GEOLOGISCHES JAHRBUCH
Reihe B Regionale Geologie Ausland
und Angewandte Geowissenschaften
Heft 103

Chemistry of Europe's Agricultural Soils

Part B

General Background Information and Further Analysis of the GEMAS Data Set

Edited by
Clemens Reimann, Manfred Birke,
Alecos Demetriades, Peter Filzmoser
& Patrick O'Connor

With 352 Pages, 121 Figures, 58 Tables and 3 Appendices

Published by
Bundesanstalt für Geowissenschaften
und Rohstoffe (BGR)

Distribution by
E. Schweizerbart'sche Verlagsbuchhandlung
(Nägele u. Obermiller)
Johannesstraße 3 A, 70176 Stuttgart

Hannover 2014

Keywords

Project, GEMAS, geochemical maps, atlas, soils, agriculture, multi-element analysis, REACH, geological surveys, international cooperation, pedogenesis, arsenic, cadmium, carbon, selenium, radioactive isotopes, boron, chlorine, fluorine, metal, metalloid, concentration, spatial distribution, monitoring, mobility, risk assessment, case studies, Norway, Sweden, Finland, Ukraine

Citation

Reimann, C., Birke, M., Demetriades, A., Filzmoser, P. & O'Connor, P. (eds.) (2014):
Chemistry of Europe's Agricultural Soils. Part B: General Background Information and Further Analysis of the GEMAS Data Set.– Geol. Jb., **B 103**: 352 pp., 121 figs., 58 Tables, 3 App.; Hannover.

Chemistry of Europe's Agricultural Soils

Part B: General Background Information and Further Analysis of the GEMAS Data Set

Edited by
CLEMENS REIMANN, MANFRED BIRKE, ALECOS DEMETRIADES,
PETER FILZMOSER & PATRICK O'CONNOR

Abstract

The GEMAS project provides a fully harmonised database of chemical elements and key parameters determining their availability for European agricultural and grazing land soil. The data set is freely available to the general public.

Part B of the GEMAS atlas provides a more detailed interpretation of the spatial distribution of selected elements (As, C, Cd, Hf, K, Se, Th, U, Zr) at the European scale than was possible to provide in Part A, the geochemical atlas. To highlight the importance of scale it also contains interpretations of the data set at a more local scale (Scandinavia and Ukraine). Possibilities for more-in-depth uses of the data set from assessing the impact of agriculture (carbon stock), through risk assessment to unravelling geological processes (loess distribution) are demonstrated. Several chapters cover the background information needed for the interpretation of a geochemical atlas (geology, soil formation, mineral deposits). Finally, data on the regional distribution of some rarely analysed elements (B, Cl, F) at the European scale are presented, based on the old FOREGS soil samples. A discussion of element concentrations, as measured in a very weak extraction (MMI®), helps us to understand the geochemical processes governing the distribution of elements at the European scale, and allows us to better detect the impact of human activities on European agricultural soil. Part B also provides Tables summarising European legislation with regard to soil, and the key statistical parameters for the GEMAS samples. |◂

Editors' addresses can be found in the list of the GEMAS project team (p. 7).

[Chemische Zusammensetzung der landwirtschaftlichen Böden Europas

Teil B: Allgemeine Hintergrundinformationen und weiterführende Auswertungen der GEMAS-Daten]

Kurzfassung

Das GEMAS-Projekt liefert im kontinentalen Maßstab eine harmonisierte und qualitätsgesicherte Datenbank von chemischen Elementkonzentrationen und weiteren Parametern, die deren Verfügbarkeit in den europäischem Acker- und Weidelandböden beeinflussen. Der Datensatz ist frei verfügbar.

Der Teil B des GEMAS-Atlasses präsentiert für ausgewählte Elemente (As, C, Cd, Hf, K, Se, Th, U, Zr) eine mehr in die Tiefe gehende Interpretation der Ergebnisse. Um die Bedeutung des gewählten Maßstabes für eine geochemische Kartierung zu veranschaulichen, wurden die Daten von Skandinavien und der Ukraine noch einmal gesondert ausgewertet. Die Daten eines geochemischen Atlasses können natürlich auch für weitergehende wissenschaftliche Studien eingesetzt werden, wie z. B. zur Untersuchung der Bedeutung der Landwirtschaft und des Klimas für den Kohlenstoffhaushalt der Böden, für die Risikobewertung der in den Böden gemessenen Metallgehalte für Bodenorganismen oder für geologische Fragestellungen, wie die räumliche Verbreitung von Lössböden in Europa. Mehrere Kapitel decken die für die Interpretation eines geochemischen Atlasses benötigten Hintergrundinformationen ab (Geologie, Bodenbildung, Mineralvorkommen). Einige Elemente (B, Cl, F) bereiten auch heute noch analytische Schwierigkeiten, ihre Gehalte in Böden werden daher selten gemessen; ein Kapitel stellt ihre Verteilung in europäischen Bodenproben aus dem FOREGS-Projekt vor. Die gesonderte Diskussion der Ergebnisse eines sehr schwachen Probenauszuges (MMI® – Mobile Metal Ion extraction) ermöglicht die Identifizierung anthropogener Einflüsse auf die Böden und liefert darüber hinaus interessante Hinweise für die Mineralexploration. Der Teil B des GEMAS-Atlasses stellt außerdem Tabellen mit den wichtigsten europäischen Bodengrenzwerten sowie statistische Übersichtstabellen der Analysenergebnisse zur Verfügung. Insgesamt zeigt der Teil B exemplarisch an Hand einiger weniger Beispiele auf, dass die Daten des GEMAS-Atlasses multivalent genutzt werden können. |◂

Editors' Foreword

by Clemens Reimann, Manfred Birke, Alecos Demetriades, Peter Filzmoser & Patrick O'Connor

Clemens Reimann

Manfred Birke

Alecos Demetriades

Peter Filzmoser

Patrick O'Connor

More than 60 organisations from around the world including government agencies, universities, contract research institutions, consultants, laboratories and the European metals industry have cooperated to make the GEMAS (**GE**ochemical **M**apping of **A**gricultural and Grazing Land **S**oil) project possible.

With an ever-growing world population, efficient food production is becoming a major challenge. Proficient agriculture depends on healthy soil. Plant and animal production require sufficient amounts of major and minor nutrients and low concentrations (or availability) of toxic elements in soil. The GEMAS project documents for the first time fully harmonised data on the concentration of almost 60 chemical elements, and of the parameters determining their availability and binding in agricultural and grazing land soil at the scale of a continent (Europe – 5.6 million km² were sampled). Key observations of the project include:

(1) Very comparable results for the two soil types sampled at the European scale.
(2) A major difference in soil composition is observed between the younger northern and the much older southern European soil for many elements.
(3) On average there is a factor 6 (and up to a factor of more than 100) difference in the median concentration of elements between the 33 participating countries.
(4) Element distributions are seen to depend on geology and climate – the anthropogenic impact is hardly detectable at the European scale. High trace element values in soil are most often related to mineral deposits and districts.
(5) Some cities (e.g., London, Paris) cause anthropogenic trace element anomalies (e.g., Au, Pb, Hg) in their near surroundings and element concentrations decrease rapidly with distance from source.
(6) Risk assessment for several metals shows that few samples have such high concentrations that they may possibly pose a toxic risk for soil organisms at the regional scale.
(7) Several important trace elements (minor nutrients, e.g., Cu, Zn) show such low levels over sizeable tracts of land in Europe that trace element deficiency is clearly of concern.

Part A is a classical geochemical atlas, describing the methodology used in the project, presenting the statistical and spatial distribution of the data and a concise interpretation of all measured parameters and elements in the survey area and providing the overall conclusions of the project. Part B starts with three chapters providing basic background information needed to interpret geochemical maps, a description of soil forming processes and the most important soil types of Europe, the Geology of Europe, and a chapter describing the distribution of mineral deposits in

Europe for the most important commodities. All raw data of the project are provided on the attached DVD (→ Part A) so that any interested reader can directly use these data, for example, as an overlay for his/her specific topic of interest.

Part B continues with a collection of articles on very different topics, from a more in-depth interpretation of the distribution of certain elements (e.g., As, C, Cd, K, Se, Th, U) already covered in Part A, describing and interpreting key results from specific analytical techniques (Chapter 13) to applications of the data in the form of a risk assessment of metals in soil (Chapter 12) or to unravel geological processes (distribution of loess deposits in Europe – Chapter 9). These chapters are only a small selection, meant to demonstrate what can be done with the GEMAS data set in addition to and beyond a classical geochemical atlas (→ Part A). Chapter 10 is special, as it does not use the GEMAS data, but presents new data obtained on soil samples from the older FOREGS project. It is included here because it describes the continental scale distribution of three important elements (B, Cl, F) in soil that are rarely reported due to analytical constraints. Chapter 11 describes the prediction of metal and metalloid partitioning coefficients in the soil samples, which are important for risk assessment. These partitioning coefficients have never before been reported on the European scale. The last two Chapters (14, 15) demonstrate what can be done with the data by going into more detail using only subsets of the data. This results in changed statistics and, of course, some patterns on the map will become more clearly visible on such a countrywide scale. Finally, landscape geochemistry, as described in Chapter 15, is the Russian way of dealing with regional geochemical data and provides a glimpse of a very different approach to present the GEMAS results. Much more could (and should) be done with the data and we hope to see many further exiting publications using GEMAS data in the coming years.

Finally, Part B closes with a Table summarising the different existing action levels for soil in Europe and Tables providing the statistical key parameters for all analysed elements and parameters.

A project of this scale would not have been possible without the backing and help from countless people and organisations. We express our sincere thanks to everybody who helped us along with the project. Despite the very large number of participants, the group really functioned as a team, and we appreciate the many enlightening discussions over the years with colleagues from industry and a wide diversity of scientific disciplines. |◄

Clemens Reimann　　Manfred Birke　　Alecos Demetriades

Peter Filzmoser　　Patrick O'Connor

GEMAS Project Team

GEMAS PROJECT MANAGEMENT

- **Clemens Reimann, NGU (Norway),**
  Chairman EuroGeoSurveys Geochemistry Expert Group
- **Manfred Birke, BGR (Germany)**
- **Alecos Demetriades, IGME (Greece)**
- **Ilse Schoeters, Rio Tinto (United Kingdom)**

Austria

Geological Survey of Austria (GBA)
Neulinggasse 38 A; 1030 Vienna; www.geologie.ac.at
- **Edith Haslinger[C/S/I]**
- **Sebastian Pfleiderer[D], Leopold Pöppel[S], Heinz Reitner[P], Albert Schedl[I]**

Vienna University of Technology (TU WIEN)
Department of Statistics and Probability Theory
Wiedner Hauptstrasse 8–10; 1040 Vienna; www.statistik.tuwien.ac.at
- **Peter Filzmoser[P/I]**

Belgium

Geological Survey of Belgium (GSB)
Department VII of the Royal Belgian Institute of Natural Sciences (RBINS)
13, Rue Jenner; 1000 Brussels
www.naturalsciences.be/institute/structure/geology/gsb_website
- **Walter De Vos[C/S/I]**

Bosnia and Herzegovina

Geological Survey of Federation of Bosnia and Herzegovina
Federal Institute for Geology (FZZG)
Ustanička br. 11; 71210 Ilidža; www.fzzg.gov.ba
- **Hazim Hrvatovic[C/S/I]**
- **Nihad Jaganjac[S], Natalija Samardzic[S], Ferid Skopljak[S]**

Bulgaria

Department of Geology and Permits for Exploration
Subsurface and Underground Resources Office,
Ministry of Environment and Water
22 Maria Louisa Blvd.; 1000 Sofia; www.moew.government.bg
- **Valeri Trendafilov[C/S/I]**
- **Vassil Belougushev[S]**

Croatia

Croatian Geological Survey (CGS)
Sachsova 2; 10000 Zagreb; www.hgi-cgs.hr
- **Josip Halamić[C/S/I]**
- **Ajka Šorša[S/I/P/D]**

University of Zagreb (UNIZG)
Faculty of Agriculture; Svetošimunska cesta 25; 10000 Zagreb; www.agr.unizg.hr
- **Stjepan Husnjak[S/I]**

Cyprus

Cyprus Geological Survey (GSD)
1415 Lefkosia; www.moa.gov.cy/moa/gsd/gsd.nsf/
- **Andreas Zissimos[C]**
- **Alecos Demetriades[S/I], Zomenia Zomeni[S]**

Czech Republic

Czech Geological Survey (CGS)
Klárov 3; 118 21 Praha 1; www.geology.cz
- **Miloslav Ďuriš** [C/S/I]
- **Jan Holák** [S], **Michal Poňavič** [S/I]

Denmark

Geological Survey of Denmark and Greenland (GEUS)
Øster Voldgade 10; 1350 Copenhagen K; www.geus.dk
- **Vibeke Ernstsen** [C/S/I]
- **Lasse Gudmundsson** [S]

Estonia

Geological Survey of Estonia (EGK)
Kadaka tee 82; Tallinn, 12618; www.egk.ee
- **Valter Petersell** [C/S/I]
- **Jaan Kivisilla** [S]

Finland

Geological Survey of Finland (GTK)
P.O. Box 96; 02151 Espoo; www.gtk.fi
- **Timo Tarvainen** [C/I]
- **Pasi Eilu** [I], **Mikael Eklund** [S/I], **Jaana Jarva** [S]

Former Yugoslav Republic of Macedonia, The

SS. Cyril and Methodius University (UKiM)
Institute of Chemistry, Faculty of Science
Arhimedova 5; 1001 Skopje; www.pmf.ukim.edu.mk/PMF/Chemistry/
- **Trajče Stafilov** [C/S/I]

France

Bureau de Recherches Géologiques et Minières (BRGM)
3 Avenue Claude Guillemin; 45060 Orléans Cedex 2; www.brgm.fr
- **Ignace Salpeteur** [C/S/I]
- **Anne Sophie Audion** [S], **Marie Branellec** [S], **Stéphanie Guignat** [S], **Hubert Haas** [S], **Jean Pierre Jegou** [S], **Gérard Longin** [S], **Jeremie Melleton** [D], **Philippe Negrel** [I], **Eric Palvadeau** [S]

Germany

Federal Institute for Geosciences and Natural Resources (BGR)
Stilleweg 2; 30655 Hannover; www.bgr.bund.de
- **Manfred Birke** [C/A/I/Ma]
- **Rainer Baritz** [I], **Irene Bitz** [A], **Reiner Dohrmann** [A], **Michael Fuchs** [S], **Levke Godbersen** [A], **Angelika Goldmann** [A], **Bodo Harazim** [A], **Rainer Hoffmann** [S], **Fabian Jähne** [I], **Christiane Kamphuis** [A], **Lars Kaste** [S/A], **Beate Keilert** [S], **Frank Korte** [A], **Cornelia Kriete** [A], **Friedrich Krone** [S], **Hans Lorenz** [A], **Heiko Oltmanns** [A], **Uwe Rauch** [I/D/Ma], **Detlev Requard** [A], **Daniela Schwenger** [A], **Jens Utermann** [S], **Elke Wargenau** [A]

Greece

Institute of Geology & Mineral Exploration (IGME)
1 Spirou Louis Street, Acharnae; 13677 Athens; www.igme.gr
- **Alecos Demetriades** [C/S/A/I/Ma]
- **Maria Kaminari** [I], **Dimitris Machaeras** [S]

Hungary

Geological and Geophysical Institute of Hungary (MFGI)
Department of Environmental Geology
Stefánia út 14; 1143 Budapest; www.mfgi.hu
- **Gyozo Jordan** [C/S]
- **Ubul Fügedi** [S/I], **Laszlo Kuti** [I]

Ireland

Geological Survey of Ireland (GSI)
Beggars Bush, Haddington Road; Dublin 4; www.gsi.ie
- **Patrick O'Connor** [C/I]
- **Vincent Gallagher** [S], **Fionnuala Ni Mairtin** [S], **Ray Scanlon** [I]

Italy

University of Napoli Federico II (UNINA)
Department of Earth, Environmental and Resources Sciences (DiSTAR)
Via Mezzocannone 8; 80134 Napoli; www.fluidenv.unina.it
- **Benedetto De Vivo** [C/S/I]
- **Stefano Albanese** [S/I], **Annamaria Lima** [S/I]

University of Sannio (UNISANNIO)
Department of Biological, Geological and Environmental Sciences (DSBGA)
Via dei Mulini 59/A; 82100 Benevento; www.sci.unisannio.it
- **Domenico Cicchella** [S/I]

University of Bologna (UNIBO)
Department of Biological, Geological and Environmental Sciences (BiGeA)
Piazza di Porta San Donato 1; 40126 Bologna; www.eng.unibo.it
- **Enrico Dinelli** [S/I]

University of Cagliari (UNICA)
Department of Civil-Environmental Engineering and Architecture (DICAAR)
Via Marengo 3; 09123 Cagliari; www.unica.it
- **Paolo Valera** [S/I]

Latvia

Latvian Environment, Geology and Meteorology Centre (LEGMC)
Maskavas Street 165; 1019 Riga; www.meteo.lv
- **Aivars Gilucis** [C/S/I]

Lithuania

Lithuanian Geological Survey (LGT)
S. Konarskio Str. 35; 03123 Vilnius; www.lgt.lt
- **Virgilija Gregorauskienė** [C/S/I]
- **Augustinas Jankauskas** [S], **Petras Pūtys** [S]

Luxembourg

Service Géologique du Luxembourg (SGL)
B.P. 17; 8005 Bertrange; www.geologie.lu
- **Robert Maquil** [C/S/I]

Montenegro

Geological Survey of Montenegro (GSM)
Cetinjski put bb; 81000 Podgorica; www.geozavod.co.me
- **Ranko Svrkota** [C/S/I]
- **Slobodan Radusinović** [S/I]

Netherlands, The

Alterra Wageningen UR Soil Science Centre
P. O. Box 47; 6700 AA Wageningen; www.wageningenur.nl
- **Gerben Mol**[C/S/I]
- **Willy de Groot**[S]

Norway

Geological Survey of Norway (NGU)
P. O. Box 6315 Sluppen; 7491 Trondheim; www.ngu.no
- **Clemens Reimann**[C/S/I/P/D]
- **Ola A. Eggen**[S], **Clea Fabian**[A], **Karl Fabian**[A/I], **Belinda Flem**[A], **Morten Jartun**[S], **Henning Jensen**[S], **Anne Nordtømme**[A], **Rolf Tore Ottesen**[I], **Henning Tiarks**[S], **Tore Volden**[S]

Norwegian Forest and Landscape Institute (skog + landskap)
P. O. 115, 1431 Ås; aha@skogoglandskap.no
- **Arnold H. Arnoldussen**[I], **Eivind Solbakken**[S], **Ragnhild Sperstad**[S]

Poland

Polish Geological Institute (PGI) Polish Geological Survey
4, Rakowiecka Street; 00-975 Warsaw; www.pgi.gov.pl
- **Anna Pasieczna**[C/I]
- **Przemyslaw Dobek**[S], **Aleksandra Dusza-Dobek**[S], **Tomasz Gliwicz**[S], **Piotr Kaszycki**[S], **Paweł Kwecko**[S], **Joanna Szyborska-Kaszycka**[S]

Portugal

National Laboratory for Energy and Geology (LNEG)
Estrada da Portela Zambujal - Alfragide, Apartado 7586;
2610-999 Amadora; www.lneg.pt
- **Maria João Batista**[C/S/I]
- **Cátia Prazeres**[S]

Serbia

Geological Survey of Serbia (GZS)
Rovinjska 12, 11000 Belgrade; www.gzs.gov.rs
- **Aleksandra Gulan**[C/S/I]
- **Božidar Đokić**[S], **Milun Jovanović**[S], **Jovan Kovačević**[I], **Dušan Nikolić**[I]

Serbian Environmental Protection Agency (SEPA)
Ruže Jovanović 27a; 11000 Belgrade; www.sepa.gov.rs
- **Dragana Vidojević**[C]

Slovakia

State Geological Institute of Dionýz Štúr (SGIDS)
Mlynská dolina 1; 817 04 Bratislava; www.geology.sk
- **Igor Slaninka**[C/I]
- **Peter Šefčík**[S], **Silvester Pramuka**[S]

Geoanalytical Laboratories (GAL) State Geological Institute of Dionýz Štúr
Regional Centre, Markušovská cesta 1; 052 01 Spišská Nová Ves
- **Daniela Mackových**[C/A/S]
- **Pavol Lučivjanský**[A]

Slovenia

Geological Survey of Slovenia (GeoZS)
Dimičeva ulica 14; 1000 Ljubljana; www.geo-zs.si
- **Mateja Gosar**[C/S/I]

Spain

Geological Survey of Spain (IGME)
Ríos Rosas, 23; 28003 Madrid; www.igme.es
- **Juan Locutura**[C/S/I]
- **Alejandro Bel-lan**[S/I], **Marta Campesino**[S], **Miguel Chamorro**[S], **Marta Domínguez**[S], **Angel García Cortés**[S], **María González**[S], **J. Mariano Martínez**[S]

Sweden

Geological Survey of Sweden (SGU)
Box 670; 751 28 Uppsala; www.sgu.se
- **Madelen Andersson**[C/S/I]
- **Anna Ladenberger**[P/I], **George Morris**[I], **Martiya Sadeghi**[P/I], **Erland Sellersjö**[S], **Jo Uhlbäck**[P/I]

Switzerland

Federal Office of Topography (swisstopo), Swiss Geological Survey (SGS)
Seftigenstrasse 264; 3084 Wabern; www.swisstopo.ch
- **Peter Hayoz**[C/I]

Research Group Swiss Soil Monitoring Network (NABO)
Research Station Agroscope Reckenholz-Tänikon (ART)
Reckenholzstrasse 191; 8046 Zürich; www.agroscope.ch
- **Reto Giulio Meuli**[S], **Peter Schwab**[S], **Daniel Zizek**[S]

Ukraine

State Geological and Subsurface Survey of Ukraine (SGSSU)
Ejen Pottier Street 16; 03680 Kyiv; www.geo.gov.ua
- **Volodymyr Klos**[C/I]
- **Grygorii Akinfiiev**[D/Ma], **Yulia Amashukeli**[I/Ma], **Yevgen Artamonov**[S], **Mykhailo Geichenko**[I], **Oleksiy Maksymenko**[S], **Borys Malyuk**[I]

United Kingdom

British Geological Survey (BGS)
Nicker Hill, Keyworth; Nottingham NG12 5GG; www.bgs.ac.uk
- **Dee Flight**[C/I]
- **Kevin Barker**[S], **Paul McDonnell**[S], **Andreas Scheib**[S/I], **Mick Strutt**[S]

International

ARCHE Consulting (Belgium)
Stapelplein 70; 9000 Gent; www.arche-consulting.be
- **Koen Oorts**[I/P]

CSIRO Land and Water (Adelaide, Australia)
Private Mail Bag No.2, Glen Osmond SA 5064; www.clw.csiro.au
- **Cathy Fiebiger**[A], **Sean Forrester**[A/I], **Les Janik**[A/I], **Jason Kirby**[A/I], **Michael J. McLaughlin**[A/I], **Jose Martin Soriano-Disla**[A/I]

SGS Minerals Services (Canada)
1885 Leslie Street, Toronto Ontario, M3B 2M3 ; www.sgs.com
- **Pierrette Prince**[A], **Nicholas Turner**[A/I]
- **Alan Mann**[A/I] (Consultant to SGS, Western Australia)

University of Otago (New Zealand)
Department of Chemistry, Forensic Chemistry
P.O. Box 56, Dunedin 9016; neon.otago.ac.nz/chemistry
- **Jurian Hoogewerff**[A/I]

A Analysis
C Coordinator
D Database
I Interpretation
Ma Maps
P Processing
S Sampling

List of Eurometaux Industry Sponsors

Eurometaux
Avenue de Broqueville 12, 1150 Brussels; Belgium; www.eurometaux.org
- **Violaine Verougstraete**[c]
- **Ilse Schoeters**[c]

European Borates Association
Twee Kerkenstraat 26 bte 2, 1000 Brussels; Belgium; www.ima-europe.eu

European Copper Institute
Tervurenlaan 168 bte 10, 1150 Brussels; Belgium; www.copperalliance.eu

European Precious Metals Federation (EPMF)
Avenue de Broqueville 12, 1150 Brussels; Belgium; www.epmf.be

The Cobalt Development Institute
167 High Street, Guildford, Surrey, GU1 3AJ; United Kingdom; www.thecdi.com

International Antimony Association (i2a)
Avenue de Broqueville 12, 1150 Brussels; Belgium; www.antimony.be

International Lead Association
Bravington House Regent Quarter, 2 Bravingtons Walk, London N1 9 AF London; United Kingdom; www.ila-lead.org/home

International Manganese Institute
17 rue Duphot, 75001 Paris; France; www.manganese.org

International Molybdenum Association
4, Heathfield Terrace, Chiswick, London W4 4JE; United Kingdom; www.imoa.info

International Tin Research Institute
Unit 3, Curo Park, Frogmore, St. Albans, Herts AL22DD; United Kingdom; www.itri.co.uk

International Zinc Association
Tervurenlaan 168 bte 4, 1150 Brussels; Belgium; www.zinc.org

The Nickel Institute
Avenue des Arts/Kunstlaan 13, 1210 Brussels; Belgium; www.nickelinstitute.org

The (REACH) Selenium and Tellurium Consortium
UMCO Umwelt Consult GmbH,
Georg-Wilhelm-Strasse 183 b, 21107 Hamburg; Germany; www.se-te-consortium.com

The (REACH) Vanadium Consortium
Auer von Welsbach Straße 1, 9330 Althofen; Austria; www.vanadiumconsortium.com

Laboratories Involved in the Analysis of GEMAS Samples

BGR (Germany)
XRF and LOI

NGU (Norway)
pH, total C, total S, Pb-isotopes, magnetic susceptibility

SGS Minerals Services (Canada)
MMI® extraction (56 elements)

Contract Laboratories

SGIDS (Slovakia)
Sample preparation and eCEC

FUGRO Consult GmbH (Germany)
TOC

ACME Analytical Laboratories Ltd. (Canada)
Aqua regia extraction (53 elements)

CSIRO Land and Water (Australia)
MIR, grain size, K_d-values

Part B

General Background Information and Further Analysis of the GEMAS Data Set

GEMAS
in Europe

Table of contents

Part B

Chapters 1 – 3
General Background Information

Chapter 1

A Short Guide to Soil Formation and the Soils of Europe

EDITH HASLINGER, CLEMENS REIMANN & MANFRED BIRKE

▸ 1.1 Introduction

GEMAS is a project about the statistical and spatial distribution of chemical elements in European agricultural and grazing land soil. When interpreting the GEMAS results, it is important to know some key facts about soil and its formation. This chapter attempts to describe some key facts of soil science in an easy-to-understand language. It is aimed at a reader with a non-soil science background, who may become interested in one or other geochemical map presented in Part A of the GEMAS atlas and is starting to wonder about the soil forming processes hidden behind the patterns on the map. Fundamental text books of soil science for further reading include BUCKMAN & BRADY 1970; ASHMAN & PURI 2002; BLUME et al. 2010) to name just a few.

The *"Soil Atlas of Europe"*, published by the European Commission, which is available online (EC 2005), is recommended for further reading about soil and the soil types of Europe, including detailed soil maps for all countries.

▸ 1.2 Soil – what is it?

Even the answer to the apparently simple question *"What is soil?"* is not an easy one. When asking students about their definition of soil, the answers range from *"We walk on it"*, *"We build houses on it"*, *"In our garden soil, we plant our vegetables"* to *"It is brown and dirty and full of earthworms"*.

Indeed, there exist several definitions of soil, and almost everybody, including scientists from different fields of Earth Sciences, has a different idea and perception of soil.

In general, soil is defined as the top layer of the Earth's crust. It is a natural substance composed of weathered rock particles (minerals), organic matter, water and air. Soil is the product of the effects of climate and living organisms on rocks, and its nature is determined by the composition of the parent rocks, the type of climate, and its age.

Numerous classifications of soil have been proposed, none of which are perfect.

The shortest definition or formula for soil is:

Soil =
Rock debris + Organic matter + Air + Water

Is that it? Yes, in essence this is it. However, there is a whole universe of variables, interactions and changes behind these four factors, so that a soil at one place can be very different from a soil just some metres away. The investigation of one or several of these factors is a science in its own right, making soil science a highly interdisciplinary branch of science. Soil is a highly complex, highly variable biomolecular sieve with an array of physical and chemical properties.

The Russian 'godfather' of soil science Vasily V. Dokuchaev (1846 – 1903) defined soil as an independent natural body developed over time under the influence of five soil-forming factors, i.e., parent rock, living organisms, climate, relief, and time (JENNY 1941):

Soil =
function of (parent rock, organisms, climate, relief, time).

These five factors are essentially responsible for the type of soil that develops in a certain area. Although this definition is over one hundred years old, it is still true and forms the basis of our understanding why there are different soils in different regions of the Earth, but also why there can be very different soil types in close proximity to one another. Because humans started to cultivate land in order to grow crops about 5000 years ago, and the human impact was strong, particularly during the last 150 years of 'industrial agriculture', the definition of Dokuchaev has to be extended by another factor – human activities.

Soil =
function of (parent rock, organisms, climate, relief, time, human activities).

Locally, the effect of human activities on soil formation, and especially soil chemistry, can be stronger than the natural factors, but on the continental scale, the natural factors still rule soil formation.

Soil is the outermost layer of the Earth's crust, which is the 'skin' of our planet. When we recall the internal structure of the Earth, the distance from the surface to the core is about 6400 km. There is an inner and an outer core, a mantle zone (lower and upper) and then - at the very surface – the Earth's crust. The Earth's crust consists of various types of rocks. Compared to the other parts, it is very thin – only 40 kilometres thick on average. Soil is the contact zone between the crust and the bio- and atmosphere. Like the crust, soil is not equally thick – it can range from just a few centimetres to several metres. Soil scientists and geologists can argue endlessly about the exact boundary between 'rocks' (including unconsolidated sediments) and 'soil'.

1.2.1 What can we do with soil?

When you imagine some elements of a 'normal' daytime routine: you go to work, sit in the office, go to the supermarket to buy food, go home, jog in the park, have a glass of water – *"do you ever think, that all of this would never be possible if soil did not exist?"*

Soil is the basis of our lives, in the truest sense of the word. Without soil, there would be no life for us human beings. On soil, we grow our food and build our houses, roads and infrastructures. Without soil, we could not jog in the park or have a picnic or a walk in a forest. In soil, organisms break down the tissues of dead plants and animals, in order to provide nutrients for the growth of new life. In soil, water and carbon dioxide are stored and made available for plants. Soil serves as a natural filter for water, as well as potential pollutants, where they are broken down or even neutralised. From soil, you can learn about the history of a landscape. Soil preserves archaeological and historical remains from damage and depletion (EC 2005; BLUME et al. 2010).

1.2.2 How does soil form?

When you imagine a boulder of solid rock - granite for instance – and this boulder is subjected to natural forces – rain, wind, snow, sun, etc., in short the 'climate' of our planet – it will not stay solid and pristine given enough time (Fig. 1.1). Within a matter of years, it will start to break up, a process that soil scientists refer to as **weathering**. One of the most important 'agents' in weathering is water, e.g., rain. When water enters a rock via cracks, which form at the rock surface due to the changes in temperature, the minerals in the rocks will react with the water in several ways. The ions (charged particles) in the water react with the ions in the minerals (the constituents of rocks). This will change the nature of the minerals and lead to rock disintegration. Some minerals, like sodium chloride ($NaCl$) – our table salt – simply dissolve in water and end up into the ocean. If sea water evaporates new salt deposits may form. Biological processes, such as the respiration or breathing of soil animals add carbon dioxide (CO_2), which in water forms (weak) carbonic acid (H_2CO_3), which also reacts with the minerals in the rocks. Plants and living organisms release organic solutions (called exudates) in order to dissolve the minerals in the rocks, and to gain access to the nutrients in the minerals. Water in cracks freezes during win-

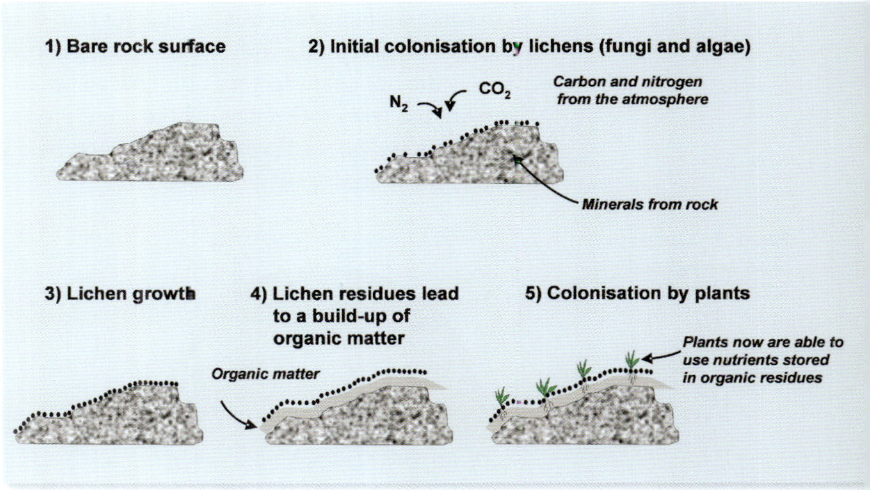

Fig. 1.1: Initial soil formation by the colonisation with lichens and the accumulation of organic matter (drawn after ASHMAN & PURI 2002: p.10).

ter and expands its volume by about 10 % during freezing. Thereby, the cracks are widened and the rock is further disaggregated.

Weathering is the main process, which gradually and eventually transforms rocks to soil. From the description above it is clear that this process can lead to substantial chemical changes from rock to soil. The parent rock from which the soil develops will have an influence on the chemical composition of the newly formed soil. However, the rock and soil chemistry can differ substantially depending on the solubility of the minerals in the rock. It is also clear that a soil will change over time. Even a soil will 'weather'. Yet, soil formation is more than the processes, which break down rocks. Apart from the mineral inputs from rock weathering, organic matter is also needed. Organic matter represents the residues of once-living organisms (plants and animals). The bare surface of rocks – even if they are already slightly weathered – do not offer very favourable conditions for plant colonisation, since the water and nutrient availability is very restricted. However, there exists a symbiosis between plants, bacteria and fungi, known as **lichens**, which are adapted to these harsh environments (Fig. 1.2). The lichens can colonise a bare rock surface and 'harvest' the nutrients from the rocks' minerals by dissolving them via the already mentioned exudates (organic acids). Therefore, lichens are called 'primary colonisers'. Strictly speaking, this colonisation of rocks by lichens is the very beginning of soil formation; however, it still takes a long time, until enough organic matter is accumulated for the colonisation of higher plants to commence (ASHMAN & PURI 2002).

When the lichens and the higher plants, which colonise the rocks die, they become 'organic matter' or – more precisely – soil organic matter. The layer of organic material, which is on top of the weathered rock becomes thicker and thicker and the amount of organic matter increases. Therefore, we have the start of the formation of a soil profile; now we have something that is very important in soil science, in order to understand the whole topic of soil formation (or pedogenesis) – the first two **soil horizons**. We will discuss this later when we talk about pedogenesis. First, we will have a closer look into the two main constituents of soil – the rocks and the organic matter.

Fig. 1.2: Initial colonisation of basalt by lichens (photograph by Edith Haslinger).

When you ask yourself *"how long does soil formation take?"* the answer is – a long time. Soil-forming processes are slow and the formation of a fully developed soil profile can take hundreds to thousands of years. For instance, it takes 100 years for 1 – 2 cm of soil to form in the temperate climate zone under permanent grassland.

When we remember our formula for soil "*Rock debris + Organic matter + Air + Water*", the '*rock debris*' is the dominant factor. From now on, we will replace the term '*rock debris*' by the more precise term '*mineral fraction*', which is explained below and in the following diagram (Fig. 1.3).

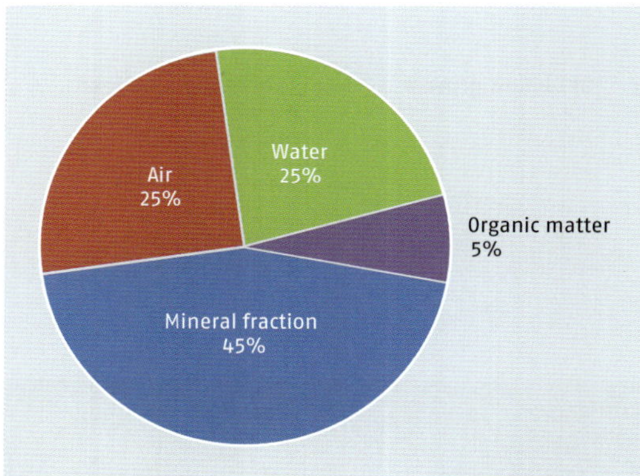

Fig. 1.3: Composition of an average soil (after Schroeder & Blum 1992).

The **'mineral fraction'** is inorganic soil material made up mostly of oxygen, silicon and aluminium (many other metals in small quantities may be included). **'Organic matter'** is derived from the residues of plants and soil organisms, and is comprised primarily of carbon, oxygen and hydrogen. The term **'air'** refers to the gaseous portion of soil composed of the same gases found in the atmosphere (oxygen, nitrogen, and carbon dioxide). **'Water'** in soil terms is the water in the pores of the soil, which is mixed with dissolved salts from the mineral fraction. Plants get their nutrients from this soil water or soil solution.

Figure 1.3 shows you the average composition of a soil. As you can see, the majority of the soil volume is inorganic, which means of non-biological origin. Roughly, half of the soil volume consists of the 'mineral fraction'. *So what is this mineral fraction?* Minerals are the constituents of rocks or – vice versa – rocks are made up of minerals. For instance, the minerals that make up granite are *quartz* (*SiO_2*), *feldspar (a silicate)* and *mica (another silicate)*; limestone is made up of the mineral *calcite* (*$CaCO_3$*), rock salt of the mineral *halite (NaCl)*. Different minerals have different weathering characteristics, i.e., from highly soluble (e.g., halite: fully water-soluble) to barely soluble at all (e.g., quartz). From weathering or the breakdown of rocks to smaller particles, we get the so-called '*mineral fraction*' of the soil, which is often made up of quite different minerals (the so-called secondary minerals that form during weathering, e.g., clay minerals are derived from mica) than the ones we find in the rocks. In soil science, the fraction < 2 mm – the so-called fine earth fraction – is used to further characterise the mineral fraction, e.g., the chemical properties, the soil texture, etc. The particles in the mineral fraction are classified according to their size (or – more precisely – to the particle diameter) (FAO 2006):

1. Sand (2 mm – 0.063 mm)
2. Silt (0.063 – 0.002 mm)
3. Clay (< 0.002 mm)

At least two of these fractions are easily understood, because we have known them since our childhood. We played with **sand** in sandpits or on the beach, and we know **clay** in the form of flower and teapots. The **silt** fraction lies between the coarse sand and the very fine clay fraction. An average agricultural soil consists of about 20 % clay.

The distribution of these three fractions in a soil is called **soil texture**. The texture is essential for the properties of a soil – "*How much water it can retain? How prone is it to erosion? How stable is it? Which plants can be grown on these soil types? What kind of yield a farmer can expect from a soil?*" etc. Since these questions are important, especially for agricultural purposes, we will focus a little bit on these soil fractions and their consequences, if a soil is, for instance, clayey or sandy – or, as gardeners and farmers would say '*heavy*' or '*light*' soil.

Sand is more or less coarse with individual grains that are easily distinguishable. The grains of sand are generally rounded, which means that they do not fit perfectly together when stacked within a soil. Therefore, in a sandy soil there are many large pores filled with air (or temporarily with water). This means that sandy soil is very well *aerated*. On the other hand, sandy soil does not retain water easily due to the large pores. This means that plants on such soil must be adapted to water stress and able to survive dry periods.

Clay is at the other end of the spectrum. We know clay as sticky, plastic and shapeable material. Clay consists of rock debris that has undergone weathering until its grain size is smaller than 0.002 mm. This fraction of the clay is called '*primary minerals*', which are the chemically and mineralogically unaltered parent rocks that have been 'ground' to this small size. Clay minerals can also be '*secondary minerals*', formed from the weathering products of other minerals, e.g., mica and feldspar. Depending on the pH-value and the conditions in the soil solution, different clay minerals can be formed. Clay minerals consist of layers, which can take up water molecules, and this is the reason for the stickiness and plasticity of the clay. When clay minerals take up water in their layers they swell, and after long periods without rain, they dry up and shrink again. This property is responsible for the typical shrinkage cracks often seen on the surface of soil rich in clay (Fig. 1.4).

A soil, which contains a high amount of clay, will behave accordingly in agriculture, i.e., it will be very sticky and highly plastic. Clayey soil is '*heavy*' soil – due to the high water storage capacity, it is often wet or even inundated after rainfall, since the water cannot drain easily. After long periods without rainfall, the clay dries up completely, leaving the soil often rock-hard. In general, clayey soil is hard to work.

Clay minerals also have another important function, which is essential for plant nutrition, but also for the filtering of rainwater, as well as assisting the 'decontamination' of soil. As mentioned above, the structure of the clay minerals allows for the take up of water molecules. However, clay minerals can also take up other molecules, such as chemical elements (ions). Plants can take up these nutrients in exchange for ions they exude from their roots. Furthermore, clay minerals can also take up and store substances, which should not be in the soil in the first place, such as organic and inorganic contaminants. Organic matter can also adsorb water and nutrients (and metals) on its surface. A high content of clay minerals and organic matter in a soil is, therefore, essential for its so-called buffering capacity, i.e., the potential to release and take up nutrients and water. This whole process is called ion exchange and the **ion exchange capacity** (cation exchange capacity, CEC) is a very important parameter to be determined in a soil.

Silt lies between the coarse sand and the fine clay fraction. Pure silt has a powdery or floury consistency. Silt is an important constituent in loamy soil. Depending on the grain size of silt (coarse or fine), it can add to the characteristics of a soil given by sand and clay. In contrast to coarse sand and fine clay, which are primarily transported by water, silt (and fine sand) is also transported and deposited by wind – the so-called *aeolian deposition*. Silt is also known as '*rock dust*' in dry, semi-arid areas, where it covers the landscape or infrastructure (houses, cars, etc.). In periglacial areas (i.e., areas at the borders of glaciated areas), silt is blown from the rocks of dry riverbeds and deposited sometimes many kilometres away, forming plateaus several tens of metres thick and many hundred square kilometres wide.

Fig. 1.4: Typical desiccation shrinkage cracks in clay-rich sediments (photograph by Jonathan Wilkins 2011).

How do we know how much sand, silt and clay is there in a soil? In the laboratory, soil samples are analysed for their soil texture by sieving and in a special apparatus, which exploits the sedimentation velocity of particles as a function of their diameter. The determination of the clay fraction is very time consuming and thus expensive. Much more fun is to determine soil texture in the field when taking samples. All you need are your hands. With a little practice this 'hand test' for soil texture is almost as good as that derived from any laboratory apparatus – and much faster and cheaper. This test for soil texture is not only useful in soil science, but it can also be used by gardeners and farmers to test how 'heavy' or 'light' their soil is, and what characteristics it has concerning water and air retention, what kind of plants will and will not grow, etc.

You take a handful of your soil (sample) and make it slightly wet, so that you can easily work it, but not so wet that water comes out of your hands when you squeeze your soil sample into a ball. Now comes the part in which you test your material with several criteria such as '*some grains to see and feel, gnashes between teeth*' and '*breaks when trying to form the wire to a ring*'. You see, soil science is a science for which you need all of your senses.

If you do not want to make the comprehensive test, there are some simple characteristics of the three main particle classes:

- **Clay**: soils fingers, is cohesive (sticky), is formable, has a high plasticity and has a shiny surface after squeezing between fingers.
- **Silt**: soils fingers, is non-sticky, only weakly formable, has a rough and ripped surface after squeezing between fingers and feels very floury (like talcum powder).
- **Sand**: cannot be shaped, does not soil fingers and feels very grainy.

However, most types of soil are a mixture of these three particle classes and, therefore, there are many subgroups and transitions. Soil scientists can have incredibly lengthy discussions about whether a soil sample is 'silty clay loam' or '(fine) sandy clay loam'. This is the point where laboratory analyses are needed for a comprehensive and comparable answer. For ordinary people, and most applications, it is enough to know whether a soil is primarily sandy, silty or clayey.

Typically, the soil texture is determined for each single layer or horizon in a soil (we will come to soil horizons later). Changes of soil texture within a soil are very interesting, since they can indicate processes within the soil, thus influencing the classification of a soil, or they can provide evidence of the changes in natural conditions during soil formation.

1.2.3 Is soil mostly 'dead'?

No, quite the contrary: *soil does not only mean life, it is also full of life*. The number of organisms that dwell in soil is utterly overwhelming. It is almost unbelievable, but in one handful of soil, there are more living organisms than there are people on Earth. A single teaspoon of garden soil may contain thousands of species, millions of individuals and a hundred metres of fungal networks (EU 2010). Some of the soil dwellers are very well known to most of us from our childhood days, when we were both fascinated and may be also a little bit scared: ants, beetles, centipedes, spiders, moles, and – eaten by the bravest in tests of courage – earthworms. This is the part of soil life that we can see – the largest share by far is invisible to our eyes. Even if one quarter of all species on Earth lives in soil, this 'universe' is still rather unknown – for instance, only 1 % of the micro-organisms in soil have been identified so far.

In Figure 1.3, we saw that the organic matter accounts for only 5 % of the total soil volume. Of this 5 %, the largest portion is, in fact, mostly dead – it primarily consists of humus, which is the decayed organic matter and accounts for 85 % (Fig. 1.5). About 10 % is comprised of plant roots, and the 'real' soil life – the soil organisms – only account for 5 % of the 5 % of total volume of organic matter in a soil. As shown in Figure 1.5, bacteria and fungi have by far the largest share of all soil organisms and – as stated before – most of the species and their functions have yet to be identified.

So, "what do all these 'animals' do all day long?" Even if they are mostly invisible, their work is essential for our lives. Depending on their size, the soil organisms chew and dissolve the dead plants, thus producing organic matter and providing the living plants with the nutrients they need to grow; by digging tunnels, burrowing and by reworking of the soil material, they aerate the soil and make the soil structure more stable; water that percolates through soil is filtered through the organic matter made by soil organisms (and also on the clay minerals); some organisms are also capable of breaking down organic pollutants converting them to non-toxic molecules. Therefore, as you can see, the life beneath our feet is working hard to sustain themselves, but also to sustain our life as well.

There are different ways to classify soil fauna, but the most common is to classify the living organisms according to their size. The smallest organisms form the group of microfauna, or micro-

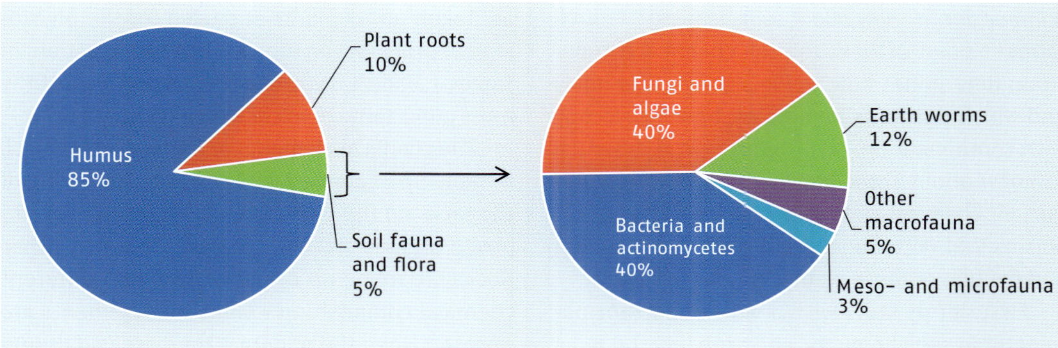

Fig. 1.5: Composition of organic matter (after SCHROEDER & BLUM 1992).

biota, and include all species smaller than 200 µm or 0.2 mm. Of this group, the bacteria are the most abundant and diverse members. If you happen to have access to a microscope, it is very interesting to watch the micro- and meso-fauna in action. You simply put a quantity of soil into a funnel, place a bottle filled with water below the funnel and place a light bulb above the funnel – this apparatus is called a Tullgren-Berlese funnel. Since the soil organisms live in the dark, they try to avoid the light, and will flee the funnel in the direction of the water-filled bottle. After a while, you can pour some water out of your bottle into a glass dish, and watch the different microanimals as they move through the water.

The following table (Table 1.1) gives an overview of the size classes for soil organisms, examples and the function of the organisms in the soil (JEFFERY et al. 2010; EU 2010; BLUME et al. 2010; SCHROEDER & BLUM 1992).

Table 1.1: Classification of soil organisms according to their size and function in soil.

Size range	Examples	Function in soil
Microfauna and Microflora **< 0.1 mm**	Bacteria, fungi, roundworms (nematodes), protozoa, wheel animals (rotifers)	They colonise freshly fallen plant litter, and start to decompose it by breaking down carbon and nitrogen in the plant litter and transforming it to carbon dioxide and nutrients for plants; they live mainly in the water-filled pores of the soil.
Mesofauna **0.1 – 2 mm**	Mites (acari), springtails (collembola)	They live on bacteria, fungi and algae, and also decompose organic matter; they mostly live in the upper parts of a soil. Their activity promotes the activity of the micro-organisms.
Macrofauna **2 – 20 mm**	Earthworms, beetles, centipedes, millipedes, snails, ants	They make tunnels in the soil, thus helping air and water to move through the soil column. By their activity, the soil material is mixed and the soil structure is loosened. They fragment the organic matter and thus prepare it for microbial decomposition. The earthworms 'eat' the soil and mix it with mucus in their digestive tract; thereby, the soil material is stabilised and better protected from erosion and aggradation.
Megafauna **> 20 mm**	Moles, rabbits, rodents	The burrowing activity of the megafauna mixes the soil material, brings organic matter, nutrients, air and water to the lower part of the soil and homogenises the soil.

▸ 1.3 Soil – an eco-friendly chemical factory

In the previous sections, we have learnt about what soil is, how it is formed, what it is made of, and what kind of organisms and animals dwell in it. Now we will enter the world of soil chemistry, where we will look at the multitude of processes and constituents that are active in the chemical composition of soil, and which are ultimately responsible for the overall result of the GEMAS project – the maps showing the spatial variation of the chemical composition of soil. *Why is it important to know about soil chemistry?* From a farmer's point of view, the soil chemical composition determines how fertile a soil is, how many and which nutrients it contains, how it can cope with contamination, etc. Therefore, it is essential to know about some key facts on the most important chemical elements in soil, and the mechanisms that determine the formation, movement and fixation of these elements.

First, let us start with a very important parameter in soil chemistry, and probably the most commonly known and measured – the **pH value**. The pH value is so critical, because it influences many other parameters in a soil to a large extent, such as the availability of nutrients and, therefore, plant growth, the types, amount and activity of soil organisms and, hence, the decomposition rate of organic matter, etc. The pH is an indicator of how **acidic** or **alkaline** a soil is (chemically, the pH is the negative log of the activity of hydrogen ions in a solution, which means, that soil pH decreases as the activity of the hydrogen ions and, therefore, the acidity increases). A soil pH of 7 is considered neutral, soil with a pH lower than 7 is acidic, and higher than 7 alkaline. In most soil, the soil pH ranges from 3 to 8. Acidic soil (low pH) forms mostly on acidic (felsic) parent material, such as granite and gneiss. Low pH occurs also in soil where the so-called base cations have been leached out due to weathering. Alkaline soil occurs mostly in the arid climate zone, where 'salt' and, therefore, Na-rich groundwater ascends through a soil profile (Na forms the strong base NaOH in aqueous solution).

Most soil is weakly to moderately acidic, because there are several processes in a soil in which acids are produced – the decomposition of organic matter, the exudation of weak organic acids by plant roots in order to take up other nutrients (cations) from the mineral fraction, as well as the 'acid rain' from emissions from industry and car traffic (the N-, P- and S-compounds in these emissions form strong acids in rainwater).

Before we consider the ideal pH-value of agricultural soil, we will have a look at the chemical elements, which can be found in a soil. The chemical elements in a soil come from the mineral and the organic fraction.

In soil terms, we often do not speak about chemical elements, but ***nutrients***, because they are taken up by plants and used for the physiological processes in a plant. In a later section, in Tables 1.5 and 1.6 you can find an overview of the most important chemical elements in a soil.

The chemical elements or nutrients that a soil is made up of are derived from the weathering of the rocks on the one hand, and from the decomposed organic matter (i.e., humus) on the other. Rocks are made up of minerals with a different chemical composition – and these elements are released when the minerals are dissolved during weathering. Accordingly, organic matter releases its chemical elements during decomposition. The longer the weathering process lasts, the more insoluble elements, such as iron and aluminium, are enriched. Old soil that has undergone a long period of weathering is, therefore, often red-coloured.

The most abundant of all chemical elements in a soil is **oxygen** (a negatively charged ion, i.e., anion), which is part of the crystal structure of minerals, but also an important constituent of the organic matter. The next most common elements are cations (i.e., positively charged ions) and are in decreasing order: **silicon, aluminium, iron, magnesium, calcium, sodium** and **potassium**. The soil organic matter is comprised mainly of carbon, hydrogen, oxygen and nitrogen. It is important to note that the most common elements are not necessarily the most essential for plants (Tables 1.5 and 1.6).

Therefore, we have two sources for nutrients in a soil – the mineral fraction and the organic matter. In the section about soil formation, we learnt that the mineral fraction is comprised of sand, silt and clay in varying amounts. Most of the nutrients in a soil are adsorbed on the inner and outer surfaces of clay minerals. As stated above, clay minerals consist of layers or sheets – they belong to the mineral class *sheet silicates*. There are different types of clay minerals – those that consist of sequences of two layers (one tetrahedral and one octahedral silicate sheet), and those that consist of sequences of three layers (one octahedral silicate sheet between the tetrahedral sheets). The types are referred to as 1:1 or 2:1 clay minerals. The 2:1 clay minerals are characterised by a large internal surface, as well as their distinct ability to take up and release water molecules and nutrients.

Plants can take up the nutrients adsorbed on the surface of clay minerals or organic matter by releasing weak organic acids. The hydrogen ions (H^+) of the acids are exchanged for other cations, e.g., calcium (Ca^{2+}), and can be used for the physiological processes in the plant. Anions can also be exchanged (anion exchange), but in soil, the cation exchange processes are dominant (Fig. 1.6), since the soil colloids (i.e., particles < 2 µm – clay minerals, humic substances, oxides and hydroxides) are mainly negatively charged.

Plants take up the chemical elements with their roots and these are accumulated predominantly in the leaves. When the leaves decay and fall to the ground (soil surface), the chemical elements are accordingly deposited on the ground with them. When the leaves start to be decomposed, the chemical elements stored in them are then enriched in the humus, which is the overall stable end product of the decomposition path of organic matter. The soluble elements, such as carbonates are washed back down into the soil with precipitation. However, the less soluble elements remain in the humus layer. Therefore, when you make a chemical analysis of your soil, including the humus layer, you will find trace elements, which typically occur only in small amounts in the soil, but are enriched in the humus layer. Therefore, in geochemical terms, the humus layer can be referred to as a 'geochemical filter' of the soil profile (Goldschmidt 1937).

Plants have individual optimum pH-ranges, but most agricultural plants thrive in soil with a weakly acid to weakly alkaline pH. This is because the soil-pH regulates the availability of chemical elements – and therefore nutrients – in a soil. Nitrogen, phosphorous and potassium, for instance, which are the dominant nutrients used in fertilisers, are less available in acidic soil. Acidic soil is generally also poor in calcium and magnesium and, therefore, acidic soil is less favourable for plant growth with the exception of plants, which are accustomed to acidic conditions (e.g., blueberry, moss, heather). In very acidic soil, iron, aluminium and manganese may be enriched to toxic levels, whereas in very alkaline soil, the availability of trace elements is decreased.

Figure 1.7 shows the pH ranges, with the corresponding availability for chemical elements, and the implication for soil organism activity, and some soil forming processes (Ashman & Puri 2002; Schroeder & Blum 1992), whereas darker colours represent high availability and light colours low availability; white fields stand for non-availability.

Agricultural soil becomes gradually more and more acidic, because the cations that counterbalance acidification (Ca, Mg and K – referred to as 'base cations') are transported away with the

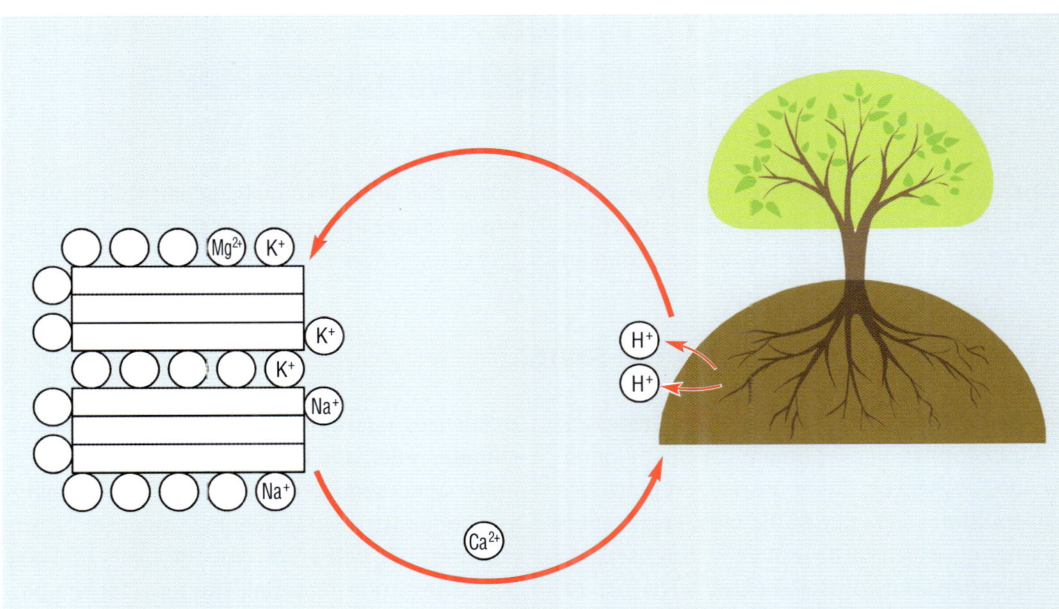

Fig. 1.6: Cation exchange. Plants exude weak organic acids (H^+) from their roots and stochiometrically exchange them for cations sorbed on the inner and outer surface of clay minerals. Note: The drawing is not true to scale since the particle diameter of clay minerals is < 2 µm, i.e., 0.002 mm.

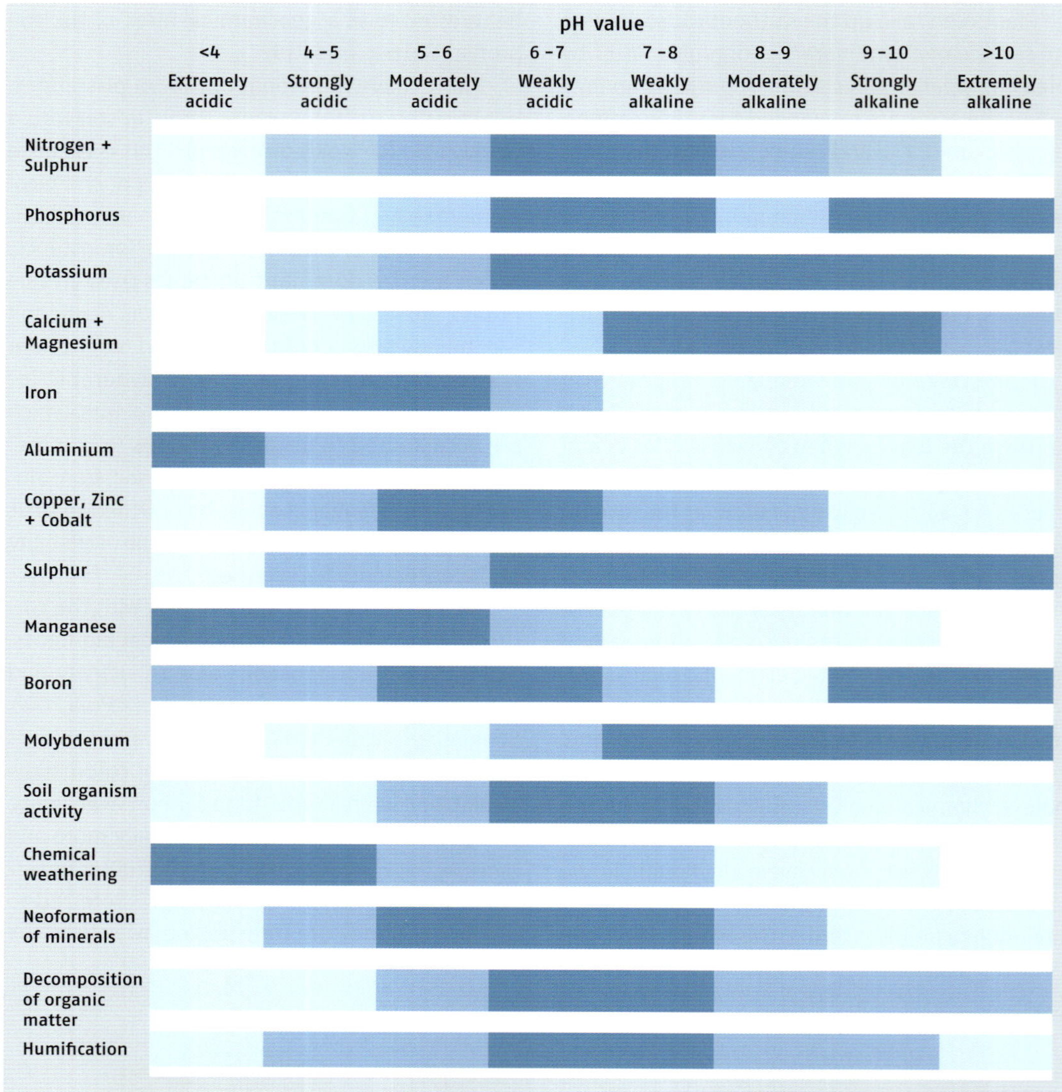

Fig. 1.7: Availability of major and trace elements based on the pH value (Ashman & Puri 2002; Schroeder & Blum 1992).

Note: Dark colours: high availability; light colours: low to no (white) availability.

plant harvest, and are not returned to the soil. In more sustainable forms of agricultural practices, the parts of the plants that are not harvested are left on the field. The remaining parts of the plants return some of the nutrients to the soil and protect the soil from erosion.

▸ 1.4 Does soil always stay the same?

No – as time passes, soil evolves. This process of soil transformation over time is called *soil formation* or *pedogenesis*. There are several processes that transform soil – one of the most important is weathering, which we have already discussed.

To understand how soil evolves, we have to recall the five soil-forming factors stated by Vasily Dokuchaev – **parent material, living organisms, climate, relief, and time**. A good starting point to understand pedogenesis is to look at the changes of one factor, while keeping the other factors constant. We will start with the basis of every soil – the parent material. The term for the parent material often used in literature is 'bedrock', but this term can be misleading, since the parent material can be solid rock – e. g., granite or limestone – but also unconsolidated sediments – e. g., glacial or river sediments. As wide is the variety of parent material, as wide is the variety of soil types that develop on them. A soil formed on a rock, such as granite, will be very different from a soil developed on limestone. Granite contains few nutrients and, therefore, plants that grow on the soil formed on granite must be adapted to these nutrient-poor conditions. Limestone consists primarily of calcium carbonate, and calcium is one of the main nutrients for plants. Therefore, a wide variety of plants can grow on soil developed in limestone areas.

The degree of weathering, and whether the parent materials are rocks or unconsolidated sediments, is also important for the type of soil, since the amount and size of pores influences the water and air regime in a soil and, therefore, the types of plants that can grow.

The third factor, which influences the soil formation, is the relief. A soil located on the flank of a hill will be constantly subjected to erosion, hindering the formation of deep soil and the colonisation by higher plants. Furthermore, water (e. g., from rainfall) will run from higher to lower areas, leaving the soil on the top and the flank of a hill often dry. At the foot of the hill material and also the organic matter from the hill will accumulate. The soil is deeper, but there is also the threat of stagnating water.

The soil aspect determines how much sun and rain a soil will get, which is essential for the temperature, the evaporation and the amount and type of soil organisms in a soil, and thus the types of plants that can grow on the soil and the mineralisation rate (and therefore the amount of nutrients).

The influence of the climate zone can result in very different – 'opposite' – soil types, when all the other factors are kept constant. Soil on granite in Scandinavia will be very different to soil on granite in Brazil. In the cold, wet climate in Scandinavia, granite will weather to nutrient-poor, sandy soil, and the soil is young due to the last Ice Age, which covered Scandinavia with a thick ice cover until about 10,000 years ago and eroded all pre-existing soil (Chapter 2). Soil in Scandinavia is thus very young. The soil in Scandinavia develops mostly under coniferous forest, the litter of which is hard to decompose, and the climatic conditions are unfavourable for soil organisms and, therefore, the vegetation litter is only slowly decomposed, leading to an accumulation of organic matter. The main soil type in Scandinavia is Podsol, which is characterised by a sandy soil texture, accumulation of organic matter, high amount of organic acids, which leach nutrients, and clay minerals from the topsoil. In Brazil, the soil is deeply weathered, since there was much more time for soil formation compared to Scandinavia. Furthermore, the higher temperatures are more favourable for soil organisms, which in turn increase the decomposition rate of organic matter. A wider variety of plants can grow on soil, resulting in increased nutrient availability, higher pH-values and less leaching of nutrients and clay minerals.

In Table 1.2, you can see some examples of the effect the five factors have on soil formation. You have to bear in mind that soil is the product, not only of one factor, but also of a combination of all five factors, and this is why we have so many different soil types on our planet.

We have now discussed the natural factors for soil formation. However, our own activities on the planet – agriculture, industry, etc. – are important factors in pedogenesis. We plough, irrigate, dig, build on, excavate, fertilise, and contaminate – in short, we use soil for our everyday life. The natural factors are the cause of the natural development or the normal pathway in the evolution of soil. The human-made or anthropogenic influences can change or disturb this pathway, and can result in a change of soil types.

Now, what is this natural pathway of soil development? We recall the very first steps of soil formation – primary colonisers, such as lichens grow on the rock surface accumulating organic matter over time (Figs. 1.1 and 1.2). Therefore, soil formation begins at the surface of rocks or on top of sediments. With time, the soil formation continues downwards such that distinct layers – the so-called **soil horizons** – are formed. Soil horizons are roughly horizontal layers of different thickness, which are distinguished by different parameters, such as organic matter content, soil texture, colour, clay content and many more. In some types of soil, the soil horizons are easily distinguishable, while in others the boundaries between the horizons are not easily detected. These soil horizons are the central point for the designation of the different **soil types** and are central in **soil classification**, which we will deal with later. You can see the soil horizons of a soil when you dig a so-called **soil profile**, which is the 'outdoor office' of every soil scientist. The soil profile is a two-dimensional vertical pit from the surface downwards. Either you can dig a soil profile or you can use a natural or human made section with visible soil horizons and parent material, for instance, where there are natural erosional channels or valleys, or where a road is being built, a road cutting

What happens next? The soil-forming factors are the cause and the starting point for the so-called **soil-forming processes** that take place within a soil. With time, the soil undergoes different processes, and which processes are active and how intense they are, depends on where the soil is located, how old it is, what the climate is like, the type of vegetation, etc. These processes differentiate the soil body and give the soil its characteristic local features or properties.

We can summarise this in the conceptual formula:

Soil-forming factors + Soil-forming processes = Soil properties

We learnt that when rocks weather and organic matter accumulates, soil starts to develop. The weathered rock and the organic matter on top are the first two **soil horizons**. These soil horizons are then the basis for the soil classification with which the soil is described and classified.

First, let us have a look at the soil horizons. We said that the first two soil horizons are the parent material (consolidated or non-consolidat-

Table 1.2: Soil forming factors and their characteristics.

Soil-forming factor	Example categories	Characteristics
Parent material	Silicate	Prone to acidification, poor soil life, low nutrient content, vegetation slowly decomposable (coniferous)
	Calcareous	High nutrient content, good buffer capacity, active soil life, vegetation more easily decomposable (deciduous)
Organisms	Active, many species	Under favourable conditions (see other factors), soil organisms decompose organic matter rapidly, providing nutrients for plants
	Poor soil life	Under unfavourable conditions (see other factors), the organic matter is only slowly decomposed and accumulates; the soil gets more and more acidic
Climate	Arid	Minimal leaching, slow dissolution of components
	Humid	Extensive leaching, rapid dissolution
	Cool	Physical weathering predominates over chemical weathering; accumulation of organic matter
	Warm	Strong chemical weathering; fast decay of organic matter (nutrient cycling)
Relief	Concave (valley, foot of a slope)	Accumulation of water and organic matter; threat of water-logging and anoxic conditions
	Flat or low inclination	Development of soil horizons according to soil-forming processes; balanced accumulation and erosion; development of deep soil profiles with time
	Convex or steeply inclined	Erosion of mineral and organic material; water run-off; loss of material and nutrients; poor and shallow soil
Time	Young	Shallow soil; parent material only slightly weathered; little or no differentiation into soil horizons; soil-forming processes only starting, no enhanced movement of particles
	Old	Deep soil; strongly weathered parent material; soil horizons well developed; profiles may contain buried soil horizons, which were eroded or formed under different conditions; many soluble chemical elements are leached and precipitated elsewhere

ed) with accumulated organic matter on top. The uppermost layer of the soil is the topsoil layer. It contains most of the organic matter mixed with the mineral fraction of the soil. This horizon is the most active layer of the soil. It contains most of the organic material and has, therefore, a darker colour than the rest of the soil profile. Young soil, but also soil in exposed areas – such as mountainous regions, consists only of these two layers. They are rather or very shallow and often very stony, so they are not suitable for agricultural purposes.

Apart from weathering, there is a wide variety of processes that can be active in the soil and which eventually determine its properties, and these in turn make it possible to group the soil into soil types. Different soil horizons develop between the topsoil layer and the parent material. Now, how many and which type of soil horizons develop depends on the five soil-forming factors.

The different soil-forming processes can be grouped into processes of translocation, transformation and mixing processes (Blume et al. 2010). The most important driver for the **translocation processes** is water. Water percolates through the soil profile and takes various compounds of the soil on its way with it – clay, humus, salts, carbonates, oxides and hydroxides. These translocation processes take place primarily downwards in the soil profile, which means that these compounds are washed from the upper parts of the soil profile, and are enriched in lower parts of the soil. Evaporation and ascending groundwater, however, can also cause translocation of mainly salts from bottom to top. The movement of these compounds and their different velocity produces the different layers or soil horizons between the topsoil layer and the parent material.

During weathering, the compounds of the soil are not only translocated, but some of them are also transformed or newly formed and, therefore, these processes belong to the group of **transformation processes**. Clay minerals are formed, which cause the soil to become loamier (heavier, finer grained). Furthermore, Fe-, Al- and Mn-oxides are formed from the parent material. These oxides – and amongst these particularly the Fe-oxides – give the soil its characteristic colour – the Fe-oxides are responsible for the yellowish-brown to darker brown colour of soil in the humid temperate climate zone. At the same time, they are able to capture many trace elements, and can thus give rise to the enrichment of other metals in certain soil layers.

The third group is the **mixing processes**. Burrowing animals, the freezing and thawing of water in the soil profile, as well as the swelling and shrinking of clay minerals during wet and dry periods, cause more or less intense reworking of the soil profile or parts of it. These mixing processes homogenise the soil material, mix the mineral fraction with the organic matter, and allow for water and air to move through the soil profile more easily by the formation of cracks, fissures, tunnels, etc.

Now, what does this mean in practice? Let us have a look at two soil types and the characteristic processes that are active.

Example 1: Podsol

The soil type Podsol typically develops in areas with a mostly cold and wet climate (the boreal climate zone), and is therefore the most common soil type found in Scandinavia. Podsol develops on felsic or acidic (low pH) bedrock (e.g., granite, gneiss, quartzite), or unconsolidated and quartz-rich sediments (e.g., sands, glacial moraines) under mostly coniferous forests with acidic litter, which is not easily decomposable. The low pH-values in this soil favours the destruction of primary (from the weathering of parent material) and secondary minerals (clay minerals), which are then washed out of the topsoil layer together with organic compounds. With time, this horizon is completely depleted of all compounds apart from quartz grains, and the characteristic pale white (ash white) layer develops, and is responsible for the name of the soil type Podsol. *'Pod'* is Russian and means *'ash'*. This process is called **podsolisation**. The compounds that are washed out from the upper layer are then 'deposited' or 'precipitated' in lower parts of the soil profile – the so-called **illuviation** zone. In well-developed Podsol, the organic and mineral compounds are illuviated in two separate layers. The dark layer contains the organic matter and the zone where the minerogenic compounds are deposited is often reddish, since Podsol develops from iron-rich sediments or rocks. The iron and aluminium compounds give

Fig. 1.8: Podsol in northern Sweden (photograph by Edith Haslinger).

the lower parts of Podsol the characteristic reddish colour. The lowermost parts of Podsol are then again lighter-coloured, since they consist of the least weathered parent material, which is quartz-rich and, therefore, rather pale (Fig. 1.8).

Example 2: Chernozem

Chernozem is a soil type that occurs in the semi-arid to continental climate, and is typical of steppe areas with loess or similar carbonate-rich sediments as parent material. As you can see in Figure 1.9, Chernozem consists of two layers. There is a very thick topsoil layer that is rich in organic matter and, therefore, dark-coloured, and a horizon with the (light) brownish parent material (loess). The continental climate zone is a rather harsh environment for soil organisms, since it is characterised by having hot summers and cold winters and only short humid periods. Due to these conditions, there is little time for the soil organisms to process and decompose the organic matter; thus, the organic matter is accumulated, and forms a very thick topsoil layer. Burrowing animals and insects make vertical tunnels of various sizes through the soil profile, and thereby mix the material of the topsoil horizon with the parent material. When the tunnels are filled with soil material, you can see dark 'spots' in the light-coloured parent material. These tunnels, filled with dark topsoil material, are known as *krotovinas* and are characteristic of Chernozem. The soil-forming process active in this soil is **bioturbation**. Furthermore, the carbonate-rich sediments of the Chernozem are decalcified – i.e., depleted of lime – with time and rainwater as agent. This process is called **decalcification** and is active in soil that develops on sediments and rocks. In these soil types – and therefore also in Chernozem – the leached carbonates precipitate then as soft powdery lime, pseudomycelium or concretions in the lower part of the soil profile.

In Table 1.3, you will find a summary of the most important soil-forming processes in European soil and their main characteristics (EC 2005; Blume et al. 2010; Schroeder & Blum 1992). The soil types are described in the following section; soil types, which are not described, are marked 'n.d.' in the Table.

Fig. 1.9: Chernozem in the Russian Steppe (photograph by Edith Haslinger).

Table 1.3: Soil forming processes and their characteristics.

Process		Description
Formation	Humus formation	Dead organic material from plants or soil organisms is decomposed by soil organisms, fungi and bacteria. After a mechanical disintegration by the soil fauna (meso- and macro-fauna), the material is colonised by microbes and fungi for further breakdown into their constituents (carbohydrates – e.g., cellulose, lignin and N-compounds). Eventually, water and carbon dioxide are released and the stable humus is formed.
	Mineral formation	Minerals can be inherited from the weathering of the parent rocks (primary minerals) or formed during pedogenesis (secondary minerals). Typical secondary minerals are oxides, hydroxides and clay minerals.
	Peat formation	Water logging inhibits decomposition of organic matter and accumulation of not fully decomposed plant material. Material with more than 30 mass-% organic matter is referred to as peat.
Decomposition, leaching	Decomposition of organic matter	Dead organic material from plants or soil organisms is decomposed by soil organisms, fungi and bacteria (see Humus formation).
	Weathering	(I) Physical weathering: mechanical disintegration of rocks by, e.g., temperature change, freezing-thawing of water in cracks, wind, pressure release, salt crystal growth, etc. – without chemical alteration. (II) Chemical weathering by (a) hydrolysis (breaking of chemical bonds of minerals by reaction with water), (b) hydration: addition of water to the crystal structure of a mineral, which causes physical stresses in the mineral, and (c) chemical oxidation of metals, e.g., Fe, which leads to the formation of Fe-oxides, such as haematite and goethite (responsible for reddish and brownish colours in the soil). (III) Biological weathering, mainly by bacteria and fungi by the release of organic acids to harvest the released nutrients of a rock.
	Leaching	Loss of soluble mineral compounds and organic matter by percolating water (rainwater). The compounds can precipitate in the lower parts of the soil profile or be washed out of the profile completely.
	Clay destruction	In soil with low pH-values, especially in cold climates, the clay minerals are irreversibly transformed and destroyed by the acid soil solution. The fragments are then leached by percolating water from the soil profile.
Translocation	Pedoturbation	In-situ mixing of soil material by animal burrowing (bioturbation), freezing and thawing processes (cryoturbation) or swelling/shrinking of clay minerals (argilliturbation).
	Clay migration	Downward movement of (fine) clay particles as well as Fe-, Mn- and Si-oxides and humus with descending water (precipitation).
	Salt and carbonate migration	Salts and carbonates are easily soluble and are leached by percolating water; in soil with groundwater, salts and carbonates can also ascend with the rising water table, and subsequently precipitated in the soil profile or even on the surface.
	Podsolisation	Leaching of organic matter, Fe- and Al-oxides in soil with low pH and in cool and humid climate with high precipitation.
	Gley formation	Fe- and Mn-oxides are reduced in the water-saturated zone of the soil and give this horizon a typical greyish-bluish colour; they are translocated with ascending groundwater to the upper parts of the profile, where they are oxidised in air-filled pores, forming Fe-/Mn-oxides and -concretions; the oxidising horizon is subsequently brownish-coloured.
	Salinisation	Enrichment of salt by saline groundwater (ascending) or artificially by irrigation or salting of roads in winter.

▸ 1.5 How many soil types are there?

The 'World Reference Base for Soil Resources' (IUSS 2006), which is the most widely used transnational soil classification system in Europe, lists 32 soil types occurring in Europe. The names of the soil types give an indication of their main features or properties. Some soil names are historical and date back to the very beginnings of soil classification by the early soil scientists like V. Dokuchaev, and often describe in simple terms what the soil looks like. For instance, the name for the soil type Chernozem comes from the Russian word **чернозём**, which simply means *black soil*.

It goes without saying that we will not discuss all the 32 soil types in an introductory chapter about soil science. For a detailed view, we selected the ten most common soil types of Europe according to the Soil Atlas of Europe (EC 2005):

Albeluvisol (15 %), Podsol (14 %), Cambisol (12 %), Chernozem (9 %), Leptosol (9 %), Luvisol (6 %), Gleysol, Histosol, Fluvisol and Calcisol (each 5 %).

In Figure 1.10, you can see a simplified map of the main soil types in Europe. A more detailed description of the major soil types of Europe can be found at the webpage of the Joint Research Centre (*http://eusoils.jrc.ec.europa.eu/Awareness/SoilTypes.html*). The following description of the 'Top Ten' soil types in Europe (Table 1.4) comprise all the soil types on the map, apart from Regosol, Kastanozem, Phaeozem, Umbrisol, Cryosol, Andosol and Vertisol. The descriptions of the soil types are summarised from the "Soil Atlas of Europe" (EC 2005), and Zech & Hintermaier-Erhard (2002).

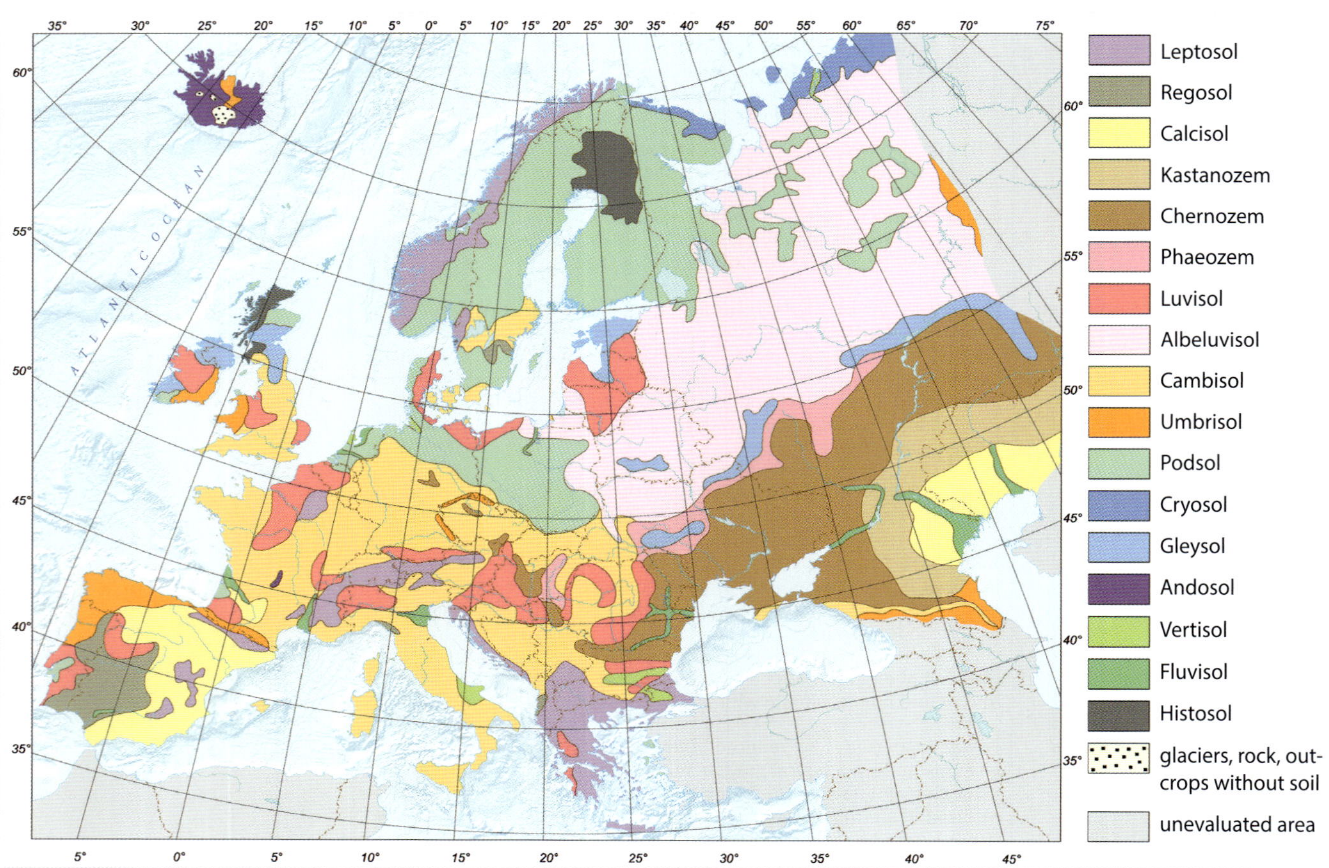

Fig. 1.10: Simplified map of the soil types of Europe (from Hrnčiarová et al. 2009).

Table 1.4: The ten most abundant soil types in Europe, their geographical distribution, parent material, use and the most important geochemical processes (photographs by Edith Haslinger).

Albeluvisol	From Latin *albus* = **white** and *eluere* = **to wash out / leach**
	Characterised by a pale brown to white layer below the organic topsoil rich in organic matter. From the pale / white, mainly clay minerals are washed out and accumulated in the layer below. These processes are referred to as eluviation (washing out) and illuviation (accumulation). The longer or the more intense the eluviation processes are, the more the colour of the layer will turn to white. The pale / white layer forms tongues reaching into the layer below, which is known as the characteristic albeluvic tonguing. In the boreal zone, the subsoil can freeze and will retain the water in the topsoil during snowmelt.
Geographical distribution	Continental boreal to temperate humid climate zone, the most common soil type in Europe (15 %), mainly in Eastern Europe (Fig. 1.10).
Parent material	Lime-free, fine-grained sediments (sands, aeolian sands, such as loess, delta sediments, etc.).
Use	Forms mediocre farmland due to low pH, low content of nutrients, tendency to water retention as well as erosion in hilly terrain. Cultivation of spring wheat, barley, sugar beets, forage plants and potatoes after application of manure and lime. Best used as grazing land and forest.
Geochemical processes	Translocation of base cations (together with clay) downwards due to eluviation processes; Fe and Mn are laterally transported above the clay-rich and dense illuviation zone and form Fe-/Mn-oxide and hydroxide concretions; clay minerals are often enriched in Al.

Podsol	From Russian *pod* = **under** and *zola* = **ash**
	As with Albeluvisol, Podsol is characterised by a pale to white layer below the organic topsoil. However, from the eluviation layer, mainly organic matter, iron and aluminium are leached and accumulated in layers below (in well-developed Podsol, there is one distinct dark layer with illuviated organic matter and one orange-brown layer with illuviated iron and aluminium). Podsol has very (or extremely) low pH-values and develops under coniferous forests. The cold climate hinders the decomposition of organic matter, leading to an accumulation and partially very thick layers of vegetation litter.
Geographical distribution	Continental boreal and temperate humid zone. The most common soil type in Scandinavia (Fig. 1.10).
Parent material	Quartz-rich, lime-free rocks, such as granite, gneiss and quartzite and their weathering debris (sandy sediment).
Use	Hardly suitable for agricultural use due to low nutrient content, low water retention capacity and low to very low pH-values. Application of lime and manure might make podsol suitable for potato cultivation.
Geochemical processes	Leaching of base cations (to groundwater); eluvation of Fe, Al, Mn, organic matter and heavy metals downwards and illuviation in the lower part of the soil profile; toxic levels of Al possible due to low pH-values.

Table 1.4: continued

Cambisol	From Italian *cambiare* = **change**
	The name of Cambisol points to the changes taking place mainly in the soil horizon below the topsoil layer. Another name for Cambisol used in several national classifications – brown soil – indicates the overall appearance of the soil profile. The brown colour comes from iron-bearing minerals of parent rocks (e.g., olivine, pyroxene, mica, amphibole), which are transformed to iron hydroxides and oxides, such as goethite (brown) and – in tropical climates – haematite (red) during weathering. During weathering of the parent rocks, they are also physically broken down to smaller particle sizes – from sand to silt and finally to clay. Clay minerals that form during weathering give the Cambisol a more or less loamy texture. Cambisol is a relatively young, moderately developed and weathered soil.
Geographical distribution	Cambisol develops on a wide variety of rocks and in different climatic zones – from continental boreal to temperate humid as well as (sub)tropical climates. Common soil type in Central and Southern Europe (Fig. 1.10).
Parent material	Calcareous and non-calcareous rocks.
Use	Highly suitable for agricultural purposes, especially on calcareous bedrock. On silicate bedrock, Cambisol is best used for grazing land or forest. This is also the case for shallow or rocky soil.
Geochemical processes	Chemical weathering of Fe-bearing silicates ⟶ base cations often leached from the soil profile, silicic acid forms clay minerals together with Al; Fe-oxides (Fe^{3+}) are formed that are responsible for the brownish colour; formation of iron-humus-complexes.

Chernozem	From Russian *tschornyj* = **black** and *zemlja* = **earth**
	Chernozem is a soil of the humid continental zone with cold winters and short, hot summers. In the dry summer months and during winter, the mineralisation is inhibited. Therefore, organic matter is accumulated, leading to the characteristic thick dark topsoil layer of the Chernozem. The biological activity is very high – burrowing animals and insects mix the organic matter with the mineral soil. The burrows that are formed by their activities – the so-called krotovinas – are filled with dark soil material.
Geographical distribution	Humid continental zone, steppe regions. Common soil type in South-east Europe (Fig. 1.10).
Parent material	Mainly loess and similar sediments.
Use	Chernozem is one of the most fertile soil types and, therefore, has a very high agricultural productivity. The dryness in the summer months and the extreme temperatures can be a limiting factor. Furthermore, Chernozem is prone to erosion in times of heavy rainfall.
Geochemical processes	Decalcification of upper parts of soil profile ⟶ precipitation of secondary carbonates and Ca-concretions in lower parts; in summer months ascending $Ca(HCO_3)_2$-rich soil ⟶ solution precipitation of secondary carbonates.

Table 1.4: continued

Leptosol	From the Greek word *leptos* = **thin**
	Leptosol is a shallow soil that develops over solid or only slightly weathered solid rock (calcareous or non-calcareous). The humus-rich topsoil layer is directly over the rock layer with no or only a slightly weathered layer in between. Leptosol indicates initial pedogenesis, or is located with inhibited pedogenesis, due to erosion or high altitude.
Geographical distribution	In mountainous areas, often on hills that are constantly subjected to erosion. Mainly in western Scandinavia, in the Alps and in the southern Balkans (Fig. 1.10).
Parent material	Calcareous or non-calcareous solid rock.
Use	Not suitable for agricultural use due to shallowness and high rock content. Use as grazing land or forest, whereas the formation of terraces helps to inhibit erosion.
Geochemical processes	Physical and chemical weathering leads to loss of base cations and subsequent disintegration of rocks.

Luvisol	From Latin *luere* = **to wash**
	The process of the eluviation of clay minerals from the topsoil to the subsoil is equal to the pedogenetic processes in the Albeluvisol, whereas the tonguing of the eluviated layer into the layer below is missing in the Luvisol.
Geographical distribution	Luvisol can be found in the temperate humid zones, often in former glacial or periglacial zones. Important soil type in Central, Western and Southern Europe (Fig. 1.10).
Parent material	Unconsolidated sediments, such as loess, aeolian sediments, sandy – silty gravel.
Use	Luvisol is highly suitable for agricultural purposes due to high nutrient content, favourable water and oxygen conditions. Due to the fine soil texture, Luvisol is prone to compaction and water retention. In hilly areas, preventive measures against erosion have to be taken (e.g. terraces).
Geochemical processes	Decalcification of upper soil parts; translocation of Fe, Mn, Al and Si downwards; leaching of base cations.

Table 1.4: continued

Gleysol	From Russian *gley* = **muddy soil mass**
	In contrast to most of the other soil types, the Gleysol is affected by groundwater. The groundwater comes near the surface and the lower part of the soil is almost constantly saturated with water. This part, therefore, shows a grey or even bluish colour, since there is no oxygen to oxidise the iron-bearing minerals, which would lead to a reddish-brown soil colour. The layer above the saturated layer is saturated with water only from time to time (according to the fluctuation of the groundwater table). Hence, this layer shows a scattered pattern of reddish-brown and grey parts, reflecting the varying conditions of saturation with water.
Geographical distribution	In lowland areas, depressions, valleys and at river, lake and sea margins. Occurs in Ireland, Scotland and Eastern Europe (Fig. 1.10).
Parent material	Fine- to medium-grained sediments.
Use	Due to the more or less constant saturation with groundwater, Gleysol is hardly of use for agricultural purposes. Intensive drainage would be needed to make this soil suitable for crop production. Forest use with trees accustomed to high water content (e.g., alder) is possible. Areas with Gleysol are often protected areas with regard to groundwater preservation.
Geochemical processes	In the oxidation zone, formation of Fe-oxides; in the groundwater saturated zone with little oxygen, Fe- and Mn-compounds (Fe^{2+}, Mn^{2+}) are mobilised and transported laterally or ascend to the oxidation zone and are oxidised. Ascending carbonate-rich groundwater can lead to precipitation of carbonates in the soil profile or on the surface.

Histosol	From the Greek word *histós* = **tissue**
	Histosol is made up nearly entirely of partially decomposed organic material; hence, its colour is almost completely dark brown to black. Commonly, the material of Histosol is known as peat or turf. Histosol develops in areas where more organic material is produced than decomposed, which is the case in concave waterlogged areas, such as in recent or former fens, lagoons or mangroves, where the organic matter gradually accumulated. Histosol is mainly found in present day cold climates, where the decomposition of the organic material is inhibited. Since there is no bedrock, which could deliver nutrients, the soil is very acidic (low pH-values) and the oxygen content is very low.
Geographical distribution	Mainly in boreal or subarctic areas and / or mountainous areas with much rainfall and low evapotranspiration – e.g., in bogs, moors and mires. Known from north-west Scotland and northern Finland (Fig. 1.10).
Parent material	Organic matter.
Use	Peat is used mainly in horticulture to improve the structure and water retention of soil, and to lower the pH-values for plants that require more acidic conditions. However, since bogs and moors are unique ecosystems, which are irretrievably lost when used, many horticulturists now refrain from using peat. The most common use of peat is for energy generation as biomass fuel, mainly in Finland and Ireland.
Geochemical processes	Availability of nutrients low, due to lack of replenishment by weathering rocks; very high CEC due to very high amount of organic matter; partially very low pH, especially in upland moors; in low-level moors, the pH-values can be higher due to ascending hydrogen carbonate ($Ca(HCO_3)_2$)-rich groundwater; anaerobic, wet and cool conditions lead to inhibited decomposition of organic matter.

Table 1.4: continued

Fluvisol

From Latin *fluvius* = **river**

Fluvisol is a soil of areas that are regularly inundated. These areas can be in river basins or in marine or lacustrine environments. Thus, the sediments deposited during inundation can be fluviatile, marine or lacustrine. The sediments reflect the respective source area. The layers of Fluvisol are not a result of pedogenesis, but consist of the deposited sediments.

The sediments reflect the geological provenance or source area. When a river flows a long distance and through various geological zones, as in the case of the river Danube, for instance, the sediments are very different in their mineralogical composition in the areas further downstream.

In areas with regular inundation (close to the water), there is not enough time for pedogenetical processes that would allow for the formation of clay minerals and/or iron oxides, which would give the soil a brownish colour. Therefore, the Fluvisol has a greyish appearance. In areas farther from the water, which are inundated only occasionally or during extensive floods, the Fluvisol is subsequently affected by pedogenesis and has a brownish colour and a loamy texture.

Geographical distribution	Along riverbanks and marine and lacustrine coasts in all climate zones.
Parent material	Sediments.
Use	Depends on regularity of inundation, the chemistry of the groundwater and the composition of the sediments. Most Fluvisol is not suitable for large agricultural machines, since its bearing capacity is limited. In case of big flood events, Fluvisol can be contaminated with material or fluids from, e.g., industrial sites.
Geochemical processes	Chemical properties dependent on type and provenance of deposited sediments; nutrient availability generally good, but threat of high contaminant contents, when, e.g., industrial areas were inundated during extensive floods.

Calcisol

From Latin *calx* = **limestone** or *calcarius* = **calcareous**

As the name indicates, the Calcisol is characterised by its high content of secondary carbonates (calcium carbonate), which comes from the parent material of this soil – calcareous sediments. The carbonates are leached from the upper part of the soil and precipitate in lower parts as so-called pseudomycelium – since it resembles the hyphens of fungi – and/or as a hard, massive layer ('calcrete').

Geographical distribution	In dry, arid and semiarid areas. Important soil type in Spain and along the coastal regions of the northern Caspian Sea (Fig. 1.10).
Parent material	Calcareous sediments.
Use	Due to the dry climate, most Calcisol must be irrigated. In this case, it is quite productive for vegetables and forage crops. Plants must be adapted to high lime-content; otherwise, there is a risk of chlorosis.
Geochemical processes	Precipitation of secondary carbonates by leaching of upper parts of the profile. Low availability of nutrients (e.g., N); deficiency in Fe and Mn; inhibition of K-uptake and proneness to chlorosis in plants due to high Ca-contents. Precipitation of carbonate leads to hard dry carbonatic horizons, which hinder root development.

Geol. Jb. B 103

▸ 1.6 Use of GEMAS data

GEMAS is a soil project, but it is also an agricultural project, since we took the samples from ploughed fields and grazing land (land under permanent grass cover). While sampling, we noted that the farmers were very interested in the project, and most of them asked whether the data would be available for them after the end of the project.

One of the main 'problems' for the use of GEMAS data by individuals is that the sampling was carried out at a very low density (1 sample per 2,500 km^2). Therefore, the samples are not directly useable for single plots of ground and not even by the farmer who owns the field where a sample was taken. Instead, they provide the continental scale overview.

The other restriction for the direct use of the data is that the chemical elements were measured in an extract of the very strong acid *aqua regia* – a mixture of nitric and hydrochloric acid. In soil science, the aqua regia extract is used to determine heavy metals in soil or other media, such as sewage sludge, which is applied on agricultural fields. This extraction with a very strong acid does not reflect natural conditions where plants or living organisms exude comparatively much weaker organic acids to dissolve the rocks in order to gain nutrients. Therefore, it cannot be said from high or low aqua regia data that the elements are also highly or poorly available for plants in the fields, or whether one element that shows high contents in one country – for instance – occurs also in potentially toxic levels for plants. Such questions would be best answered, when using a very weak extraction (see Chapter 13 on MMI® extraction). The problem here is that no generally accepted extraction procedure that is suited for all the elements of interest in the GEMAS project exists. Furthermore, quality issues (and detection limit problems for many elements) become more and more important the weaker the extraction used.

However, the data of the GEMAS project do give a good overview of the variation of the chemical elements in agricultural and grazing land soil on a continental scale. They reflect the overall geology combined with the soil forming processes of countries or larger regions; therefore, they can be used to get primary information, whether one element, for instance, is enriched in a region, or if there are deficiencies in another.

Certain elements are essential for plants; some can be problematic or even toxic when they occur in higher concentrations.

Tables 1.5, 1.6 and 1.7 may be of interest, particularly for farmers, who always want to know, what the elements in their fertilisers do in the soil, why they are important and why some elements might be toxic when enriched in soil. The information in the Table was compiled from Bowen (1979), Schroeder & Blum (1992), Kabata-Pendias (2001), Kabata-Pendias & Mukherjee (2007) and Blume et al. (2010).

In addition to the geochemical situation in a region or country, farmers or other interested persons can consult soil and geological maps of the area or field that they are interested in. These maps are available at the national Geological Survey of each country and some of them may even be available online. Some communities and municipalities also offer maps with detailed information about the soil types in a certain area. For individual plots, it is important to know about the texture as well as the water and nutrient contents of a soil. The GEMAS data, together with the soil characteristics of individual fields, can help farmers with land management with respect to choice of crop, tillage and fertilisation. |◂

Table 1.5: The most important macronutrients in a soil, their source, their function, deficiency and toxicity symptoms, and their concentration in a typical soil.

Macronutrients in Soil					
				Content range in soil	
			Deficiency (D) / Toxicity (T) Symptoms	Range	Global median
Nutrient	Source	Function in soil		[%]	
Calcium (Ca)	Limestone (calcium carbonate); Ca-bearing silicates (feldspar, augite, hornblende); anhydrite and gypsum (Ca-sulphates); Ca-containing phosphates. Lime is often used in agriculture and horticulture to increase the pH-value of the soil. Highest contents in soil developed on shell bearing limestone.	Plant and root growth, stabilisation of cell wall, activation of membranes	**(D):** Root growth defects, brown spots on fruits.	0.07 to 50	1.5
Magnesium (Mg)	Mg-silicates (serpentine, chlorite, augite, hornblende, olivine, biotite); Mg-carbonate (dolomite, magnesite)	Magnesium is a constituent of important plant compounds (chlorophyll, enzymes, etc.). Magnesium is important for ion-transport systems; it is taken into plant roots by mass adsorption.	**(D):** Chlorosis, necrosis, inhibited growth. Intensive fertilisation with ammonium, potassium and calcium reduces the uptake of Mg.	0.04 to 0.9	0.5
Nitrogen (N)	Organic matter, nitrogen of the atmosphere, oxides, ammonia	Nitrogen is essential for metabolic and structure functions in plants and for the function of enzymes, chlorophyll, nucleic acids, amino acids, proteins and other compounds; essential nutrient for biological growth and activity (atmospheric N is the source for plant growth)	**(D):** Inhibited plant growth.	0.02 to 0.5	0.2
Phosphorus (P)	Phosphate-bearing minerals (Ca-, Al-, Fe-phosphates) and phosphate rocks	Constituent of different compounds, e.g., proteins. The phosphorus anions are rapidly metabolised into organic compounds; the main function is in energy transfer as adenosine triphosphate (ATP).	**(D):** Inhibited or delayed growth of leaves and roots. Deficiency only in sandy soil; in humid-temperate climate, and especially in areas with industrialised agriculture, P-excess due to intensive fertilisation.	0.0035 to 0.53	0.08
Potassium (K)	Sheet silicates (biotite, illite) and K-feldspar; salt deposits (potassium chloride and sulphate, potash salt deposits)	Potassium activates enzymes; it is essential for the water balance of the plants, inhibits the evapotranspiration, thus protecting the plant from dehydration; stimulation of photosynthesis; strengthens the flow of carbohydrates from the leaf tips.	**(D):** Inhibited growth and plant stability; fruits and vegetables have a poorer taste and have decreased storage stability. Potassium is adsorbed on clay minerals and organic matter; therefore, there are deficiencies in soil rich in clay or humus, as well as in sandy soil.	0.008 to 3.7	1.4
Sulphur (S)	Fe-sulphides, Ca-sulphates (e.g., gypsum); concentrated in coal, gas, sulphide ores, in crude oil and elemental forms (several allotropic forms)	The S allotropic forms mainly consist of the amino acids cystine, cysteine and methionine. In soil containing a high proportion of organic matter, the sulphur content is mainly bound in organic molecules. Soil acidification by acid precipitation can lead to a loss of plant nutrients, and the accumulation of toxic aluminium, both of which may affect vegetation. Sulphur supplementation of plants largely depends on the uptake of sulphates by the roots where they are reduced to sulphides, and assimilation of reduced sulphur into sulphur containing amino acids and peptides.	**(D):** Plants deficient in sulphur are smaller than normally developed plants and show the following symptoms: branchless roots, shorter stems, smaller leaves, and overall yellowing of new leaves. Influence of sulphur excess: direct damage caused by stack gases near towns and industrial sites.	0.003 to 0.16	0.07

Table 1.6: The most important micronutrients in a soil, their source, their function, deficiency and toxicity symptoms, and their concentration in a typical soil.

Micronutrients in Soil					
				Content range in soil	
				Range	Global median
Nutrient	Source	Function in soil	Deficiency (D) / Toxicity (T) Symptoms	[mg/kg]*	
Boron (B)	In B-minerals (borates), in silicates (tourmaline, mica) and in some salts; accumulation in granite and pegmatite; affinity to organic matter. Borates (boracite, borax).	Boron is an essential micronutrient for plants; phytotoxic in higher concentrations. Regulates the lipid composition of the plasma membrane; can affect total boron uptake by a plant, accumulated in old leaves.	B-deficiency in acidic and sandy soil, especially in warm and dry years, and in arid climate. In humid climate, toxic contents are due to fertilisation, application of sewage sludge and in waste water (washing powder); in arid climate, due to B-accumulation in salt, but also due to irrigation with B-rich water. Plants with high B-need: maize, sugar beet, rape, apples, wine, cotton. B-fertilisation is delicate, since B can be easily under- and over-dosed. **(D):** Dry rot, cork formation, dying of young leaves. **(T):** Loss of photosynthetic capacity, necrosis on leaves; chlorosis (yellowing) of the leaf tip.	5 to 50	20
Chlorine (Cl)	Chlorides, highest concentrations in mafic and some calcareous rocks; most occurrence in its organic form or as mineral salt; fixed in solid matrix of volcanic activity, concentrated as organic-bond Cl_2 in oils and coals.	Chlorine is absorbed by plant leaves and is considered an essential micronutrient for higher plants; regulation of water balance and cation-anion-equilibrium in plants; central role in the osmotic system of plants (influence of acid-base balance).	Halophilic plants: sugar beet, spinach, celery. Salt-sensitive plants: potatoes, fruit, tomatoes, wine, tobacco, etc. **(T):** Abiotic damage of leaves	8 to 1800	100
Cobalt (Co)	Ultramafic rocks; minerals enriched in Fe- and Mn-hydroxides (e.g., biotite, Fe-/Mn-oxides); in sulphidic copper ores, in arsenides and arsenosulphides; high concentrations in heavy loamy soil.	Essential for the N_2-fixation by rhizobia, cyano- and other bacteria.	**(D):** In acidic, sandy soil (Podsol) **(T):** White, dead leaf margins and tips, chlorosis of new leaves (is linked with Fe-chlorosis)	0.05 to 65	8
Copper (Cu)	In Cu-minrerals (sulphides, sulphates, oxides, carbonates). In soil, primarily adsorbed on organic matter and pedogenic oxides. Copper is used in many industrial products; processing of copper ore; application of sewage sludge. Partially excessive use in viniculture and pomiculture.	Essential for plants, humans and animals; as part of the chloroplast Cu is essential for the photosynthesis; important for the water balance of plants.	**(D):** In soil rich in organic matter (Cu is immobilised), such as Podsol and Histosol. Plants sensitive to Cu-deficiency: wheat, oat, legumes, fruit trees. **(D):** Chlorosis and white colouring of leaves. Toxic in high concentrations; leads to decreases in farm yield	2 to 250	30
Iron (Fe)	In Fe-minerals (Fe-oxides and -hydroxides, Fe-sulphides); in Silicates (augite, hornblende, biotite, olivine); accumulation in concretions and nodules, forming of chelates with humic and non-humic organic substances.	Essential element for plants, humans and animals; key metal in energy transformation needed for synthesis and other life processes of the cells; plant roots may reduce Fe^{3+} to Fe^{2+}; influencing of chlorophyll formation; organic Fe-complexes are involved in the photosynthesis; iron is directly implicated in the metabolism of nucleic acids.	Deficiency in $CaCO_3$-rich soil (or in limed soil), soil in arid climate or in very acidic soil (moors); deficiency leads to chlorosis symptoms in plants. **(T):** Root damage	0.2 to 55 %	4 %

Table 1.6: continued

Micronutrients in Soil					
				Content range in soil	
				Range	Global median
Nutrient	Source	Function in soil	Deficiency (D) / Toxicity (T) Symptoms	[mg/kg] *	
Manganese (Mn)	In Mn-minerals (manganite, pyrolusite), in silicates (as accessory); high contents often in soil derived from mafic rocks, in Fe and/or organic rich soil, and in soil from arid to semiarid climatic regions.	Essential element for all organisms; water balance in plants, synthesis of chlorophyll, formation of amino acids, etc.; most important function in plants is related to oxidation-reduction processes.	Deficiency occurs in sandy soil (e.g. Podsol), fens and marshes in humid climate; in arid climate in carbonate-rich soil. Plants sensitive to Mn-deficiency: peas, oats, beets and potatoes; lack of Mn leads to growth disturbances. **(D):** Grey staining in plants or death of the plant.	10 to 10.000	1000
Molybdenum (Mo)	In Mo-minerals (sulphide); in silicates, oxides and hydroxides (as accessory). High Mo-contents in areas with ore smelting, coal and crude oil processing	Constituent of enzymes; needed for nitrogen fixation.	Mo-deficiency in sandy soil (Podsol). Mo is needed for the N-fixation – deficiency leads to N-depletion symptoms in legumes and delays the formation of chlorophyll. High Mo-contents in fodder leads to poisoning symptoms, especially with cattle.		
Selenium (Se)	Constituent of a wide variety of rocks; enriched in Se-bearing ores (Cu_2Se), which occur together with sulphides; high concentrations in volcanic rocks, sandstone, uranium deposits and carbonaceous rocks. Se-enrichment by incineration of coal, application of sewage sludge and coal ash, industrial emissions; volcanic activities; highest contents often in organic rich, calcareous and volcanic soil.	Not essential element, but beneficial for plants; essential for humans and animals. Increasing of soil acidity reduces the rate of biotransformation of Se compounds to dimethyl selenide, which results also to reduce movement of soil Se to plants.	Threshold range between deficient and toxic concentration very narrow; **(D/T):** symptoms mainly in cattle. **(T):** Some plants are selenium-tolerant, in non-tolerant plants selenium compounds may impair germination and growth and lead to chlorosis, damage of plants (replacement of sulphur in proteins by selenium); causes so-called alkali disease in cattle.	0.005 to 3.5	0.4
Silicon (Si)	Second most common element (after oxygen) in the Earth's crust; in silicates and oxides.	Silicon is a common mineral constituent of plants and is not defined to be essential. Plants take up silicic acid (H_4SiO_4), which is formed by weathering of silicates. Some plants accumulate and can contain Si up to 10 % (e.g., rice). Important element for increasing the tolerance of plants against environmental stress; necessary for plant growth and stabilisation of plant tissue (e.g., stalks); Si increases the sought tolerance of plants and provides them with a higher resistance against diseases and insects.	**(T):** Only known from very acidic forest soil, where the Si-uptake in needles of coniferous trees is increased, leading to inhibited water and nutrient flow. In arid areas, Si leads to a cementation of soil particles and decreased root propagation. Fertilisation with phosphorus leads to an increased availability of Si.	25 to 41 %	33 %
Zinc (Zn)	In Zn-minerals (sulphides, phosphates, carbonates, hydroxides), in silicates (accessory); highest concentrations in calcareous and organic soil. In industrial areas, Zn occurs in toxic concentrations (zinc coatings), smelting of zinc ore; emissions from car traffic, application of sewage sludge. Occurs often together with other heavy metals (Cd, Pb, Cu).	Essential element for plants, humans and animals. It influences the permeability of membranes and stabilises cellular components; stimulates the resistance of plants to dry and hot weather, and to bacterial and fungal diseases.	Deficiency occurs only rarely, sometimes in $CaCO_3$-rich soil, in semi-humid to arid climate zone or in phosphate-rich soil. Plants sensitive to Zn-deficiency: apples, vine, lemon, maize, beans, potatoes, onions. Symptoms: chlorosis, smaller leaves. **(T):** Inhibited growth; impairment of micro-organism activity. Phytotoxicity is reported especially for acid and heavily sludged soil. Liming and application of Fe-oxides may meliorate Zn-contaminated soil.	1 to 900	90

* unless otherwise stated

Table 1.7: Potentially toxic elements in a soil, their source, their behaviour, toxicity symptoms and their concentration in a typical soil.

Potentially toxic elements in soil					
				Content range in soil	
				Range	Global median
Element	Source	Behaviour in soil	Toxicity symptoms	[mg/kg]	
Cadmium (Cd)	Often occurs together with Zn, e.g. in the main Zn-mineral sphalerite; mineral greenockite (CdS); in industrial waste, sewage sludge, air, phosphate fertilisers; high concentrations in organic-rich shale, phosphorite and carbonates, kaolin sand, coal.	Cadmium is chemically similar to Zn; relative Zn/Cd-ratio in rocks (500) and soil (100) – Cd is enriched during soil weathering; Cd is adsorbed to the organic matter of soil.	Toxic for humans and animals already in very low concentrations. Very toxic to plants; chlorosis, necrosis, growth inhibition; Cd accumulation in plants causes disorders in enzyme activities, metabolism transpiration and photo-synthesis; plants sensitive to Cd-contamination: beans, peas, beets, spinach	0.01 to 2.0	0.35
Chromium (Cr)	In serpentinite and ultramafic igneous rocks; chromite ores ($FeCr_2O_4$); Cr(III) can also be constituent of Fe-oxides; higher contents in soil derived from mafic rocks and argillaceous sediments. Hexavalent Cr is widely used in industry (steel, paintings, glass, ceramic, leather); in (fertilisers from) industrial waste and wastewater; sewage sludge.	Enriched mainly in plant roots, in mineral fraction and in organic matter of soil. Cr availability strongly depends on Cr valency. Cr (III) minerals almost insoluble under natural conditions. Only essential for humans and animals for glucose metabolism, enzymes; stabilisation of proteins and nucleic acid.	Hexavalent Cr(IV) is extremely toxic for human beings, animals and plants. Cr(III) is toxic for plants only in extremely high concentrations (inhibition of growth).	5 to 1500	70
Fluorine (F)	Emissions from chemical and metal industry; waste incineration; fossil fuel power plants, P-fertiliser factories, Al-smelters; highest contents in heavy clay soil and in soil derived from mafic rocks.	Fluorine not needed by plants; for humans and animals essential for dental enamel.	Toxic for all plants, humans and animals at high concentration. Necrosis on leaves; retarded growth inhibited reproduction, yield reduction.	20 to 700	200
Lead (Pb)	Accumulated in Fe-Mn-nodules and concretions; higher contents in calcareous and organic soil (concentrated in carbonate or phosphate particles). In areas with Pb-ores and in industrial and/or contaminated areas; use of Pb-bearing petrol. Pb carbonate precitpitation in Ca-rich mining waters.	Less toxic than Cd and Hg. In soil with $pH > 5$ immobile; $pH < 4$ very mobile. Lead is immobilised in organic matter, which can hinder mobilisation also in acidic soil. Adsorption on Fe-, Mn- and Al-oxides, accumulated in topsoil due to sorption on the soil organic matter. In phosphate-rich soil formation of Pb-phosphates; in $CaCO_3$-rich soil formation of Pb-carbonate.	Plants are usually well protected against uptake of Pb, however, on top of mineralisation, plant samples have returned values over 1000 mg/kg Pb. Toxic in high concentrations; leads to decreases in farm yield.	2 to 300	35
Mercury (Hg)	Hg-bearing ores, cinnabar; native Hg; volcanic activities; highest contents in organic soil. Incineration of fossil fuels; application of Hg-bearing fungicides; application of sewage sludge; industrial processes.	Very insoluble in soil, not plant-available; organic matter immobilises Hg, and also pedogenic oxides and clay minerals. Hg^{2+} species in soil undergo biological and chemical reactions forming different inorganic and organic compounds, which are controlled by factors as temperature, pH and soil organic matter content.	Extremely toxic for humans and animals; for plants only in very high concentrations due to the immobilisation in soil (above all in the organic matter); Hg-salts, or metallic Hg, cause hazardous effects on plants and soil biota. Stunting of growth of seedlings and roots, inhibition of photosynthesis, inhibition of K uptake, reduction of yield.	0.01 to 0.5	0.06
Nickel (Ni)	Ultramafic rocks with high amount of olivine; higher contents in clay rich and calcareous loam soil; emissions from metal industry, fossil fuels, vehicles.	Beneficial for plants and animals in low concentrat-ions; adsorbed on Mn-, Fe- and Al-oxides and clay minerals; higher availability in acidic soil; in surface soil Ni occurs mainly in bound organic forms. Nickel behaviour and its phytoavailability are controlled by soil properties (pH, clay content, soil organic matter); very high Ni concentration in soil solutions; Ni is readily taken up by plants (e.g., from polluted soil).	Increased Ni levels in fodder plants are associated with health hazards. Inhibited growth; impairment of micro-organism activity; chlorosis and necrosis on leaves.	2 to 750	50

Edith Haslinger
AIT Austrian Institute of Technology GmbH; Tulln, Austria
E-mail: edith.haslinger@ait.ac.at

Clemens Reimann
Geological Survey of Norway (NGU); Trondheim, Norway

Manfred Birke
Federal Institute for Geosciences and Natural Resources (BGR); Hannover, Germany

Chapter 2

Geology of Europe

FABIAN JÄHNE

▸ 2.1 Introduction

According to the Glossary of Geology (BATES & JACKSON 1980) Geology is: "The study of the planet Earth – the materials of which it is made, the processes that act on these materials, the products formed, and the history of the planet and its life forms since its origin".

This chapter provides an overview of the different rock types that make up the continent and provides a summary of the Historical Geology of Europe. Historical Geology focuses on geological processes that change the Earth's surface and subsurface; and it involves the use of stratigraphy, structural geology and palaeontology to tell the sequence of events throughout geological time (LEVIN 2010).

A concise description of the Geology of Europe can be found in the *"FOREGS – Geochemical Atlas of Europe"* (Chapter 2 by PLANT et al. 2005), in *"The Geology of Central Europe"* (MCCANN 2008) and in *"A Continent Revealed – the European Geotraverse"* (BLUNDELL et al. 1992). In the past few decades, several atlases of the deeper subsurface (e.g., DOORNENBAL & STEVENSON 2010; EVANS et al. 2003), palaeogeographical map compilations (e.g., DERCOURT et al. 2000; ZIEGLER 1990) and several plate-tectonic reconstructions (e.g., SCOTESE 2010; STAMPFLI & KOZUR 2006) have been published, which help to understand the complex geological history of Europe and the evolution of today's structural patterns.

Figure 2.1 provides a Geological Map of Europe, which was especially drafted for the scale and the projection used in this atlas. The age span of geological formations at the surface extends from Archaean rocks (3.75 – 2.5 Ga), extensively exposed, in the north of Europe and in the Ukraine to Quaternary and recent deposits in the lowlands of Europe and young volcanic deposits in the Southern and Eastern Mediterranean. The age distribution of rocks (Fig. 2.1) reflects in parts the main structures of Europe too (compare Fig. 2.5). Most of Southern and Western Europe is covered by Mesozoic (251 – 65.5 Ma) to Caenozoic (65.5 Ma – recent) sediments.

The oldest rocks are found in outcrops on the East European Craton (fragment of Baltica) in the Ukraine Massif (Figs. 2.1, 2.5) and are mid-Archaean (*ca.* 3700 Ma) in age (GEE & STEPHENSON 2006). In Scandinavia, Archaean and Proterozoic rocks outcrop over large areas. It is a good place to study processes of crustal genesis in the early history of the Earth on the basis of some rare igneous rocks like komatiites, which are mainly found along greenstone belts in continental masses of Archaean age. Greenstone belts (Figs. 2.1, 2.3) are Archaean to Proterozoic metamorphosed mafic to ultramafic volcanic sequences with associated sedimentary rocks that have been interpreted as ancient oceanic spreading centres and island arc terranes (DE WIT & ASHWAL 1997).

Palaeozoic terranes form most of the crustal basement of Western Europe. The Palaeozoic basement outcrops over large areas in the Bohemian and Rhenish Massif, in the Massif Central, in the Armorican Massif of western France, in Western Iberia, in Ireland and northern Great Britain (Fig. 2.5). In the other parts of Western Europe and in the Mediterranean region, the Palaeozoic bedrocks are covered by Mesozoic to Caenozoic terrestrial to marine sediments. The greatest thicknesses of these sediments are attained in the sediment basins shown in Figure 2.8.

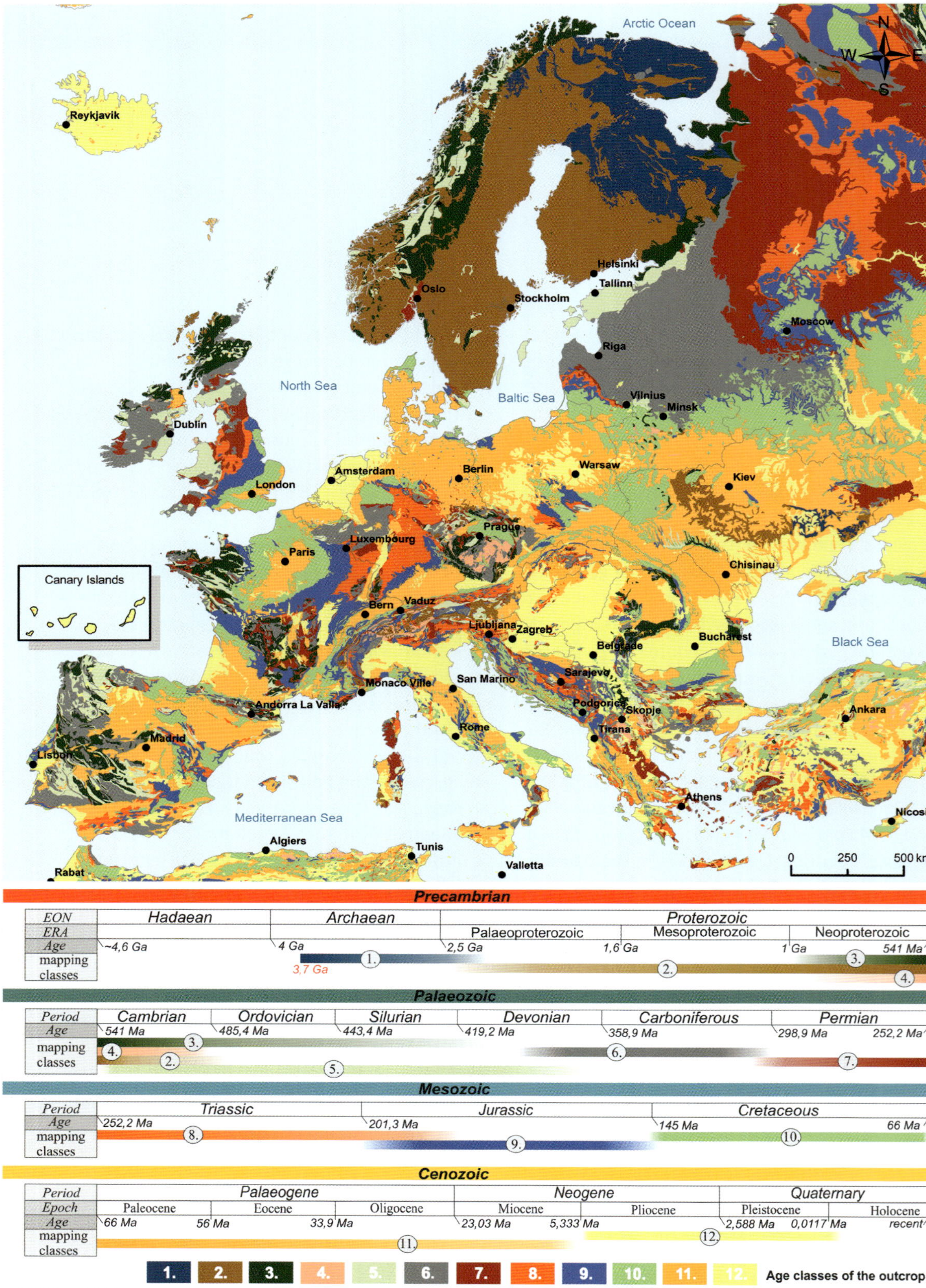

Fig. 2.1: Simplified geological map of Europe. The legend shows that there is an overlap between some of the age classes chosen; this is an effect of inaccuracy in dating and stratigraphical classification of the rock units (edited by F. Jähne; the map is based on the IGME 5000 by ASCH (2003); the time scale of the legend is based on the International Chronostratigraphic Chart by COHEN et al. 2013).

▸ 2.2 The European plate tectonic evolution in the Phanerozoic

European Phanerozoic tectonic history was dominated by several orogenic events with accretion of terranes, phases of extensive rifting and continental breakup. Throughout most of the Early Phanerozoic, Baltica, that is now part of the East European Craton (EEC) represented a more or less stable continent. During this time, it acted as a nucleus for a lot of smaller microcontinents and island arcs that form the crystalline basement of Central and Western Europe. Especially in Late Neoproterozoic and Palaeozoic times, subduction with accretion of terranes and microcontinents and the adjacent formation of massive orogenies had taken place along all borders of Baltica. The Phanerozoic orogenies with the greatest impact on the structural pattern of today's Europe are the Cadomian Orogeny, the Caledonian Orogeny, the Variscan Orogeny, the Cimmerian and the Alpidic Orogeny.

The Phanerozoic evolution of Europe started with the Cadomian Orogeny that commenced in the mid-Neoproterozoic (*ca.* 750 Ma) and extended into the earliest Cambrian. It was the last in a series of events that formed the crystalline basement rocks of Europe (LINNEMANN et al. 2008). The Cadomian Orogeny was driven by amalgamation of the 'Vendian supercontinent', which comprised Laurentia, Gondwana and Baltica (MCCANN 2008a). By breakup of the Vendian Supercontinent in the Ediacaran (600 – 550 Ma), the formation of the Iapetus Ocean was initiated (Fig. 2.2a). From the Late Neoproterozoic up to the Ordovician, most European terranes were situated around the southern polar region (Figs. 2.2a, b). Baltica as the greatest terrane of Europe started to drift northward in the Early to Mid-Ordovician and crossed the Equator in the Early Carboniferous. The Caledonian orogenic cycle occurred between Late Cambrian and earliest Devonian times. The collision of Baltica and Laurentia with Avalonia (Fig. 2.5), a fragment of Gondwana, induced the complex closure of the Iapetus Ocean and of the Tornquist Sea (Fig. 2.2b). However the main closure events happened in the Silurian (*ca.* 425 Ma) and collision-related deformation continued into the Devonian (*ca.* 390 Ma), followed by post-orogenic wrench faulting, back-arc rifting processes in the Caledonides and Variscan-related extension (Rhenohercynian Basin). The Caledonides (Fig. 2.5) form most of the crust of Great Britain, Ireland, Norway and the basement beneath the Central European Basin System in The Netherlands, Germany and in parts of Poland (LITTKE et al. 2008). The transition from the Caledonian Orogeny to the Variscan Orogeny (Middle Devonian – Late Carboniferous) is almost continuous. Most of the crust in Western and South-Western Europe, but also parts of the crystalline basement of the Alps, is characterised by Variscan deformation (Fig. 2.5). The Variscan Orogeny marks the final amalgamation of Laurasia (*Laurentia*, *Baltica* and other continents that make up today's continents of the northern hemisphere) in the North, and Gondwana (*Antarctica, South America*, *Africa*, *Madagascar*, *Australia-New Guinea*, and *New Zealand*, as well as *Arabia* and the *Indian subcontinent*) in the South to form the supercontinent of Pangaea, by closure of major oceans, such as the Rheic Ocean and the

Fig. 2.2: Reconstruction of important stages of European plate tectonic evolution based on MCCANN (2008a). a.: Tectonic setting and plate positions during the Late Ediacaran; b.: The plate tectonic setting during the Caledonian Orogeny; c.: The end of the Variscan Orogeny and formation of the supercontinent Pangaea; d.: The break-up of Pangaea and the opening of the Central Atlantic.

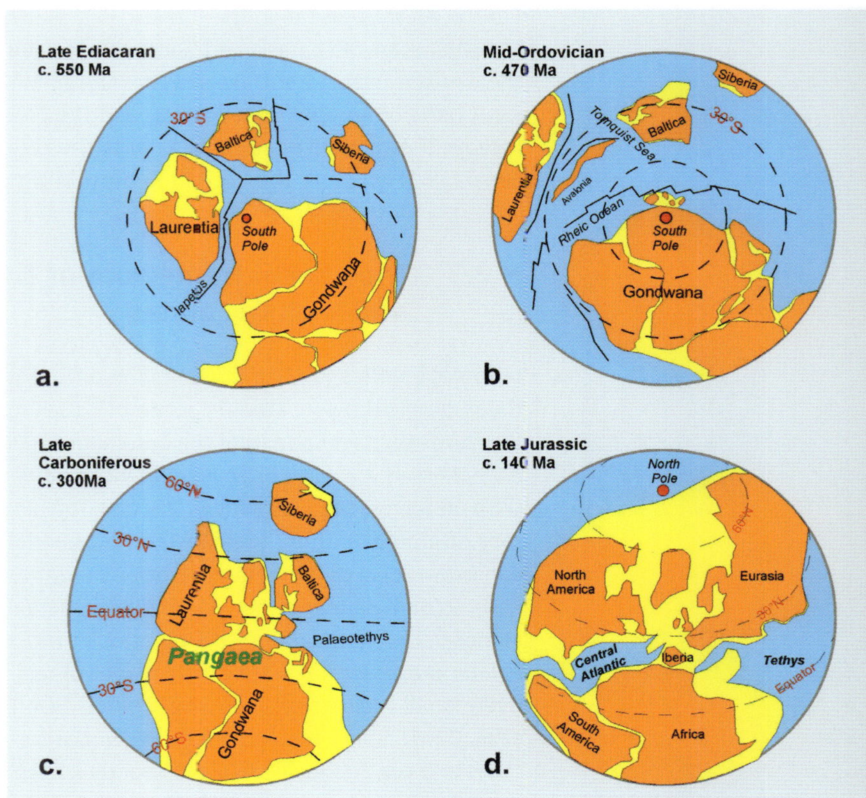

Proto-Tethys (Fig. 2.2c). Variscan deformation continued into the Late Carboniferous (McCann et al. 2008), with widespread thrust and wrench faulting, folding, post-tectonic granite emplacement and extensive sedimentation in foreland basins (North-west Carboniferous Basin). The change of convergence between Gondwana and Laurussia (Laurentia, Baltica and Avalonia – the Old Red Continent), in the late stages of the Variscan Orogeny, to an east-west orientation formed continental-scale dextral shears (Ziegler & Dezes 2006). One of the most prominent is the Tornquist-Teisseyre Zone (TTZ; Figs. 2.4, 2.5, 2.9). The late stages of the Variscan orogen were characterised by emplacement of granitoids in many parts of Variscan Europe and by a widespread late orogenic volcanism in the Oslo Rift and in north-east Germany (Timmerman et al. 2009; McCann et al. 2006). The beginning of the collapse of Pangaea started in the Late Permian to the Mesozoic with extensive rifting, also in the Southern Permian basin and in the North Sea, which resulted in significant crustal thinning. At first North America and Gondwana were separated by the opening of the Central Atlantic Ocean in the Jurassic (Fig. 2.2d). The separation of Laurasia and Gondwana formed the western Tethys, a shallow sea covering significant parts of Europe in the Jurassic and the Cretaceous (Pieńkowski et al. 2008). Prior to the Late Mesozoic (Fig. 2.2d), the Iberian Peninsula was joined to Armorica (northern France). In the Early Cretaceous, the Bay of Biscay (Figs. 2.4, 2.5) started opening around 126 Ma and separated Iberia from Western Europe (Stampfli & Borel 2004). Up to the Santonian, Iberia rotated anticlockwise relative to Eurasia. In the Late Cretaceous to Palaeogene, Europe began to separate from Greenland by a northward opening of the North Atlantic.

The Alpine Orogeny is the last main event for Europe's structural genesis, which has formed the mountain belts in Central Europe and around the Mediterranean Sea (Figs. 2.4, 2.5). The former deformation and accretion of terranes, during the Cimmerian Orogeny, are well preserved in the south-eastern Alps and in Turkey (Figs. 2.3, 2.5). The formation of the Alps resulted from the collision of the composite Adria microplate (Fig. 2.5), and the stretched southern margin of the European plate, including the subduction of several strips of intervening oceanic lithosphere (Kley & Voigt 2008). The Alps show two main collisional stages (Eisbacher & Brandner 1996; Ring et al. 1988; Schmid et al. 1996). The first stage, from Middle to Late Cretaceous, is only recorded in the Austroalpine units of the Adria plate (Kley & Voigt 2008). The second stage formed the shape of today's Alps and resulted from N–S directed plate motions between Africa and Europe in the Eocene, and NW – SE directed motions since the Miocene (McCann 2008a).

▸ 2.3 The crustal framework of Europe – distribution of terranes and tectonic plates

Inhomogeneities in structure and the different rheological behaviour (the response of a rock to the forces that act upon it) of the basement rocks (Figs. 2.6b – d, 2.7) have often an influence on subsequent tectonic evolution (Figs. 2.5, 2.9). The history of amalgamation, breakup, compressional and extensional deformation of Europe's terranes and tectonic plates formed a complex structural mosaic. In places, the reworking of older rocks in later orogenies has resulted in collages of relatively small shear-zone-bounded terranes (Plant et al. 2005). Some boundaries between the terranes and continents are in places poorly known, because of a lack of data or they are concealed by younger rocks (Figs. 2.1, 2.8). The distribution of ophiolites in Europe, as well as their ages, gives hints for the reconstruction of terrane and continent boundaries (Figs. 2.3, 2.5).

An ophiolite is a section of oceanic crust, and sometimes parts of the upper mantle, that has been tectonically emplaced on top of continental crust and is often exposed within the present continental crust (Dilek & Newcomb 2003). Ophiolite occurrences are widespread in Europe and, because these rocks have a very special geochemistry (often strongly enriched in elements like Mg, Cr, Ni, Cu, Co), they can significantly influence the chemistry of soil formed in their vicinity. Their distribution and age is shown in Figure 2.3. Note that they often only cover very small areas, so that they tend to become invisible in a geological overview map (e.g., Fig. 2.6c), while they often have a major importance in the interpretation of a geochemical map.

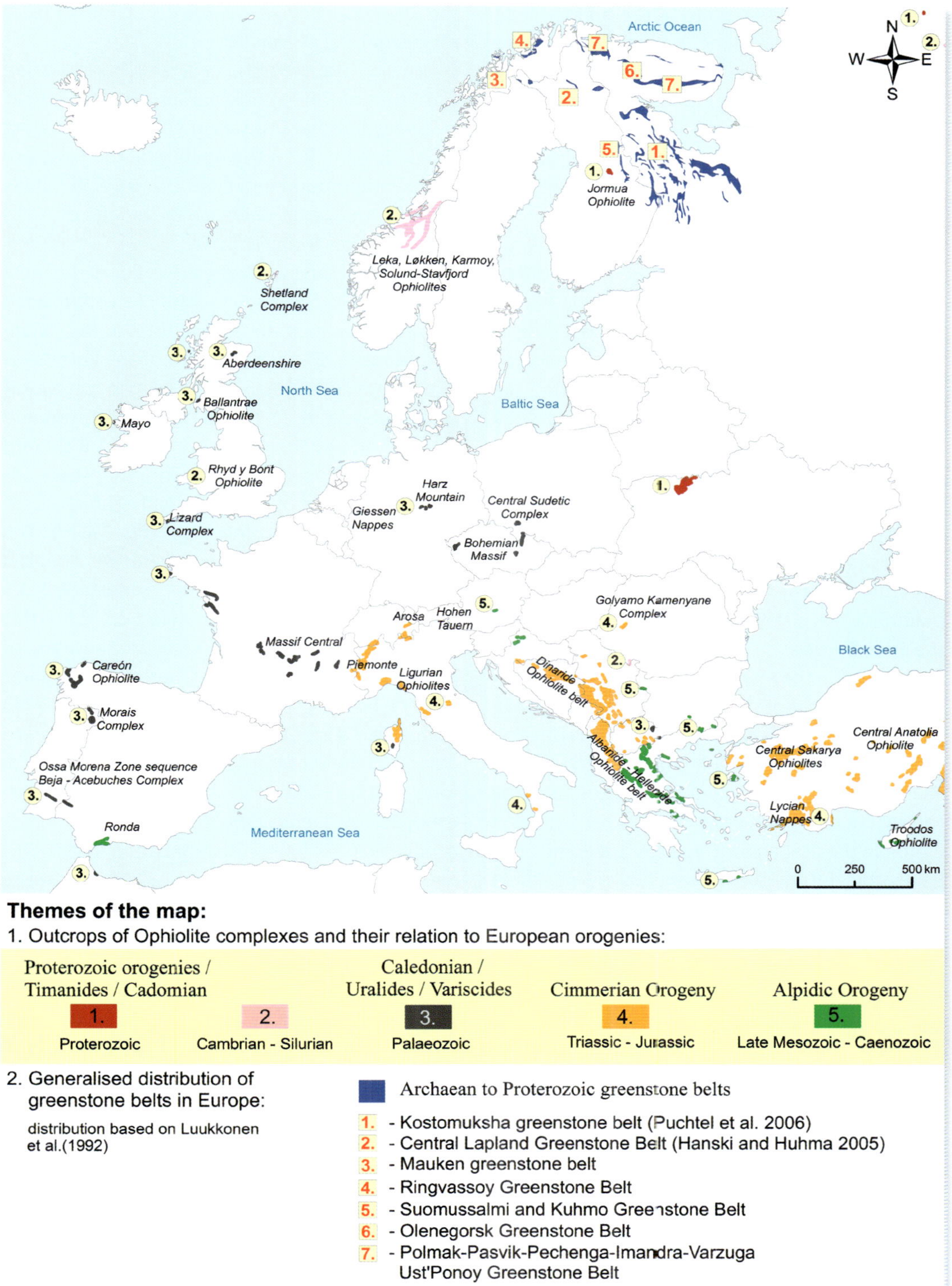

Fig. 2.3: The distribution and ages of ophiolite complexes in Europe provide hints for the reconstruction of ancient terrane and continental boundaries. Greenstone belts are named for the green hue of most of their rocks, which is caused by newly formed green minerals (chlorite, actinolite, amphibole), developed during the alteration of these rocks (DE WIT & ASHWAL 1997). By their larger degree of heterogeneity, greenstone belts are tectonic markers far more distinct than the surrounding homogeneous granite-gneiss units (edited and compiled by F. Jähne; the spatial data based mainly on ASCH 2003).

Europe is today a part of the Eurasian continent, and extends from the Ural Mountains in the east to the Iberian Peninsula and the passive continental margin of Norway in the west, and from the partly active plate border in the Mediterranean Sea to the Arctic Sea in the north. Note that in parts of Europe the boundaries between terranes and their affinity to other structural elements are still under discussion.

The **Mohorovičić discontinuity**, often simply referred to as the **MOHO**, is a major seismic discontinuity at the transition from the crust (the outermost solid shell of the earth) to the mantle (a highly viscous layer which is chemically different from the crust). Because of a change in rock composition and rock density, the seismic velocity increases along this discontinuity. Major geological processes like mountain building and rifting can have an impact to the depth trend of the MOHO. The map view in Figure 2.4 with the depth of the MOHO shows clearly that the European lithosphere (crust and uppermost solid mantle) can be differentiated in two parts, a western and an eastern to northern part. Very old Precambrian structures like the Svecofennides or former plate boundaries (e.g., Mid-Lithuanian Suture Zone) of the East European Craton are preserved over a long period in the internal structure of the MOHO. These structures are often highlighted by a lowered MOHO. Most of the basins and rifts with oceanic or thinned crust are highlighted by a raised MOHO. The most eye-catching features are the contrast in depth between the old East European Craton in the east and the younger predominantly Palaeozoic terranes of Western Europe, as well as the contrast along the transition from continental crust to the oceanic crust of the Atlantic Ocean or in the Mediterranean Sea (Fig. 2.4).

Figure 2.5 shows the main geological terranes and the ages of their genesis and accretion to Europe. The oldest Archaean to Proterozoic terranes are parts of the East European Craton (EEC) in the north and the north-east of Europe. One major segment of the EEC is Fennoscandia (syn. Fennoscandian Shield), where the crystalline basement is covered by Palaeozoic and Mesozoic sediments only in the Baltic Sea area (Figs. 2.1, 2.8). The other two major segments are the Volga-Uralia segment and Sarmatia, which outcrops in the Ukrainian Massif (with the oldest rocks of Europe) and the Voronezh Massif. The segments of the EEC, assembled in the Late Palaeoproterozoic, show little intraplate deformation from the Mesoproterozoic to the Phanerozoic (Figs. 2.5, 2.8, e.g., Central Russia rift system, Pachelma rift).

The crustal structure and the Precambrian evolution of the Fennoscandian Shield are relatively well understood. Four NW to SE trending orogenic belts form the main structural pattern of the Fennoscandian Shield (Plant et al. 2005). There is an overall decrease in the age of rocks and structures from NE to SW.

The Kola-Karelian Orogen (Fig. 2.5) is the most north-eastern one and consists of five Archaean terranes amalgamated by collision in the Palaeoproterozoic (2 – 1.9 Ga). South of the Kola-Karelian orogen, the Svecofennian orogen consists of rocks younger than 2.2 Ga, which accreted and underwent collisions between 2.0 – 1.8 Ga and were reworked by crustal melting between 1.8-1.54 Ga (Plant et al. 2005). No Archaean rocks are described in the Svecofennides (Figs. 2.4, 2.5). The structural pattern of the basement was finalised with the deposition of the Jotnian sandstone at 1.5 Ga in a large elongate basin (Figs. 2.7, 2.8) – this sandstone is a remarkable feature on many of the geochemical maps (see Chapter 11, e.g., Section 11.16 Ca(XRF) map; Section 11.51 Si(XRF) map → Part A). Situated along the south-western margin of the Svecofennian orogen is the 1600 km long, 150 km wide and 1.84 – 1.75 Ga old Transscandinavian Igneous Belt (TIB), also called Transscandinavian Batholith (Blundell et al. 1992; Bingen et al. 2008; Högdahl et al. 2004). Below the main batholith area of central Finland (Luosto 1991) the MOHO is depressed to 60 km (Fig. 2.4). The western border of the Transscandinavian Batholith (Fig. 2.5) and the remainder of the Gothian orogen are heavily reworked by Sveco-Norwegian deformation (1.05 – 0.9 Ga, Greenville age orogenesis). The Gothian orogen comprises rocks accreted and formed between 1.77 – 1.5 Ga. A small part of the Gothian orogen is well preserved in south-east Sweden and on Bornholm Island, because it has escaped Sveco-Norwegian reworking (Johansson & Larsen 1989). The other Archaean to Proterozoic remnant of Europe outcrops along the shelf in the north-west of Scotland and in the

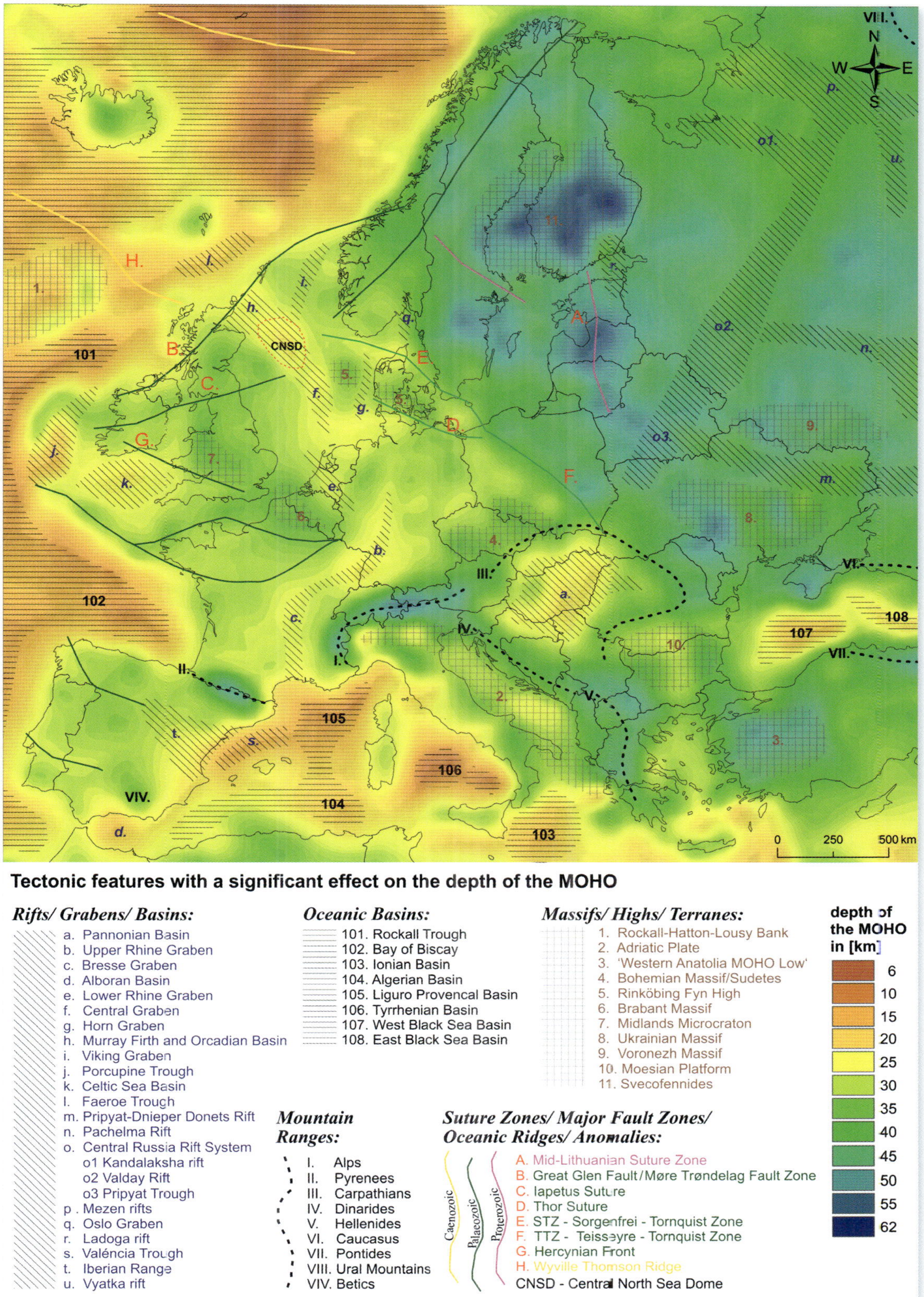

Fig. 2.4: MOHO depth map (circa crustal thickness) and tectonic features with a significant effect on the depth of the MOHO. Geological processes and very old crustal remnants can have an impact on the trend of the MOHO beneath Europe (edited and compiled by F. Jähne; background map of the MOHO based on Grad & Tiira (2009)).

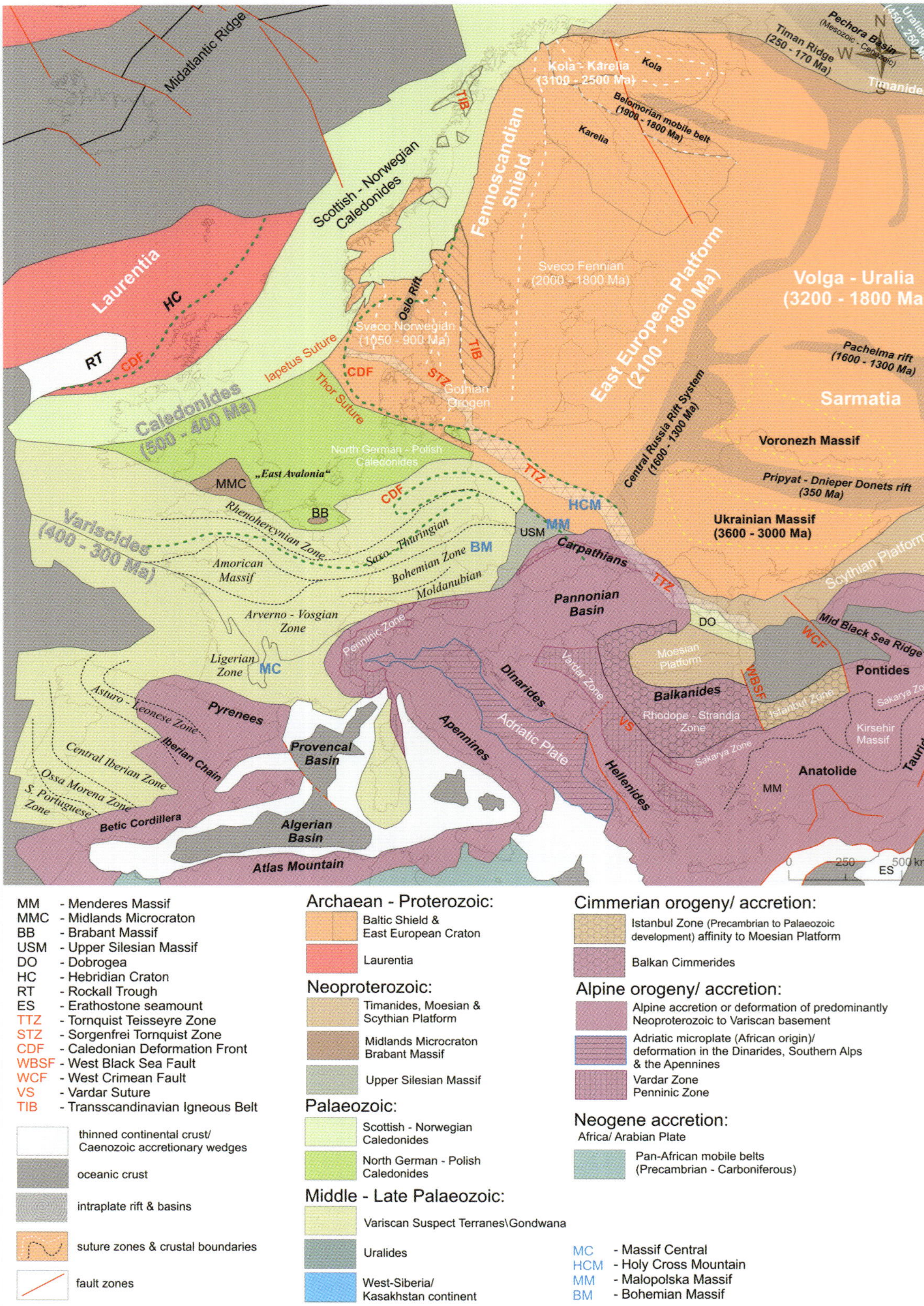

Fig. 2.5: Terrane map of Europe shows the main geological terranes and tectonic plates. The major internal structures of the terranes and continents are labelled (compiled and edited by F. Jähne; based on PHARAOH (1999), OCZLON (2006), OKAY et al. (2006), BLUNDELL et al. (1992)).

north of Ireland (Fig. 2.5). It is a fragment of the continent of Laurentia, initially part of a North American-Greenland landmass, which was separated by the formation of the North Atlantic Ocean in the Early Caenozoic.

In Europe there is evidence for two Neoproterozoic orogenies, the Timanian Orogeny along the eastern flanks of proto-Baltica and the Cadomian Orogeny described in Western Europe and Iberia (Fig. 2.5). Late Neoproterozoic accretionary or collisional Timanides are preserved in the north-east of Fennoscandia and beneath the Phanerozoic sediments of the Pechora Basin (Fig. 2.8). From the southern Uralides in the east to the Scythian Platform in the south, there is indirect evidence indicating Timanian-age orogenesis (Gee & Stephenson 2006). The age, the boundaries and the crustal geometry of the Scythian Platform are not well known and still under discussion. The basement complexes of Western and Central Europe contain several smaller continental crustal blocks with an origin or deformation during the Cadomian Orogeny (e.g., Brabant Massif, Armorican Massif, Bohemian Massif and Malopolska Massif, Western Iberia). However, they were mostly overprinted by later Caledonian and Variscan deformation processes. The Midlands Microcraton (MMC) and the Brabant Massif in Belgium are parts of the Anglo-Brabant Massif. Their crust comprises of Late Proterozoic parts of 'Avalonia' affinity, formed by accreted oceanic arcs and marginal-basins (Pharaoh & Gibbons 1994). The MMC has a relatively thin cover of Palaeozoic sediments, which were not strongly deformed during the Caledonian Orogeny. Soper et al. (1987) proposed that the MMC has acted as a 'rigid indenter', around which the Lower Palaeozoic rocks of the Caledonian belts were deformed.

The Caledonides of North-West Europe (Figs. 2.3, 2.5) define the collision zone of Laurentia with Baltica and the suturing of Laurussia (Dewey & Strachan 2005; Gee 2005). They are subdivided in the NE-SW trending Scottish-Norwegian Caledonides north of the Iapetus Suture, between Laurentia and Baltica, and in the Avalonian microcontinent, south of the Iapetus and Tornquist Suture. In Norway, the Caledonides are well exposed and four major compressive/transpressive events are recognised (Finnmarkian (Late Cambrian), Trondheim (Early Arenig), Taconian (Mid-Late Ordovician) and Scandian (Mid Silurian – Early Devonian)). The Caledonides, south of the Iapetus Suture, are largely concealed by a cover of late Palaeozoic, Mesozoic and Caenozoic strata. Major Neoproterozoic terranes in the west, and an unknown number of smaller Neoproterozoic to Palaeozoic terranes in the east, are within the assemblage of East-Avalonia. The Anglo-Brabant Massif in the west of East-Avalonia formed a structural high for much of subsequent Phanerozoic history (Pharaoh et al. 1997; Pharaoh 1999). The nature of the crust underlying northern Germany and the Polish Trough is uncertain overall. To the east and south, the reworking by the later Variscan Orogeny increases (Fig. 2.5). The Palaeozoic Massif of the Holy Cross Mountains (south-east Poland) is one of the most south-eastern outcrops of the Caledonides and shows an Early Palaeozoic succession of predominantly shallow water facies with a presumed provenance in the EEC (Pharaoh et al. 1997; Pharaoh 1999). The Variscan orogen on Iberia shows indications of a former Caledonian evolution too.

Most of Western and Southern Europe is composed of heterogeneous Variscan basement, which is overprinted by Mesozoic to Caenozoic intraplate extension and by the Alpidic Orogeny (Figs. 2.5, 2.9). The Variscan Orogeny had an unusually complex evolution and many aspects are the subject of active debates (Franke 2000). The extent of the Variscan fold belt in Europe is close to 3000 km long and 700–800 km wide (Matte 1991).

The central to western European Variscides are subdivided into the Rhenohercynian Zone, the Saxothuringian Zone and the Moldanubian Zone (McCann 2008a). The Variscan fold belt shows great lateral continuity, which is often interpreted as an effect of bilateral symmetry (Franke & Engel 1986; Matte 1991). Regardless of the many uncertainties, the tectonostratigraphic zones in the Iberian Peninsula can be correlated with confidence with equivalent zones in Central and Western Europe. The South Portuguese Zone is comparable with the Rhenohercynian Zone; the Ossa Morena Zone with the Armorican Massif and with the Saxothuringian Zone and the Iberian Zone is comparable with the Massif Central and the Moldanubian Zone (Fig. 2.5). The characteristics and geometry of the Variscides in South-Eastern Europe are not well defined because of extensive overprinting by Alpine deformation.

The central and southern European crustal structure is affected by the Alpine Orogeny and the previous Cimmerian Orogeny. The Anatolian – Balkan region consists of several small continental fragments or terranes bearing evidence of various periods of deformation, metamorphism and magmatism before the Alpine Orogeny (OKAY et al. 2006). The Late Triassic to Early Cretaceous Cimmerian Orogeny had affected the Sakarya Zone, the Istanbul Zone and the Rhodope-Strandja Zone (Fig. 2.5). These terranes were slightly affected by the Alpine Orogeny, exhibiting Laurasian affinities and were finally amalgamated in a single terrane by Mid-Cretaceous times. The Istanbul Zone consists of a Neoproterozoic crystalline basement overlain by Lower Ordovician to Eocene sedimentary rocks (GÖRÜR et al. 1997). Before the Late Cretaceous opening of the Black Sea, the Istanbul Zone was situated east of the Moesian Platform and adjacent to the Scythian Platform (OKAY et al. 1994). The intensely deformed and partly metamorphosed southern Anatolide-Tauride Block shows Gondwana affinities. The Late Cretaceous Kirşehir Massif consists of metamorphic and plutonic rocks. An affinity to the Anatolide-Tauride Block is assumed (OKAY & TÜYSÜZ 1999). The Alpine Orogeny formed mountain belts all around the Mediterranean Sea and the Black Sea. The Adriatic or Apulian Plate is a spur of continental crust that broke up in the Cretaceous from northern Africa along a large transform fault. The Adriatic Plate collided as an indenter with the European plate and formed the mountain ranges of the Alps. The ophiolitic Vardar and Penninic Zones (Figs. 2.3, 2.5) mark the eastern boundary of the Adriatic Plate.

▸ 2.4 Igneous rocks of Europe

Figure 2.6a shows the distribution of different outcrops of igneous rocks. Igneous rocks are one of the three main types of rocks as defined by geologists. The other two are sedimentary and metamorphic rocks. According to the derivation of the word igneous from the Latin word igneus/ignis, which means fire (LE MAITRE 2002), these rocks result from cooling and solidification of a magma (hot melt). Over 700 types of igneous rocks have been described, which can be divided into volcanic and plutonic rocks.

Volcanic or extrusive igneous rocks are formed on the Earth's surface or on the sea floor.

Their composition and texture depend on the depth of origin and composition of the initial magma in the earth's interior, on the interaction of rising magma with wall rocks, on the site of eruption and cooling, whether extruded on the sea floor or the terrestrial surface.

Regarding terrestrial volcanism, everyone has seen pictures from catastrophic eruptions like the eruption of Pinatubo (Philippines) in the early 1990s, which produced more than 10 km^3 of volcanic ashes and tephra. The eruption of Eyjafjallajökull in 2010 on Iceland was much smaller, but the ashes produced by the eruption and blown into the atmosphere, were sufficient to have a strong impact on European air traffic for days. Another kind of eruption can be observed recurring at irregular intervals on Sicily at Mount Etna, one of Europe's largest active volcanoes. Volcanoes like Mount Etna rarely show explosive eruptions, but produce large amounts of lava flows. In the oceanic environment, most of the upper crust is formed by volcanic rocks, mainly extruded along submarine mid-oceanic ridges as lava flows with predominantly basaltic composition. Recent subaerial volcanic activity is observable along the Mid-Atlantic Ridge. Iceland, as an example, is a volcanic centre at the Mid-Atlantic Ridge (Figs. 2.4, 2.5, 2.9), which formed by extensive activity in the Neogene caused by an underlying mantle plume (hot-spot).

In principle, fast cooling of extruded magma at the surface or on the sea floor means that the solidification of the lava happens quickly. For this reason, volcanic rocks are characterised by a fine-grained or glassy matrix.

The other class of igneous rocks are the plutonic or intrusive igneous rocks.

They are formed from partial melting of the mantle or crustal material, but the generated melts did not reach the surface of the earth. These rocks are more variable in composition and texture than volcanic rocks (LE MAITRE 2002). This variety of plutonic rocks depends on many variables, which have an influence on their composition and the process of their formation. The resulting com-

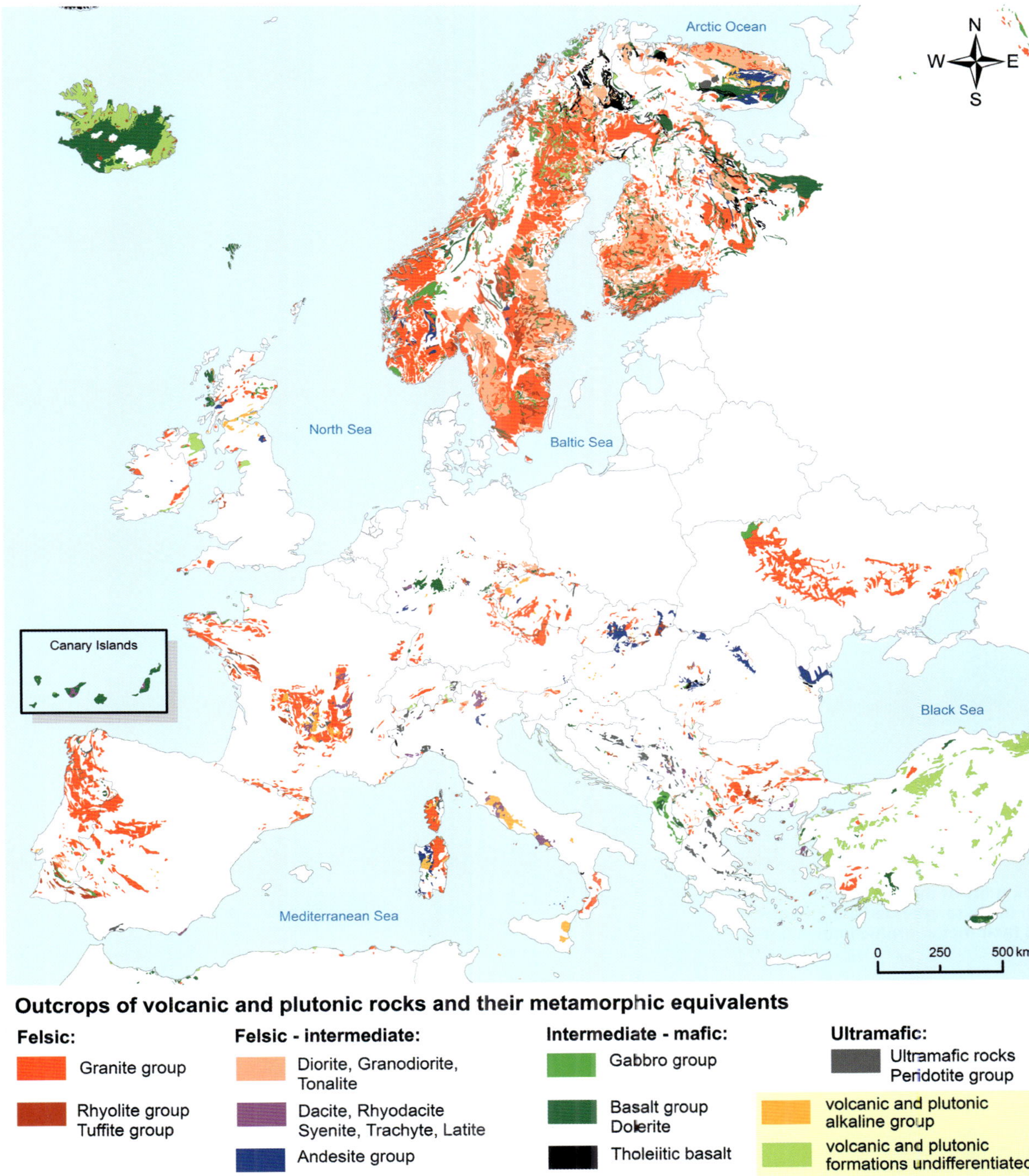

Fig. 2.6a: Map of igneous rocks at the surface reflecting the overall distribution of outcrops of old crystalline basement in Europe. The igneous rocks on the map are classified according to their absolute content of SiO_2 (edited by F. Jähne; based on the IGME 5000 by ASCH (2003)).

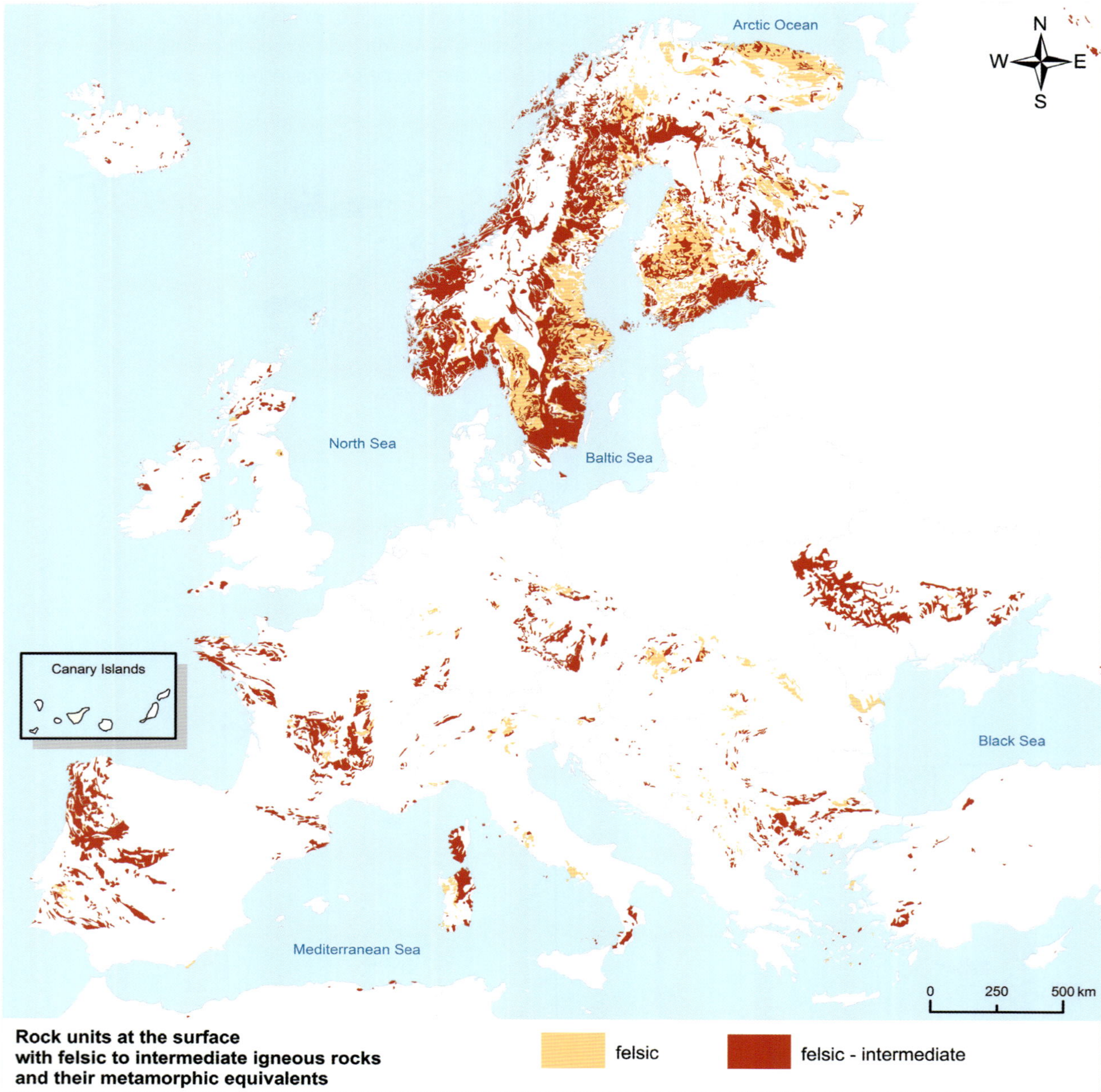

Fig. 2.6b: Generalised map of felsic (syn. acid) to intermediate igneous rocks at the surface. On the map, only the primary rock type of the different outcrops is displayed (edited by F. Jähne; based on the IGME 5000 by Asch (2003)).

position and texture of a plutonic rock is, amongst others, a function of the amount and composition of the original magma, of the temperature and pressure conditions prevailing over time, of the nature of the country rocks and circulating fluids. Due to the slow cooling of the magma and crystallisation over millions of years in crustal reservoirs these rocks are in general coarse-grained.

Feldspar, *quartz* or *feldspathoids*, *olivine*, *pyroxene*, *amphibole*, and *mica* are important minerals in the formation of almost all igneous rocks, and they are essential for the classification of these rocks (Le Maitre 2002). In general, most plutonic rocks have volcanic equivalents and thus both groups show chemical and mineralogical similarities. Only the textures and sites of final emplacement are essentially different. Due to this fact, plutonic and volcanic rocks can be classified as in Figures 2.6b–d according to their chemical and mineralogical composition.

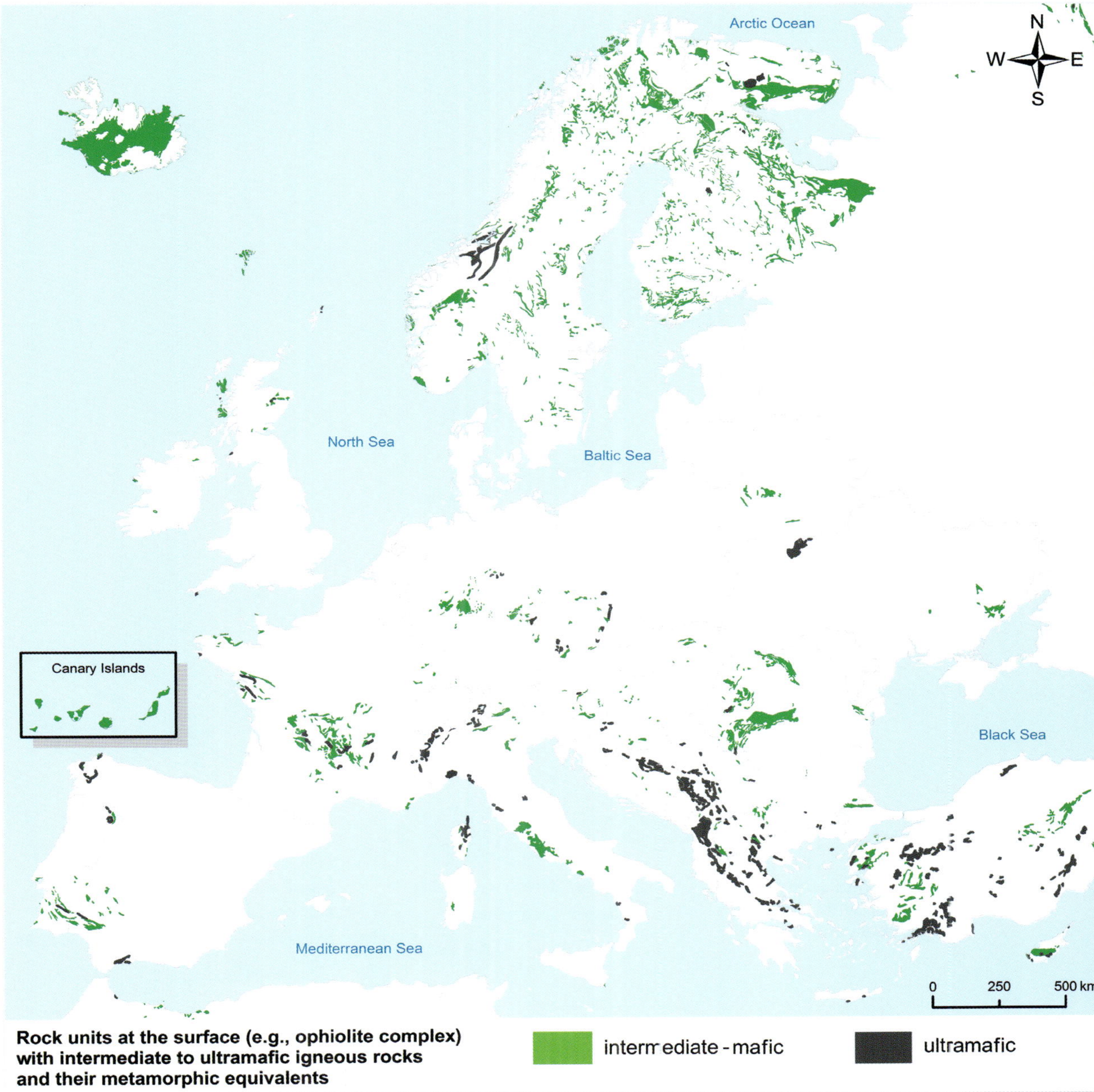

Fig. 2.6c: Generalised map of intermediate, mafic (syn. basic) to ultramafic igneous rocks at the surface (edited by F. Jähne; based on the IGME 5000 by Asch (2003)).

Felsic (syn. acid) igneous rocks (Figs. 2.6a, b) contain high amounts of silica (greater than 63 % SiO_2). An example of a felsic plutonic rock is *granite* and its volcanic equivalent *rhyolite*. Igneous rocks with silica contents between 52–63 % SiO_2 are classified as intermediate (Fig. 2.6a). An example of an intermediate plutonic rock is diorite and its volcanic equivalent andesite. *Mafic* (syn. basic) igneous rocks have low silica 45–52 % and typically high iron-magnesium content. An example of a mafic plutonic rock is gabbro and its volcanic equivalent *basalt*. *Ultramafic rocks* (Figs. 2.3, 2.6c) show a silica content less than 45 %. An example of an ultramafic plutonic rock is *peridotite*; examples of ultramafic volcanics are *picrite* and *komatiite*.

Alkaline igneous rocks show 5–15 % alkali (K_2O + Na_2O) content or a molar ratio of alkali to silica greater than 1 : 6 (*phonolite* and *trachyte*). Alkaline rocks are generally characterised by an unusual

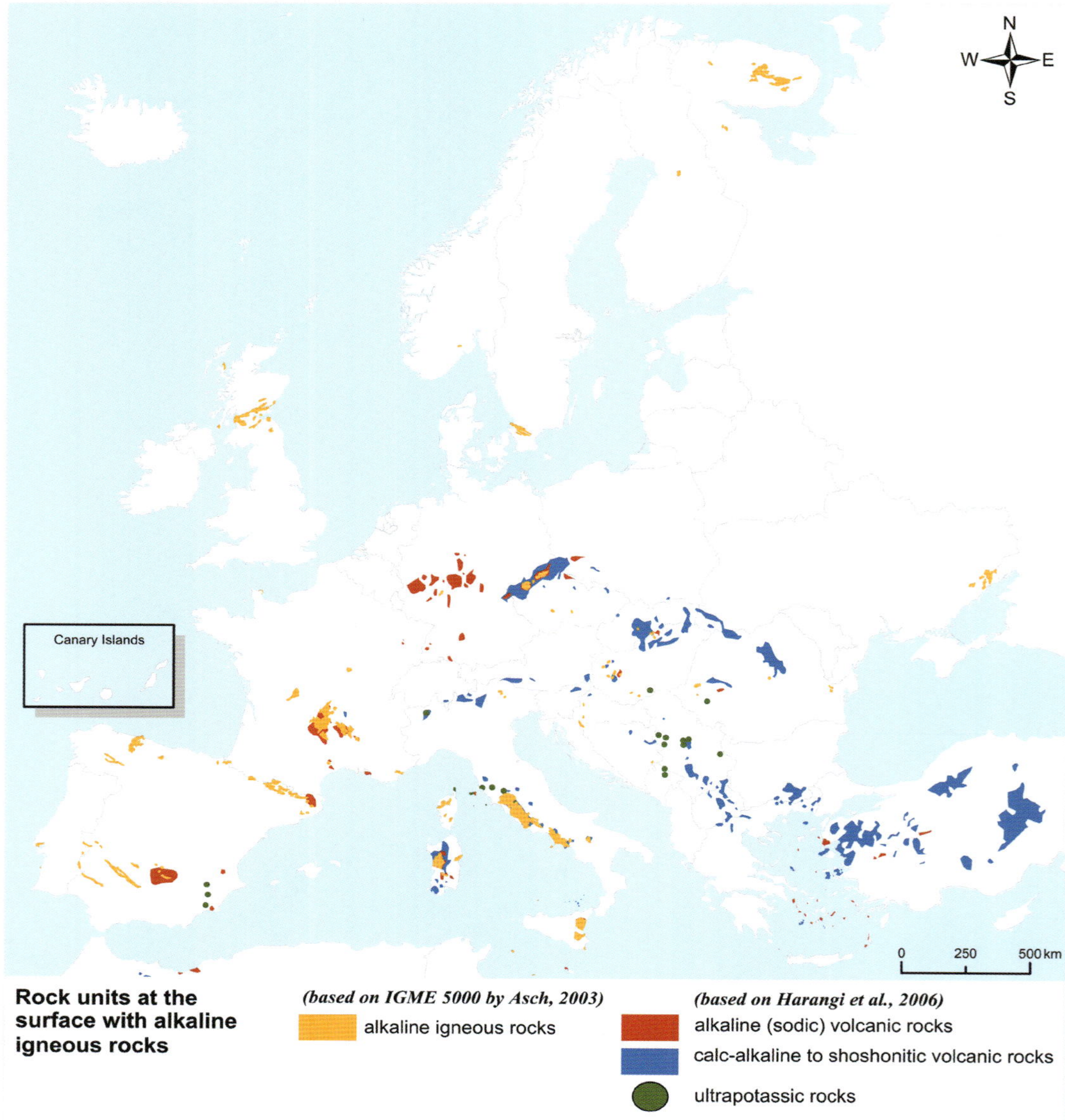

Fig. 2.6d: Generalised map of alkaline igneous rocks at the surface. The map visualises two different data sets with different focus on data and with different spatial resolution. The data set, based on Harangi et al. (2006), focuses on the distribution of alkaline, calc-alkaline and ultra-potassic rocks in Western and Central Europe and around the Mediterranean region. The data set by Asch (2003) shows the Europe-wide distribution of rock units with alkaline igneous rocks as primary rock type.

geochemistry, and, just like ophiolite, they can be expected to have a substantial impact on regional surface soil chemistry, notwithstanding their rather minor spatial distribution (Fig. 2.6d).

Most of the igneous rocks in Europe are related to Archaean to Palaeozoic basement outcrops (compare Fig. 2.1 and 2.6b–d). In large parts of Western and Central Europe the distribution of igneous rocks in the upper crust is uncertain, because they are buried beneath deep Palaeozoic to Caenozoic sediment basins (Fig. 2.8). Figure 2.6a shows a worldwide trend that the continental crust is generally dominated by felsic to intermediate igneous rocks. In contrast, Iceland which is built of oceanic crust is dominated by mafic rocks (Fig. 2.6c). The ultramafic rocks are mainly related to outcrops of former oceanic lower crustal or mantle fragments (ophiolites), which are now aggregated to the European continental mass (Fig. 2.3). Most of the volcanic rocks at the surface of Central and Southern Europe as shown in Figure 2.6a are of Caenozoic age. For example, the volcanic rocks in Germany and in the Bohemian Massif were formed by Caenozoic rifting and the presence of a postulated mantle plume underneath the Eifel. The Massif Central in France shows similar features. The flood basalts on Iceland and Greenland, as well as the basalts in north-east Ireland, are related to ocean spreading of the Northern Atlantic Ocean. Quaternary and present day volcanism is observed on Iceland and from Italy to the Eastern Mediterranean.

▸ 2.5 Sedimentary basin evolution in Europe from the Palaeozoic to the Caenozoic

Most of offshore and onshore Europe, with the exception of most of Scandinavia and Eastern Europe, is covered with Palaeozoic to Caenozoic sediments (Figs. 2.1, 2.7, 2.8). Their deposition is linked to the evolution of different types of sedimentary basins, rifts and failed rifts (aulacogen). The area between these sedimentary basins (platform or structural highs) is often characterised by major unconformities and reduced thicknesses of the Palaeozoic to Mesozoic sedimentary cover. Sedimentary basins of Europe can be classified by their differences in evolution and their different lithologies and sedimentary sequences. In some regions, for example in Western Europe, the deposited rocks of former sediment basins are covered with the basin fill of a posterior to recent sediment basin (Fig. 2.8). There, the crystalline basement can be buried under more than 15 km thick Palaeozoic to Caenozoic sediments. From Poland to Britain a broadly east-west trending basin (North-West Carboniferous Basin (NWCB)) was formed during the final amalgamation of Pangaea in the Carboniferous as a foreland basin of the Variscan mountain belt. The North-West Carboniferous Basin can be regarded as a precursor of the Late Palaeozoic to Mesozoic Southern Permian Basin (DOORNENBAL & STEVENSON 2010) but both basins show a fundamentally different tectonic setting. The NWCB developed in front of an actively deforming mountain belt and is shaped by a compressional regime, whereas the Southern Permian Basin developed in a mainly extensional regime (Fig. 2.9). Recent foreland basins extend along the mountain belts of the Alpine orogen (e.g., North Alpine Molasse Basin, Carpathian foredeep, Po basin) and are filled with Late Mesozoic to Tertiary rocks (Figs. 2.1, 2.8). Foreland basins are often asymmetric wedge shaped basins and the thickness of their sedimentary fill increases towards the orogen. An idealised stratigraphy of these basins shows a change from deep marine to marine conditions with turbidite and hemipelagic mudstone to carbonate sediments (in prior European publications named as flysch) to terrestrial conditions with siliciclastic deposits (in prior European publications named as molasse) caused by the erosion of the uplifted orogenic belt. The thick coal beds of the North-West Carboniferous Basin, which were important for the industrialisation of Europe, and are one of the source rocks for natural gas in the Southern Permian Basin, are molasse deposits of the Variscan foreland basin.

In most of Western Europe, and in parts of the North Sea, this former basin is covered by up to 10 km thick succession of Permian to Tertiary sediments (e.g., Central Graben, Glückstadt Graben) of the Central European Basin System (CEBS), an intra-continental basin system (LITTKE et al. 2008). The evolution of Europe's intra-continental basins is linked to thermal instabilities and Palaeozoic to Mesozoic rifting events, because of the breakup of the supercontinent Pangaea (Figs. 2.2c, d; MCCANN 2008a). Most of the sediments in the CEBS were deposited in shallow marine conditions or show an arid terrestrial provenance (Fig. 2.7). The infill of intra-continental basins and grabens also form the Mesozoic sedimentary cover of France and Spain, in the North Sea, as well as parts within the Alpine orogen (Fig. 2.8). However, in the Alps, the Mesozoic basin evolution is often obscured by later mountain building from the Late Mesozoic to Caenozoic.

The evolution of the Central European Basin System started in the Late Carboniferous to Early Permian with the deposition of primarily volcanic rocks (calc-alkaline, SiO_2 rich lava and ignimbrite, with some associated basaltic lava), which reaches thicknesses up to 3000 m (PLEIN 1995; BENEK et al. 1996). The accumulation of the Pangaea supercontinent at the end of the Variscan Orogeny in the Late Carboniferous had an enormous influence

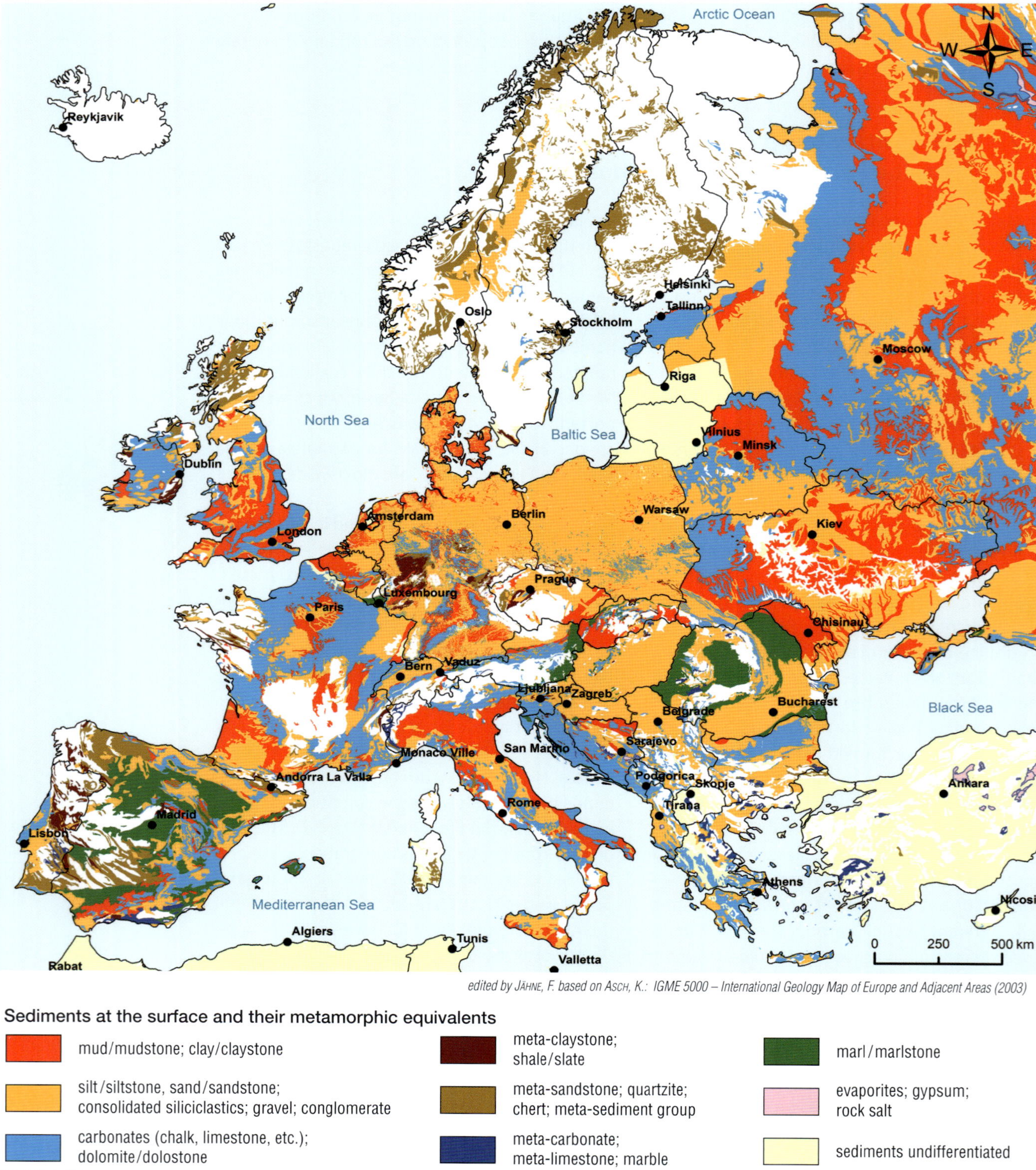

Fig. 2.7: Most of Europe is covered by Mesozoic to Caenozoic sediments. The different types of sediment can have an influence on the geochemical fingerprint of a region. For example, the young siliciclastic and unconsolidated sediments in European sedimentary basins show low concentrations for most of the elements (e.g., U, Sn, Pb, Sr, Rb, Th, Ti, V) (edited by F. Jähne; based on Asch (2003)).

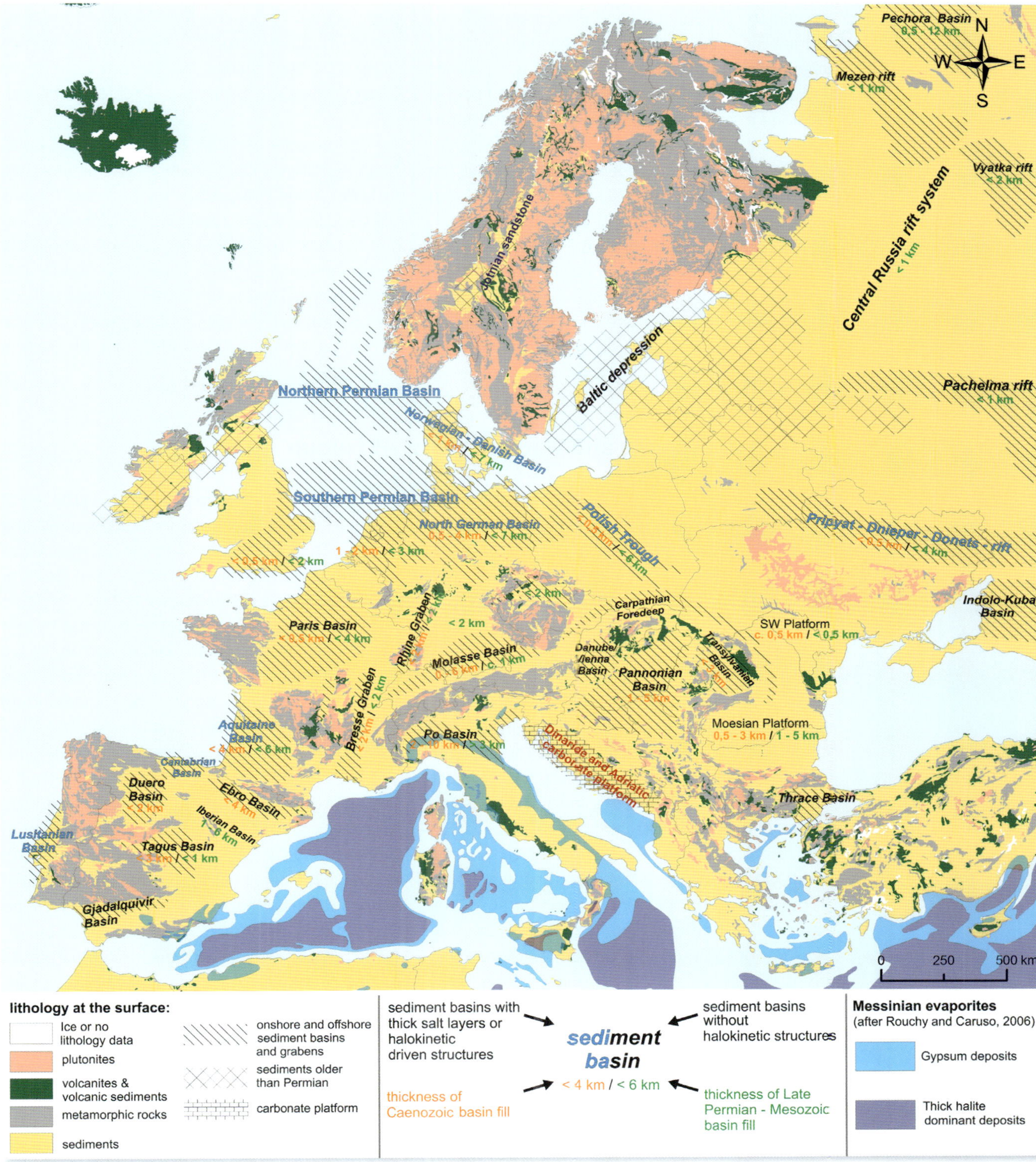

Fig. 2.8: Distribution and thickness of the Mesozoic to Caenozoic basin fill of the major sedimentary basins of Europe, (compiled by F. Jähne), combined with the lithology at the surface (based on Asch 2003).

on climate, the sea level and geodynamic effects. During the northward drift of Pangaea, a dramatic change in climate, from humid in the Carboniferous to arid in the Permian and in the Triassic, resulted in a massive sedimentation of evaporite in shallow epicontinental seas along two major intracontinental basins, generally referred to as the Northern and Southern Permian basins (McCann et al. 2008a).

The tectonic mobilisation of Permian (in Western and Central Europe) to Triassic rock salt (Iberia) and the resulting rise of salt diapirs had a major impact on basin subsidence and the structural pattern of the sedimentary cover in European Mesozoic basins (Fig. 2.8). The Jurassic to Cretaceous deposits of European continental basins and platforms primarily consist of shale and carbonates (Figs. 2.7, 2.8) of shallow marine facies, but at the basin borders the amount of sandstones increases.

Compared to Western Europe, the East European Platform is covered only by relatively thin, little deformed Phanerozoic platform successions (Gee & Stephenson 2006). The thickness of the sedimentary cover only increases significantly along Palaeozoic to Mesozoic rifts, like the Pripyat-Dnieper Donets Rift or along Meso- to Neoproterozoic broken rifts (aulacogenes), like the Pachelma Rift or the Central Russia Rift System (Figs. 2.8, 2.9).

▸ 2.6 Major faults and fracture zones of Europe

Roughly, a fault can be described as a planar or curved discontinuity in a volume of rock which shows significant vertical or horizontal displacement along the given fracture. The classification of faults is based on their sense of slip and additionally in more detailed classifications on their geometry. Strike-slip faults are characterised by predominantly horizontal movement of rock bodies, and reverse or thrust faults as well as normal faults are characterised by predominantly vertical movement. There are, however, a couple of exceptions in this classification. Most of these are special cases, which describe inversion or reactivation of faults or describe faults that show both horizontal and vertical movement (oblique-slip faults). For interpretation of geochemical patterns at the surface, faults are interesting, because they can provide pathways for fluid movement. It is thus not uncommon that certain fault directions are directly visible on geochemical maps. For a striking example see Sæther et al. (2005).

Figure 2.9 shows the major faults and fault zones of Europe. A fault zone is a zone that is characterised by a network of interconnected fractures. In Figure 2.9, the Mesozoic to Caenozoic sedimentary cover (Fig. 2.8) was ignored in a few cases, to image buried fault zones, which are important for the evolution of the European structural framework. Most of the fault zones near the coast continue into the offshore areas. For example, the evolution of complex rifts in the North Sea (e.g., Central Graben, Horn Graben and Viking Graben) from the Mesozoic to the Caenozoic has a strong impact on the thicknesses and facies of sediments in this region (Figs. 2.8, 2.9).

The orientation, geometry, kinematics and evolution of the displayed faults/ fault zones are important parameters for plate reconstruction, palaeo-stress analysis, and for the understanding of sedimentation in basins and the distribution of igneous rocks. However, *what is the impact of faults on the geochemical pattern at the surface?* Faults, their genesis and rheological influences on the host rocks can influence the geochemical signature at the surface. Faults and related fissures can have an important influence on the distribution of springs at the Earth's surface and the underground flow of fluids. Major deep rising fault zones are often pathways for fluids with chemical signatures of mantle or crustal origin. But only some faults leak. The orientation of a fault in the recent stress-field, the lithology, in which the fault is hosted, and the time that has passed since last activity are important parameters for the flow of fluids along faults. Because of interconnection of faulting and syn-sedimentary processes and the relationship of faulting with the intrusion of salt or igneous bodies, faults often define lithological changes in the subsurface. It can be assumed that in salt-influenced basins of Europe (Baldschuhn et al. 1996, Littke et al. 2008)

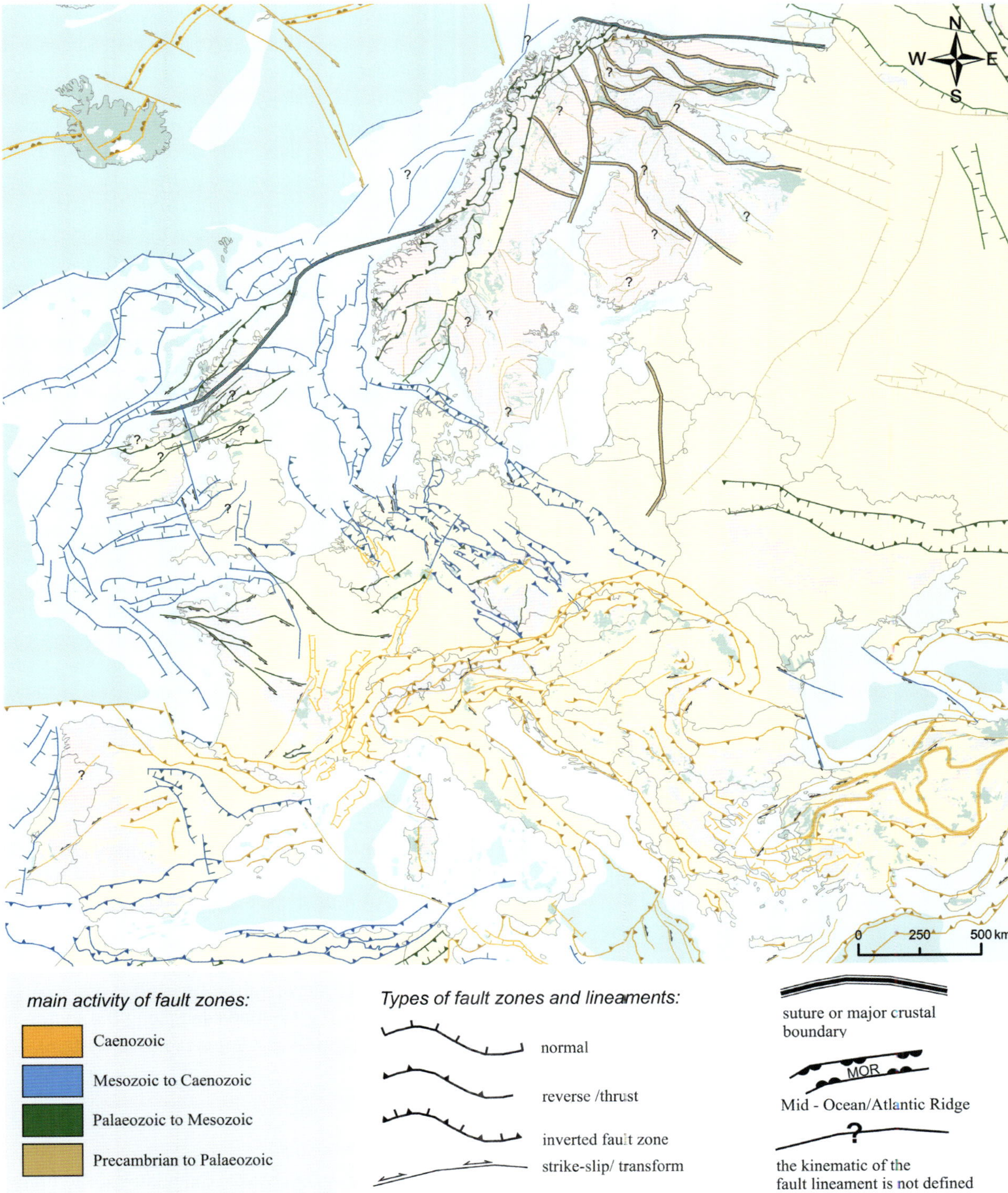

Fig. 2.9: Distribution, ages and the orientation of fault zones onshore and offshore (compiled and edited by F. Jähne). Background colours (based on Asch 2003) provide an idea about the main lithological units at the surface (young sediments (light yellow), intrusive rocks (pink), extrusive rocks (dark green), metamorphic rocks (grey)). In a few cases, the Mesozoic to Caenozoic sedimentary cover was ignored. For readability of the fault pattern at the mapping scale, some faults with similar evolution were generalised to one lineament.

there exists, for some structures, a causal connection between faulting and diapiric rise.

The map in Figure 2.9 with the distribution and estimated ages of fault zones indicates that onshore Europe is divided into a tectonically relatively stable region in the north and north-east of Europe, and in a region around the Mediterranean Sea and also parts of Western Europe, which shows increased tectonic activity and faulting in the Caenozoic up to the present (Figs. 2.5, 2.9). In the last decade, major earthquakes occurred in the Mediterranean region (e. g., L' Aquila in Italy (2009) and Izmit in Turkey (1999)), which have their origin in movement along major fault zones like the North Anatolian fault.

The oldest structural events and related faults are well exposed in Scandinavia, whereas the Mediterranean Sea is surrounded by Caenozoic alpidic mountain ranges (e. g., Pyrenees, Alps, Betics, Appennines, Atlas Mountains, Taurus Mountains), which are dominated by thrust tectonics and especially in the case of the Alps by strike-slip tectonics, too (Fig. 2.9). West of the Alps, a Caenozoic graben system can be traced from the western Mediterranean Sea to The Netherlands. It is supposed that the orientation of parts of the Caenozoic rift system in Germany (Rhine Graben) is controlled by Variscan basement structures (Edel et al. 2007). The Caenozoic volcanism in the Massif Central, in the Upper Rhine Graben, in the Rhenish Massif or around the Bohemian Massif is a result of these rifting processes.

The Proterozoic intraplate rifts (Figs. 2.5, 2.8, 2.9) on the Eastern European Platform are covered by relatively thin Palaeozoic to Mesozoic sediments and their structural pattern and structural evolution are only in parts well described (Bogdanova et al. 1996, 2008; Nikishin et al. 1996; Artemieva 2007). The NW-SE trending Palaeozoic fold and thrust belts of the Uralides (Russia) separate the East European Craton (EEC) from West Siberia and the Kazakhstan-Continent in the east (Figs. 2.5, 2.9). North-east striking Palaeozoic (Caledonian) thrust nappes are exposed in most of Norway. The basement outcrops in central France and in the Armorican Massif are dominated by Hercynian strike-slip movement and related structures. In contrast, in Belgium, Germany and Poland the Variscides show predominantly north-vergent thrusts. In southern France, and in Western Iberia, the Variscan basement shows mainly south-vergent thrusting.

The most prominent border-zone of the EEC in Western and South-Eastern Europe is the NW-SE trending Tornquist -Teisseyre Zone (Figs. 2.4, 2.5, 2.9), also termed Trans European Suture Zone (TESZ; Pharaoh 1999, Pharaoh et al. 2006). The TESZ extends approximately 2000 km from the North Sea to the Dobrogea region of the Black Sea. It is clearly visible on the MOHO map that its depth (Fig. 2.4) increases across the TESZ from *ca.* 30 km beneath Phanerozoic Europe to *ca.* 45 km beneath the EEC (Pharaoh 1999; Pharaoh et al. 2006). The Palaeozoic structures of the TESZ are largely obscured by Mesozoic and younger strata of the Central European Basin System (Pharaoh 1999). It appears that parts of the Central European Basin System developed by reactivation of Caledonian to Variscan basement structures. Because of multiple reactivations in different stress fields in the Mesozoic, the TESZ played a major role in the evolution of the Northern and Southern Permian Basin (SPB), especially in the Polish and Danish parts of the Permian basins. Reactivation of former structures and basin inversion are not only limited to the TESZ (Fig. 2.9). Late Mesozoic inversion structures are also observed in other parts of Europe (e. g. NE Iberia and along the southern border of the Central European Basin System).

▸ 2.7 Impact structures of Europe

Europe's recent topography shows some circular structures up to 50 km in diameter. Most of them are of volcanic origin or due to erosional processes. In a few cases, however, these structures are interpreted as impacts of comets or meteorites (Fig. 2.10). Most of the impact craters all over the world, proven or possible candidates are accurately described in internet databases (Rajmon 2009; EID 2013).

Table 2.1 provides some details of prominent impact structures in Europe. In the crystalline basement of Scandinavia remnants of old impact events with estimated crater size greater than 50 km are preserved (Siljan crater in Sweden, approx. 380 Ma old). Through erosion over time, deeper levels of these structures are exposed and often show a characteristic multi-scale deformation pattern of the host rocks. The forces that act during impact events can result in partial melting and shock metamorphic effects at the surface and in deeper crustal levels too. The degree and style of the deformation in the subsurface depends, among other things, on the diameter, density, composition, velocity and entrance angle of the cosmic object. An impact can also induce hydrothermal activity and the crust fractured by an impact provides good pathways for the rise of fluids from deeper crustal levels. In a few cases, mineral deposits are related to impact structures. The Siljan Impact Structure (Sweden) shows impact-related hydrothermal deposits with economic importance (Pb, Zn).

The mineralisation in the Siljan Structure occurs in the form of veins, as well as fracture and breccia fillings, in Ordovician limestone (Reimold et al. 2005). It is thought that these traps for ore fluids formed as a consequence of tectonic disturbances resulting from the Siljan impact event (Reimold et al. 2005). Meteorites are often enriched in elements like iridium (Ir), which are rare in the upper crust. In a few cases, impact events are the reason for geochemical anomalies in sediments (Grieve 1997). The most prominent example is an Ir-anomaly in sediments, dated to the Cretaceous-Tertiary (K-T) boundary, caused by the Chicxulub structure in Mexico. Figure 2.10 shows that most of the known impact structures in Europe are situated in southern Scandinavia (Sweden and Finland) and on the East European Craton. Some of the crater structures, especially in Scandinavia, are partially under water. The greatest known confirmed impact structure in Europe is the Puchezh-Katunki structure (Fig. 2.10), a large impact crater at the Volga River, near the city of Nizhniy Novgorod in Russia; measuring 80 km in diameter, it is the fifth largest known terrestrial impact crater in the Phanerozoic (Grieve et al. 1995; Grieve 1997; Pálfy 2004).

In Western Europe, a few large craters are also known. For example, the prominent Nördlinger Ries and the Steinheim Craters in southern Germany are relatively young Caenozoic crater structures (Fig. 2.10, Table 2.1).

Table 2.1: The biggest confirmed impact craters in Europe's Phanerozoic history and supplementary information. Extract from the Earth Impact Database (*http://www.passc.net/*).

Crater Name	Location	Latitude Longitude	Diameter [km]	Age [Ma]*	Exposed	Drilled	Target Rock**	Bolide Type***
Puchezh-Katunki	Russia	N 56° 58' E 43° 43'	80	182.5 ±15	Y	Y	M	-
Siljan	Sweden	N 61° 2' E 14° 52'	52	376.8 ± 1.7	Y	Y	M	-
Kamensk	Russia	N 48° 21' E 40° 30'	25	49.0 ± 0.2	N	Y	S	-
Boltysh	Ukraine	N 48° 45' E 32° 10'	24	65.17 ± 0.64	N	N	C	Chondrite
Nördlinger Ries	Germany	N 48° 53' E 10° 37'	24	15.1 ± 0.1	Y	Y	M	Achondrite
Lappajärvi	Finland	N 63° 12' E 23° 42'	23	73.3 ± 5.3	Y	Y	M	Chondrite
Rochechouart	France	N 45° 50' E 0° 56'	23	214 ± 8	Y	N	C	Chondrite? Iron?
Obolon	Ukraine	N 49° 35' E 32° 55'	20	169 ± 7	N	Y	M	Iron?
Dellen	Sweden	N 61° 48' E 16° 48'	19	89.0 ± 2.7	Y	N	C	Stone?
Kaluga	Russia	N 54° 30' E 36° 12'	15	380 ± 5	N	Y	M	-
Logoisk	Belarus	N 54° 12' E 27° 48'	15	42.3 ± 1.1	N	Y	M	-

* pre-1977 K-Ar, Ar-Ar and Rb-Sr ages recalculated using the decay constants of Steiger & Jäger (1977). Ages in millions of years (Ma) before present. Age range and supplementary information of the Puchezh-Katunki event derived from Pálfy (2004).

** Abbreviations: **C** - Crystalline Target; **M** - Mixed Target (i.e., sedimentary strata overlying crystalline basement); **S** - Sedimentary Target (i.e., no crystalline rocks affected by the impact event). From Osinski et al. (2008).

*** From Koeberl (1998, 2007) and PASSC Files (IAB, IIIAB, IIIB, IIID - Iron Meteorite).

Fig. 2.10: Top map: Diameter and ages of impact craters in Europe. In the crystalline basement of Scandinavia remnants of very old impact events >1 Ga are preserved. One of the most prominent impact structures in Europe is the Caenozoic Nördlinger Ries (Germany). The deformation of rocks by impact forces form a typical structural pattern and can induce hydrothermal activity. Resulting mineral deposits or element anomalies in synsedimentary sequences can have an influence on the geochemical fingerprint in a region. Bottom map: Distinction of impact structures in plausibility classes. The status for some of Europe's impact structures is still under discussion. It is reasonable to take into account only objects with confirmed 'most probable' status.

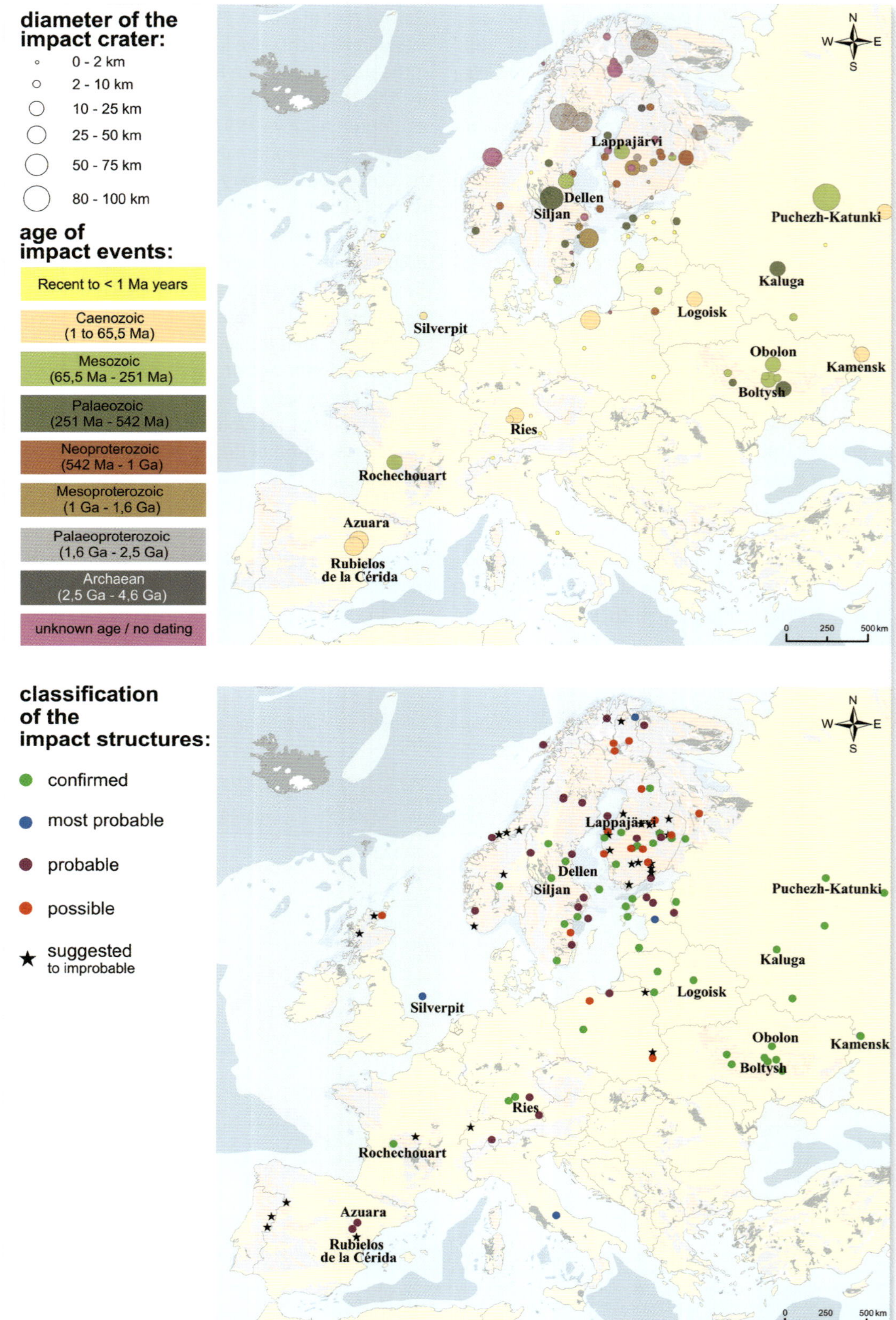

▸ 2.8 Quaternary development of Europe

Plant et al. (2005) point out the effect of deposits and geological and climatic processes of the Quaternary on the geochemical signature at the surface. The processes mentioned and deposits formed often affected the first hundreds of metres beneath the present surface. Quaternary deposits blanket most of Europe's bedrock geology. During the last 2.6 Ma, glacio-fluvial and aeolian processes transported huge masses of detritus, partly over thousands of kilometres, and deposited the sediments in a complex pattern all over Europe's lowlands (Fig. 2.11). *"The consequence of this complex erosional and depositional genesis, is that material from localised mineral sources is highly likely to be dispersed in the direction of net resultant transportation, giving elongate geochemical signature anomalies that extend for considerable distances away from the source area, and potentially 'smearing' geochemical features for considerable distances away from the source area"* (Plant et al. 2005: p.40). Igneous or metamorphic erratic boulders and gravels at the surface along the lowlands of Germany and Poland, which are correlated with Scandinavian bedrock geology, show the major impact of glaciogenic processes on the recent topography in these areas. Glacial morphology of U-shaped deep valleys along European mountain ranges and the fjords of Scandinavia and of Britain and Ireland are spectacular examples of the driving glacial forces. In the Mediterranean region, Quaternary karst development, especially around the Adriatic area, produced distinctive landforms (Figs. 2.8, 2.11).

Superficial deposits of the Quaternary range from soil and weathered bedrock to a variety of

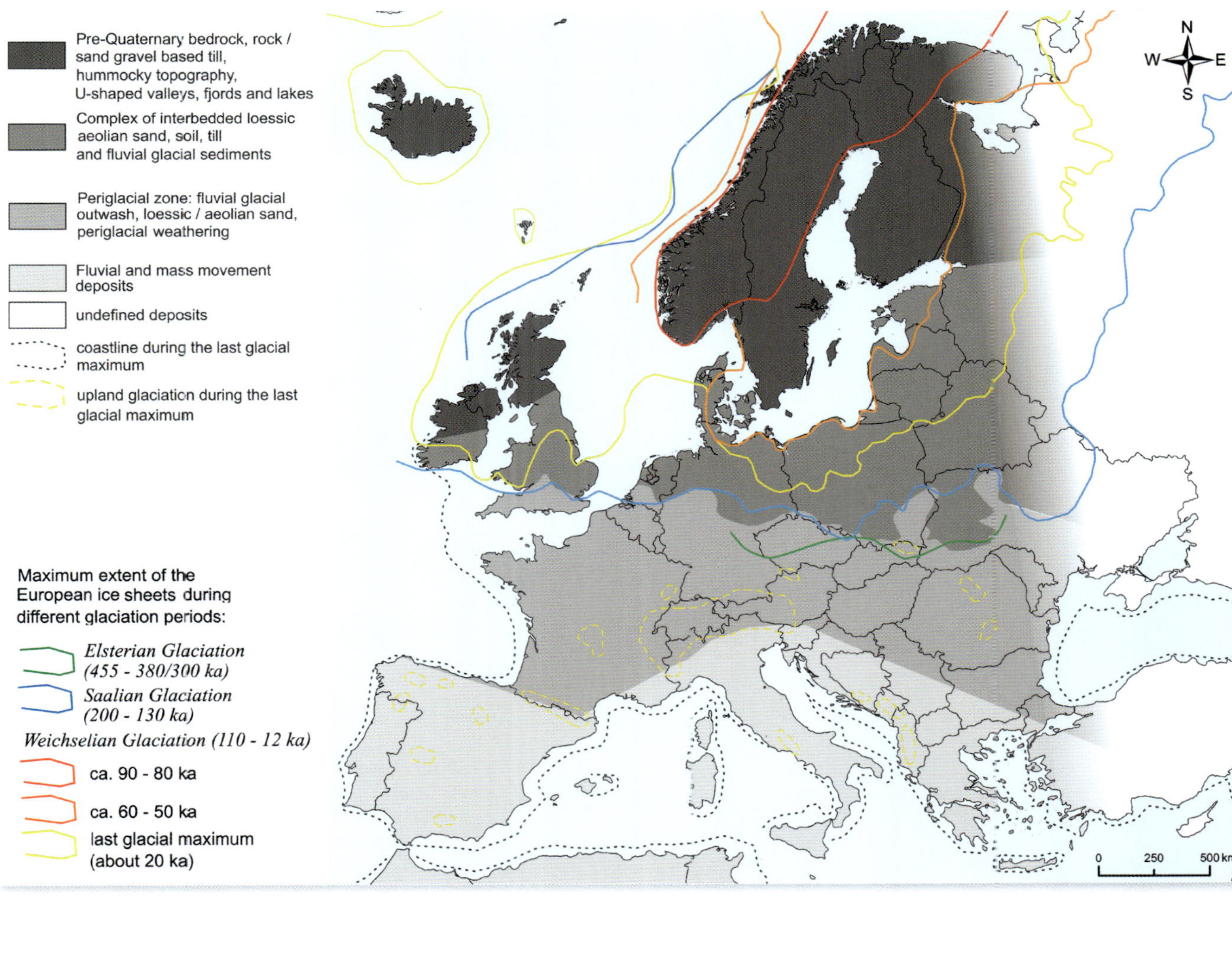

Fig. 2.11: The map shows the maximum extent of the European Quaternary ice sheets (from Anderson & Borns 1997; Spielhagen et al. 2004; Svendsen et al. 2004; Plant et al. 2003a: Fig. 3, p.B230) and the Quaternary domains with different morphological features and deposits (from Plant et al. 2003a: Fig. 2, p.B230 and Plant et al. 2005).

sediments with different composition and genesis. The Quaternary period is defined by frequent and marked climatic oscillations. In marine sediments, up to 25 glacial cycles are recognised, which are eroded or obscured by subsequent events on land (Plant et al. 2005). Figure 2.11 shows the widest extent of the four major glacial events, and gives a rough view of the distribution of sediments, resulting from a range of climate-determined processes.

Broecker & van Donk (1970) point out the saw tooth nature of some glacial cycles, where a slow cooling (glacials) is followed by rapid warming (interglacials). During the warm periods, the temperatures were similar to present day conditions. The former ice sheets retreated or disappeared in these epochs. Currently, the Earth is in such an interglacial period. In the periods of cooling (glacials), ice sheets spread from the Fennoscandian Shield to the Mid-European lowlands and locked away large amounts of freshwater. By bonding the surface water in up to 2.5 km thick ice-sheets and glaciers, the global sea level was depressed during the last glacial maximum (Weichselian) by over 120 m and exposed parts of the European continental shelf, which is now about 90 m below sea level (Plant et al. 2005). During this glacial maximum, about 18,000 years ago, the resulting land bridges connected the major landmass of Europe with the present day islands of Ireland and Great Britain (Fig. 2.11). By the end of this last glacial period, approximately 10,000 years ago, the present day distribution of land and sea was established.

The last 10,000 years until our present time are defined as the Holocene epoch. The Holocene in Europe is characterised by extensive human settlement caused by a warmer and more stable climate. Human changes to the environment have a strong influence on the evolution of the surface morphology and on soil development too. *"Subsequently, social development following the mastery of fire, the development of settled agriculture, which has become gradually industrialised, together with industrialisation and urbanisation, have increasingly contaminated the environment, leaving a significant anthropogenic geochemical imprint"* (Plant et al. 2005: p. 40). The latter statement is certainly true at the local scale, but *"what will we see when mapping agricultural soil at the continental scale?" Will we detect a major human impact on soil chemistry or that 'Nature rules' still today?* |◄

Fabian Jähne
Federal Institute for Geosciences
and Natural Resources (BGR);
Hannover, Germany
E-mail: fabian.jaehne@bgr.de

Chapter 3

Mineral Deposits of Europe

Alecos Demetriades & Clemens Reimann

▸ 3.1 Introduction

A mineral deposit is "*a mass of naturally occurring mineral material*, e.g., *metal ores or non-metallic minerals, usually of economic value* ..." (Neuendorf et al. 2011: p. 415). Such a mineral deposit will normally result in a geochemical anomaly in a large variety of different sample materials in its surroundings, the basis of geochemical exploration (e.g., Hawkes & Webb 1962). Knowledge about the distribution of mineral deposits is thus of paramount significance when attempting to interpret any geochemical map, especially at the scale of a whole continent.

For detailed information on the formation and occurrence of mineral deposits and mineralisation relevant textbooks should be consulted (e.g., Lindgren 1933; Bateman 1950, 1955; Open University 1974; Park & MacDiarmid 1975; Smirnov 1976; Walker 1976; Wright 1977; Evans 1980; Mitchell & Garson 1981; Skinner 1981; Amstutz et al. 1982; Hutchison 1983; Sawkins 1984; Misra 2000; Hedenquist 2005; Guilbert & Park 2007). In addition, the mineral deposits of Europe are described in a number of textbooks (e.g., Bowie et al. 1978, 1989; Dunning et al. 1982, 1989; Schneider 1983; Dunning & Evans 1986; Boissonnas & Ometto 1988; Friedrich & Herzig 1988; Blundell et al. 2005).

The mineral deposit and mineralisation maps presented here are based on the ProMine (2013) mineral deposit database. ProMine is a project under the EU Seventh Framework Programme. Among its objectives are (a) to develop the first pan-European GIS-based database containing the known and predicted metalliferous and non-metalliferous resources, which together define the strategic reserves (including secondary resources) of European Union member countries, and (b) to calculate the volumes of potentially strategic metals (e.g., cobalt, niobium, vanadium, antimony, platinum group elements and REE) and minerals that are currently not extracted in Europe. The ProMine project has developed a mineral deposit interactive database, which can be viewed at *http://ptrarc.gtk.fi/ProMine/*.

▸ 3.2 Mineral deposit and mineralisation maps

In total, 64 mineral deposit and mineralisation maps (Table 3.1) were prepared based on the ProMine database (Cassard et al. 2012); (most of these can be found on the attached DVD → Part A). Only ten maps are shown in print here as examples (Figs. 3.1 to 3.10), i.e., silver (Ag), aluminium (Al), gold (Au), copper (Cu), iron (Fe), molybdenum (Mo), nickel (Ni), lead (Pb), uranium (U) and zinc (Zn). The maps highlight large differences between the elements, the distribution of Al deposits, for example, is related to climate and these deposits occur exclusively in certain areas of southern Europe, while the distribution of occurrences of other elements (e.g., Cu, Pb) is much more widespread. For all elements, it is possible to discern certain areas in Europe where clusters of numerous deposits occur, i.e., a mineral district or belt or province.

The maps were plotted using version 7 of Golden Software's MapViewer. The mineral deposit maps consist of two layers. The first (top) layer is the major element deposit layer, which is represented by a different colour for each element, and the growing dots correspond to the tonnage size of the deposit as indicated in Table 3.1, i.e., very large deposit, large deposit, medium deposit and small deposit; since there are many unclassified deposits with respect to tonnage size, these are indicated by a rhombus. The second (bottom) layer represents the occurrence of the mapped element in other mineral deposits as an associated element; this relationship is presented on all maps with light green growing dots. For example, silver is associated with deposits of arsenic, gold, baryte, cobalt, copper, iron, manganese, nickel, lead, antimony, tin and zinc. |◂

Note: The maps presented in this chapter (Figs. 3.1–3.10) are highlighted (the mineral deposit maps are on the attached DVD → Part A).

Table 3.1: Commodities and mineral deposit classification according to ore reserves in metric tonnes from the ProMine project (CASSARD et al. 2012).

		Class A	Class B	Class C	Class D
Element	**ProMine Commodity**	**Very large deposit (>X)**	**Large deposit (>X)**	**Medium deposit (>X)**	**Small deposit (>X)**
Ag	Silver (metal)	10,000	3,000	500	100
Al	Aluminium (Bauxite ore)	1,000,000,000	100,000,000	10,000,000	1,000,000
As	Arsenic (metal)	200,000	20,000	2,000	200
Au	Gold (metal)	300	50	10	1
B	Borates (B_2O_3)	30,000,000	10,000,000	100,000	10,000
Ba	Baryte ($BaSO_4$)	5,000,000	1,000,000	200,000	50,000
Be	Beryllium (BeO)	50,000	5,000	500	50
Bi	Bismuth (metal)	20,000	2,000	200	2
Cd	Cadmium (metal)	10,000	2,000	500	100
Ce	Cerium (Ce_2O_3)	300,000	30,000	3,000	300
Co	Cobalt (metal)	100,000	10,000	2,000	200
Cr	Chromium (Cr_2O_3)	30,000,000	5,000,000	1,000,000	200,000
Cs	Caesium (Cs_2O)	1,000	100	10	1
Cu	Copper (metal)	10,000,000	1,000,000	100,000	10,000
F	Fluorite (CaF_2)	5,000,000	1,000,000	200,000	50,000
Fe	Iron (metal)	1,000,000,000	100,000,000	10,000,000	1,000,000
Ga	Gallium (metal)	100	50	10	1
Ge	Germanium (metal)	500	100	20	5
Hf	Hafnium (metal)	10,000	1,000	100	10
Hg	Mercury (metal)	50,000	5,000	500	100
In	Indium (metal)	500	100	30	5
K	Potash (sylvite, carnallite) (K_2O)	500,000,000	50,000,000	5,000,000	500,000
Li	Lithium (Li_2O)	5,000,000	500,000	50,000	5,000
Mg	Magnesium, magnesite ($MgCO_3$)	100,000,000	10,000,000	1,000,000	100,000
Mn	Manganese (metal)	100,000,000	10,000,000	1,000,000	100,000
Mo	Molybdenum (metal)	500,000	50,000	5,000	1,000
Na	Rock salt (NaCl)	500,000,000	50,000,000	5,000,000	500,000
Nb	Niobium-Columbium (Nb_2O_5)	300,000	50,000	10,000	2,000
Ni	Nickel (metal)	2,000,000	200,000	20,000	2,000
P	Phosphate (P_2O_5)	2,000,000,000	200,000,000	20,000,000	2,000,000
Pb	Lead (metal)	1,000,000	100,000	10,000	1,000
Pb+Zn	Lead + Zinc (metal)	2,000,000	200,000	20,000	2,000
Pd	Palladium (metal)	300	50	10	1
Pt	Platinum (metal) & Platinoids	300	50	10	1
Rb	Rubidium (Rb_2O)	1,000	100	10	1

Table 3.1: continued

Element	ProMine Commodity	Class A	Class B	Class C	Class D
		Very large deposit (>X)	Large deposit (>X)	Medium deposit (>X)	Small deposit (>X)
Re	Rhenium (metal)	5,000	500	50	5
REE	Rare Earth Elements (RE_2O_3)	300,000	30,000	3,000	300
S	Sulphur (substance)	20,000,000	2,000,000	200,000	20,000
Sb	Antimony (metal)	50,000	10,000	2,000	1,000
Sc	Scandium (metal)	1,000	100	10	1
Se	Selenium	5,000	1,000	300	50
Si	Silica, silica sand	10,000,000	3,000,000	500,000	100,000
Sn	Tin (metal)	100,000	10,000	1,000	100
Sr	Strontium ($SrCO_3$ or $SrSO_4$)	1,000,000	100,000	10,000	1,000
Ta	Tantalum (Ta_2O_5)	3,000	500	100	20
Te	Tellurium (metal)	500	100	20	5
Th	Thorium (metal)	100,000	10,000	1,000	100
Ti	Titanium, general & Ilmenite (TiO_2)	20,000,000	2,000,000	200,000	20,000
Tl	Thallium (metal)	5,000	500	50	5
U	Uranium (metal)	50,000	5,000	500	50
V	Vanadium (metal)	10,000,000	1,000,000	100,000	10,000
W	Tungsten (WO_3)	50,000	5,000	500	50
Y	Yttrium (Y_2O_3)	300,000	30,000	3,000	300
Zn	Zinc (metal)	2,000,000	200,000	20,000	2,000
Zr	Zirconium (ZrO_2)	1,000,000	100,000	10,000	1,000
Additional resource maps					
Coal	Coal, lignite	10,000,000,000	1,000,000,000	100,000,000	5,000,000
Dolomite	Dolomite	500,000,000	50,000,000	5,000,000	500,000
Evaporite	Evaporite	500,000,000	50,000,000	5,000,000	500,000
Feldspar	Feldspar	100,000,000	10,000,000	1,000,000	100,000
Gabbro	Gabbro	100,000,000	20,000,000	5,000,000	1,000,000
Granite	Granite	100,000,000	20,000,000	5,000,000	1,000,000
Gypsum	Gypsum, anhydrite	500,000,000	50,000,000	5,000,000	500,000
Olivine	Olivine	100,000,000	20,000,000	5,000,000	1,000,000
Pyrite	Pyrite (FeS_2)	100,000,000	20,000,000	5,000,000	200,000

Source of European Mineral Deposit data

The mineral deposit data were compiled by an EU co-funded project under the 7th Framework Programme entitled *"Nano-particle products from new mineral resources in Europe"* with the acronym "ProMine" (www.promine.gtk.fi). The ProMine team was led by Daniel Cassard (BRGM), and the mineral deposit information was compiled by people from many countries (Guillaume Bertrand, Fréderik Maldan, Gabor Gaàl, Juha Kaija, Soile Aatos, Jean-Michel Angel, Nikos Arvanitidis, Dimitris Ballas, Mario Billa, Christos Christidis, Dimitrina Dimitrova, Pasi Eilu, Augusto Filipe, Emmy Gazea, Carlos Inverno, Esa Kauniskangas, Timo Mäki, João Matos, Maël Meliani, Costas Michael, Vassilka Mladenova, Javier Navas, Mateusz Niedbal, George Perantonis, Jacek Pyra, Helena Santana, Todor Serafimovski, Jean-Jacques Serrano, Juha Strengell, Goran Tasev, Fernando Tornos and George Tudor). Yves Husson (BRGM) provided the ProMine mineral deposit database in a format that could be used for plotting the GEMAS mineral deposit maps presented in this chapter. Christos Christidis and Costas Michael (IGME) provided important clarifications about the ProMine mineral deposit database.

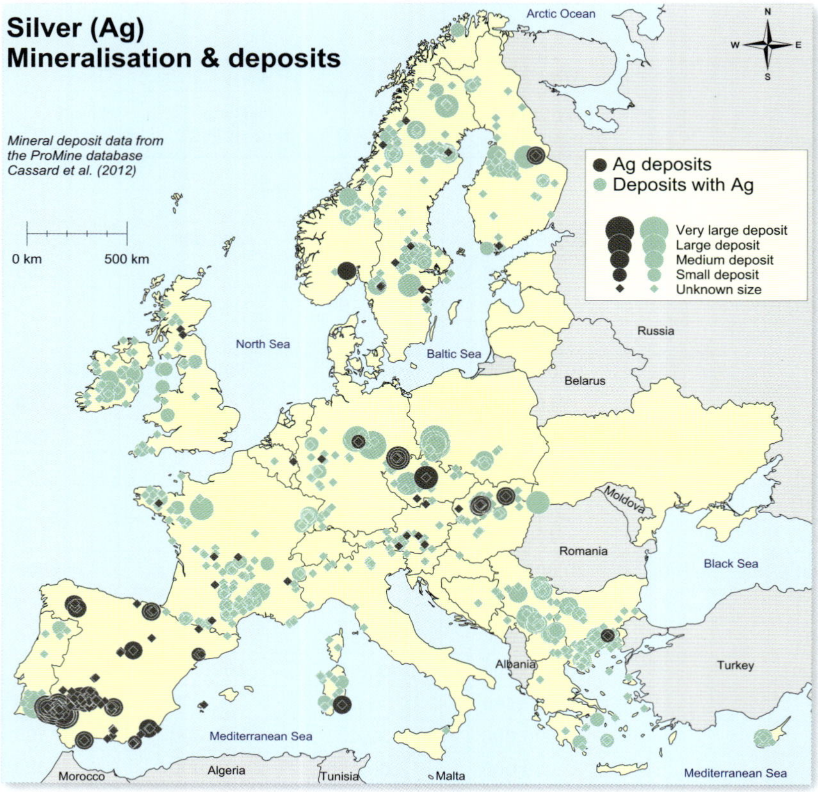

Fig. 3.1: Distribution of silver (Ag) mineral deposits and mineralisation in Europe (source ProMine mineral deposit database, Cassard et al. 2012).

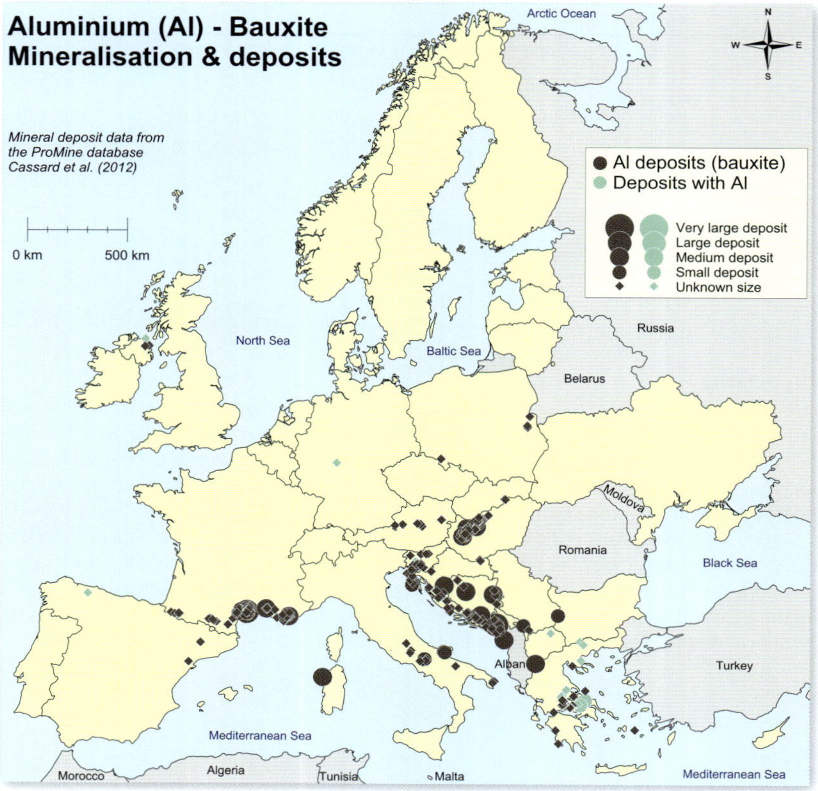

Fig. 3.2: Distribution of bauxite - aluminium (Al) mineral deposits and mineralisation in Europe (source ProMine mineral deposit database, Cassard et al. 2012).

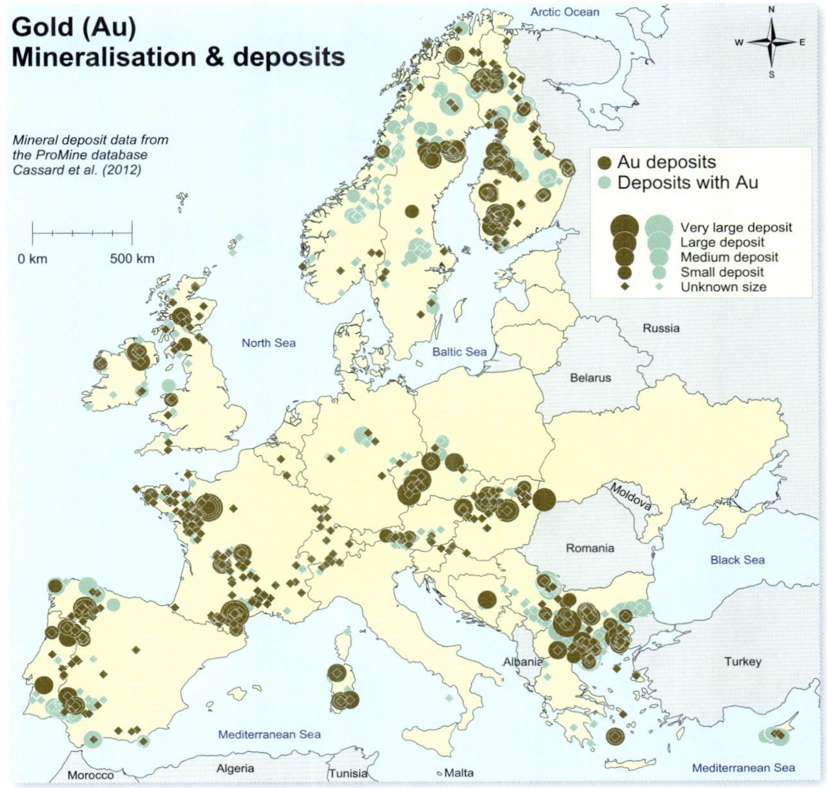

Fig. 3.3: Distribution of gold (Au) mineral deposits and mineralisation in Europe (source ProMine mineral deposit database, CASSARD et al. 2012).

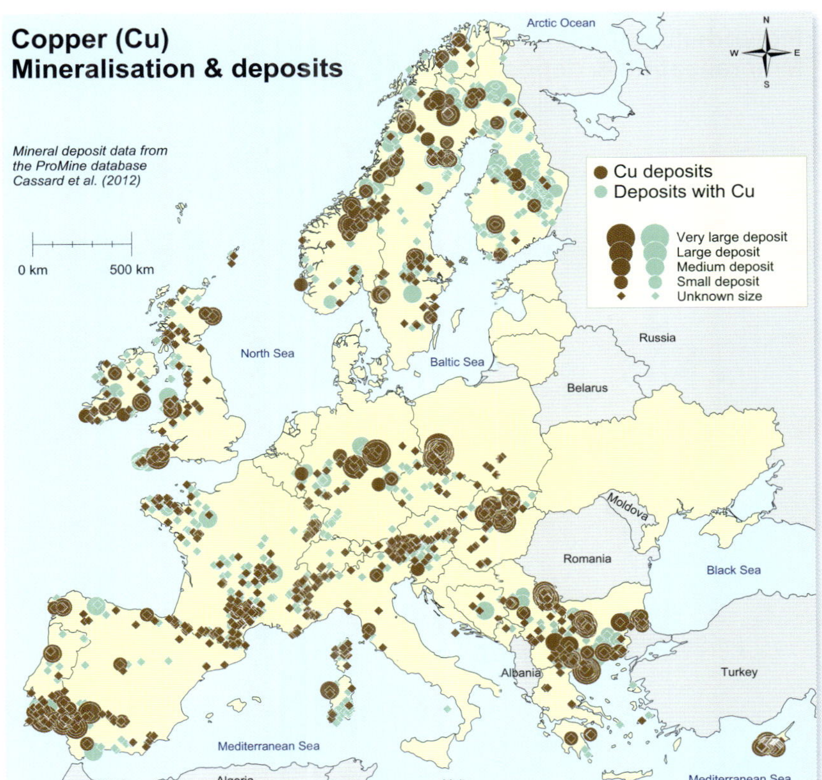

Fig. 3.4: Distribution of copper (Cu) mineral deposits and mineralisation in Europe (source ProMine mineral deposit database, CASSARD et al. 2012).

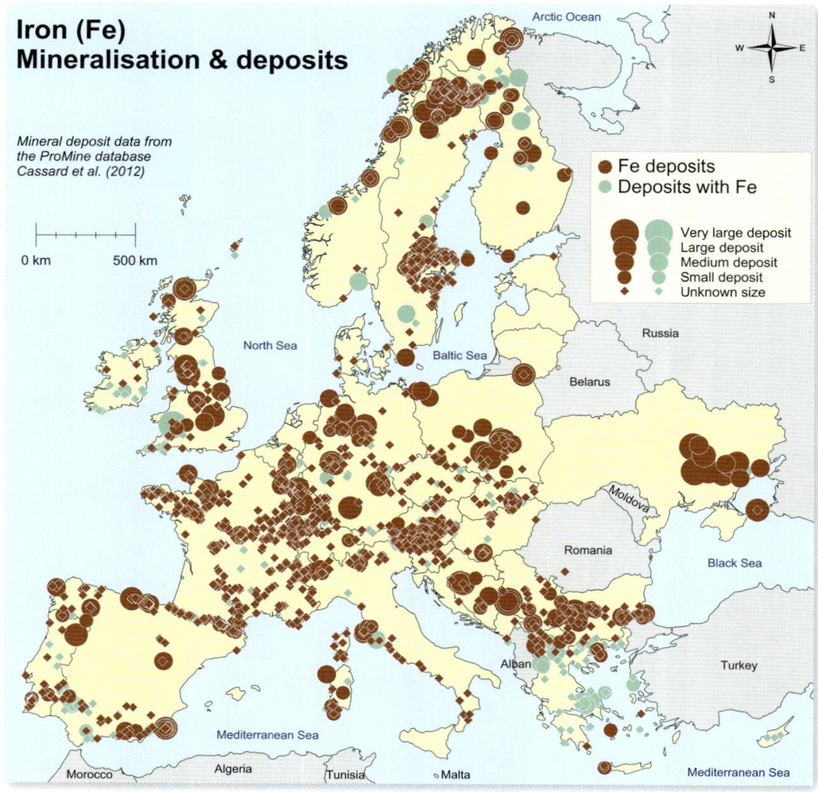

Fig. 3.5: Distribution of iron (Fe) mineral deposits and mineralisation in Europe (source ProMine mineral deposit database, CASSARD et al. 2012).

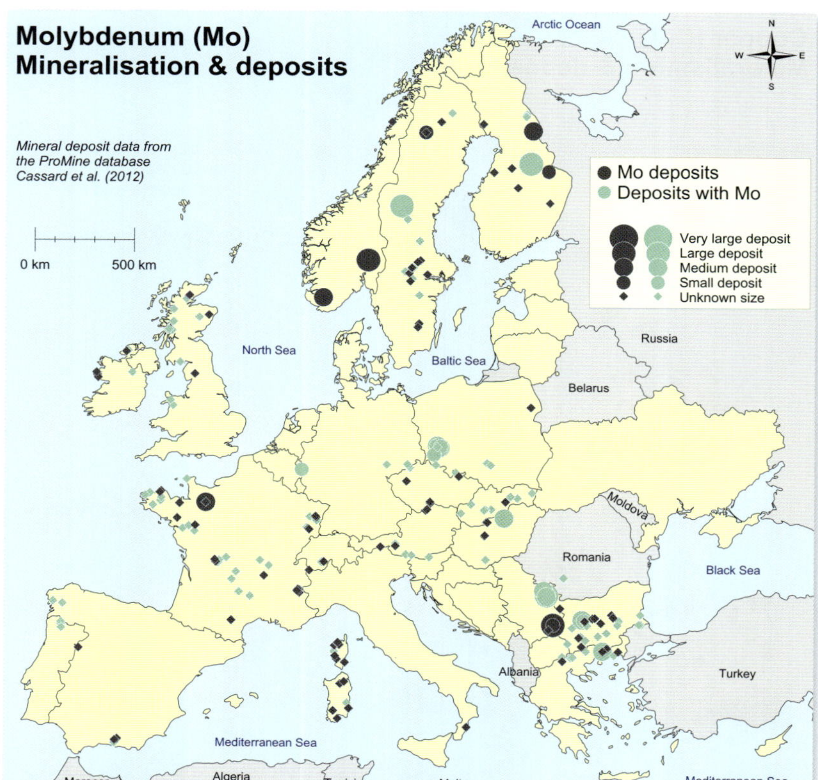

Fig. 3.6: Distribution of molybdenum (Mo) mineral deposits and mineralisation in Europe (source ProMine mineral deposit database, CASSARD et al. 2012).

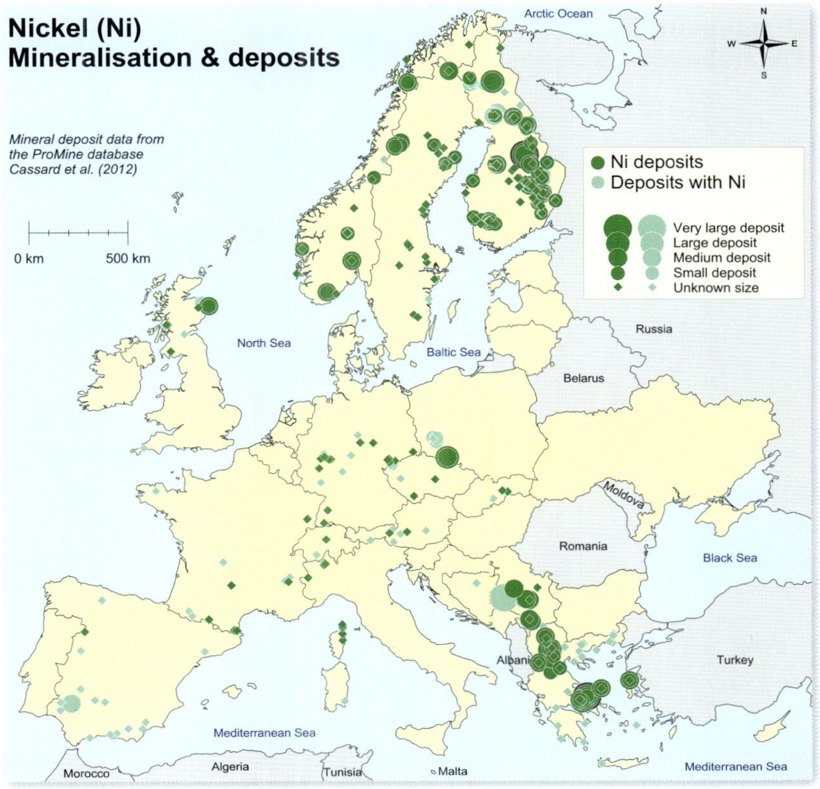

Fig. 3.7: Distribution of nickel (Ni) mineral deposits and mineralisation in Europe (source ProMine mineral deposit database, CASSARD et al. 2012).

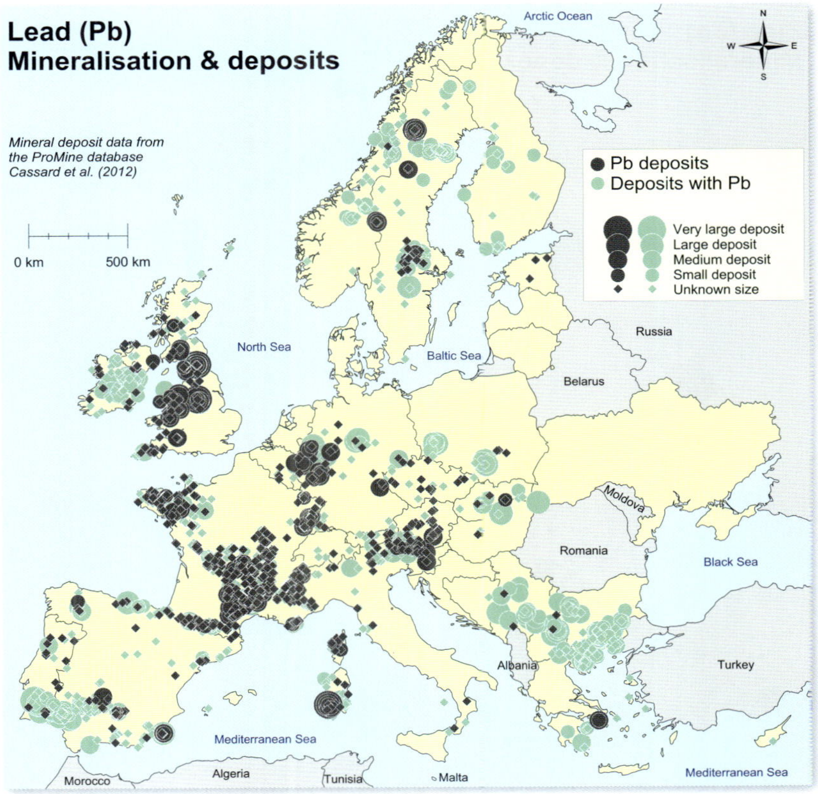

Fig. 3.8: Distribution of lead (Pb) mineral deposits and mineralisation in Europe (source ProMine mineral deposit database, CASSARD et al. 2012).

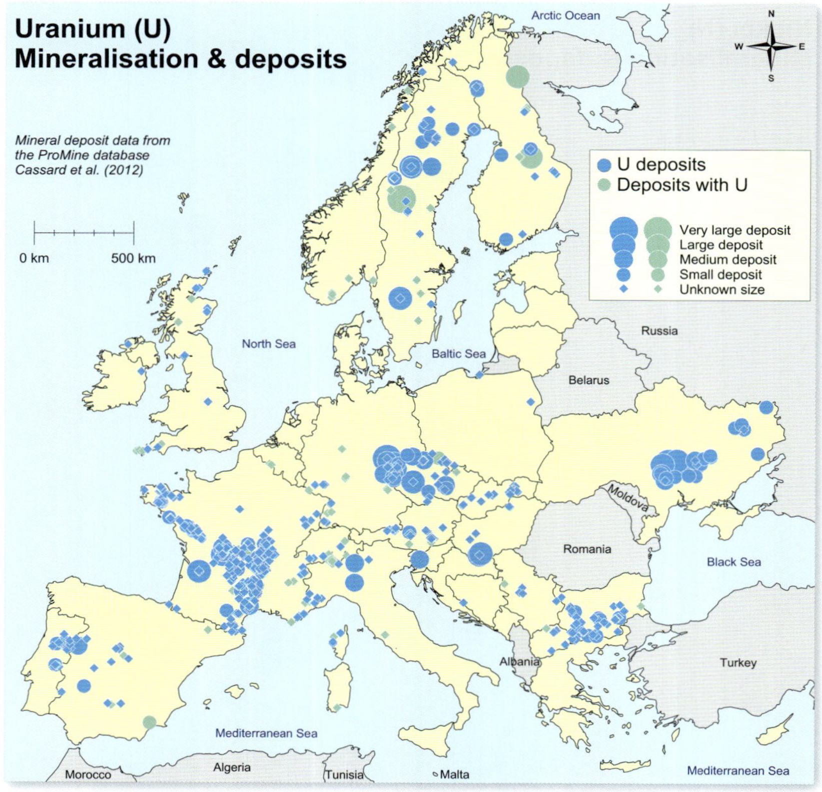

Fig. 3.9: Distribution of uranium (U) mineral deposits and mineralisation in Europe (source ProMine mineral deposit database, Cassard et al. 2012).

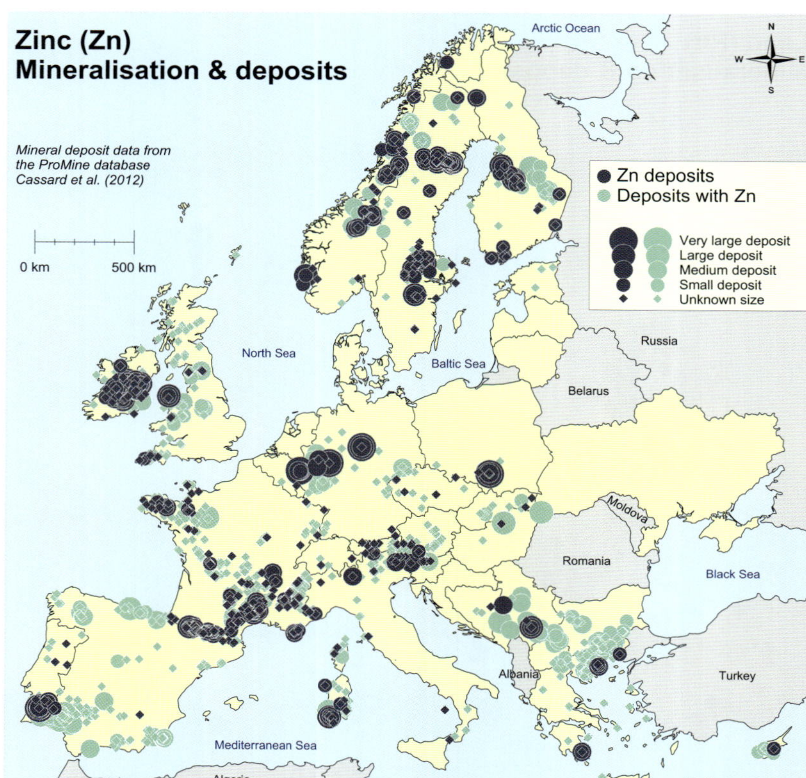

Fig. 3.10: Distribution of zinc (Zn) mineral deposits and mineralisation in Europe (source ProMine mineral deposit database, Cassard et al. 2012).

Alecos Demetriades
Institute of Geology and Mineral Exploration (IGME); Athens, Greece
E-mail: alecos.demetriades@gmail.com

Clemens Reimann
Geological Survey of Norway (NGU); Trondheim, Norway

Part B

Chapters 4 – 10

Applied Geochemistry of Selected Elements

Chapter 4

Arsenic Anomalies in European Agricultural and Grazing Land Soil

Timo Tarvainen, Manfred Birke, Clemens Reimann,
Michal Poňavič & Stefano Albanese

▸ 4.1 Introduction

The GEMAS data set provides a homogenised overview of arsenic distribution in agricultural (Ap) and grazing land soil (Gr) of Europe. Similar large-scale geochemical maps of agricultural soil have earlier been available only from the Baltic Sea Rim countries (Reimann et al. 2003). Tarvainen et al. (2013) reported the median values of the GEMAS arsenic data and studied the overall regional distribution on the continental scale. Median As concentrations in an aqua regia extraction analysed by inductively coupled plasma emission mass spectrometer (ICP-MS) are 5.7 mg/kg for the Ap samples and 5.8 mg/kg for the Gr samples. The median for the total As concentration as analysed by X-ray fluorescence spectrometry (XRF) is 7 mg/kg in both soil materials.

Maps of the As distribution for both land use types (Ap and Gr) showed a very similar geographical distribution. The dominant feature on both maps is the southern margin of the former glacial cover seen in the form of a sharp boundary between northern Europe with lower arsenic concentrations and southern Europe with more than 3-fold higher As concentrations. The border between northern and southern European As concentrations follows almost exactly the maximum extent of the last glaciation, and occurs near one of the main tectonic borders in Europe called the Trans-European Suture Zone.

According to Tarvainen et al. (2013), most of the As anomalies displayed on the maps can be directly linked to geology. However, some features have an anthropogenic origin. This chapter gives a more detailed insight into individual arsenic anomalies found by the GEMAS mapping project.

▸ 4.2 Materials and methods

This study on arsenic anomalies in European agricultural and grazing land soil is based on the GEMAS samples. Soil samples from agricultural land were taken from the topmost 20 cm (A_p horizon), while samples from grazing land, represent the topmost 10 cm of soil. Detailed description of the sampling protocol is given in EGS (2008), (see also summary in Chapter 3 → Part A).

Soil samples were air-dried and sieved through a 2 mm nylon screen in the laboratory of the State Geological Institute of Dionýz Štúr, Slovakia (Chapter 4 → Part A). Total (XRF) arsenic concentrations were determined using milled samples at the Federal Institute for Geosciences and Natural Resources (BGR) in Germany by wavelength dispersive X-ray fluorescence spectrometry on fused beads (Reimann et al. 2011; Chapter 5 → Part A). Aqua regia (AR) extractable arsenic concentration was determined by ICP-MS at ACME laboratories in Vancouver, Canada (Reimann et al. 2009; Chapter 5 → Part A). The XRF detection limit of As at 3 mg/kg is too high for many of the soil samples, thus the interpretation of anomalies is based on the aqua regia (AR) extractable concentrations.

The regional distribution of arsenic in agricultural and grazing land soil is presented using colour surface maps. Kriging was used to convert the values from irregularly distributed sampling sites to a regular grid for map production (Chapter 8 → Part A).

▸ 4.3 Results

The regional distribution of As (aqua regia extraction of Ap samples) in European agricultural soil is shown on Figure 4.1. According to TARVAINEN et al. (2013), the spatial distribution of arsenic in the Ap samples (0 – 20 cm) and the grazing land soil (Gr) samples (0 – 10 cm) are very similar. The most important anomalies are numbered on Figure 4.1 (Ap results). These anomalies are discussed in more detail in the following section.

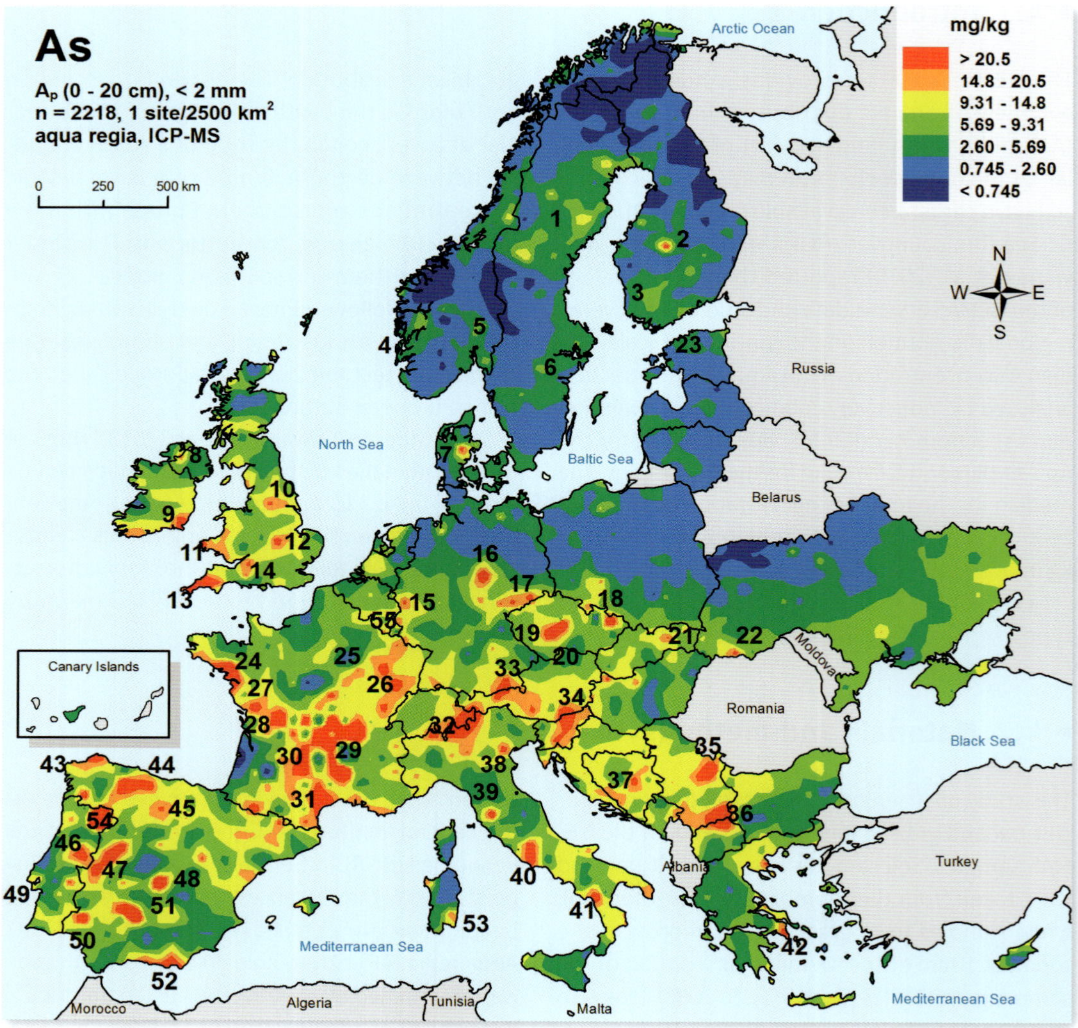

Fig. 4.1: Spatial distribution of arsenic in European agricultural topsoil (0 – 20 cm). Aqua regia extraction on the < 2 mm size fraction and determination by ICP-MS.

▸ 4.4 Discussion

4.4.1 Northern Europe

The Ap arsenic map (Fig. 4.1) shows a clear boundary between northern and southern Europe. This boundary coincides with the maximum extent of the last glaciation and occurs near the Trans-European Suture Zone (TARVAINEN et al. 2013). As a consequence of their glacial history, soil in northern Europe is often coarse-grained and less weathered compared to that in southern Europe. In general, arsenic concentrations in northern European soil are lower (AR median Ap 2.3 mg/kg) than in southern European soil (AR median Ap 8 mg/kg). This large difference in soil As concentrations between northern and southern Europe also has statistical consequences on map presentation. If the data set is subdivided into two halves along this boundary and mapped with different class limits for each half, many of the low-level anomalies in northern Europe appear much more prominent at the Scandinavian scale, while some anomalies in southern Europe are subdued. This can be clearly seen when comparing the GEMAS As map with an older map presented in the BSS atlas (Fig. 4.2; REIMANN et al. 2003). The BSS project only covered northern Europe, and here anomaly 1 on Figure 4.1 (discussed in more detail below) is the most prominent As anomaly in the whole survey area. It is in this area that the Geological Survey of Sweden identified several groundwater wells with As concentrations far above the drinking water action level (PITEÅ KOMMUN 2005). Even from a cursory glance at these two maps, it is clear that it is not possible to define one soil background level for As that is valid for all of Europe.

On a more local scale, some of the As anomalies on the map coincide with known mineral belts or ore deposits, where As is a frequent by-product, e.g., Boliden/Skellefte ore district (anomaly 1) and southern Bergslagen district (anomaly 6) in Sweden. The elevated and anomalous arsenic concentrations in Scandinavia are most often caused by sulphide mineralisation in granitic bedrock areas. In Bergslagen (anomaly 6), some of the high values may also be due to the presence of marine clay. In Norway, the elevated values (anomalies 4 and 5) could also be due to the occurrence of arsenopyrite-rich black shale or even to concentrations of arsenopyrite in Quaternary sediments.

Although As concentrations are generally much lower in northern Europe, quite a number of local As anomalies still occur when the sam-

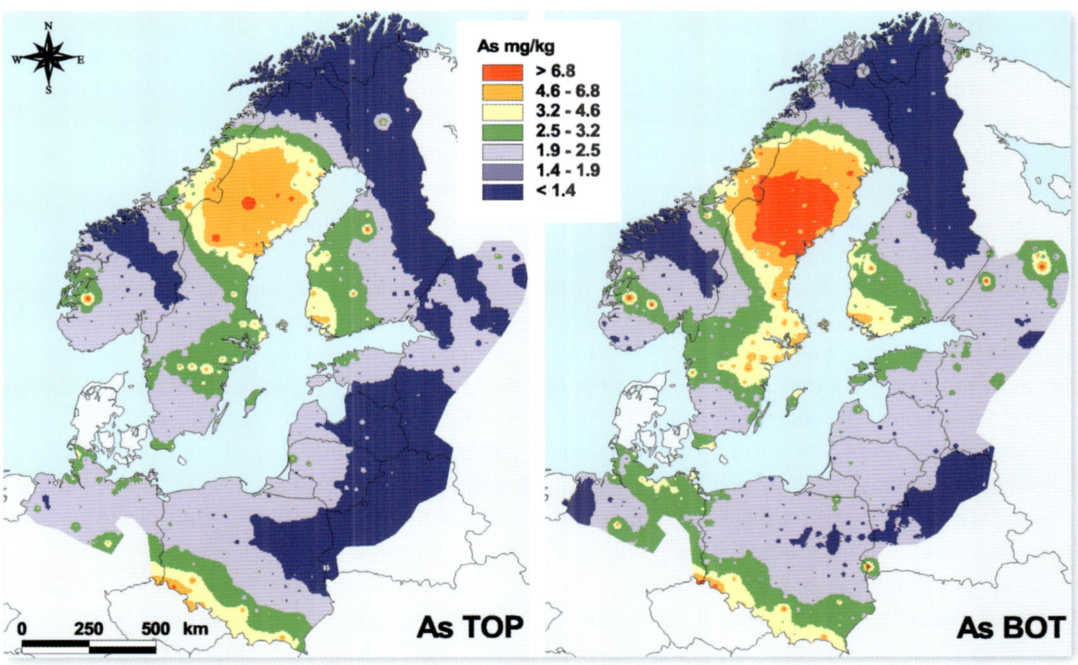

Fig. 4.2: Spatial distribution of arsenic in agricultural soil (Topsoil: 0–25 cm; Bottom soil: 50–75 cm). Aqua regia extraction on the <2 mm size fraction and determination by AAS (from REIMANN et al. 2003: p.63) showing the more detailed patterns of anomaly 1 on Figure 4.1.

ple sets are separated and northern Europe is mapped individually. For example, during the Baltic Soil Survey a major As anomaly covering almost 200 000 km^2 was detected in central Sweden (Fig. 4.2; REIMANN et al. 2003). More than ten new sulphide ore deposits have been found along the fringes of this anomaly during the last 10 years. In Finland, the well-known arsenic anomaly of the Tampere-Hämeenlinna region (LOUKOLA-RUSKEENIEMI et al. 2007) is shown as a local anomaly in south-west Finland (anomaly 3). The cause of the relatively high arsenic content in a single sample in central Finland (anomaly 2) remains unresolved.

Anomaly number 7 in Denmark is unexplained, but according to TARVAINEN et al. (2013), it has most probably an anthropogenic origin.

In Estonia, soil with a very high content of organic matter occurs near Viimsi (anomaly 23). In the area around Maardu, the high concentrations of As are most likely due to originally elevated contents in the Dictyonema Shale or to the presence of sulphides (BIRKE et al. 1998). The arsenic anomaly in northern Estonia, however, is most likely caused by industrial emissions (high industrial density, industrial dusts, metal processing and engineering). In the area around Tallinn, anomalous As concentrations in topsoil vary between 60 and 70 mg/kg (BITYUKOVA et al. 2000; BITYUKOVA & BIRKE 2011).

4.4.2 Central and Southern Europe

In southern and central Europe many of the arsenic anomalies coincide with known mineral belts, such as the Rhenish Massif, Harz Mountains and Erzgebirge ore districts (Ore Mountains) in Germany (anomalies 15 – 17); most vein-type As mineralisation in France (e.g., anomaly 26), the metallogenic province on the border of Serbia/Bulgaria/FYROM (anomaly 36) and the ore deposits in the Iberian Pyrite Belt, Spain (see As mineral deposits map on attached DVD → Part A). Because most of the known deposits have been mined, local industrial sources often developed in the same areas and some anthropogenic anomalies can coincide, therefore, with the geological ones. More detailed scale geochemical mapping can reveal the difference between geogenic and anthropogenic anomalies. For example, the Geochemical Atlas of Saxony (RANK et al. 1999) covers the Erzgebirge (Ore Mountains) area at a density of 1 sample/16 km^2 and demonstrates that although most of the anomalies can be traced back to known ore bodies, high As concentrations have been spread over a much larger area via anthropogenic activities (mining, smelting; Fig. 4.3).

In south-east Ireland, the elevated As levels (anomaly 9) can be linked to Ordovician metavolcanic formations that host volcanic massive sulphide and gold deposits locally. Also, anomalies in Northern Ireland (8) and in England (10 – 13) can be linked to geological sources (TARVAINEN et al. 2013). Anomaly 14 may be connected to processing of As-rich metal ores.

In the data set of agricultural soil, by far the highest concentration is observed at a site in north-east Portugal (666 mg/kg As; anomaly 54), which was sampled approximately in the alignment of an important E – W gold mineralised vein and 500 m north of an abandoned gold mine (Maria João Batista, pers. comm.). The next highest value of 126 mg/kg is found in France and, in fact, all the Ap samples with more than 70 mg/kg As originate from France. This may be an indication of As-based herbicides having been used in France on ploughed fields. One of these high As values came from a vineyard, while the others were observed on fields where the last crop sown was maize. For the grazing land soil, the maximum value is 292 mg/kg (a sample from France). The Gr samples with values above 70 mg/kg As are not exclusively from France, but from many other European countries.

Arsenic anomaly number 49 in Lisbon, Portugal, is the best example of arsenic enrichment in urban surroundings. Other examples of potential anthropogenic anomalies include numbers 18 and 19 in Poland and Czechia, respectively. These anomalies are related to coal-fired electricity power plants. One anomaly in Belgium (55) is also most probably anthropogenic. A sample from grazing land has 72.7 mg/kg As, while the adjacent sample from agricultural soil contains 18.4 mg/kg As. The sampling site is located in an area underlain by Jurassic sandstone, Hettangian to Sinemurian in age, with generally low trace element contents, and it is thus likely that the grazing land site is contaminated.

Regions with relatively high arsenic concentrations in agricultural and grazing land soil include the Variscan basement in Portugal and the western parts of Spain (Iberian Massif (anomalies 44, 46, 47), Iberian Pyrite Belt (50), Galicia (43)) , the central part of the Betic Cordillera in southern Spain (52), the Pyrenees, most of France (Armorican Massif (24), Massif Central (29 – 30) and Vosges (26)), the southern Alps and the Po-Adriatic Basin (32 and 38), the western Alps and Switzerland, the Pannonian Basin in Hungary, Croatia, Serbia (Sumadija; mining: Rudnik, Babe), on the border between Serbia and north-eastern FYROM (anomaly 36, mineralised zones and mining: Lece-Haldiki, Besna Kobila-Osogovo, Zletovo-Sasa-Tornica), as well as the Harz Mountains (16), the Black Forest and the Ore Mountains (17) in Germany. Samples were not available from Romania, but anomalies 22 and 35 most probably extend into this country. Areas with high As concentrations are also found in south-west England (Cornwall, SW England mining field, central Wales mining field, Conway valley mineralisation), in Czechia (Bohemian Massif, Ostrava, anomaly 20) and on the Attica peninsula in Greece. Anomalies in Slovakia (21) and south-western Ukraine (22) are linked to mineralisation. Anomaly 37 in Bosnia-Herzegovina remains unexplained.

In Spain, arsen c anomalies are located in the same areas as the arsenic anomalies seen in the FOREGS geochemical mapping of Europe (Salminen et al. 2005; de Vos & Tarvainen 2006). High As concentrations are controlled by geology and mineralisation, i.e.. in Galicia (anomaly 43) by regional shear zones that control the intrusion of some granitic bodies that contain Au-As-(W) mineralisation. In Asturias (44), several small granite masses are accompanied by epithermal Au-As-(W) mineralisation. In the remaining Iberian massif, elevated As concentrations in soil are most likely associated with Sn-W and precious metal mineralisation (de Vos & Tarvainen 2006). Anomaly 45 in the Castilla-León Tertiary Basin, anomaly 48 at

Fig. 4.3: Distribution of arsenic in topsoil, Free State of Saxony, showing the anomalous patterns in the Erzgebirge (Ore Mountains) area at a density of 1 sample/16 km^2 (from Kardel & Rank 2010) showing the more detailed patterns of anomaly 17 on Figure 4.1.

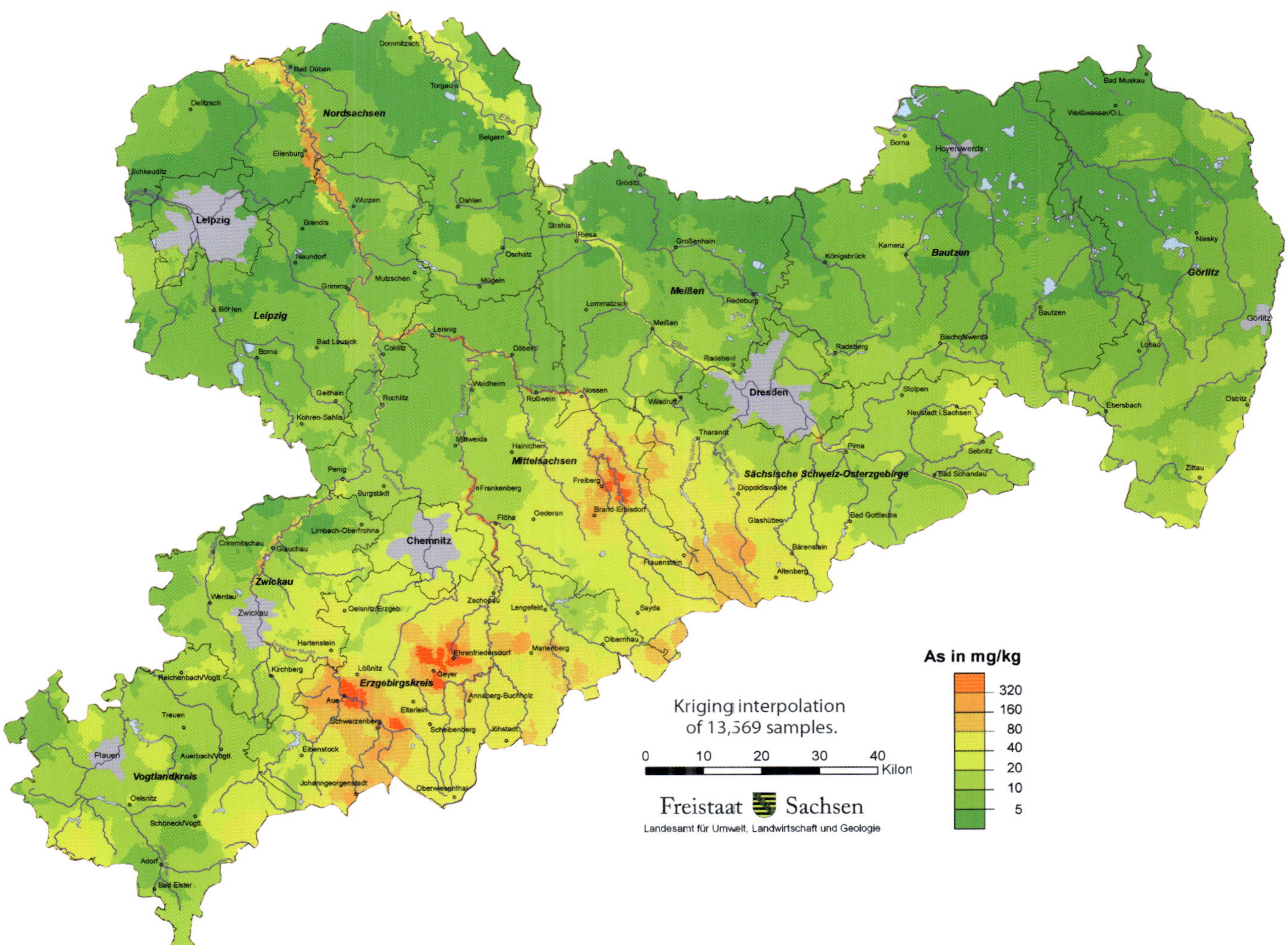

Toledo Mountains and 51 at Pedroches batholith can also be linked to geological sources.

In north-west France (anomalies 24 and 27), high As concentrations in Ap and Gr samples delineate the south-western Armorican shear zone, which extends from Brittany to the Massif Central (de Vos & Tarvainen 2006), with associated Sb-As-Au mineralisation. Further, anomaly 28 in France has most probably a geological source, i.e., Hercynian granite in Jurassic black marl. The arsenic-rich Cambrian black shale in the Pyrenees has an impact on the arsenic distribution in topsoil. In the southern Massif Central, the anomalous As values are associated with primary mineralisation (Ales Sb-As district). The north-west part of anomaly 31 in France is clearly related to the major gold-arsenopyrite deposit of Salsigne (mesothermal gold), while the south-east part of the same anomaly may be related to pesticides having been used in orchards and vineyards. Anomaly 25 can be linked to geological sources (Tarvainen et al. 2013).

The local As maximum in south-eastern Greece mainland around Vavrona (42) is possibly related to polymetallic sulphide mineralisation, but there may be a contribution from As-based herbicides used in the vineyard. The As anomaly in north-eastern Greece in the Chalkidhiki area (Palaeochori) is correlated with anomalous Au concentrations and attributed to polymetallic mineralisation (Alecos Demetriades, pers. comm.).

The Swiss-Austrian-Italian border area shows elevated As concentrations related to natural mineralisation along complex tectonic structures (anomaly 32). However, the known arsenic anomalies in southern Switzerland between Lugano and the Aar Massif are not well represented in the GEMAS data set (Peter Hayoz, pers. comm.). Anomaly 33 can be linked to fahlore deposits in Schwaz-Brixlegg. Soil in the area of anomaly 34 in the Styrian Basin has elevated As concentration.

In Italy, As concentrations are mostly elevated in association with the Po River plain (anomaly 38), some areas along the Tyrrhenian Sea coastline (between southern Tuscany and southern Campania) and in Sardinia. In the Po River plain, soil is mostly developed on Quaternary alluvial and glacio-fluvial deposits characterised by a clayey texture. In an alluvial/lacustrine environment, As is generally adsorbed on the surface of Fe hydroxides. In the Po River plain, the Fe hydroxides following the deposition process of the fine-grained sediments (clay and silt), due to their similar crystalline dimensions, could also have been able to generate the actual enrichment of As in soil. At the same time, it is not easy to determine the primary source of As, which could be derived from the use of CCA-preserved wood, for example, treatment of vineyard posts, the past use of arsenic-containing fertilisers or arsenic-based herbicides or from the sediments of arsenopyrite-bearing metamorphic rock (gneiss) from the Alps. Natural accumulations of As in the sediments of river basins are a worldwide phenomenon and may cause many problems for drinking water quality in these areas.

In Tuscany, As enrichments in the alluvial plain of Scarlino (Piana di Scarlino) opposite Elba Island are mostly of natural origin (anomaly 39). The primary source of As is the geological materials from the middle valley of the Pecora River (upstream of Scarlino), where As-enriched phases have been mobilised toward the alluvial plain from surface alterations of Cu-Pb-Zn and Fe deposits. Arsenic could also be associated with travertine outcrops generated by thermal waters originating from many springs widely occurring in the middle valley of the Pecora River.

Arsenic enrichments along the Tyrrhenian coastline are generally related to the volcanic nature of the sampled soil (anomaly 40). This soil has been developed by the pedogenesis of the volcanic materials (mostly pyroclastics) deposited by the Quaternary activity of the 'Roman-Neapolitan Comagmatic Province (RNCP)'. Arsenic enrichments can also be found in areas far from the volcanic centres of the RNCP (southern Campania region), where carbonatic sedimentary rocks have been covered by the fallout of wind-transported pyroclastics (mostly ashes) from volcanic eruptions (anomaly 41).

Arsenic enrichments in Sardinia (53) are probably primarily due to the presence of arsenopyrite veins in areas where base metal sulphide mineralisation occurs (see As mineral deposit map on attached DVD → Part A). Mining has been going on in these areas for a long time, causing a general alteration in the chemistry of the surrounding soil, especially in the downstream areas, by means of both dispersed dust and fluvial transport.

In Czechia, the As distribution maps for both types of soil are almost identical. The highest As values (Ap soil 61.3 mg/kg; Gr soil 131 mg/kg) were recorded on the north-western edge of Czechia. Again, the relative enrichment of As in these soil samples can have several possible causes: (1) anthropogenic contamination due to brown coal combustion, chemical and other industries (e. g., Sucharova et al. 2011), and (2) it could be associated with As-Ag-U vein mineralisation in the large mining district of Jachymov in the Erzgebirge region, and thus be a primary natural anomaly indicating the presence of mineralisation. The median total As content is 39.8 mg/kg in brown coal from the North Bohemian Brown Coal Basin, with a range varying from 2.92 to 387 mg/kg (Śkorová et al. 2007).

The second highest As values in Czechia (Ap soil 61.1 mg/kg; Gr soil 51.5 mg/kg) occur in the central part of the country, near the Kutná Hora Ag–Pb–Zn mining district. This could indicate the mineralised area or be caused by anthropogenic emissions from the mining area or both. Detailed mineralogical investigations of the soil sample would be necessary to unravel the true source of the anomaly.

In Germany, the NNE-SSW-striking Lower Harz Ore Zone (anomaly 16 on Fig. 4.1) is characterised by geogenic anomalous As concentrations of more than 30 mg/kg. It is evident that, even with the given low sample density, it is possible to predict the existence of hidden intrusive bodies of geochemically specialised granitoid magmas (Ramberg granite) at depth, as indicated by an As halo. The reason may be that the magmas were intruded into anchimetamorphic slate (clay schist) and that the mineralisation in the roof region of the granite has a clear structural control. In the Gr samples, the As distribution is similar to the zinc, copper and lead distribution in the Upper Harz Mountains, caused by the ore deposits and old mines at Bad Grund, St. Andreasberg and Rammelsberg near Goslar. Pyrite-baryte ores with Zn, Pb and Cu sulphides are characteristic of the Rammelsberg ore deposit. The high arsenic concentrations in the Harz Mountains result from the ore deposits in this area.

A major As anomaly is observed in the German part of the Ore Mountains (17) along the border with Czechia. This area is world famous for its ore occurrences (predominantly polymetallic quartz sulphide veins). A local geochemical atlas (Rank et al. 1999; Kardel & Rank 2010) demonstrates that almost all of Saxony is characterised by highly elevated As concentrations in its soil (Fig. 4.3). The many mines, mine waste heaps, and associated industrial activities, have led to widespread contamination in this area, as can be demonstrated in more detailed investigations using stream sediment (Birke et al. 1995, 2006) and terrestrial moss (Siewers & Herpin 1998). It is almost impossible to determine how much of the As in the agricultural and grazing land soil, collected for the GEMAS project, is due to the highly elevated background in this area, and how much is actually due to contamination of single sites. The detailed investigations cited above demonstrate that severe contamination is usually observed at a much more detailed scale than the continental scale used here.

Distinctly elevated concentrations (up to 50 mg/kg) in the grazing land soil samples characterise the granite and the sulphide mineralisation of the southern Black Forest.

The elevated As values to the north-west of Saarbrücken in the Saar-Nahe Valley are caused by Pb-Zn-Cu sulphide mineralisation. The anomalous concentrations around Aachen-Stollberg and in other areas of the Rhenish Slate Mountains are linked to the sulphide mineralisation and ore deposits in this region. Anthropogenic influences (mining and smelting) are superimposed on the natural sources causing the elevated background values (median values) in the Ruhr district.

The elevated As concentrations in the topsoil of the Bavarian Molasse basin may be caused by groundwater from Tertiary aquifers containing high As (Weigand et al. 2010). Weigand et al. (2010) focused on redox processes induced by water table fluctuations as a possible route of As enrichment in the topsoil. The slightly elevated As concentrations in the Franconian and Swabian Jura regions (especially in the Franconian and Swabian Keuper-Lias country) are associated with known sulphide mineralisation. The elevated arsenic values in Gr samples in north-eastern Germany (Mecklenburg-Vorpommern) are correlated with other elevated trace metal contents (Cd, Cu, Ni, Hg, Pb, Tl, Zn) and have been attributed to the industrial area around Schwedt (oil refinery) at the Oder-Neisse border with Poland.

The arsenic anomalies in the Ap samples in northern and southern Italy, in the south of Ukraine, south-east of Switzerland and in the west and south of Austria are correlated with elevated and anomalous gold values. In central and south-west England, arsenic anomalies are also coincident with anomalous gold contents, and thus provide indications of mineralisation. In Germany, the anomalous As levels around the old mining and smelting district of Aachen-Stollberg correlate with an anomalous gold content in agricultural soil. The anomalous As concentrations in Ap samples in the eastern Alps, on the border between Germany and western Austria, correlate with elevated gold values. In the grazing land samples, a clear correlation between anomalous arsenic and gold values can additionally be detected in eastern Serbia, in the north-west of Italy, in south and south-western Austria, and especially in north-eastern Greece.

▸ 4.5 Conclusions

Most of the arsenic anomalies in European agricultural soil can be linked to geological sources. In some cases, the geological anomaly pattern has been overprinted by anthropogenic emissions from the ore processing and metal industry in the same district. More detailed geochemical mapping would be needed to separate anthropogenic and geogenic anomalous patterns in mineralised districts. Only very few anomalies could be explained by other human activities such as urbanisation and energy production.

The general level of arsenic concentrations is lower in northern Europe. Some anomalies might be more clearly delineated if a different colour scale was used to construct the arsenic map of northern Europe. |◂

Timo Tarvainen
Geological Survey of Finland (GTK);
Espoo, Finland
E-mail: timo.tarvainen@gtk.fi

Manfred Birke
Federal Institute for Geosciences and Natural Resources (BGR);
Hannover, Germany

Clemens Reimann
Geological Survey of Norway (NGU);
Trondheim, Norway

Michal Poňavič
Czech Geological Survey (CGS);
Praha, Czech Republic

Stefano Albanese
University of Napoli Federico II (UNINA);
Napoli, Italy

Chapter 5

Distribution of Cadmium in European Agricultural and Grazing Land Soil

MANFRED BIRKE, CLEMENS REIMANN, UWE RAUCH, BENEDETTO DE VIVO, JOSIP HALAMIĆ, VOLODYMYR KLOS, MATEJA GOSAR & ANNA LADENBERGER

▸ 5.1 Introduction

Cadmium has a relatively low abundance in materials of the Earth's upper continental crust. The average concentration of Cd in the upper continental crust (UCC) is rather low, around 0.1 (0.06 – 0.102) mg/kg (REIMANN & DE CARITAT 1998; RUDNICK & GAO 2003). A comprehensive overview of Cd distribution in the geosphere is given by REEDER et al. (2006) and by PAN et al. (2010). It is a relatively rare soft metal that occurs in the environment typically in association with zinc ores and, to a lesser extent, with lead and copper ores. Some inorganic Cd compounds are soluble in water, while Cd oxide and Cd sulphide are almost insoluble. Cadmium rapidly oxidises in air. Wet and dry deposition transfers Cd from the air to soil, where it is adsorbed by plants and enters the food chain. This process can be influenced by acidification, which increases the availability of Cd in soil.

Cadmium has eight naturally occurring isotopes, of which ^{114}Cd and ^{112}Cd are the most abundant at 28.73 % and 24.13 % of its mass, respectively. Furthermore, about 35 radioisotopes are known. Its geochemical properties are close to zinc, mercury and indium, and it has a strong affinity for sulphur (ALLOWAY 1995; KABATA-PENDIAS & MUKHERJEE 2007). Cadmium forms compounds with oxygen and it can be present in chlorides, nitrates and carbonates (GREENWOOD & EARNSHAW 1997). In aqueous solution, Cd often forms coordinate bonds with many ligands, including halides, hydroxide, cyanide and nitrate (ALLOWAY 1995; GREENWOOD & EARNSHAW 1997). Organocadmium compounds are rather reactive in both air and water.

Cadmium has locally elevated abundance due to anthropogenic activities. Atmospheric levels of Cd range up to 0.005 µg/m^3 in rural areas, from 0.005 µg/m^3 to 0.015 µg/m^3 in urban areas, and up to 0.06 µg/m^3 in industrial areas (WHO 1992; WORLD BANK GROUP 1998). Freshwater bodies typically contain levels of Cd between 0.01 µg/L and 3.0 µg/L. In freshwater, Cd has a tendency to form complexes with organic compounds. In European bottled mineral water, Cd concentrations vary between < 0.003 µg/L and 1.1 µg/L, and in European tap water from < 0.003 µg/L to 1.4 µg/L (REIMANN & BIRKE 2010).

There are very small differences in the overall abundance of Cd in mafic and granitic rocks. Sedimentation processes tend to concentrate the element as evidenced by the high concentrations (up to 11.0 mg/kg) in sedimentary rocks. Elevated Cd levels are found in organic-rich shale, oceanic and lacustrine sediments, and in marine manganese nodules (5.0 – 8.4 mg/kg), in phosphorite, and in carbonates, kaolin and coal (WEDEPOHL 1968; FERGUSSON 1990; REEDER et al. 2006). Marine zinc-bearing phosphorite also has high Cd concentrations (60 – 340 mg/kg; GOLDSCHMIDT 1958; AUER 1977). Cadmium in phosphorite and black shale is present as sulphide and in organic complexes. Sphalerite is its principle carrier (GARRET et al. 1996). Cadmium concentrations of commercial interest are only found in zinc sulphide deposits (up to 1.3 % Cd, more commonly within the range of 0.1 to 0.5 % Cd; USGS 2011), as well as in lead and copper sulphide deposits. Therefore, it is invariably recovered as a by-product of these ores and not as the principal product of any mine. Cadmium can be used in geochemical exploration as a typical pathfinder element for sulphide mineralisation.

Cadmium is a very toxic element to plants and animals. Foods contain Cd as a result of uptake from soil and bioaccumulation in terrestrial and aquatic animals. As an environmental hazard, Cd was identified as being responsible for the occurrence of itai-itai (painful-painful) disease in Japan in the 1950s, although the disease was known since the early 1900s. Itai-itai is a form of acquired Fanconi's syndrome and is characterised by severe kidney damage, chronic calcium loss and the development of brittle and easily broken bones. The Cd had leached from a waste heap at a nearby lead-zinc mine and contaminated the water supply of the village Fuchu. Over a period of about 15 years, the inhabitants of the Toyama Prefecture had ingested Cd in drinking water and in rice that had been grown in paddies irrigated by the same polluted water of the River Jinzu (Nirei et al. 2010).

▸ 5.2 Occurrence and sources

Cadmium forms sulphides like greenockite (CdS), which is a common component of polymetallic ores of other chalcophile elements; it is of little importance in other minerals like cadmoselite (CdSe), monteponite (CdO), otavite ($CdCO_3$), hawleyite (CdS), and the amorphous form of CdS. Cadmium occurs in trace concentrations in sphalerite, biotite, and amphibole. Its content tends to increase in the following order of rock types: igneous (0.1 – 0.3 mg/kg) < metamorphic (0.3 – 10 mg/kg) < sedimentary (0.3 – 11 mg/kg).

In uncontaminated light sandy and heavy-loamy soil Cd contents were observed from 0.22 to 0.51 mg/kg (Kabata-Pendias 2001). Depending on the parent material and soil type, concentrations vary from 0.06 mg/kg to about 10 mg/kg. Cadmium is often associated with soil colloids, where it may be present in several types of complex ions and organic chelates. Taylor & Percipal (2001) found that between 55 and 90 % of Cd in soil solution is present as the free metal ion Cd^{2+}. According to Welch & Norvell (1999), Cd also occurs in complexes with various organic acids in the soil solution (Gottschalch et al. 2007). The type of organic matter is very important for the Cd and Pb adsorption. Under anaerobic conditions, the amount of Cd in soil solution is governed by sulphide precipitation (Kabata-Pendias & Mukherjee 2007). Cadmium is bound to organic-matter-rich raw materials, such as crude oil (0.01 – 10,000 mg/kg) and coal (0.01 – 300 mg/kg; Alloway 1995). As the organic matter content in soil increases, the retention of Cd in soil also increases (e. g., podsol 0.07 mg/kg, loess 0.2 mg/kg, chernozem 0.35 mg/kg).

However, Cd concentration in soil is controlled by pH, content of humus and iron and manganese oxides, and concentration of phosphates and chlorides. Cadmium solubility increases with decreasing pH, and very little Cd adsorption by soil colloids, hydrous oxides, and organic matter takes place. At pH values > 6, Cd is adsorbed by solids in soil or is precipitated, and the concentrations of dissolved Cd are greatly reduced (Adriano et al. 2005). Cadmium forms soluble complexes with inorganic and organic ligands, especially with chloride ions. Thornton (1991) determined that Cd mobility and bioavailability are higher in non-calcareous soil than in calcareous ones. Liming of soil raises the pH, increasing Cd adsorption by the soil and reducing bioavailability (Thornton 1992). Cadmium may be adsorbed by clay minerals, carbonates or hydrous oxides of iron and manganese or may be precipitated as Cd carbonate, hydroxide or phosphate.

Cadmium is widely distributed in mineral deposits and is found in shale and igneous rocks, coal, sandstone, limestone, lacustrine and marine sediments, and soil. The Cd content of typical zinc ores can be as high as 1.3 %, but is more commonly in the range of 0.1 to 0.5 % (USGS 2011). It is also found in phosphorite, hydrothermally mineralised rocks and some black shale deposits. Garret et al. (2008, 2010) described geogenic Cd enrichment in Jamaica in connection with marine phosphorite. Extensive analysis of surface soil has determined substantial Cd enrichment (up to 200 mg/kg; Lalor et al. 1998) above many phosphate ore deposits. Cadmium is never found as native metal, and only a few rare Cd minerals are known, none of which are found in sufficient concentration to be mined alone as an ore. Greenockite (CdS, hexagonal crystal form and brown colour), hawleyite (CdS, cubic crystal form and yellow colour), and otavite ($CdCO_3$) are often found occluded in the zinc minerals wurtzite, sphalerite, and smithsonite. Cadmium of commercial interest occurs only in the sulphide deposits of zinc, lead and copper, and is recovered as a by-product of these ores. The zinc-to-cadmium ratios in typical zinc ores range from 200 : 1 to 400 : 1 (USGS 2011). For every tonne of refined zinc about 3 kg of Cd are produced.

Cadmium production is, therefore, much more dependent on zinc production than on demand for Cd. The amount of Cd relative to zinc present in zinc ores depends on the type of deposit. Because Cd is more chalcophile than zinc, it tends to remain in solution in a magma more readily than zinc does and, consequently, is more enriched relative to zinc in igneous rocks than in other types of deposits. Hydrothermal zinc ores form in the later, low-temperature stages of magmatic differentiation. Consequently, Cd is even more enriched relative to zinc in low-temperature reducing environments, such as those in which black shale is formed.

Cadmium reserves are calculated as a percentage of zinc reserves. Identified world resources of Cd in 2010, which are based on zinc resources, are estimated to be nearly 660,000 tonnes (Singer & Menzie 2010; USGS 2011). This is around 145 times world production in 2010 (USGS 2011). India, China, Australia, Kazakhstan, Mexico, Peru, United States, Poland, Russia, and Canada have more than 85 % of global Cd resources; India and China alone have about 34 % (USGS 2011); less than 10 % of the resources are in Europe. Reserves data are, however, dynamic. They may be reduced as ore is mined, or more commonly, they may continue to increase as additional deposits are developed, or currently exploited deposits are more thoroughly explored.

Cadmium emission to the environment may arise from both natural and human-made sources. Natural sources of Cd, mainly volcanic emissions, transport of wind eroded material (aeolian particles from soil erosion), forest fires, marine aerosols and weathering of parent rocks, are estimated to release annually about 300 tonnes of Cd.

Mining, storage and production of Cd metal and Cd oxide, as well as Cd chemicals, result in emissions of Cd to water, air and soil. Similarly, the manufacturing of Cd products – and the generation of scrap, solid waste, hazardous waste, wastewater and sludge containing Cd – result in additions to the surface environment. Cadmium is often a by-product during extraction of zinc, lead, or copper from their sulphide ores. The mined sulphide ore is first broken, crushed and ground to liberate the zinc sulphide particles from the host rock and treated with water and reagents, followed by the concentration process, which involves flotation. Annual anthropogenic sources of Cd in the environment may be about 5500 tonnes.

Sources of soil contamination by Cd are the mining of non-ferrous ores, lead and zinc smelters, metallurgical plants (iron and steel mills), ashes from fossil-fuel combustion, waste disposal from cement plants, municipal refuse, incineration of plastic containers and batteries (Kazantzis 1987), production of ink, paints and plastics, electroplating, sewage sludge application to land, and especially phosphate fertilisers and atmospheric deposition (deposition of aerosol particles from urban and industrial air pollution, 1 – 50 mg Cd/m^3). Phosphate fertilisers are the most widespread cause of Cd contamination, but application of sewage sludge to soil has caused also significant local problems.

Anthropogenic sources of Cd include materials to which Cd has been added to impart specific properties, or in which Cd is naturally present as a residual or impurity element.

According to Morrow (2001), Cd amounts from the intentional uses of Cd (NiCd batteries, pigments, alloys and stabilisers) are far less than the amounts of Cd released into the environment from the production and use of zinc and steel, fertilisers, cement, coal and oil, waste incineration and wastewater treatment). The author describes an example of this where even though the Cd level in NiCd batteries (7 – 25 % Cd) is higher than the Cd levels in phosphate fertilisers (average content of 3 – 90 mg/kg); the amounts of applied fertilisers are many times higher than the amount of NiCd batteries produced.

The Cd concentrations found in phosphorite fertilisers vary greatly (0.2 – 345 mg/kg; Kabata-

Pendias & Mukherjee 2007) depending on the origin of the phosphorite (0.1 – 584 mg/kg; Alloway 1995). It is sometimes found in high concentrations in phosphate fertilisers, since some marine phosphates and phosphorites have been reported to contain Cd levels as high as 500 mg/kg (Cook & Morrow 1995). Other causes for elevated Cd concentrations in soil are the use of Cd-contaminated sewage sludge fertilisers and the deposition of floodplain sediments by polluted streams and rivers. Municipal sewage sludge contains Cd in the range of < 1 to 3500 mg/kg, with medians in the order of 5 to 20 mg/kg (Nriagu & Pacyna 1988). In compost, Cd concentrations range from 0.26 to 11.7 mg/kg, and in fly ash between 1.5 and 13 mg/kg (Kabata-Pendias & Mukherjee 2007).

▸ 5.3 **Production and use**

In the past, Cd metal was recovered during the smelting of zinc ores by fractional distillation under reducing conditions to increase its purity (Alloway 1995). The current practice in most European countries involves the electrolytic treatment of Cd sulphate solutions collected during production of zinc (ECB 2007).

Cadmium is produced in countries where zinc is refined, not necessarily in the countries where zinc ore is mined. China, Japan, and Korea are the world's largest producers, followed by Mexico, the United States, The Netherlands, India, the United Kingdom, Peru and Germany (USGS 2011). Since most of the world's NiCd batteries are produced in China, Japan, and Korea, it is not surprising that Asia has also come to dominate world Cd consumption as well as Cd production.

Cadmium production is closely linked to the production of other non-ferrous metals. At the end of the last century, about 80 % of Cd output worldwide was derived from the mining, smelting and refining of zinc ores, and the remaining 20 % was obtained from secondary sources, such as the recycling of Cd products, and from the production of other primary metals (copper and lead smelting). Global refinery production of Cd was estimated to increase in 2010 up to 22,000 tonnes as a result of production increases at zinc smelters that also recovered Cd as a by-product. Most of the world's primary Cd metal was produced in Asia (China, Japan, and Republic of Korea), followed by North America, central and western Europe (USGS 2011). At the end of the 1990s, the largest European producers were Finland, The Netherlands, Italy, and Germany. Belgium, which used to be one of the world's largest Cd producers, ceased primary Cd metal production in 2002 (Cooper & Kuck 2007). Between 2000 and 2004, primary Cd metal production in Germany increased from 458 tonnes per year to 640 tonnes per year; in 2010 German refinery production was 440 tonnes (Cooper & Kuck 2007; USGS 2011).

Cadmium, its alloys and compounds are used in a variety of different industrial and consumer products. About two-thirds of all Cd produced is used for alloying iron and copper and other metals to protect them from corrosion. Other products containing Cd include solders and electrical parts, pigments, plastics, rubber, pesticides, galvanised iron, etc. Currently, most Cd is used in the production of NiCd batteries (7 – 25 % Cd), although alternative products are available, such as lithium ion batteries (ATSDR 2008). Cadmium telluride is also used as a semiconductor and photoconductor in solar cells and other electronic devices (ATSDR 2008). A wide variety of Cd sulphide minerals are used for making yellow, orange and red pigments (about 1 % Cd), for plastics, glasses, enamels, ceramics and artist's paints. Cadmium pigments are particularly well suited for applications requiring high-temperature or high-pressure processing, or applications where other pigments will readily degrade. One of the more promising non-cadmium colourants is a cerium sulphide (Cooper & Kuck 2007).

Cadmium coatings (about 0.2 % Cd) are utilised for corrosion protection of iron and steel, aluminium and titanium, especially in the aircraft and car industry. Restrictions exist on Cd products in the EU (EC 1991), but not outside the EU.

Cadmium-based stabilisers (about 1 % Cd, Cd laurate and Cd stearate) have been used extensively in the past to provide ultraviolet light and weathering resistance to PVC. However, it has been found that other Cd-free compounds (Ca-Zn, Ba-Zn and organo-Sn stabilisers) can be used equally well. Similarly, many different types of Cd-containing alloys (1 – 20 % Cd) have been used in the past for brazing and soldering applications that have now been generally replaced by Cd-free compositions with equal performance.

Cadmium is present in other minor applications, e. g., photography, photocopying, dyeing, calico printing, vacuum tube manufacture, lubricants, ice-nucleation agents, and in the manufacture of special mirrors (Alloway 1995; Kabata-Pendias & Mukherjee 2007; ATSDR 2008). Because Cd strongly absorbs neutrons, it is also used in nuclear reactor control rods to dampen the nuclear reaction and keep the fission reactions under control.

▸ 5.4 Vegetation and aquatic biota

Cadmium is accumulated by all plants. Soil pH is the most important factor controlling Cd uptake by plants, with higher uptake at lower pH. Other soil factors that increase the uptake of Cd by plants are high salinity, high total Cd concentrations, low organic-matter content, low cation exchange capacity, low clay content, low Fe and Zn concentrations, and high temperature (de Meeûs et al. 2002). Transfer from soil to plant is particularly effective in the following crops: wheat, spinach, celery, lettuce, linseed and certain wild mushrooms. Phytotoxicity of Cd is manifested by chlorosis, reduction in photosynthesis, wilting, and necrosis (Herber 2004). Toxicity and bioavailability of Cd are influenced by soil characteristics (ECB 2007).

Vegetation contains Cd at the amounts 0.1 – 2.5 mg/kg, enriched in wheat, celery, lettuce, cabbage, spinach and carrots. As a rule, underground parts accumulate more Cd than above ground parts. Nevertheless, some plant species (e. g., *Thlaspi caerulescens*) can accumulate Cd in amounts exceeding 10 mg/kg (Kapusta et al. 2006; Lombi et al. 2001). Toxic accumulation of Cd in plants causes many disorders in enzyme activity, metabolism, transpiration, photosynthesis, etc. leading to death (Khan & Khan 2006). Increasing soil zinc is known to reduce Cd availability to plants (ECB 2007), because Zn inhibits Cd uptake (Wallnoefer & Engelhardt 1995).

Aquatic biota also accumulate Cd and other trace elements (Hans et al. 2010; Jöst & Zauke 2008). These studies indicate Cd accumulation in many polar organisms (2 – 133 mg/kg), e. g., crustaceans, polychaetes, and sea spiders. These results from the unpolluted Antarctic environment are of great importance for differentiating human impact and natural variability of Cd and other trace metals (Hans et al. 2010).

▸ 5.5 Health implications and ecotoxicology

Cadmium is highly toxic to plants and animals and no evidence has been found that it is essential for any biological function. The effects of Cd on human health have been discussed by Herber (2004), Hodgson (2004) and Pan et al. (2010). Among the sources of exposure to Cd mentioned above, direct exposure through airborne Cd is minimal to the general population, with the exception of tobacco smokers. Cadmium, especially inorganic Cd, is one of the most toxic metals known to humans, and it is a recognised carcinogen and teratogen (Kalberlah 1999). It is noted that the half-life of Cd in the human body is estimated to be between 10 – 30 years (Hodgson 2004).

In the environment, inorganic Cd compounds are found in drinking water, soil, air, dust, in food and tobacco smoke (Ferguson 1990; Chaney et al. 1996; Hill 1996). For most humans, Cd exposure comes from its ingestion in food – and most of the Cd uptake by plants (crops) originates from fertilisers, sewage sludge, manure, and atmospheric deposition (van Asche 1998). The contribution from other pathways to total human uptake is overall small.

Studies have estimated that 90 % of the ingested Cd comes from terrestrial foods (UNEP 2006; UNEP 2010a). Under normal circumstances, cadmium intake through drinking water and beverages, the air and soil and dust is of minor importance relative to dietary Cd intake.

The major hazard to human health from Cd is its accumulation in the kidneys, where it can cause dysfunction if its concentration in the kidney cortex exceeds 200 µg/kg (WHO 1987; Alloway 1995). Kidney damage has long since been described to be the main problem of patients chronically exposed to Cd (Barbier et al. 2005; Baselt 2011; Arndt 2012). It can be caused even at very low Cd levels of environmental exposure (Järup et al. 1998, 2002).

Acute or chronic exposure to Cd may also cause bronchitis, renal damage, dysfunction of the kidneys, hypertension and bone softening. High Cd dietary intake coupled with low availability of calcium, iron and zinc, may give rise to itai-itai bone disease (osteomalacia), which was associated with exposure to Cd as early as the 1950s and 1960s in Japan, or even earlier. Research has shown that Cd transfer in the rice food chain is unique, and that other food chains do not pose as great a risk (Chaney et al. 1996).

As a carcinogen, Cd may cause lung and prostate cancer and is suspected of initiating cancer of other organs (Godt et al. 2006; Benbrahim-Tallaa et al. 2007; Il'Yasova & Schwartz 2005).

The mean daily intake of Cd in Europe varies between 9 µg/day in The Netherlands and 35–64 µg/day in Italy (Council of Europe 1994). The average dietary intake of Cd around the world ranges between 25 and 75 µg/day (Page et al. 1981). The European Commission has estimated the current tolerable weekly intake (TWI) of 2.5 µg/kg body weight for Cd, but a provisional tolerable monthly intake of 25 µg/kg body weight was established by the Joint FAO/WHO Expert Committee on Food Activities (EFSA 2011). The lethal dose is 250–500 mg.

Only a small proportion (5–10 %) of ingested Cd is adsorbed by humans, and large variations exist among individuals. Deficiencies of iron, zinc, and calcium in the human body generally facilitate Cd adsorption (Chaney et al. 1999, 2004).

Less than 50 % of inhaled Cd is adsorbed by the lungs. Acute and chronic exposure to Cd dust and fumes can result in Cd poisoning. Acute respiratory effects can be expected at a concentration above 1 mg/m^3 (Environment Agency 2009a, b). Considering the various sources of exposure and applying a safety factor, WHO (1987) estimated that 0.2 µg/m^3 was a safe level of atmospheric Cd concentration with regard to renal effects through inhalation.

The effect of dermal exposure to Cd is not expected to be significant. Dermal adsorption appears to be lower than oral absorption. The contributions of oral, dermal and inhalation exposure are collectively taken into account in a risk assessment.

▸ 5.6 Survey area

Topographic and land use maps of Europe can be found in almost any world atlas. A number of maps covering different themes (topography, geology, tectonics, fault and fracture zones, distribution of different rock types, distribution of the main sedimentary basins, precipitation and population density) at about the scale of the GEMAS project can be found in Chapters 2 and 3 and in Reimann & Birke (2010). An excellent source of information on land use in Europe is the CORINE Land Use Map of Europe (GLC 2000 database 2003). A detailed geological map of Europe is provided by Asch (2005), concise descriptions of the geology of Europe can be found in Blundell et al. (1992) and McCann (2008). The Soil Atlas of Europe provides a wealth of information about the different soil types of Europe but also contains maps of average precipitation, temperature, land use, population density, extent of the last glaciation and soil texture (Jones et al. 2005).

▸ 5.7 Results

Geochemical data for Cd concentrations in European soil are presented in three separate sections. Firstly, the Cd concentrations in agricultural (Ap) and grazing land (Gr) soil samples are discussed in comparison with European limits. Secondly, a statistical comparison of the aqua regia extractable, total extractable (FOREGS samples) and MMI® extractable Cd concentrations is presented to indicate the differences in the variability of several digestions.

Thirdly, the spatial distribution of Cd concentrations across Europe in Ap and Gr soil samples, and in FOREGS samples, and the main distribution patterns are shown in colour shaded contour maps and their primary sources are analysed (see EDA maps on attached DVD → Part A).

5.7.1 Cadmium in soil and statutory guideline values

Bowen (1979) gave the global median value of Cd in world soil as 0.35 mg/kg with a range of < 0.005 – 8.1 mg/kg. Vinogradov (1954) reported a world mean concentration in soil of 0.54 mg/kg. Kabata-Pendias & Mukherjee (2007) conclude that mean Cd contents reported from most surveys of background levels of metals in soil lie between 0.06 – 1.1 mg/kg, with a calculated global mean of 0.53 mg/kg. The soil survey of England and Wales estimated the mean Cd concentration in topsoil (0 – 15 cm; N = 5,691) to be 0.8 mg/kg, with a median value of 0.7 mg/kg and a range of < 0.2 – 40.9 mg/kg (McGrath & Loveland 1992). The same samples, used in this atlas (determined by ICP-AES following an AR extraction), have recently been reanalysed for total concentrations by XRFS at BGS and the mean is 0.52 mg/kg, with a median of 0.33 mg/kg and a range between 0.14 – 48 mg/kg Cd. Finally, a recent investigation by BGS, with the objective to determine the normal background levels in soil for a number of contaminants, analysed systematically 14,310 topsoil samples by XRFS and the mean of Cd is 0.85 mg/kg, with a median of < 0.5 mg/kg and a range of < 0.5 – 165 mg/kg (Defra 2012). In the agricultural soil of northern Europe, Reimann et al. (2003) determined a median value of 0.13 mg/kg (aqua regia extraction), with a range of < 0.005 to 1.1 mg/kg Cd. Table 5.1 gives an overview of the Cd mean values in agricultural soil worldwide. The Cd mean value in European agricultural soil varies between 0.08 mg/kg (Portugal) and 0.39 mg/kg (The Netherlands; EEA 1998). Concentrations of Cd in topsoil in the USA range from 0.005 to 2.4 mg/kg, with mean and median values of 0.27 mg/kg and 0.20 mg/kg, respectively (Page et al. 1987). The maximum Cd levels were found in the soil of coastal regions of central and southern California (Holmgren et al. 1993).

The European Union sets a limit on Cd content only for sewage sludge (20 – 40 mg/kg of dry matter) to be used on agricultural land and for soil exposed to sludge (EC Directive 86/278/EEC (EC 1986); see Appendix A). In Austria, Belgium, Luxembourg, Denmark, Finland, Germany, The Netherlands, Portugal, Sweden, Norway, UK, France, Switzerland, Czech Republic and Lithuania, the limits for Cd in soil are below the EU limit (Table 5.2). When the European limits (1 – 3 mg Cd/kg) are compared with the values found in topsoil (Ap and Gr samples) in the European countries

Table 5.1: Mean, minimum and maximum values of cadmium (Cd) (mg/kg) in agricultural soil of selected countries (UNEP 2010a).

Country	mean	min.	max.
Czech Republic	0.30	-	-
Denmark	0.25	0.05	0.60
	0.26 (1)	-	-
Canada	-	0.11	0.43
	0.90 (1)	0.40 (1)	1.7 (1)
Japan	0.30 (3)	-	-
Moldova	0.40 (1)	0.28	0.81
Poland	0.21	0.01	49.73
USA	0.27	0.01	2.0
England and Wales	0.80 (1)	0.20 (1)	40.9 (1)
Scotland	0.77 (2)	0.30 (2)	1.5 (2)
Australia	0.19 (2)	0.013 (2)	0.56 (2)
Sweden	0.22 (2)	0.03 (2)	2.3 (2)
Finland	0.16 (4)	0.105 (4)	
	0.06 (5)	0.01 (5)	0.227 (5)
	0.07 (6)	0.02 (6)	0.238 (6)
	0.17 (7)	0.04 (7)	0.144 (7)
Mongolia	0.02 (2)	-	-
Norway	0.22 (8)	0.025 (8)	1.80 (8)
	0.17 (9)	0.14 (9)	2.0 (9)
Switzerland	0.41 (10)	0.081 (10)	2.028 (10)
	0.33 (11)	0.195 (11)	0.577 (11)
	0.27 (12)	0.100 (12)	0.971 (12)
	0.43 (13)	0.005 (13)	23.0 (13)
	0.50 (14)	0.010 (14)	32.40 (14)
Russia	0.45	-	-

(1) unspecified topsoi (0 – 15 cm)
(2) unspecified topsoi
(3) uncontaminated soil
(4) vertic cambisol
(5) dystric cambisol
(6) haplic podsol
(7) fibric histosol
(8) agricultural soil (0 – 5 cm)
(9) agricultural soil (> 40 cm)
(10) pasture
(11) 'intensive grassland'
(12) agricultural soil of Swiss soil monitoring network site
(13) all agricultural soil
(14) all 'grassland' soil

(Fig. 5.1), it is evident that most of the GEMAS samples have Cd concentrations below the EU limits.

For agricultural (Ap) soil, only five samples exceed the highest European limit of 3 mg/kg, and a total of 41 samples (1.93 %) are above the lowest limit of 1.0 mg/kg Cd. For grazing land soil, 13 Gr samples (0.58 %) exceed the highest limit and 77 samples (3.6 %) are above the lowest European limit (Fig. 5.1).

A number of countries have implemented legislation to manage and control releases of Cd and limit use of and exposure to Cd within their territories. Table 5.2 gives an overview of soil guidelines and action levels in different countries worldwide.

Some countries (e.g., UK, Germany, The Netherlands, Canada) have established remediation criteria for Cd soil concentration at contaminated sites depending on land use (Table 5.2).

Table 5.2: National guideline and/or limit values of cadmium (Cd) (mg/kg) in soil of selected countries.

Country	Guideline and / or limit value	Land use or medium	Reference
Argentina	3.0	agricultural	UNEP (2010a)
	5.0	residential	
	20.0	industrial	
Austria	1.0	agricultural and horticultural	ÖNORM L 1075 (2004)
	1 – 2	unspecified land use	Marmo (2002)
	0.5	agricultural	
	1.0	pasture	
Belgium: Flanders	1.2	unspecified land use	Marmo (2002)
Belgium: Wallonia	2.0	unspecified land use	Marmo (2002)
Brazil	3.0	agricultural	UNEP (2010a)
	8.0	residential	
	20.0	industrial	
Canada	1.4	agricultural	Canadian Soil Quality Guidelines (1999)
	10.0	residential and parkland	
	22.0	commercial and industrial	
China	≤ 0.3	agricultural (pH < 7.5)	UNEP (2010a)
	≤ 0.6	agricultural (pH > 7.5)	
Croatia	0.5 (1) – 1.0 (2)	agricultural and grazing land	Official Gazette (2010)
Czech Republic	1.0	agricultural and grazing land	MoE (1994)
Denmark	0.5	agricultural and grazing land	Miljøstyrelsen (2010), UNEP (2010a), Marmo (2002)
Ecuador	2.0	agricultural	UNEP (2010a)
	5.0	residential	
	10.0	industrial	
EU	1 – 3	unspecified soil (when sewage sludge is spread on agricultural soil)	EU Directive 86/278/EEC (EC 1986)
Finland	0.5	unspecified land use	Marmo (2002)
France	2.0	unspecified land use	Marmo (2002)
Germany	0.1	agricultural (after ammonium nitrate extraction)	BBodSchV (1999)
	1.5	unspecified land use	Marmo (2002)
	20.0	grazing land	BBodSchV (1999)
	10.0	playground	
	20.0	residential	
	50.0	parkland	
	60.0	commercial and industrial	
	1.0	agricultural soil (light soil, clay content < 5 %, 5.5 < pH < 6.0)	AbfKlärV (2012)
	1.5	agricultural and horticultural soil	
	5.0	sewage sludge (light soil, clay content < 5 %, 5.0 < ph < 6.0)	
	10.0	sewage sludge (agricultural and horticultural soil)	

Table 5.2: continued

Country	Guideline and / or limit value	Land use or medium	Reference
Greece	1 – 3	unspecified land use	Marmo (2002)
Ireland	1.0	unspecified land use	Marmo (2002)
Italy	1.5	unspecified land use	Marmo (2002)
Lithuania	3.0	agricultural and grazing land	Decree No. V-114 (2004)
Luxembourg	1–3	unspecified land use	Marmo (2002)
Madagascar	3.0	unspecified soil	UNEP (2010a)
Moldavia	0.5	unspecified land use	UNEP (2010a)
Norway	3.0	most sensitive land use	UNEP (2010a)
Poland	4.0	agricultural and grazing land	Decree of the Polish Ministry (2002)
Portugal	1.0	unspecified land use (pH < 5.5)	Marmo (2002)
	3.0	unspecified land use (5.5 < pH < 7.0)	
	4.0	unspecified land use (pH > 7.0)	
Republic of Korea	4.0	agricultural and residential	UNEP (2010a)
Romania	1.0	unspecified soil	UNEP (2010a)
Russia	0.5	sandy and clay-sandy soil	UNEP (2010a)
	1.0	loamy and clay soil (pH_{KCl} ≤ 5.5)	
	2.0	loamy and clay soil (pH_{KCl} > 5.5)	
Slovakia	0.4	agricultural, sandy and clayey-sand soil	UNEP (2010a)
	0.7	agricultural, sandy-clay and clay soil	
	1.0	agricultural, clay soil	
Spain	1.0	unspecified land use (pH < 7)	Marmo (2002)
	3.0	unspecified land use (pH > 7)	
Sweden	0.4	unspecified land use	Marmo (2002)
Switzerland	0.8 (3)	agricultural and grazing land	VBBO (1998)
	0.02 (4)	agricultural and grazing land	
	2.0 (5)	agricultural land	
Thailand	37.0	agricultural	UNEP (2010a)
The Netherlands	0.8	unspecified soil	Marmo (2002)
Trinidad and Tobago	0.7	unspecified soil	UNEP (2010a)
United Kingdom	1.8	agricultural and grazing land and allotment	Environment Agency (2009 a, b)
	10.0	residential	
	230	commercial	
	3.0	unspecified land use	Marmo (2002)
Uzbekistan	0.5	unspecified soil	UNEP (2010a)

(1) guideline value for sandy and silty soil
(2) guideline value for loamy soil
(3) guideline value for total Cd concentration
(4) Cd limit for soluble concentration
(5) Cd limit for total concentration

Moreover, a few countries have introduced limits for Cd input to agricultural soil and/or content of Cd in unspecified soil that are below the EU limits (EC Directive 86/278/EEC; EC 1986).

Limits on the content of Cd in phosphate fertilisers have been established only in Austria (75 mg Cd/kg P_2O_5), Finland (21.5 mg Cd/kg P_2O_5) and Sweden (43 mg Cd/kg P_2O_5). Based on ECB data (2007), Cd concentrations in fertilisers used in Austria, Belgium, Denmark, Finland, Germany, Norway, Sweden and UK in the period 1985-1995 ranged between 2.5 and 80 mg Cd/kg. Hutton & de Meeûs (2001) found that fertilisers with very low Cd content (< 7 mg Cd/kg P_2O_5) were used in Finland and Sweden.

Many countries have legal standards for the maximum concentration of Cd in drinking water (Reimann & Birke 2010) and quality standards for surface water bodies, etc., and have established Cd threshold limits for air exposure in the working environment (UNEP 2010a), which are not discussed in this contribution on European soil.

5.7.2 Statistical distribution of Cadmium in agricultural and grazing land soil

Cadmium concentrations in GEMAS samples (aqua regia extracts) of agricultural (Ap) soil range from <0.01 to 7.5 mg/kg, with a median value of 0.181 mg/kg, and in grazing land (Gr) soil from <0.01 to 6.02 mg/kg, with a median value of 0.202 mg/kg (Table 5.3; Fig. 5.1). In general, the median values of Cd, measured in aqua regia extracts of Ap and Gr samples, are very similar. In both sample types, Cd concentrations have a log-normal distribution with high positive skewness values. The highest background values for Cd were determined in agricultural soil over calcare-

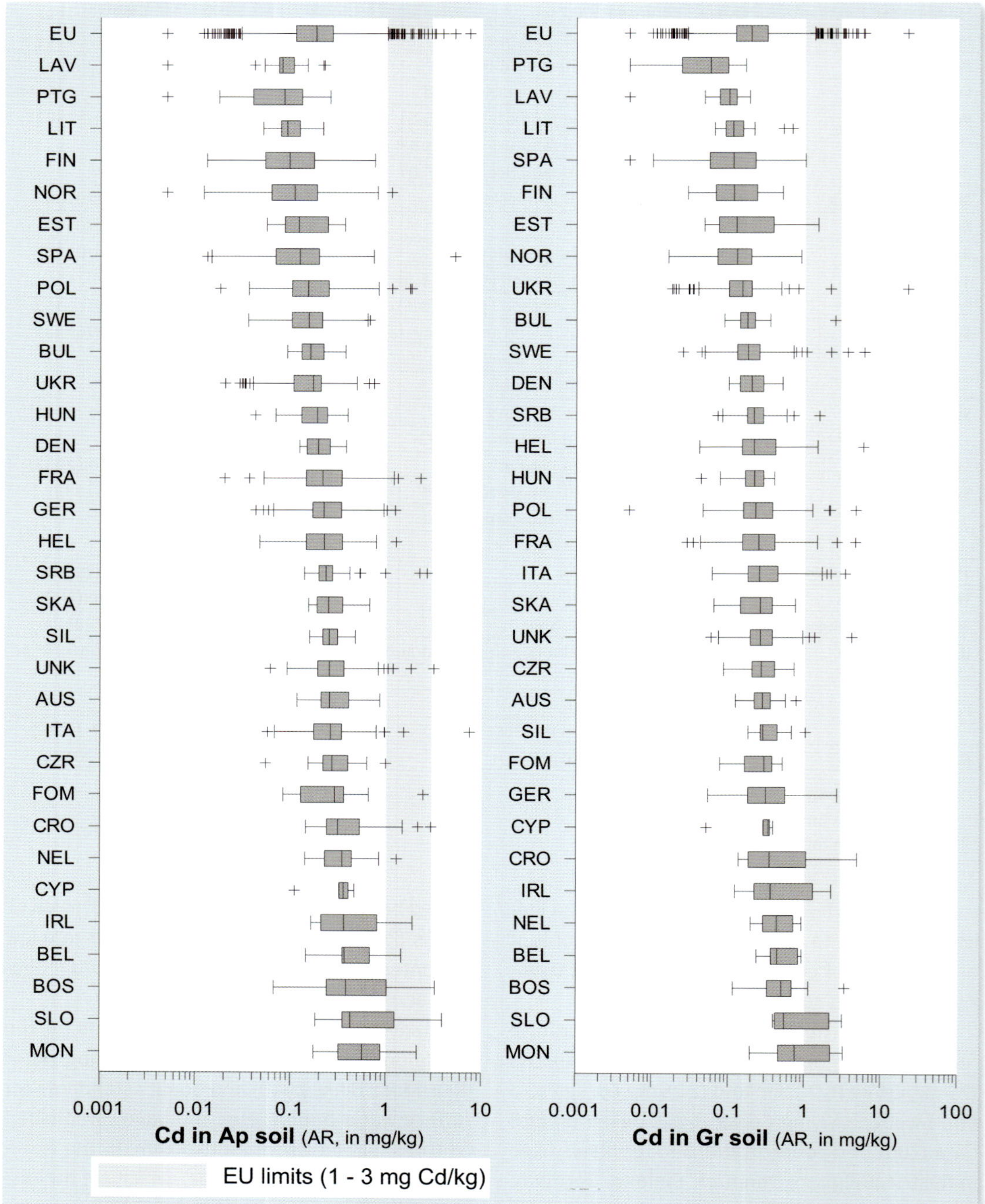

Fig. 5.1: Boxplot comparison of cadmium (Cd) in aqua regia extracts of Ap and Gr soil samples of European countries.
AUS: Austria;
BEL: Belgium;
BOS: Bosnia and Herzegovina;
BUL: Bulgaria;
CRO: Croatia;
CYP: Cyprus;
CZR: Czech Republic;
DEN: Denmark;
EST: Estonia;
EU: European Union;
FIN: Finland;
FOM: The Former Yugoslav Republic of Macedonia;
FRA: France;
GER: Germany;
HEL: Greece (Hellas);
HUN: Hungary;
IRL: Republic of Ireland;
ITA: Italy;
LAV: Latvia;
LIT: Lithuania;
MON: Montenegro;
NEL: The Netherlands;
NOR: Norway;
POL: Poland;
PTG: Portugal;
SIL: Switzerland;
SKA: Slovakia;
SLO: Slovenia;
SPA: Spain;
SRB: Serbia;
SWE: Sweden;
UKR: Ukraine;
UNK: United Kingdom.

Table 5.3: Statistical parameters of analytical results for cadmium (Cd) (mg/kg) from different continental scale soil geochemical surveys in Europe.

Media	N	min.	P25	median	mean	P75	P90	max.	skewness
FOREGS (Salminen et al. 2005; de Vos & Tarvainen 2006)									
Topsoil (tot.)	840	< 0.01	0.080	0.145	0.284	0.260	0.480	14.1	11.8
Baltic Soil Survey (Reimann et al. 2003)									
Topsoil (AR)	744	< 0.005	0.088	0.133	0.165	0.203	0.298	1.10	3.04
GEMAS, all of Europe									
Ap (AR)	2108	< 0.01	0.110	0.181	0.247	0.278	0.446	7.50	9.72
Gr (AR)	2024	< 0.01	0.122	0.202	0.306	0.329	0.574	6.02	6.43
Ap (MMI®)	2108	< 0.001	0.032	0.054	0.071	0.085	0.127	2.81	14.3
GEMAS, northern Europe									
Ap (AR)	814	< 0.01	0.079	0.122	0.148	0.184	0.260	1.14	2.80
Gr (AR)	711	< 0.01	0.096	0.152	0.217	0.234	0.385	6.02	10.3
Ap (MMI®)	814	< 0.001	0.028	0.043	0.054	0.068	0.097	0.404	2.94
GEMAS, southern Europe									
Ap (AR)	1294	< 0.01	0.148	0.224	0.309	0.338	0.532	7.50	8.45
Gr (AR)	1312	< 0.01	0.151	0.235	0.354	0.382	0.689	5.79	5.54
Ap (MMI®)	1294	< 0.001	0.037	0.061	0.082	0.096	0.145	2.81	12.9

ous rocks (Table 5.4). In the MMI® extract of the Ap samples, the median value of Cd is 0.054 mg/kg, more than one third of the aqua regia extractable Cd (Table 5.3; Fig. 5.2; Appendix B).

The median value of Cd in the Ap and Gr samples are just below the global median value (0.35 mg/kg) of Bowen (1979) and significantly below the global mean concentration (0.54 mg/kg) of Vinogradov (1954). The median values of the GEMAS project samples are similar to the median for Cd in the USA topsoil of 0.20 mg/kg (Holmgren et al. 1993; Burt et al. 2003). They slightly exceed the median values for Cd in topsoil of the Baltic Soil Survey (Reimann et al. 2003) and the FOREGS project (Salminen et al. 2005; de Vos & Tarvainen 2006; Table 5.3). In comparison with the upper continental crust (0.06 mg/kg, Hu & Gao 2008), Cd (aqua regia extracts) is significantly enriched in the GEMAS soil (Table 5.3).

In the GEMAS aqua regia and MMI® extraction results, the major division between northern (low Cd) and southern Europe (high Cd) can be observed in the Ap and Gr median values (Table 5.3). Figure 5.3 presents a cumulative probability plot (CP plot) of the Cd results for the GEMAS Ap and Gr samples (aqua regia extracts) in comparison with MMI® values in Ap samples, and the results from aqua regia extracts of the Baltic Soil Survey

Table 5.4: Median values of aqua regia extractable cadmium (Cd) (mg/kg) in European agricultural soil (Ap) according to parent material (after Hartwich et al. 2005).

Parent material	median	N
Quaternary marine deposits	0.221	20
Quaternary marine deposits (undifferentiated), overlying igneous and metamorphic rocks	0.175	9
Fluvial deposits	0.193	204
Moraine deposits	0.133	369
Glaciofluvial deposits	0.130	111
Glaciolacustrine deposits	0.147	25
Aeolian sand	0.211	24
Loess	0.196	213
Peat	0.0735	75
Tertiary sedimentary rocks	0.134	106
Tertiary molasse	0.197	22
Mesozoic, partly Tertiary rocks	0.240	230
Mesozoic calcareous sedimentary rocks	0.406	98
Flysch with Tertiary and Mesozoic sedimentary rocks	0.305	63
Palaeozoic, partly Mesozoic and Tertiary sedimentary rocks	0.247	137
Volcanic rocks	0.221	5
Igneous and metamorphic rocks	0.152	392

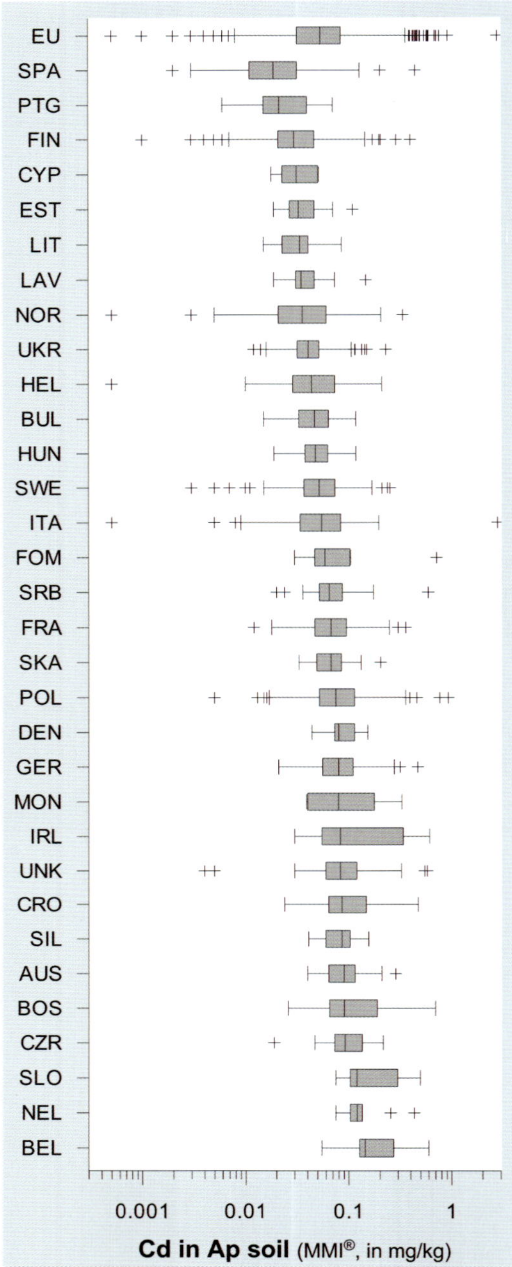

Fig. 5.2: Boxplot comparison of cadmium (Cd) in MMI® extracts of Ap soil samples of European countries. For country notation see Figure 5.1.

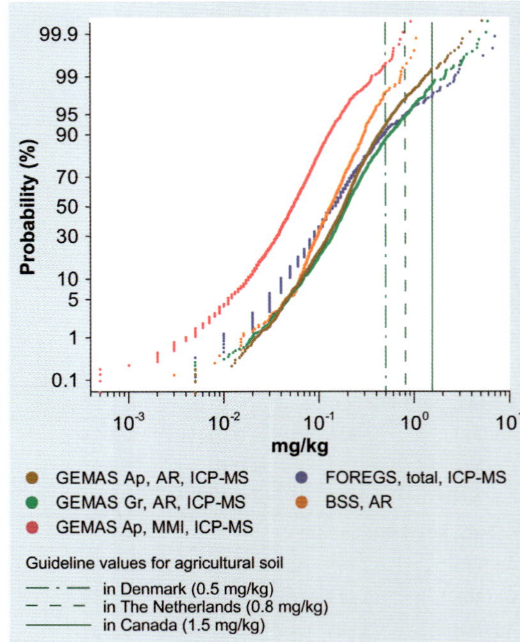

Fig. 5.3: Cumulative probability plot of cadmium (Cd) in Baltic Soil Survey (Reimann et al. 2003), FOREGS topsoil (Salminen et al. 2005) and GEMAS Ap and Gr samples. Soil guideline values from Denmark, The Netherlands and Canada are indicated on the plot.

in topsoil (Reimann et al. 2003), as well as total concentration values of the FOREGS topsoil (nitric acid extracts).

The CP plots for Cd (Fig. 5.3) show only small differences between aqua regia and total extraction used for GEMAS and FOREGS samples. The differences between aqua regia and total extraction for soil samples measured by ICP-MS are also very small (Fig. 5.3). Further, the boxplots for the soil samples from the different European geochemistry soil projects (GEMAS with Ap-AR, Gr-AR and Ap-MMI®, FOREGS, BSS) show slightly higher Cd concentrations in the GEMAS grazing land (Gr) soil (Fig. 5.4). The boxplots display the difference between aqua regia (GEMAS, BSS) and total extraction in comparison to the MMI® extraction (low concentration levels) of Ap samples. Additionally the boxplots in Figure 5.4 highlight the low interquartile variation of the main body of data (boxes) for all data sets. The boxplots for the Ap and Gr soil samples in aqua regia extraction of all participating European countries (Fig. 5.1) show the highest concentration levels in Montenegro, Slovenia, Bosnia and Herzegovina, Belgium and Ireland caused by different sources (Tables 5.5, 5.6).

Comparison of Cd concentrations with soil type and clay content reveals that with increasing clay content of the European agricultural soil (Ap samples) Cd concentration increases. Cadmium contents significantly increase from light to medium soil, with only a slight increase from medium to heavy soil (clay content > 25 %) (Fig. 5.5).

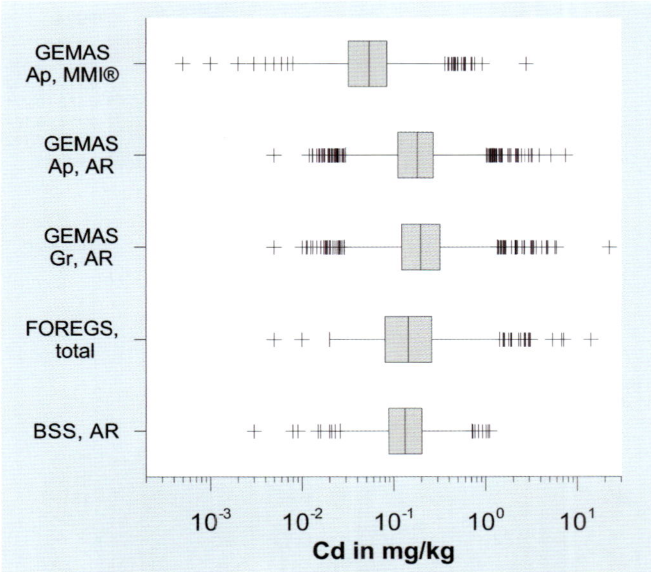

Fig. 5.4: Boxplot comparison of cadmium (Cd) in Baltic Soil Survey (Reimann et al. 2003), FOREGS topsoil (Salminen et al. 2005) and GEMAS Ap and Gr samples.

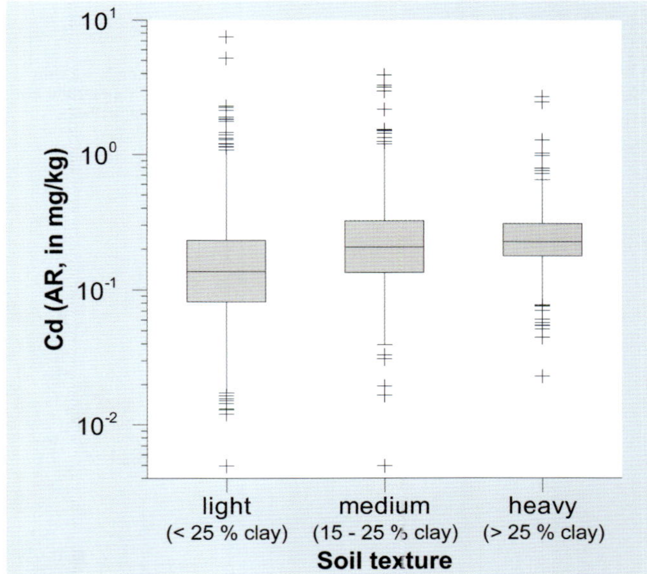

Fig. 5.5: Boxplot comparison of cadmium (Cd) in agricultural (Ap) soil according to soil texture (see Chapter 1).

5.7.3 Spatial distribution of Cadmium in agricultural and grazing land soil across Europe

The distribution of Cd concentration in the aqua regia extraction (AR) of the Ap samples is illustrated on Figure 5.6 and in the Gr samples on Figure 5.7. The distribution of Cd concentration in the Ap samples based on ICP-MS analysis after MMI® extraction is shown on Figure 5.8. The Cd results, after total extraction with nitric acid from topsoil, were interpolated by the same geostatistical procedure (kriging) as the other elements, and are presented on Figure 5.9 for comparison with the FOREGS results. Firstly, the Cd distribution in the Ap and Gr samples (aqua regia extraction) appears similar (Figs. 5.6, 5.7). The most intense Cd anomalies are mapped in both types of samples (Ap and Gr) in Croatia, Slovenia, Germany (e.g., Rhine-Ruhr region, eastern Ore Mountains), UK (England), southern Poland, Greece, central and eastern France, Ireland and south-western Ukraine. The Cd distribution map for the Gr samples (Fig. 5.7) shows, however, many more small local anomalies, the sources of which can be explained by the different land use. The main Cd anomalies on the distribution maps (Ap and Gr) are numbered and an explanation for the most likely source of the Cd anomalies is provided in Tables 5.5 and 5.6.

On the geochemical maps of Cd distribution (Figs. 5.6, 5.7), the boundary between northern and southern Europe coincides very well with the maximum extent of the last glaciations, especially in central and eastern Europe. This distribution pattern (north-south concentration gradient) was determined for most of the major and trace elements (Chapter 11 → Part A). The Trans-European Suture Zone (Fig. 5.7; Reimann et al. 2012: Fig. 18.1) is more prominent in the Cd distribution of the Gr samples than in the Ap samples. The same pattern is seen in the FOREGS topsoil (Fig. 5.9; Salminen et al. 2005; de Vos & Tarvainen 2006). The increase of Cd content in agricultural and grazing land soil from northern to southern Europe also reflects the different climate zones (Figs. 5.10, 5.11; Table 5.7; Appendix C).

The Cd concentrations in agricultural and grazing land soil (Ap and Gr), as well as in the FOREGS topsoil (Figs. 5.6, 5.7, 5.9), reflect the natural background variation and the difference between the young, often coarse-grained and less weathered soil in northern Europe and the older, finer grained and more weathered soil of southern Europe.

Fig. 5.6: Spatial distribution of cadmium (Cd) in European agricultural soil (in mg/kg, aqua regia extraction of the <2 mm size fraction).

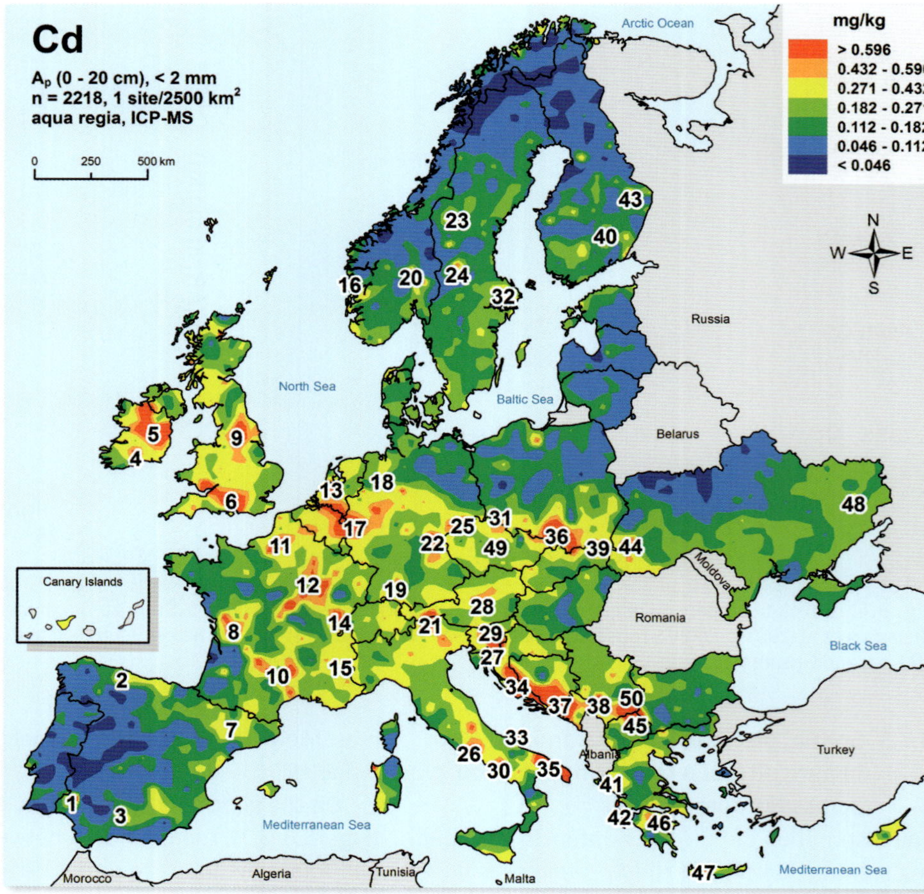

Table 5.5: Explanation of the cadmium (Cd) anomalies in agricultural soil samples (Ap) as indicated on Figure 5.6.

No.	Location of GEMAS Cd anomalies	Assumed primary reason	Possible other sources at site; remarks
		Ore deposits and mineralisation districts	
4	**S Ireland**	Pb-Zn mineralisation, Mallow district	
5	**Central Ireland**	Zn-Pb-(Cd) mineralisation in carboniferous rocks	Mining
9	**UK**	Peak district, Pennine Pb-Zn ore field	Mining, Pb smelting, industries, urban areas
32	**E Sweden, Uppland**	Cu-Zn-Fe mineralisation, Bergslagen district	Mining
24	**Central Sweden (near Älvdalen)**	Ore deposits, Pb-Ag-Zn mineralisation	Mineralisation in sedimentary and metavolcanic rocks
23	**Sweden, Central Jaemtland**	Cd-Pb-U-Mo mineralisation in Lower Allochthon, along Scandinavian Caledonides	Sedimentary rocks of Lower Allochthon of the Caledonides, occurrences of alum shale, Precambrian sandstone
36	**S Poland**	Silesian-Cracovian Zn-Pb district, Zn-Pb deposits	Mining, smelting, industrial area: steelworks, power cable factories, metal-working plants, non-ferrous metals smelter
28	**Austria, Gurktaler Alps**	Pb-Zn occurrences	Small scale mining
21	**SW Austria, Southern Tirol, Stubaital**	Fe-Cu-Pb-Zn deposits of Mühlbach/Telfes, Fe-Pb deposit of Kampl	Mining
	N Italy, S Tirol	Pb-Zn-Ag occurrences, Schneeberg mining ore	Mining, smelting, metal processing
38	**SW Serbia**	Pb-Zn mineralisation, Kopaonik district	Mining, smelting, metallurgical complex (Kasovska Mitrovica)
50	**S Serbia, Kosovo**	Pb-Zn occurrences, metallogenic Serbo-Macedonian Province	Mining, smelting, metal processing
45	**NE FYROM**	Pb-Zn deposits	Mining (Toranica, Sasa and Zletovo mine), smelting, metal processing

Table 5.5: continued

No.	Location of GEMAS Cd anomalies	Assumed primary reason	Possible other sources at site; remarks
		Ore deposits and mineralisation districts	
41	**Greece, Epirus**	Phosphorite mineralisation	Use of phosphate fertilisers
46	**Greece, Corinthia-Argolidha**	Sulphide mineralisation	Use of phosphate fertilisers
47	**Greece, Crete, Rethimno**	Pb-Zn mineralisation	Use of phosphate fertilisers
49	**Czech Republic, Kutna Hora**	Pb-Ag-Cu sulphide mineralisation	Old mines
2	**N Spain, Asturias and Cantabric region**	Pb-Zn occurrences	Mining, processing, use of fertilisers (manure)
7	**W Spain, Lérida-Aragón, N of Ebro Basin**	Pb-Zn sulphide mineralisation	Soil developed over limestone
1	**SW Spain, Sierra de Aracena**	Pb-Zn mineralisation	Old Pb-Zn-Ag mines
18	**N Germany, S of Bremen**	Oolitic Fe ores near Damme	
17	**W Germany (Rhenish Slate Mountains)**	Aachen-Stollberger and Bensberg ore districts	Mining, smelting
	and E Belgium	La Calamine district, Moresnet district	Mining, smelting, metals industry
25	**Germany, Ore Mountains**	Freiberg region, Lichtenberg-Roethenbach district, quartz-polymetallic mineralisation (Zn-Sn-Cu, Fe-Cu)	Old mines, smelting
19	**S Germany, Swabian Alb**	Stratiform Fe-Mn mineralisation	Mining near Gutmadingen
22	**Germany, Frankenwald**	Pb-Zn sulphide mineralisation near Wallenfels-Dürenwaid	
		Geology / Parent material	
6	**UK, S England**	Chalk bedrock with phosphatic nodules	Historical zinc smelting S Wales and Avonmouth. Mendips (Shipham) sulphide mineralisation
40, 43	**Finland**	Organic-rich soil	Use of fertilisers
16	**S Norway**	Organic-rich soil	Bergen urban area
20	**S Norway**	Black shale in the Hamar area	
39	**SE Poland**	Sedimentary rocks of the Carpathian nappes	Sandstone with layers of carbonates
44	**SW Ukraine**	Pb-Zn mineralisation and deposits; sedimentary rocks in folded area of the Carpathian nappes	Sandstone with layers of carbonates
31	**N Czech Republic**	Palaeozoic plutonic rocks of the Krkonose-Jizera Composite Massif	
27	**W Croatia, Gorski Kotar region**	Organic-rich soil developed over carbonate rocks	
34	**S Croatia, Lika and central Dalmatia**	Organic-rich soil over karstified areas (Red soil), Triassic pyroclastics (Pietra Verde)	
37	**S Bosnia and Herzegovina, S Croatia, W Montenegro**	Soil developed over carbonate rocks, Red soil	
26	**Italy, Latium-Campania**	Alkaline volcanic rocks of the Roman Province	Urban areas of Naples, industries
30	**Italy, Naples, Campania**	Vesuvius and alkaline volcanic rocks	Urban areas, industries
33	**Italy, Apulia-Gargano**	Red soil ('terra rossa')	
35	**Italy, Apulia-Salento**	Red soil ('terra rossa')	
11	**N France**	Eocene Fe-rich alterite over Cretaceous rocks	
12	**Central France (in N of Massif Central)**	Fe-rich soil over calcareous rocks	Cd enrichment in Plio-Quarterny formations of the Forez Graben
14	**France (in E of Massif Central)**	Clay-rich soil over calcareous oolitic rocks	
15	**France, Prealps area**	Soils developed over calcareous rocks	
10	**S France**	Fe-rich alterite	Pb mining (?)
8	**W France**	Eocene Fe-rich alterite over Cretaceous rocks	Use of fertilisers (?)
		Contamination	
48	**E Ukraine**	Donetsk region, Donetsk Basin, coal mining, processing industries, SbHg mineralisation	Urban and industrial areas
29	**Central Slovenia**	Use of fertilisers (pesticides), intensive agriculture	Detection of pesticides in soil samples
42	**Greece, Lefkadha-Kefalonia**	Use of phosphate fertiliser	
3	**S Spain, Guadalquivir River Basin**	Use of fertilisers, intensive agriculture	
13	**Amsterdam**	Urban area, industries	

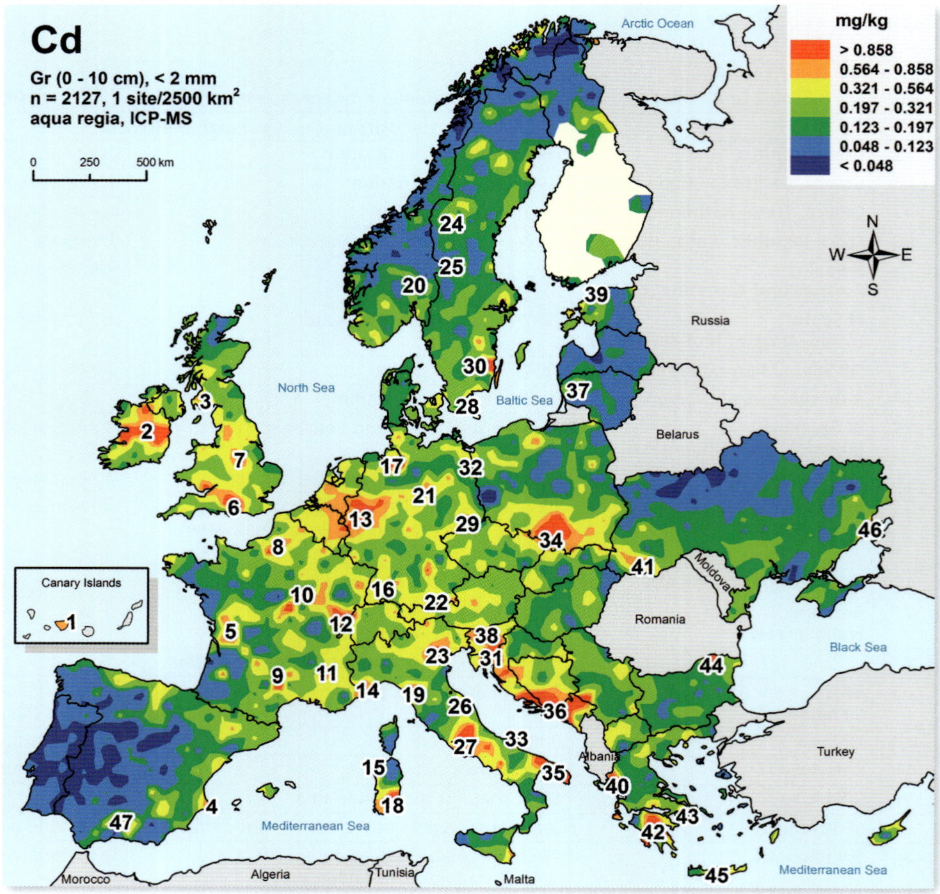

Fig. 5.7: Spatial distribution of cadmium (Cd) in European grazing land soil (in mg/kg, aqua regia extraction of the <2 mm size fraction).

Table 5.6: Explanations about the cadmium (Cd) anomalies in grazing land soil samples (Gr) as indicated on Figure 5.7.

No.	Location of GEMAS Cd anomalies	Assumed primary reason	Possible other sources at site; remarks
	Ore deposits and mineralisation districts		
2	**Central Ireland**	Pb-Zn mineralisation in the Mallow district	Mining, extensive deposits of fluviodeltaic and marine shale, siltstone and coal
7	**UK**	Historical Pb-Zn mining area of the southern Pennine Orefield,	Mineralised carboniferous rocks, industries, urban areas
24	**Sweden, central Jaemtland**	Pb mineralisation	Sedimentary rocks of Lower Allochthon of the Caledonides, Cd enrichment in Precambrian sandstone, occurrences of alum shale (S of Oestersund)
25	**Sweden, near Älvdalen**	Pb-Ag-Zn mineralisation and ore deposits	Mining, sedimentary and metavolcanic rocks
30	**SE Sweden, Oskarshamn**	Polymetallic ore deposits: Vätland (Cu-Au, Ni, Cu, Au), Gladhammer-Västervik (Cu-Co, REE-U, Fe, Cu)	
28	**S Sweden, Skåne**	Pb-Zn-Ag mineralisation in Precambrian-Cambrian sedimentary rocks	Cd enrichment in Cambrian sandstone
34	**S Poland, Upper Silesia**	Silesian-Cracowian Zn-Pb district, sulphide Zn-Pb ores (Cd, Ag, Tl, As)	Mining, smelting, metals industry, Zn metallurgy, urban areas
41	**SW Ukraine**	Pb-Zn mineralisation and deposits	Sedimentary rocks of the Carpathian nappes, sandstone and shale with carbonate layers
42	**Greece, Corinthia-Argolidha**	Sulphide mineralisation	
43	**Greece, Attiki**	Sulphide mineralisation	
45	**Greece, Rethimno, Crete**	Pb-Zn mineralisation	
14	**N Italy, Liguria**	Polymetallic Hercynian mineralisation	
19	**Italy, Tuscany-Liguria**	Small scale mineralisation	

Table 5.6: continued

No.	Location of GEMAS Cd anomalies	Assumed primary reason	Possible other sources at site; remarks
		Ore deposits and mineralisation districts	
15	**Italy, N Sardinia**	Polymetallic mineralisation	
18	**Italy, S Sardinia**	Polymetallic mineralisation	
47	**S Spain, Betic Range, (Baza, Granada)**	Pb-Zn mineralisation district	Pb-Zn mining
17	**NW Germany, N of Bremen**	Fe deposit near Achim, diatomite deposit near Hetendorf S of Hamburg	Pb smelter in Norderham
13	**W Germany, Rhenish Slate Mountains**	Aachen-Stollberg and Bensberg districts, Ramsbeck, Meggen, Ruhr district; Pb-Zn and Zn-Pb-Ba occurrences, Fe sulphides	Mining and smelting, mining disposals
16	**S Germany, Swabian Alb, (S of Stuttgart)**	Stratiform Fe-Mn mineralisation near Gutmadingen-Blumberg	Mining
29	**SE Germany, Ore Mountains**	Freiberg region, quartz-polymetallic (Zn-Sn-Cu, Fe-Cu, Pb-Zn) deposits of Lichtenberg-Roethenbach	Old mines, smelting, mining waste heaps
		Geology / Parent material	
6	**UK, S England**	Baseline enrichment over Chalk bedrock, phosphatic nodules in calcareous rocks	Historical zinc smelting S Wales and Avonmouth. Mendips (Shipham) sulphide mineralisation
20	**S Norway**	Black shale in the Hamar area	
39	**E Estonia**	Outcrops of Dictyonema Shale (black shale) and Maardu phosphorite deposit	Ordovician limestone (Tallinn landfill?)
37	**W Lithuania**	Organic-rich soil	Meliorated former peatlands, use of fertilisers (herbicides, pesticides)
31	**W Croatia, Gorski kotar**	Organic-rich soil developed over carbonate rocks	
36	**Bosnia and Herzegovina, S Croatia, W Montenegro**	Red soil, organic-rich soil developed over karstified areas	Bauxite deposits
23	**Italy, Veneto**	Tertiary alkaline volcanics of the Lessini Mountains	Lead smelter (Eurometalli)
27	**Italy, Latium (S of Rome)**	Soil developed over volcanic rocks	
33	**S Italy, Apulia-Gargano**	Red soil ('terra rossa')	
35	**SE Italy, Apulia-Salento**	Red soil ('terra rossa')	
8	**N France**	Eocene Fe-rich alterite over Cretaceous rocks	
10	**NE France**	Fe-rich alterite over Cretaceous rocks	
12	**E France**	Clay-rich soil developed over calcareous oolitic rocks	
5	**W France**	Eocene Fe-rich alterite over Cretaceous rocks	Use of fertilisers (?)
9	**S France**	Eocene Fe-rich alterite	Pb mining (?)
11	**SE France, Prealps area**	Soil developed over calcareous rocks	
22	**S Germany, Subalpine young moraine area**	Organic-rich soil (bogs), Tertiary molasse	Palaeogene coal deposits
21	**Germany, N Saxony-Anhalt**	Organic-rich soil	Wetlands (bogs), may be dredged material from the Midland Canal
		Contamination	
1	**Tenerife Island (Canary Islands)**	In neighbourhood of airport	Urban area
46	**E Ukraine**	Iron and steel plant in Mariupol, metallurgical complex (Azovstal)	
44	**NE Bulgaria**	Possible anthropogenic sources, metallurgical industry in Calvasi (Romania), near Danube river	Soil developed over Pleistocene aeolian deposits
38	**Slovenia**	Intensive agriculture, use of fertilisers (pesticides)	
40	**Greece, Epirus**	Intensive agriculture in past, use of fertilisers	
26	**Italy, Umbria-Marche**	Urban area	
4	**SE Spain, Cabo de La Nao**	Urban areas (Benidorm, Javea), industrial urban areas (Alcoy, Cocentaina)	
32	**NE Germany, near Schwedt**	Industrial (chemical plants, refinery) and urban areas	
		Unexplained	
3	**NW United Kingdom, W of Southern Uplands**	Possible use of fertilisers (high P, U and Cd concentrations indicate phosphatic material)	

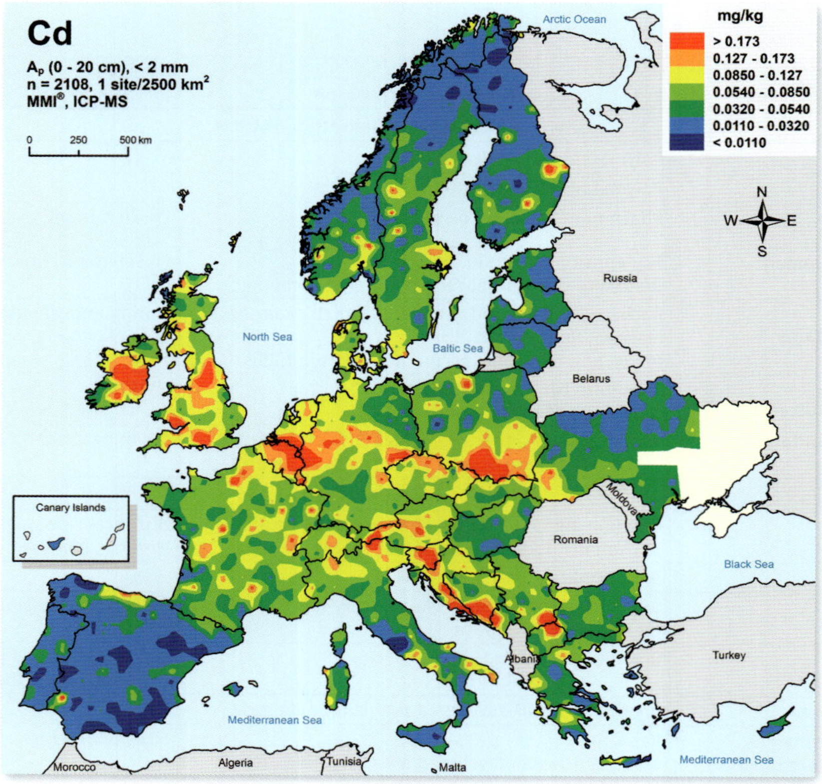

Fig. 5.8: Spatial distribution of cadmium (Cd) in European agricultural soil (in mg/kg, MMI® extraction of the <2 mm size fraction).

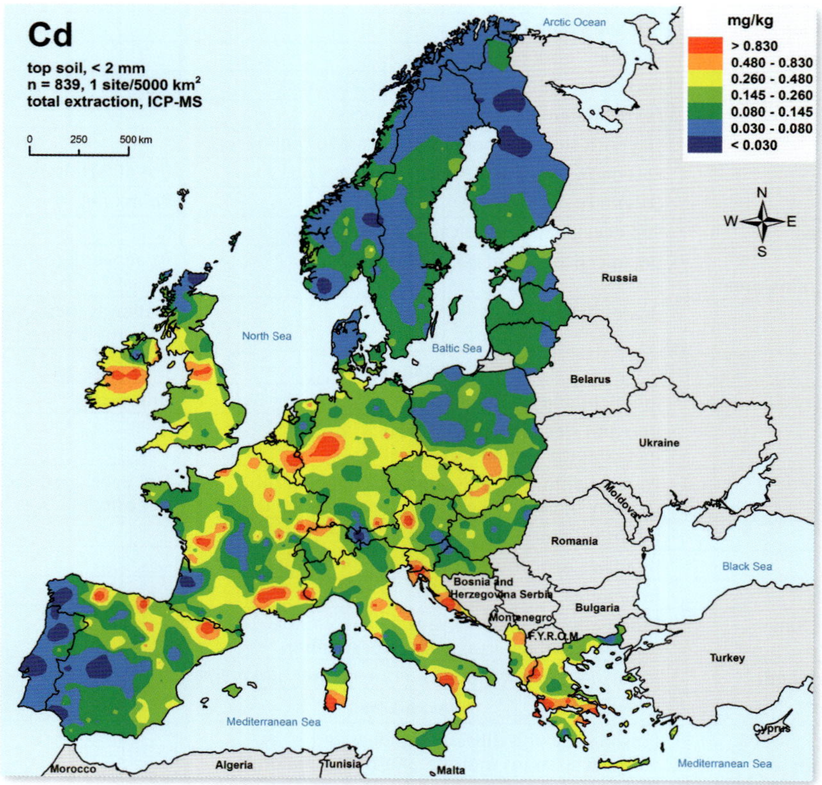

Fig. 5.9: Spatial distribution of cadmium (Cd) in the FOREGS topsoil (in mg/kg, total (nitric acid) extraction of the <2 mm size fraction).

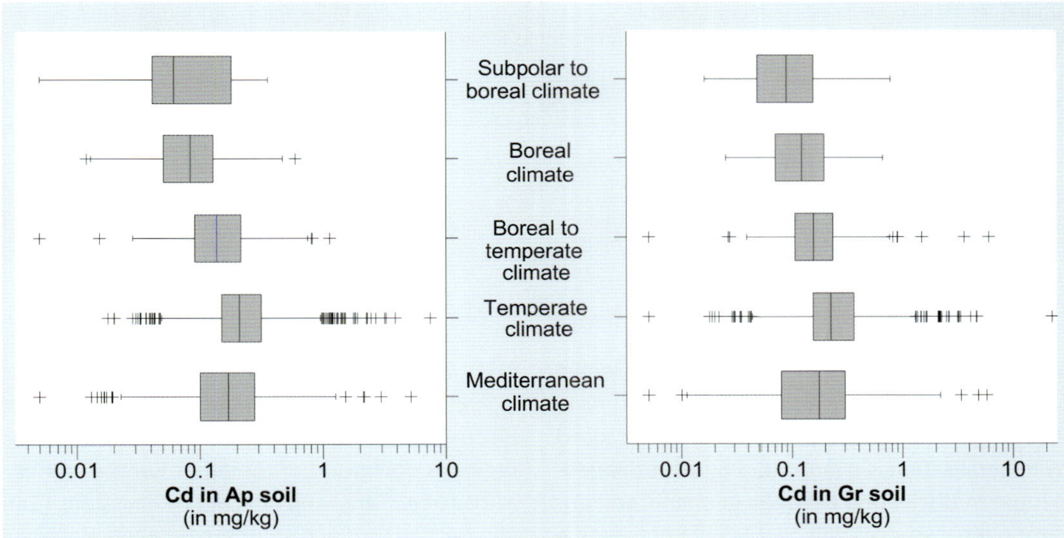

Fig. 5.10: Boxplot comparison of cadmium (Cd) in agricultural (Ap) and grazing land (Gr) soil according to European climate regions (see Fig. 5.11).

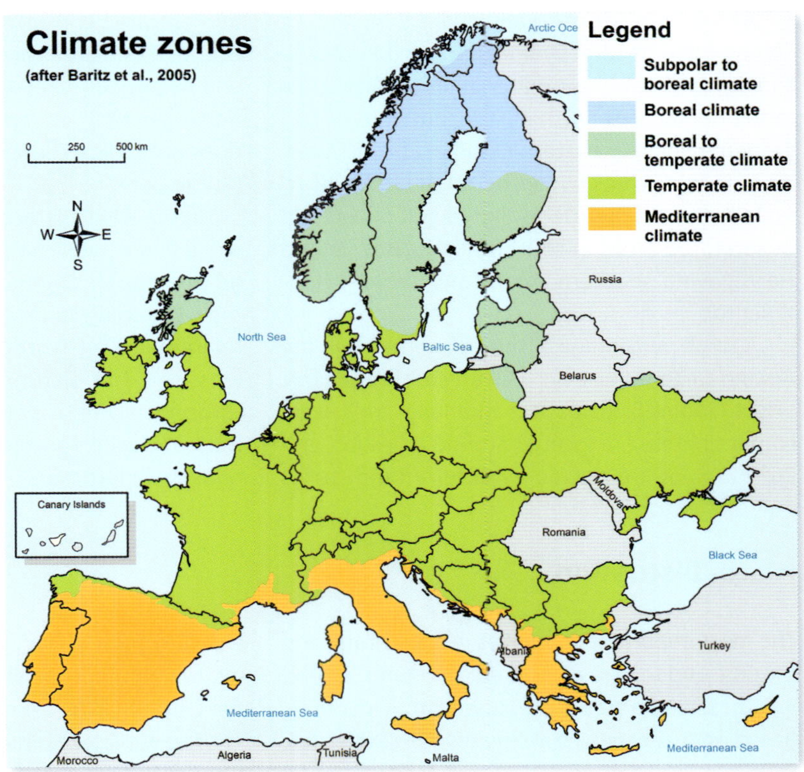

Fig. 5.11: Climate regions of Europe (after BARITZ et al. 2005).

About 50 % in Ap samples and 40 % in Gr samples of the most intense Cd anomalies (Tables 5.5, 5.6) coincide with known mineralisation and/or mineral deposits (e.g., central Ireland, Peak district in the UK, Rheinish Slate Mountains and Ore Mountains in Germany, Moresnet district in Belgium, Silesian – Cracovian district in southern Poland, Bergslager district in Sweden, southern Serbia, northern FYROM, sulphide mineralisation in Greece, Czech Republic, Spain, Austria), which also have been mined, and local industries (especially metallurgical, metal working and chemical

Table 5.7: Median values of cadmium and other selected elements (in mg/kg) in European agricultural (Ap) soil after aqua regia extraction and pH according to climate zones (after BARITZ et al. 2005; see also Figs. 5.10 and 5.11).

Climate zones	N	Cd	As	Pb	Zn	Sb	pH
Subpolar to boreal	22	0.062	1.60	5.14	35.5	0.048	4.77
Boreal	182	0.084	1.82	6.44	31.8	0.072	4.80
Boreal to temperate	353	0.137	2.36	11.3	40.8	0.116	5.21
Temperate	1107	0.215	6.84	19.0	47.8	0.307	6.01
Mediterranean	439	0.170	7.46	16.7	46.3	0.306	7.23

plants) often developed in the same areas (see Cd mineral deposit map on attached DVD → Part A). In these regions, the natural high Cd content caused by mineralisation is overprinted by anthropogenic input. A distinction between geogenic and anthropogenic sources will be possible only with a detailed geochemical survey.

A large number of Cd anomalies in Ap and Gr samples can be directly linked to the bedrock lithology and/or parent material (e.g., black shale in the Hamar area north of Oslo; outcrops of Dictyonema Shale in eastern Estonia; Eocene alterite in France; chalk in southern England; calcareous rocks; organic-rich soil; red soil developed in karstified areas in southern Italy, Bosnia and Herzegovina, southern Croatia and western Montenegro).

A small percentage of the Cd anomalies in the Ap and Gr samples have a clearly anthropogenic origin (Tables 5.5, 5.6). The anomalies indicating contamination are located around cities (4^{Gr} and 26^{Gr}; 13^{Ap}) or are caused by different industries (32^{Gr}, 44^{Gr} and 46^{Gr}; 48^{Ap}) and the use of fertilisers (3^{Ap}, 29^{Ap} and 42^{Ap}; 3^{Gr} and 38^{Gr}). It has often been discussed in the literature that the Rhine-Ruhr region is downwind of the other European megacities (e.g., Paris) and affected by emissions from The Netherlands and southern England. However, the distribution pattern and form of Cd maxima and elevated concentrations in this region indicate that emissions likely originate from the local and neighbouring areas. Further, the Cd distribution in the MMI® extraction of Ap samples (Fig. 5.8) does not indicate long-range transport from distant regions.

In the MMI® extraction (Fig. 5.8), a zone of elevated and anomalous concentrations can be seen on the map throughout almost all of central Europe. Especially in central Germany, the zone correlates with the distribution of black and brown loess soil. This distribution pattern suggests anthropogenic input of MMI®-available Cd concentrations due to agricultural land use and input via phosphate fertilisers.

▸ 5.8 Discussion

The most important anomalies in agricultural and grazing land soil are numbered on Figures 5.6 and 5.7. These anomalies are discussed in more detail, based on main sources in this section (Tables 5.5, 5.6).

5.8.1 Geogenic anomalies

Most of the geogenic Cd anomalies (Tables 5.5, 5.6) can be directly linked to geology (parent materials) and mineralisation.

In general, the Cd concentrations tend to be higher in southern and south-eastern Scandinavia with increasing clay content in soil (Fig. 5.12). The distribution pattern also shows a relationship to the Caledonian lineaments.

The elevated and anomalous Cd concentrations in Sweden are often caused by Pb-Cu-Zn±Fe, Ag sulphide mineralisation in areas of granitic and sedimentary rock. In eastern Sweden (Uppland, in northern Mälaren) the elevated Cd values (32^{Ap}, Fig. 5.6) are derived from marine sediments (high clay content) and the Pb-Zn-Fe mineralisation of the Bergslagen district. Near Älvdalen (Dalarna) anomaly 24^{Ap} in agricultural soil samples can be attributed to known Pb-Ag-Zn mineralisation, mostly in sedimentary and metavolcanic rocks. The anomalous Cd concentration in Härjedalen between anomalies 23^{Ap} and 24^{Ap} in Ap samples is possibly related to the mineralisation in the Vassbo region (e. g., the Vassbo mine), where sulphide (Pb) ore deposits are known. In central Jämtland (23^{Ap}), the anomalous concentrations are associated with Lower Allochthon sedimentary rocks of the Caledonides (also known Pb, U and Mo mineralisation) and occurrences of alum shale south of Östersund. The elevated Cd values might be related to Precambrian sandstone along the Caledonian front. The anomalies in the Gr samples (Table 5.6; Fig. 5.7) near Älvdalen (25^{Gr}) and in central Jämtland (Östersund region) are also shown. Anomalous concentrations were also found in southern Sweden (28^{Gr}) between Tomelilla and Simrishamn caused by Pb-Zn-Ag mineralisation in Cambrian sandstone. This sandstone is well known for its high Cd contents. Anomaly 30^{Gr} in south-eastern Sweden (Table 5.6) near Oskarshamn coincides with known polymetallic ore deposits between the Vetlanda (Cu-Au, Ni, Cu, Au) and Gladhammar-Västervik (Cu-Co, REE-U, Fe, Cu) ore districts.

In Norway, anomaly 16^{Ap} in agricultural soil samples in the vicinity of Bergen occurs in an area with very high precipitation and organic-rich soil (Fig. 5.6). It is undoubtedly a geogenic (natural) anomaly due to binding of Cd to organic matter in soil. However, due to the closeness to urban Bergen, an anthropogenic input cannot be completely excluded. Anomaly $20^{Ap\&Gr}$ in Ap and Gr samples from southern Norway is caused by black shale in the Hamar area (Figs. 5.6, 5.7). It is quite well known in Norway that the wheat grown in this area has very high Cd concentrations (Sæther et al. 2010).

In Finland, Cd concentrations tend to be higher in clay soil in south-western Finland than in the northern and eastern parts of the country (Fig. 5.6). The two anomalies (40^{Ap} and 43^{Ap}) in Ap soil (Table 5.5) are related to organic-rich soil, and they could also reflect Cd input from fertilisers used over a long time.

In the Baltic States, two anomalies are observed in Gr samples (Fig. 5.7). The anomalous

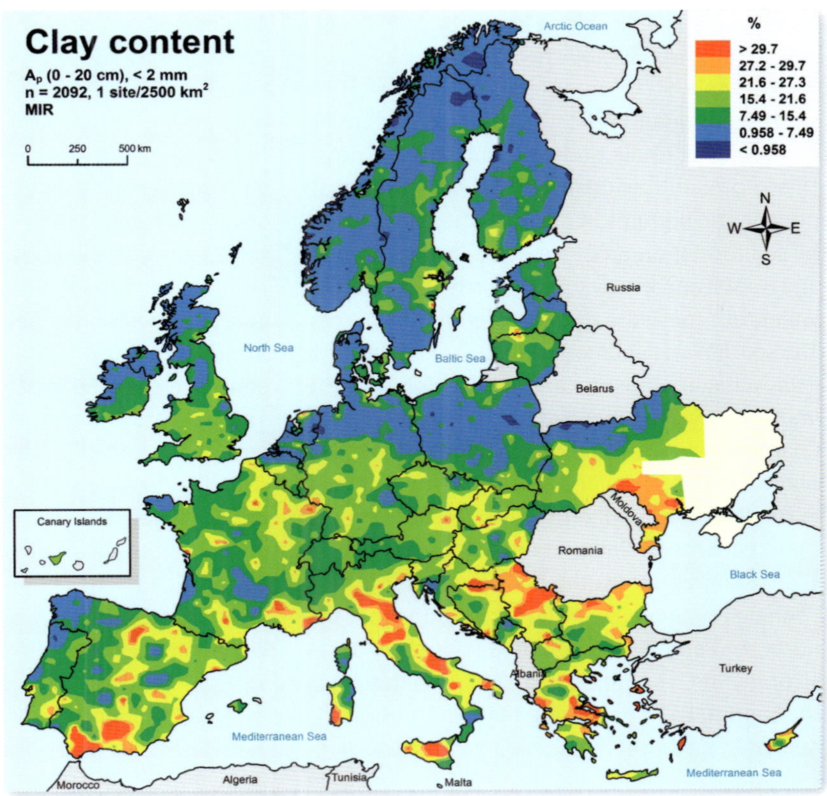

Fig. 5.12: Distribution of clay content in European agricultural soil.

concentrations in western Lithuania (37^{Gr}) can be linked to organic-rich soil (high sorption capacity of organic material). These former peat lands, and current agricultural soil, have also been enriched by several orders of magnitude in trace elements originating from the long use of fertilisers. In Estonia (39^{Gr}), soil with a very high content of organic matter is associated with fine-grained sedimentary rocks (Dictyonema shale) enriched in U, Mo, V, Ag and Cd (Petersell 1997; Petersell et al. 1997). The Dictyonema shale belongs to the extensive Cambrian-Ordovician black shale and correlates with the upper part of the alum shale in central and southern Sweden. The elevated and anomalous concentrations of Cd are most likely due to the outcrops of Dictyonema shale and the presence of fragmental and fine fraction phosphorite in Quaternary sediments (Raudsep 1997).

The anomalies in England (6^{Ap} and 9^{Ap}; 6^{Gr} and 7^{Gr}) and Ireland (4^{Ap} and 5^{Ap}; 2^{Gr}) can be also linked to geological sources (Tables 5.5, 5.6; Figs. 5.6, 5.7). The anomalies in Ap and Gr samples from Ireland both coincide with extensive deposits of Namurian rocks (fluviodeltaic and marine shale, siltstone and coal), and also overlie areas where Pb-Zn mineralisation (and operating mines) occurs in the Carboniferous host rocks.

Anomaly $6^{Ap\,\&\,Gr}$ in Ap and Gr samples from southern England reflects baseline enrichment over parts of Chalk bedrock. One theory is that this represents widespread Cd enrichment in phosphatic nodules in chalk, perhaps enhanced by soil formation processes. The anomalies in Gr and Ap soil apparently extend north-west to south Wales, but these are due to different Cd sources. The south Wales anomaly around Swansea is most likely a result of historical zinc smelting, and the impact of the Avonmouth smelter near the Bristol Channel may also be seen. Further, there is sulphide mineralisation in the area south of Bristol in the Mendips (Shipham) that may contribute to these enhanced values. Anomaly 9^{Ap} in agricultural soil samples marks the northern and southern Pennine Pb-Zn ore fields. It is interesting that the northern Pennine ore field does not show up as strongly in Gr samples (Fig. 5.7). This indicates that anomaly 7^{Gr} in grazing land soil samples is related to historical Pb-Zn mining in the southern Pennine ore field and as such, it is a combination of geogenic (mineralisation) and anthropogenic dispersion of strongly mineralised Carboniferous rocks. The source for anomaly 3^{Gr} in Gr samples in the western part of the Southern Uplands, Scotland, is unknown.

The Variscan basement in Portugal and the western parts of Spain (Iberian Massif) are marked by low Cd concentrations in agricultural and grazing land soil. The distribution of elevated and anomalous concentrations in Spain mirrors the sediment distribution. In total, the Cd distribution in Spain is very similar to that from the FOREGS mapping (Fig. 5.9). The single anomaly (1^{Ap}) in the Aracena area is caused by old Pb-Zn-Ag mines and is associated mainly with Cambrian limestone. The Gr sample was taken several kilometres from the Ap sample and does not show an anomalous signature (except for Mo). In northern Spain (Asturias and Cantabria regions), the anomalous belt (2^{Ap}) from Oviedo-Gijón (Asturias) to Santander and to Basque Country (Bilbao and Nervión river catchment area) is accompanied by Pb-Zn mineralisation, and mines occur in this belt (Reoci, close to Santander). The geogenic anomalous concentrations are overprinted by anthropogenic influences (old mining and industrial activities). This anomaly (Fig. 5.6) is also seen with lower intensity in Gr samples (Fig. 5.7), which may be associated with agricultural activity (use of fertilisers). In eastern Spain (Lérida-Aragón and north of the Ebro basin), the widespread Cd anomaly (7^{Ap}) is also caused by several different sources (Pb-Zn mineralisation in the Pre-Pyrenees sedimentary cover, intensive agriculture with use of fertilisers). The anomalous Cd concentrations in the Betic Range (Baza, Granada) in southern Spain (47^{Gr}) may be related to mining sources (old Pb-Zn mines within a Pb-Zn district). The elevated concentrations in grazing land soil east of anomaly 47^{Gr} are close to the Caravca river (Murcia province) and can be related to the old Fe mines (Cehegin).

Most of the Cd anomalies in France are related to iron-rich soil developed from alterite over Cretaceous rocks (8^{Ap}, 11^{Ap} and 12^{Ap}; 5^{Gr}, 8^{Gr} and 10^{Gr}). The anomalies in south-eastern France (14^{Ap} and 15^{Ap}; 11^{Gr} and 12^{Gr}) are in clay rich soil over calcareous (oolitic) rocks with elevated background values of Cd between 1.3 and 1.5 mg/kg (Dubois et al. 1996). The anomalous concentrations in southern France (10^{Ap}; 9^{Gr}) can be linked to iron-rich alterite.

In western Germany (Rhenish Slate Mountains) and eastern Belgium the widespread Cd anomalies (17^{Ap}; Gr: 13^{Gr}) are related to well known Pb-Zn ore districts (Aachen-Stolberg and Bensberg districts; Morestnet district, La Calamine) and smelting and mining (Figs. 5.6, 5.13).

Anthropogenic Cd enrichments of up to 8 mg/kg were determined in the floodplain soil of the Elbe and Mulde rivers in Germany (Rank et al. 1999; Beuge & Ulique 1997).

Polymetallic mineralisation (Zn-Sn-Cu, Fe-Cu) in the Lichtenberg-Roethenbach district in Germany is the source of anomalous Cd concentrations in the Freiberg region (25^{Ap}; 29^{Gr}). The anomalous Cd contents caused by the mineralisation also contain contributions by smelter emissions and dust blown from the mine waste heaps. Cadmium concentrations up to 2.9 mg/kg (median value) were found throughout the immediate vicinity of mine and smelter sites (Geochemical Atlas of Saxony, Rank et al. 1999; Kardel & Rank 2010; Barth et al. 1996).

In southern Germany (19^{Ap}), the anomalous Cd concentrations can be related to Fe-Mn mineralisation. In Frankenwald, anomaly 22^{Ap} in the Ap samples is caused by Pb-Zn sulphide mineralisation. Anomaly 17^{Gr} in Gr samples (Table 5.6, Fig. 5.7) in north-western Germany can be related to the iron deposit near Achim, but emissions from the Nordenham lead smelter cannot be excluded. In northern Saxony-Anhalt, anomaly 21^{Gr} in Gr samples is associated with organic-rich soil.

In Austria, the Cd anomalies (21^{Ap}) are caused by Pb-Zn-Ag mineralisation (Schneeberg mining district). Fe-Cu-Pb-Zn (Mühlbach-Telfes ore deposit) and Fe-Pb (Kampl deposit) mineralisations in the Stubai valley are the sources of anomalies in southern Tirol. The last two occurrences are at the base of the Brenner Mesozoic in which sphalerite containing 0.06 – 0.41 % Cd, and the limonite weathering crust outcrops have high molybdenum contents.

The biggest anomaly in Poland (36^{Ap}; 34^{Gr}) is related to the Upper Silesia mining district with Carboniferous black coal (hosted by sandstone) and Pb-Zn ore deposits in Triassic dolomite. Today, exploited sulphide Pb-Zn ores contain on average 4.2 % Zn and 1.4 % Pb as well as Cd, Ag and other minor elements (Co, Cu, Tl, Ge). Cadmium content in sphalerite of the Silesian-Cracowian deposits reaches 5,000 – 10,000 mg/kg (Viets et al. 1996; Cabala 2009). This geogenic anomaly is very intense, and is overprinted by anthropogenic emissions (mining areas and heaps, metal smelters, steelworks, rolling mills, metalworking plants, power cable factories, recycling companies). A certain proportion of Cd can be of an anthropogenic origin, since this area has been severely polluted. Additionally, the main rock type in this region – Jurassic carbonates (limestone, marl), results in a high carbonate content in soil, which may help to fix Cd in Ap and Gr soil. Anomaly 39^{Ap} in the Carpathian Mountains cannot be explained by known mineralisation. However, the sedimentary rocks of

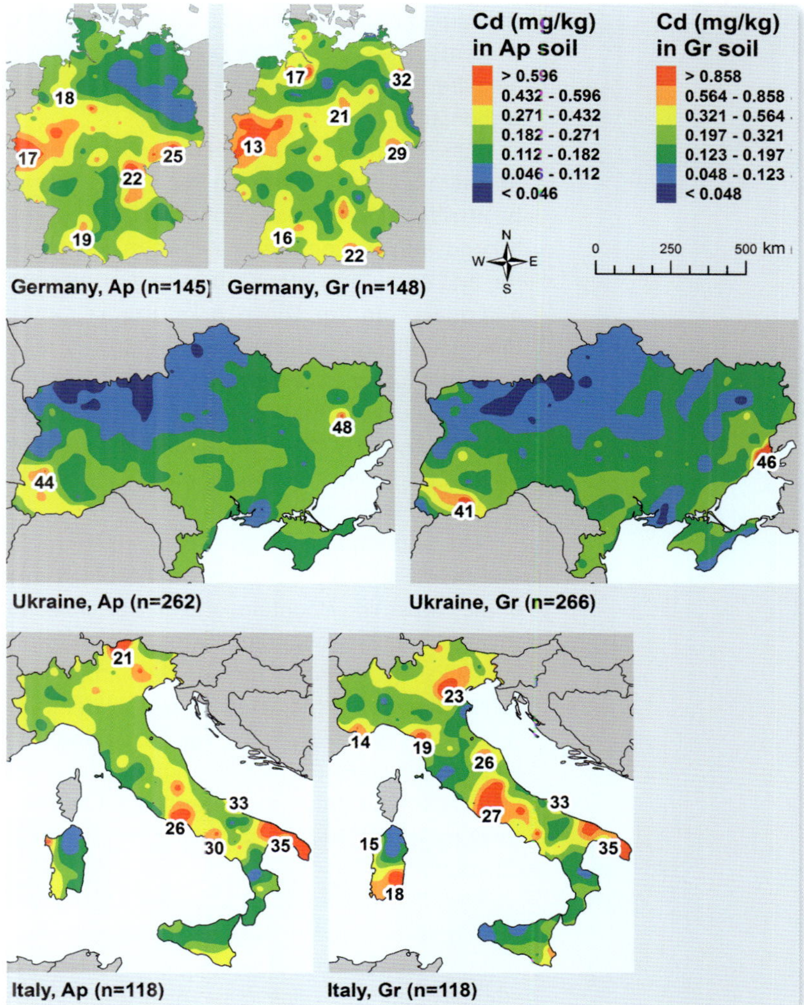

Fig. 5.13: Cadmium (Cd) (mg/kg) distribution maps of Germany, Ukraine and Italy in agricultural and grazing land soil (aqua regia extraction on the < 2 mm fraction).

the Carpathian nappes may have natural higher Cd contents. The typical lithologies occurring in the area are sandstone and shale with local layers of carbonates.

The elevated Cd concentrations along the northern border of the Czech Republic (31^{Ap}) can be attributed to Palaeozoic plutonic rocks of the Krkonose-Jizera Composite Massif (mainly granite, and some hornblende-biotite granodiorite and quartz diorite). These rocks (Jizera Granite, Krkonose-Jizera Granite) have a relatively high amount of other trace elements (Cr, Cu, Ni, Pb, Ta, Th, REE, Ga, W, Rb, Sr; Klominský 1969; Lorenc et al. 1998). Anomaly 49^{Ap} in Ap samples is caused by Pb-Ag-Cu sulphide mineralisation (old mining) in the Kutna Hora region.

The anomalous concentrations in Ukraine (44^{Ap}; 41^{Gr}; Figs. 5.6, 5.7, 5.13) are in the folded area of the Carpathian Mountains and may be linked to the geochemistry of parent materials (soil-forming rocks). Sedimentary rocks in the Carpathians within Ukraine are characterised by elevated concentrations of elements like Hg, Cd, Ba and F. In general, the Cd distribution in Ukraine (Fig. 5.13) reflects the boundary between the Palaeozoic cover rocks and the Mesozoic platform sediments (Chapter 2).

The distribution of Cd anomalies along the western (south-eastern Italy) and eastern (Dalmatian Croatia, Slovenia, Montenegro, Greece) Adriatic coast may be explained by the distribution of red soil ('terra rossa') in southern Europe (Muhs et al. 2010: Fig. 5). This red Mediterranean soil is typical in the karst areas around the Adriatic Sea.

In south-eastern Italy (Apulia-Gargano, Apulia-Salento), the anomalous concentrations of Cd in the Ap and Gr samples (33^{Ap} and 35^{Ap}; 33^{Gr} and 35^{Gr}) occur in soil developed over limestone, often with development of red terra rossa soil (rich in iron and aluminium hydroxides). Most of the Cd anomalies are recorded in soil developed on the Dinaride Carbonate platform along the Adriatic coast of Croatia. The soil there is developed mainly on Jurassic, Triassic, and Cretaceous carbonate rocks. Anomalies 34^{Ap} (Lika and central Dalmatia, Fig. 5.14) and 37^{Ap} (southern Dalmatia) in the Ap samples, as well as anomaly 36^{Gr} in the Gr samples, may be due to red terra rossa soil developed on karst. Anomalies 27^{Ap} and 31^{Gr} in the Gorski Kotar region are linked mainly to humus rich soil (organic element bonding). The anomalous Cd concentrations in this Mediterranean region also correlate with other elements (Al, Be, Bi, Ca, Ce, Cu, Fe, Ga, Hf, In, La, Mn, Mo, Nb, Re, Te, Th, Tl, V, Y).

In Italy, the agricultural soil (Ap) in the areas of anomalies 26^{Ap} (Latium – Campania) and 30^{Ap} (Campania), developed over a volcanic substratum, might reflect an elevated Cd background related to alkaline volcanic rocks of the Roman Province. Additionally, anthropogenic influences from the urban area of Naples are possible. In the grazing land soil samples, anomalies 14^{Gr} (Liguria), 15^{Gr} (northern Sardinia) and 18^{Gr} (southern Sardinia) are caused by known polymetallic mineralisation. In general, the Cd anomalies (26^{Ap} and 30^{Ap}; 27^{Gr}) in southern Italy (Campania, Latium, Apulia) may be mostly related to tephra occurrences in the soil profile, also shown on maps of other elements that are known to be enriched in alkaline volcanic rocks of the Roman-Neapolitan Volcanic Province (Vesuvius, Phlegraean fields, Campanian ignimbrite, and the Roman volcanoes). The elevated and anomalous Cd concentrations (23^{Gr}) in the Veneto area are related to the Tertiary alkaline volcanics of the Lessini Mountains.

The anomalous concentrations in agricultural soil of south-western Serbia (38^{Ap}, Kopaonik district), southern Serbia (50^{Ap}, Serbo-Macedonian metallogenic province) and in northern FYROM (45^{Ap}) indicate known Pb-Zn mineralisation, and they may be overprinted, however, by anthropogenic emissions from the nearby mining and smelting activities.

The Serbian Pb-Zn deposits in the Kopaonik district contain replacement ores related to Tertiary volcano-plutonic events, and are generally hosted by carbonates in contact with dioritic intrusions. Other deposits are dispersed throughout Serbia and they are also replacement-type ores and skarn deposits related to Tertiary volcano-plutonic events (Monthel et al. 2002).

In the north-eastern part of FYROM (45^{Ap}) there are three Pb-Zn mines (the Toranica, Sasa and Zletovo mines) with flotation plants producing Pb and Zn concentrates. In addition to Pb and Zn, the ore concentrates contain Ag, Au, Cd, In, Sb and other elements. Until 2004, a Pb-Zn smelter in the city of Veles was operational using the

Fig. 5.14: Agricultural soil sampling site (CRO 027 Ap) and profile near Radošić village, 16 km north-west of Split, Red soil in Croatia, cadmium (Cd) anomaly 34 (see Fig. 5.6) (photograph by Josip Halamić).

concentrates from these mines, producing about 100,000 t/year of Pb and Zn (Stafilov et al. 2010). Besides these metals, about 300 t/y of Cd was produced as well as some Ag and Au (Stafilov et al. 2010). Due to environmental problems, this smelter was closed.

The biggest mines and smelter plants in the Kopaonik district are in Kosovo around the Kosovska Mitrovica (Kosovska Mitrovica metallurgical complex). Several surveys of soil pollution in this region were completed and interpreted by Stafilov et al. (2010a) and Aliu et al. (2010, 2011, 2011a).

The anomalous Cd concentrations in the grazing land soil of Greece are mostly due to sulphide and Pb-Zn mineralisation (Corinthia-Argolidha: 42^{Gr}, Attiki: 43^{Gr}, Rethimno, Crete: 45^{Gr}). Geogenic anomaly 47^{Ap} in the agricultural soil samples near Rethimno on the island of Crete (Pb-Zn mineralisation) may be also influenced by the use of phosphate fertilisers.

5.8.2 Anthropogenic anomalies

Anthropogenic element enrichments are caused by certain industrial emissions, mining, fertiliser use in agriculture, as well as urbanisation. Very often, a combination of geogenic and anthropogenic sources is observed (Tables 5.5, 5.6). These complex anomalies have a relatively high anthropogenic component (e.g., Freiberg region, Ruhr district, Kosovska Mitrovica, Donetsk region, Upper Silesia/Katowice region, north-eastern FYROM), and cover widespread areas with elevated background values.

Some anomalous Cd concentrations in agricultural and grazing land soil can be due to the use of phosphate and other fertilisers (Slovenia: 38^{Ap}; 29^{Gr}, Fig. 5.15; Greece: 40^{Ap}; 42^{Gr}; Spain: 3^{Gr}). For several anomalies in France (5^{Ap}), Lithuania (37^{Ap}), Finland (40^{Gr}) and Greece (41^{Ap}, 46^{Ap} and 47^{Ap}), the anomalous Cd concentrations are partly due to mineralisation or parent material, and partly to intensive agricultural land use (fertilisers).

The anomalous concentrations in agricultural and grazing land soil of Slovenia are not located in typical karst areas with specific soil development (terra rossa). This soil is developed on carbonate rocks and it is usually enriched in many trace elements. The maximum Cd concentration in the agricultural soil is as high as 3.9 mg/kg, and in the grazing land soil reaches up to 3.13 mg/kg. These Cd values correlate with anomalous phosphorous concentrations in Ap and Gr samples. A GC-MS screening of moderately volatile organic compounds showed the presence of tributyl phosphate (TBP, organophosphorous compound used in herbicides and fungicides) in all Ap and Gr samples in Slovenia. PAH and PCB compounds were not found, however, in the Slovenian samples. A strong agricultural influence (pesticides/fungicides) is present in anomalies 29^{Ap} and 38^{Gr}.

In eastern Ukraine, anomalies with distinct industrial emissions were observed in Mariupol

Fig. 5.15: Grazing land soil sampling site (SLO 008 Gr) and profile near Goli Vrh, 6 km north-west of Novo mesto, Slovenia, cadmium (Cd) anomaly 38 (see Fig. 5.7) (photograph by Mateja Gosar).

(46^{Gr}, iron and steel plant) and in the Donetsk region (48^{Ap}, coal mining and coal processing industries).

The Cd anomaly in north-eastern Bulgaria (44^{Gr}) is in the area of the metallurgical industry near the Danube River at Calvasi, Romania.

In north-eastern Germany, the chemical industry near Schwedt causes local Cd enrichment (32^{Gr}). The Nordenham lead smelter also appears in the data as a Cd anomaly (17^{Gr}).

In north-eastern FYROM, the geogenic Cd concentrations from known mineralisation and ore deposits are overprinted by an intense anthropogenic component from the mining and processing industries (45^{Ap}).

Anthropogenic influences are observed around urban areas in Italy (26^{Gr}, Umbria-Marche; 26^{Ap} and 30^{Ap}, Latium-Campania, Campania) and in Spain (4^{Gr}, Cabo de La Nao). The Italian anomalies in agricultural soil samples, 26^{Ap} and 30^{Ap} (Fig. 5.13), border the largely urbanised area of Naples and its suburbs. On the Canary Islands, there is an anomaly (1^{Gr}) near the Tenerife airport.

The Cd anomaly in the greater Amsterdam area (13^{Ap}) is also caused by urbanisation and industrial plants.

▸ 5.9 Conclusions

The pattern dominating the Cd maps is the southern limit of the last glaciation, as well as the distribution of clay content in the European agricultural soil.

In Scandinavia, a relationship to the Caledonian structural lineaments can be recognised in the distribution pattern; in eastern Europe, it is the boundary between the Palaeozoic cover and the Mesozoic platform sediments that stands out.

The data for the aqua regia/HNO_3 extracts on the maps establish the natural geochemical background of Cd concentrations in European agricultural and grazing land soil; they demonstrate the existence of two major natural background regimes, a northern and a southern European, separated by the Trans-European Suture Zone.

Most of the Cd anomalies in European agricultural and grazing land soil can be linked to geological (parent material and mineralisation) sources. In several cases, the natural anomaly pattern is overprinted by anthropogenic emissions from old mining, and the ore processing and metal industries in the same region. For a more discrete separation of geogenic and anthropogenic Cd anomalous contents, a more detailed geochemical mapping or a multivariate statistical data analysis would be needed.

Only a few Cd anomalies could be attributed to mainly anthropogenic sources, such as urbanisation and the use of fertilisers. The Cd distribution pattern of MMI® extracts shows a zone of relatively high values throughout central Europe, which can be linked to agricultural land use (fertilisers). |◂

Manfred Birke
Federal Institute for Geosciences
and Natural Resources (BGR);
Hannover, Germany
E-mail: manfred.birke@bgr.de

Clemens Reimann
Geological Survey of Norway (NGU);
Trondheim, Norway

Uwe Rauch
Federal Institute for Geosciences
and Natural Resources (BGR);
Hannover, Germany

Benedetto De Vivo
University of Napoli Federico II (UNINA);
Napoli, Italy

Josip Halamić
Croatian Geological Survey (HGI-CGS);
Zagreb, Croatia

Volodymyr Klos
State Geological and Subsurface Survey
of Ukraine (SGSSU);
Kyiv, Ukraine

Mateja Gosar
Geological Survey of Slovenia (GeoZS);
Ljubljana, Slovenia

Anna Ladenberger
Geological Survey of Sweden (SGU);
Uppsala, Sweden

Chapter 6

Carbon Concentrations in European Agricultural and Grazing Land Soil

RAINER BARITZ, VIBEKE ERNSTSEN & DIETMAR ZIRLEWAGEN

▸ 6.1 Introduction

Carbon in soil represents a key indicator for the functioning of soil. Carbon in soil appears mainly in two different chemical variants, in mineral and organic form. Mineral forms of carbon in soil are lime, gypsum, ash, charcoal; organic forms are derived from decomposition of organic matter due to biological activity. The GEMAS database has determined the total organic carbon (TOC) in topsoil samples of agricultural (regularly ploughed) and grazing (permanent grassland) land. Even though the database largely concentrates on trace elements, which are mainly of geogenic origin, soil organic carbon represents one of the key indicators for the fertility and productivity of soil. Soil organic matter – about half of it is carbon – improves the structure and aggregation of soil, thus it serves to preserve site stability; it also serves as habitat for soil fauna, and this way affects the available source for decomposition.

Soil also acts as a selective filter for trace elements (metals typically at low concentrations) due to its high reactive surface. This surface largely depends on the clay content, but also soil organic matter. Because of the high absorption capacity, soil immobilises some metals thus keeping them from being washed out or taken up by plants too easily. Soil organic matter and clay host more than 80 % of the microbial biomass in soil (MASSFELDT 2011). These micro-organisms are responsible for more than 90 % of the decomposition, producing about 2/3 of the CO_2 in topsoil (TERYTZE et al. 2000). These data demonstrate the enormous value of organic matter for soil functioning.

MONTANARELLA et al. (2004) have studied the availability of soil carbon data in Europe. They also referred to the database of the FOREGS Geochemical Atlas of Europe (SALMINEN et al. 2005; DE VOS & TARVAINEN 2006) as a good example to produce harmonised soil chemical data for a large area, with only one laboratory involved in the determination of TOC for reasons of comparability and minimised error budget. The GEMAS database has in fact extended this inventory with improved representativeness with a higher number of sampling sites for agricultural and grazing land soil across all of Europe. SCHILS et al. (2008) have also stressed the need for improved regional soil carbon data. They also discuss the challenges to develop comparable data for Europe, given the great variability of soil and national methodologies. In contrast to the homogenous inventory design for GEMAS, the European Forest Level I inventory suffers from many national deviations from the main protocol for sampling and analysis (BARITZ et al. 2010). Thus, systematic error is introduced, which reduces the capacity for developing a homogenous baseline map, and to conduct trend assessments based on future sampling repetitions.

This chapter presents and discusses two upscaling approaches to evaluate the TOC data from the GEMAS database. First, a straightforward geostatistical upscaling presents an overview about the regional distribution of carbon in European agricultural and grazing land soil. In addition, a regression-based regionalisation is conducted in order to study further the main drivers for potential soil organic matter accumulation and loss across Europe. Spatially explicit high-resolution data about carbon storage, such as those produced by the regression-based approach, is also important for defining baselines based on measurements, e.g., to investigate the role of climate change on soil organic matter by feeding such baseline data into soil biophysical models. Similar to the approaches presented here for TOC, pH and phosphorous were also evaluated, however, with deviating regional stratification.

▸ 6.2 Materials and methods

The analytical procedures related to the GEMAS database can be found in Chapter 5 (→ Part A) and in REIMANN et al. (2011), (see also analytical data on DVD → Part A). Figure 6.1 presents the distribution map of agricultural soil sample sites (ploughed topsoil; Ap horizon, classified according to FAO 2006). The grazing land (Gr) sites were independently sampled in close proximity to the Ap sites. In total, 2211 agricultural (topsoil 0 – 20 cm), and 2118 grazing land sites (topsoil 0 – 10 cm) were sampled (Chapter 3 → Part A). The resulting sample density is 1 site/2500 km^2.

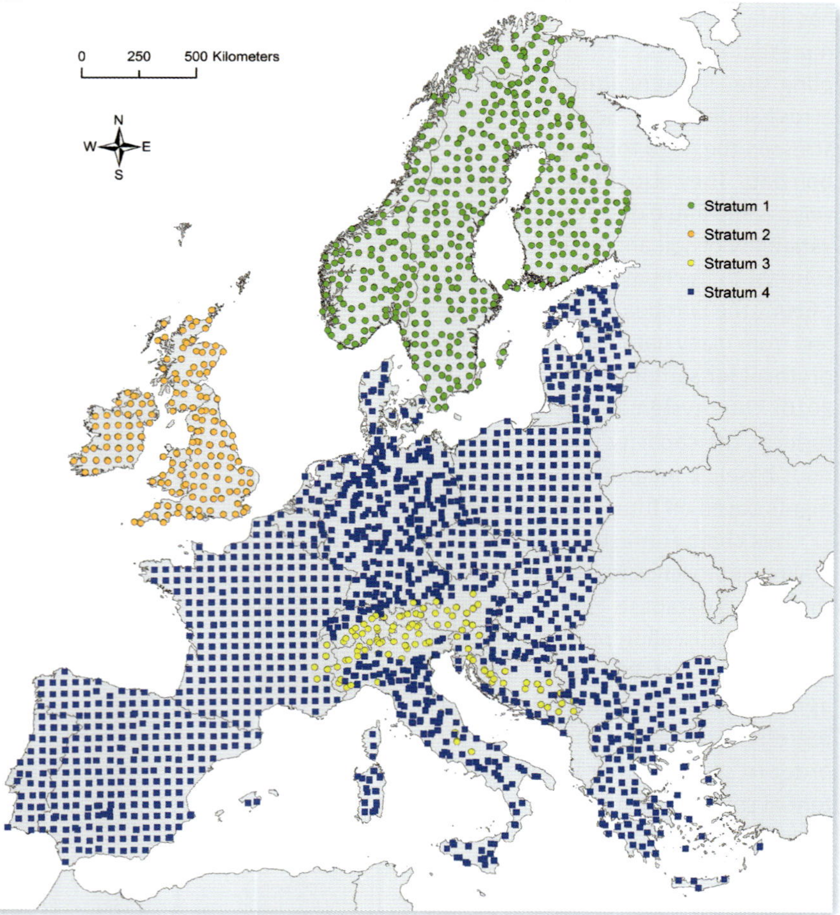

The data analysis followed two main approaches:

- ▸ Geostatistics, and
- ▸ Regression kriging.

6.2.1 Regression kriging

Regression kriging is a hybrid technique of geostatistical and regression techniques (MCBRATNEY et al. 2000, 2003). It combines a regression for a response variable (here: TOC) and auxiliary variables (such as terrain parameters, soil type, etc.) with simple kriging of the regression residuals (ODEH et al. 1995). Recent digital terrain analysis helps to identify a landscape zonation based on geomorphological, climatic and hydrological processes in a fairly high spatial resolution (MOORE et al. 1991). Combining land use related parameters along with climate and soil maps, these data were applied as auxiliary information in the regression models.

After exploratory data analysis, multiple linear regression with stepwise selection was applied using the REG procedure of the SAS statistical package (SAS 2004). Descriptive statistics, boxplots and probability plots were used to test for normal distribution of the residuals. The residual plots indicated some skewness for the TOC values. An approximate Gaussian distribution of the response variables could be achieved by transforming them to natural logarithms. The back-transformation from the logarithmic values to arithmetic units was done by applying a correction for bias (MILLER 1984).

In order to find the 'best' prediction model, stepwise regressions were performed using the default SAS-significance level for parameter entry. Statistical performance can be affected by collinearity between regressor variables. A collin-

Fig. 6.1: GEMAS sample sites classified according to stratification for TOC (Stratum 1: Boreal; Stratum 2: United Kingdom and Ireland; Stratum 3: High mountainous, and Stratum 4: European continent).

earity problem occurs when a regressor is nearly a linear combination of other regressors in the model. Since there were many collinearities to expect among the large number of auxiliary variables tested, a preliminary step of exploratory data analysis (e.g., through pair-wise scatter plots of variables), and the stepwise regression techniques were used to eliminate ineffective predictors. Multicollinearity of regressors was examined using the options COLLIN and TOL of the REG procedure and the covariance matrix as well. The plot of studentised residuals (residual divided by its standard error), and Cook's D influence statistic (measures the change to the estimates that results from deleting each observation) were plotted in order to exclude outliers and to reduce their effect on the least squares methods used in the regressions. The plot of studentised residuals was also used to test the models for homogeneity of variance.

6.2.2 Stratification

One of the main aspects, during the development of transfer regression models, is the selection of the proper stratification, since a single model cannot represent all the different ecological site factors in Europe. It can be expected that due to the extensive environmental variation across Europe, different explanatory variables dominate in strata or geographical regions. Consequently, the original regression model was split into several models (regional transfer models). While expert knowledge is an important basis for stratification, we also applied cluster analysis by using the Ward's Minimum-Variance Method. In this case, observations for combinations of variables are grouped, based on the ANOVA sum of squares between different clusters added up over all the variables.

The final stratification for TOC is presented on Figure 6.1:

- **Stratum 1**: Sweden, Norway and Finland (*'Boreal'*)
- **Stratum 2**: United Kingdom and Ireland
- **Stratum 3**: Eco-Regions (DMEER) with Code-Numbers 56, 27, 10 (*'High mountainous'*)
- **Stratum 4**: the remaining European target area (*'European continent'*).

These clusters (strata) are also used for the descriptive study of bivariate relationships and trends.

6.2.3 Validation

If the statistical requirements for regression analysis are not fully satisfied, and if the coefficient of determination is not close to 100 %, then the separation of a subset of data for validation is important, given sufficient number of samples. Even though the number of samples for all Europe is fairly low (1 sample site/2500 km^2), the GEMAS data were subdivided into a training and a validation data set. During the regression analysis, the regional models are derived from the training data, and the prediction error is then compared to results from modelling with the validation data. This prediction error on the validation data can also be used to decide when to terminate the selection process, or to decide what effects to include as the selection process proceeds. Hastie et al. (2009) note that it is difficult to give a general rule on how many observations should be assigned to each subset. We used a typical split with 50 % for training and 50 % for validation. The split of the data was based on a combination of (1) a spatial balancing using large-scale squares, and (2) a stratification using geo-ecological categories.

6.2.4 Auxiliary variables

It was shown in earlier studies (e.g., Zirlewagen & von Wilpert 2010) that soil carbon development also depends on position in the relief. Therefore, a Europe-wide digital elevation model was needed for this study. We used the digital elevation data that are available only south of 60 degrees North (SRTM 2013), which were combined with contour data for areas north of 60 degrees North that were prepared by the GIS Experts Co. (Digital Elevation Data 2013), contracted for the EU project MASCAREF *"Study under EEC 2152/2003 Forest Focus regulation on developing harmonised methods for assessing carbon sequestration in European forests"* (Cienciala et al. 2010).

The following relief parameters were calculated using ArcGIS 9 and SAGA-GIS 2.07:

- **Aspect, slope, curvature,** and **upstream and downstream distance** along the flow path for each cell (ArcGIS).
- **Topographic-Wetness-Index** TWI (Beven & Kirkby 1979), **Stream-Power-Index** SPI and

Slope-Length-Factor SLF (Moore et al. 1991; Moore & Wilson 1992) were calculated according to Schmidt & Persson (2003) using different mean filters for different circle neighbourhoods (1, 10, 35, 100 m).

- **Topographic Position Index** TPI, **Slope Position** and **Landform Classifications** were calculated according to Jenness (2005); TPI as circle neighbourhood with 300, 500, 750, 1000 and 3000 m distance, Slope Position categories based on TPI 1000 m, and Landform Classifications on the basis of TPI 500 and 1000 m.
- **Annual solar radiation** was assessed after McCune & Keon (2002).
- The **Convergence Index** was calculated after Koethe & Lehmeier (1996) in SAGA-GIS 2.07.
- **Slope Height, Valley Depth, Normalised Height, Standardised Height and Mid-Slope Position** were all derived from the digital elevation data with SAGA-GIS 2.07.
- The **Terrain Ruggedness** was calculated as Vector Ruggedness Measure (VRM) as described in Sappington et al. (2007) (neighbourhood size-variants: 3, 7, 11, 13, 19, 41 and 81 cells) as stand-alone Python-script, and as Terrain Ruggedness Index TRI according to Riley et al. (1999) in SAGA-GIS 2.07.

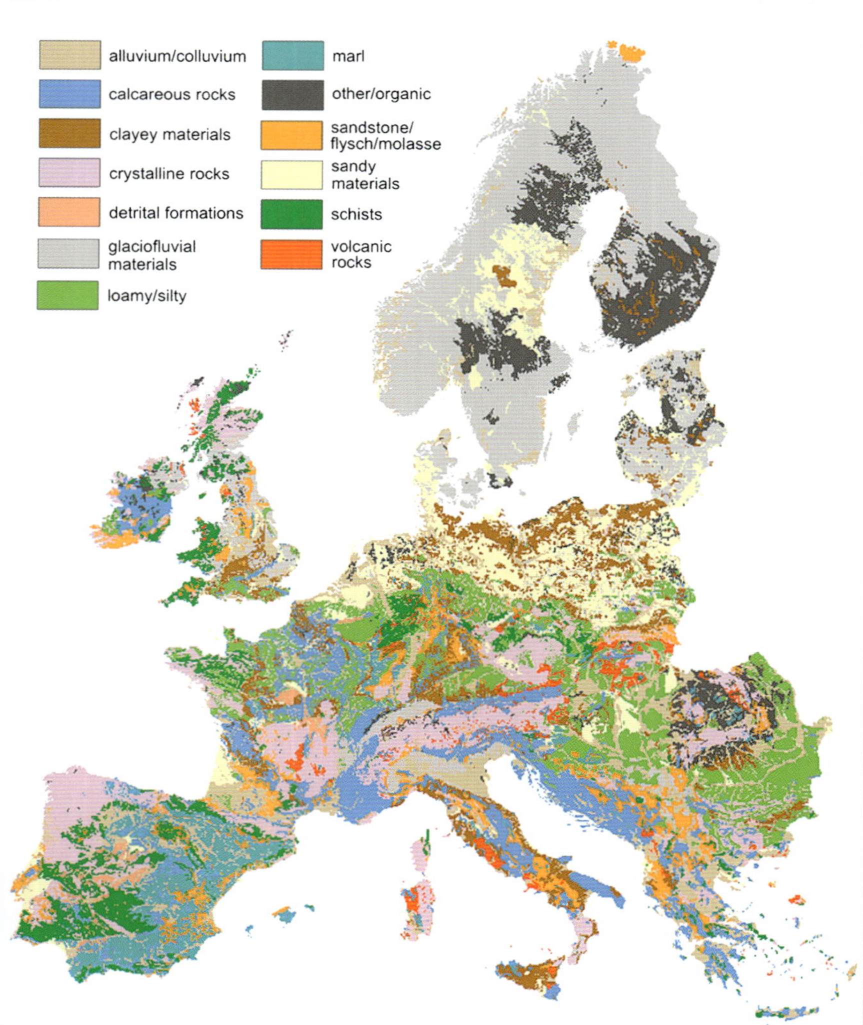

Fig. 6.2: Parent material groups derived from the European Soil Map (Günther et al. 2013: Fig. 2, p. 299).

Climate data were derived from the WorldClim data set (Hijmans et al. 2005). They are based on interpolated weather station data of average monthly climate values on a 30 arc-second resolution grid (often referred to as '1 km²' resolution).

Soil and parent material information was derived from the European Soil Data Base (ESDB 2004; Panagos 2006; Panagos et al. 2012). For example, water regime, texture and soil type-related parameters were used in this analysis. While binary features were derived from the level of soil mapping units (SMU), metric attributes were calculated using the values and area proportions for soil typological units (STU). Parent material was regrouped into 13 classes according to Günther et al. (2013); (Fig. 6.2).

Land use related parameters for the regression analysis were derived from CORINE land cover version 2000 (Norway, parts of southern Europe) and 2006 (rest of Europe); the following land cover attributes were considered, after grouping different grid codes, arable, crops, pasture, irrigated land, wetland, open vegetation and shrub land.

In order to compare the spatial extent of crop types at GEMAS sites, the LUCAS soil survey 2003 was evaluated. The LUCAS soil inventory represents a spatially explicit land use inventory (CEC 2001). The land cover was recorded in an 18 × 18 km grid across Europe, and for the 10 sub-sample points at each grid cell. Figure 6.3 presents the spatial distribution of arable (agricultural) land according to PELCOM (Mücher 2000), the Pan-European Land Cover Monitoring project, which developed a pan-European land cover database (1996 – 1999) with 1 km resolution.

The data source for stratification (biogeographical regions) is the European Soil Regions Map (BGR

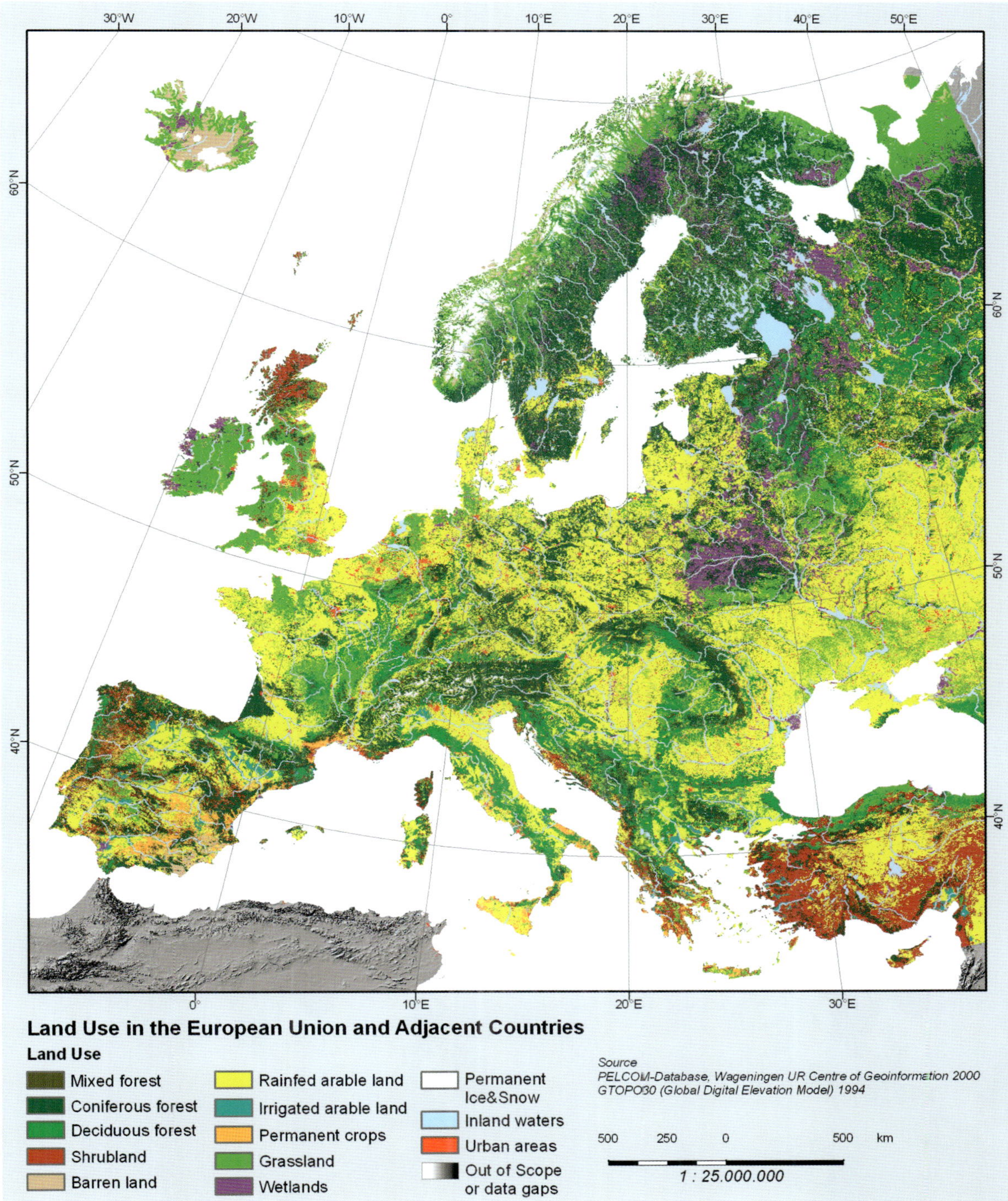

Fig. 6.3: Land use (based on the PELCOM 1 km data set, prepared by HARTWICH et al. 2005).

2005; HARTWICH et al. 2006), Digital Map of European Ecological Regions (DMEER, EEA/ETC BD 2000) and interpolated EMEP data of the year 2005 (EEA/EMEP website, nitrogen oxides; EEA 2005).

In total, N = 227 attributes could be derived from these data sources. For each layer, the local information for each GEMAS site was derived using point-polygon analysis.

▸ 6.3 Results

6.3.1 Spatial distribution of TOC

The kriging results provide a first overview about the distribution of TOC in Europe (Section 11.15 → Part A). The highest values are found in Norway, Sweden, Finland, the United Kingdom and Ireland, and the Alpine and Dinarides region; the lowest values occur in central Spain (Fig. 6.4 B). While the spatial pattern between agricultural and grazing land is similar for most of Europe, TOC values for grazing land completely contrast those for agricultural land in the northern European Pleistocene lowlands (Germany, Poland, Lithuania). Even though sandy soil dominates in this area for both land cover types, grazing land sandy soil is often found in areas with hydromorphic conditions, while agricultural sandy soil occurs in upland areas. For the rest of Europe, it can be hypothesised that climate is the main driver for soil organic matter storage.

Because soil organic matter is a key indicator for many filter and transport processes related to water, nutrient and heavy metals, the controlling factors for TOC in agricultural and grazing land soil are of interest. For example, sites depleted of carbon are considered less productive and less stable than carbon-rich sites under comparable climate. If the main drivers for TOC storage were known, its spatial distribution and sensitiveness for management-induced degradation would be known too. If agricultural practices can achieve high yields with low TOC values (as in central Europe), then it is the carbon-rich soil in northern Europe, the Alps, and grassland soil, which are likely to lose more carbon after intensive management, thus become net CO_2-sources. Therefore, in this study, we collected spatial data about environmental covariates for soil organic matter storage, and statistically searched for trends via regression analysis.

The two maps of Figure 6.4 A present the results of the spatial regression analysis. The interpretations are easier if the land use map in Figure 6.3 is used for orientation about the coverage of agricultural and grazing land. The areas with the lowest TOC values in agricultural soil are found in the northern part of eastern Europe and central Spain. The values are slightly higher in western France and in the continental steppe-like soil in eastern Europe. Further, higher TOC values are found in agricultural soil in low mountainous ranges (e.g., central part of Germany) and under

Fig. 6.4 A: Spatial distribution of TOC in European agricultural and grazing land soil interpolated by regression analysis (left Gr; right Ap).

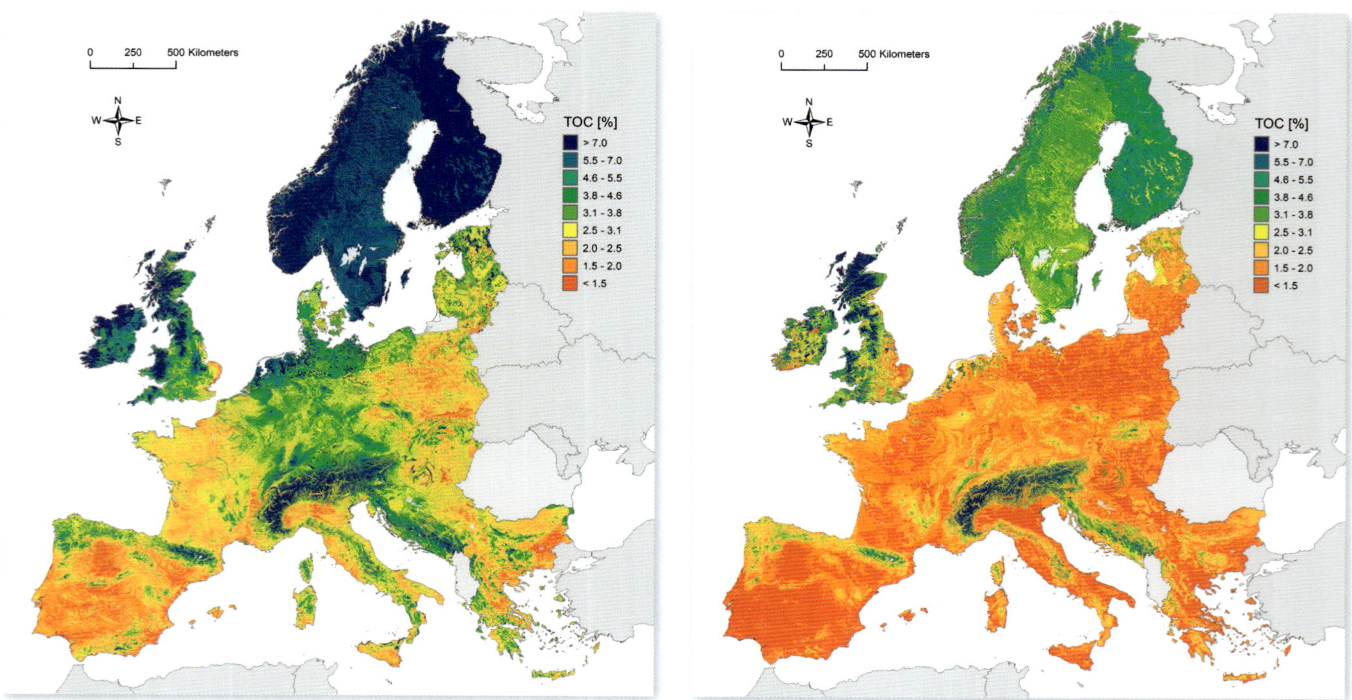

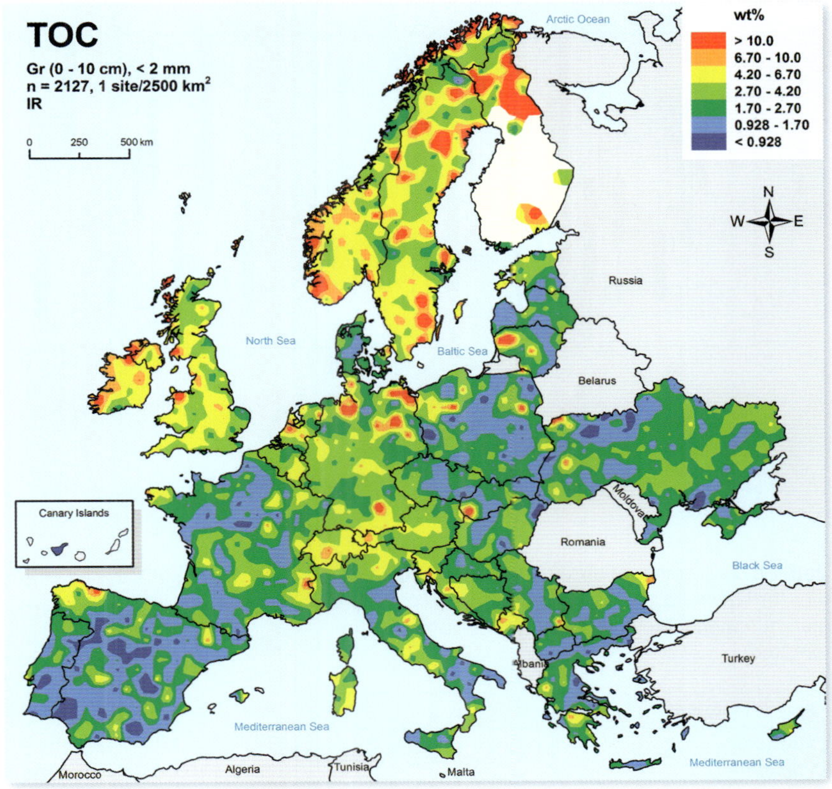

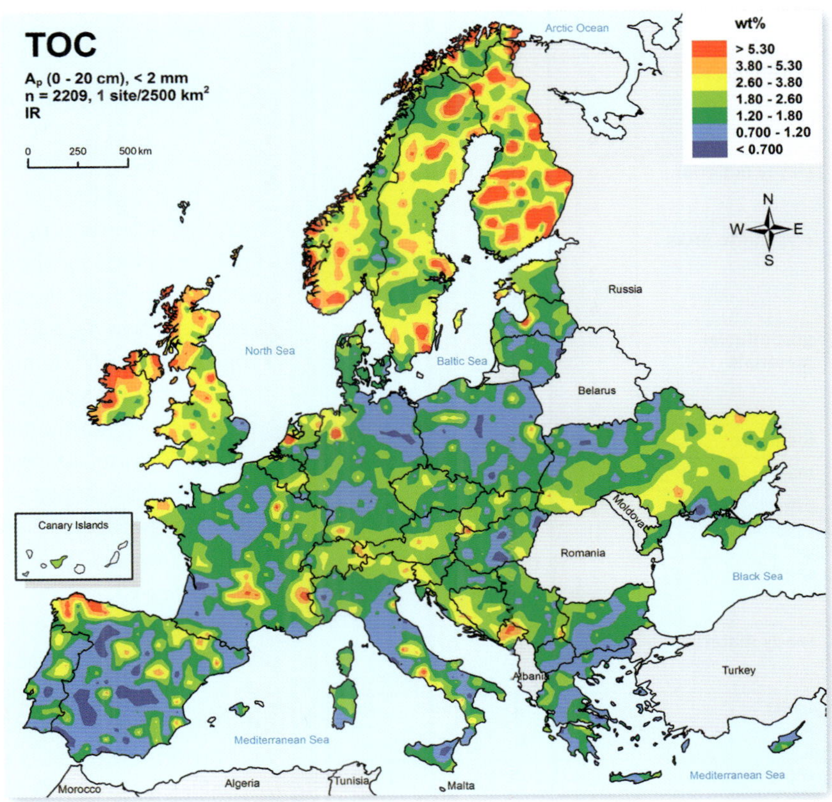

Fig. 6.4 B: Spatial distribution of TOC in European agricultural and grazing land soil interpolated by kriging (top Gr, bottom Ap).

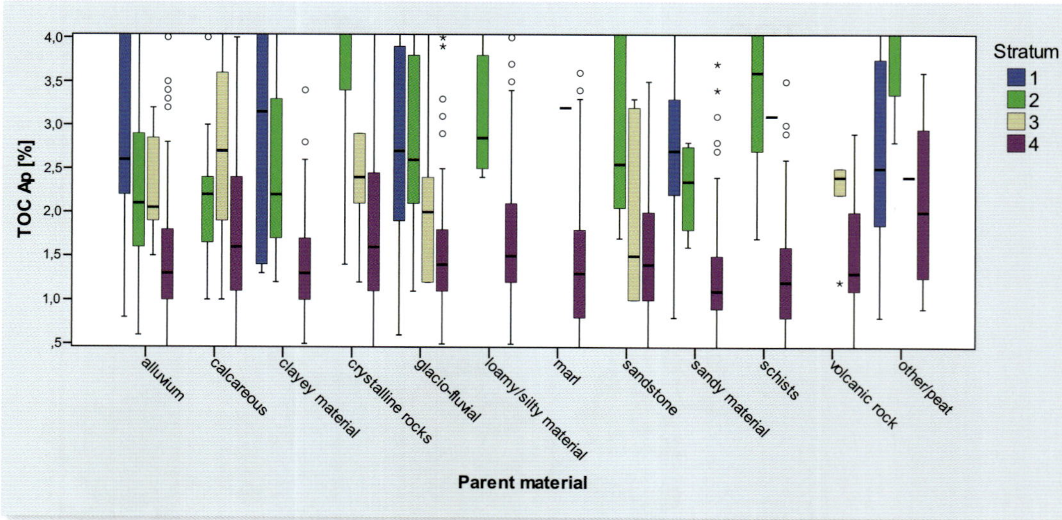

Fig. 6.5: Statistical distribution of TOC in Ap soil samples according to parent material groups (Stratum 1: Boreal; Stratum 2: United Kingdom and Ireland; Stratum 3: High mountainous; Stratum 4: European continent).

Table 6.1: Number of observations and median values for TOC in Ap and Gr topsoil samples according to parent material group (strata with N ≤ 5 samples are omitted).

Parent material	Stratum	N	Ap	N	Gr
alluvium/ colluvium	1	33	2.60	32	4.85
	2	20	2.10	20	4.00
	3	16	2.05	13	2.70
	4	264	1.30	220	2.05
calcareous rocks	2	12	2.20	12	4.75
	3	21	2.70	27	4.30
	4	123	1.60	128	2.40
clayey material	2	10	2.20	10	4.45
	4	102	1.30	118	2.20
crystalline rocks	2	13	4.80	14	4.25
	3	9	2.40	18	4.00
	4	123	1.60	129	2.40
glaciofluvial materials	1	243	2.70	196	4.90
	2	33	2.60	37	4.50
	4	77	1.40	72	2.05
loam/silt	4	170	1.50	149	2.50
marl	4	87	1.30	87	1.60
other/ organic	1	107	2.50	56	4.05
	4	23	2.00	24	2.40
sandstone/flysch	2	20	2.55	18	4.35
	4	65	1.40	75	2.20
sandy materials	1	57	2.70	55	4.20
	4	136	1.10	133	2.00
schist	2	30	3.60	32	5.05
	4	70	1.20	75	1.60
volcanic rock	4	27	1.30	30	2.95

loess influence. The regression maps also help to differentiate more clearly between the agricultural and grazing land. Grazing land soil has clearly higher TOC content, especially in northern and central Europe. Most likely, effects of hydromorphic soil conditions in the north, and higher altitudes in central Europe, are the factors that enhance the combined macro-climatic north-south and west-east gradients.

6.3.2 TOC and parent material

Figure 6.5 presents the distribution of TOC according to parent material. It can be seen that the range of TOC values would be huge if it was not stratified according to climatic strata. TOC variability is highest for the boreal zone; the highest medians are found in samples rich in clayey material (3.15 % in Ap, 6.6 % in Gr soil); however, the range of values (25 and 75 percentiles) is similar for all parent material groups. The differentiation according to parent material is best expressed in stratum 2 (United Kingdom and Ireland) and 3 (high mountain ranges: Alps, Dinarides, and Apennines). In the United Kingdom and Ireland stratum, the highest values are found at sites with crystalline rock and on 'Other/Organic Materials' (4.8 % and 4.1 % in the Ap, respectively), but also on sites having Schist as parent material, the values are quite high (3.6 % in the Ap) (Table 6.1). The stratum of mountainous ranges is predominantly represented by sites in the Alps; here, highest values are found at sites with calcareous rock

(2.7 % in Ap; 4.3 % in Gr soil samples). All medians of stratum 3 are clearly lower than those for strata 1 and 2. However, the lowest values are found for stratum 1. Compared to Ap sites, differences for parent materials are slightly better expressed for grazing land soil sites.

With respect to stratum 4 (central European continent), the lowest values for grazing land are found on marl (1.6 %) and schist (1.6 %); the highest median and range occur on sites with volcanic rock (3.0 %). In general, the TOC values are clearly higher for grazing land compared to agricultural soil; however, the general distribution pattern according to climate and parent material is fairly similar for both land cover types.

6.3.3 TOC and crop type

Figure 6.6 presents the distribution of TOC values for the Ap soil according to crop type. For 423 sites, there was no information about the crop type, and for 548 sites, the category 'Other' was used; corn was the crop type at 294 sites and wheat at 369 sites. Therefore, for almost half of all Ap sites, no reliable information about the crop type was recorded so that the stratification according to crop type and climate in some cases leads to very low numbers of observations in strata 1, 2 and 3. This is the reason why median values have to be discussed with great care, because it cannot be expected that the range of values is fully representative. Values for wine, energy crops, forage crops and rye are at a slightly lower level than for sugar beet, wheat and corn. Otherwise, there is no clearly pronounced trend across crop types.

Similar to parent material, climate seems to be the most dominant site factor for explaining TOC variability (for example, see values for corn, Fig. 6.6) with median values of 2.7 %, 2.8 %, 2.4 % and 1.0 % for strata 1, 2, 3 and 4, respectively. The

Table 6.2: Number of observations and median values for TOC (%) in agricultural soil (Ap) according to crop type (strata with N ≤ 5 samples are omitted).

Crop	N	TOC Ap [%]	Crop	N	TOC Ap [%]
Stratum 1: Boreal			**Stratum 4: European continent**		
Corn	37	2.70	Cereal	6	1.00
Other	325	2.70	Corn	220	1.50
Potato	17	2.20	Energy crop	48	1.20
Wheat	31	2.20	Forage crop	26	1.20
Stratum 2: UK and Ireland			Ochard	36	1.30
Corn	22	2.80	Other	191	1.10
Other	25	3.80	Potato	35	1.30
Potato	14	4.30	Rye	52	1.20
Wheat	40	2.40	Sugar beet	10	1.40
Stratum 3: High mountainous			Vegetable	15	1.20
Corn	15	2.40	Vine	15	1.05
Other	7	2.70	Wheat	284	1.50
Potato	9	2.80			
Wheat	14	2.90			

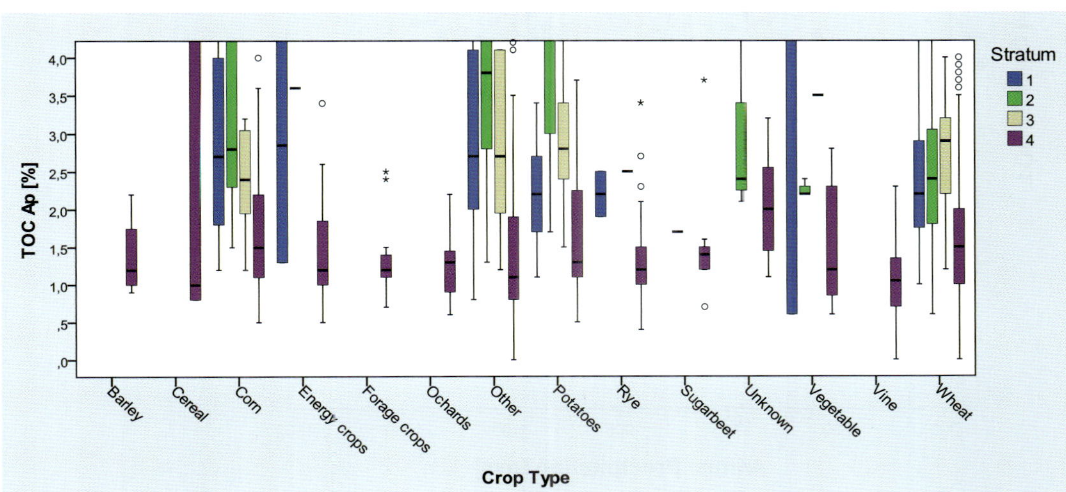

Fig. 6.6: Statistical distribution of TOC in Ap soil according to crop type (Stratum 1: Boreal; Stratum 2: United Kingdom and Ireland; Stratum 3: High mountainous; Stratum 4: European continent).

difference between strata 1 to 3 versus stratum 4 is very pronounced. In stratum 4, TOC values are generally at a very low level, between 1.0% and 1.5%.

6.3.4 TOC and climate

An attempt has been made to test the effect of climate after deriving long-term averages for temperature and precipitation from a 1 km raster climatic data set. Here we used annual climate averages. Figure 6.7 shows the results for grazing land topsoil. There is only a very slight trend for TOC increase with increasing precipitation (see stratum 4), which may be expected as these areas often are located in areas with wet soil. For the same stratum, there is also a slight trend for TOC decrease with increasing annual average temperature.

6.3.5 TOC and texture

Besides climate, TOC has been proven to increase with increasing clay content. This correlation has been found in many studies. Not so with the GEMAS data. Across the large gradient for all of Europe, no trend is found of TOC with respect to clay (see quantile plots on attached DVD → Part A). In order to investigate the plausibility of this observation, the LUCAS soil survey points with known land cover type was overlaid with texture class abstracted from the European Soil map 1:1,000,000 (ESDB 2004; ESBN 2005). The results are presented in Table 6.3. It can be seen that many crops are cultivated across a spectrum of two or three different texture classes. Some crops, such as potatoes, corn and rye (but also shrub lands and grazing land) are mostly abundant on sites with coarser texture, while sugar beet, wheat and barley cover the medium textured spectrum with a tendency towards slightly heavier soil. On finer grained soil, many Mediterranean crops are found; this is probably due to their abundance on heavy soil from calcareous residuum (typically clay-rich), combined with water shortage. With regard to TOC and soil texture, descriptive statistics with the GEMAS data base did not provide clear results.

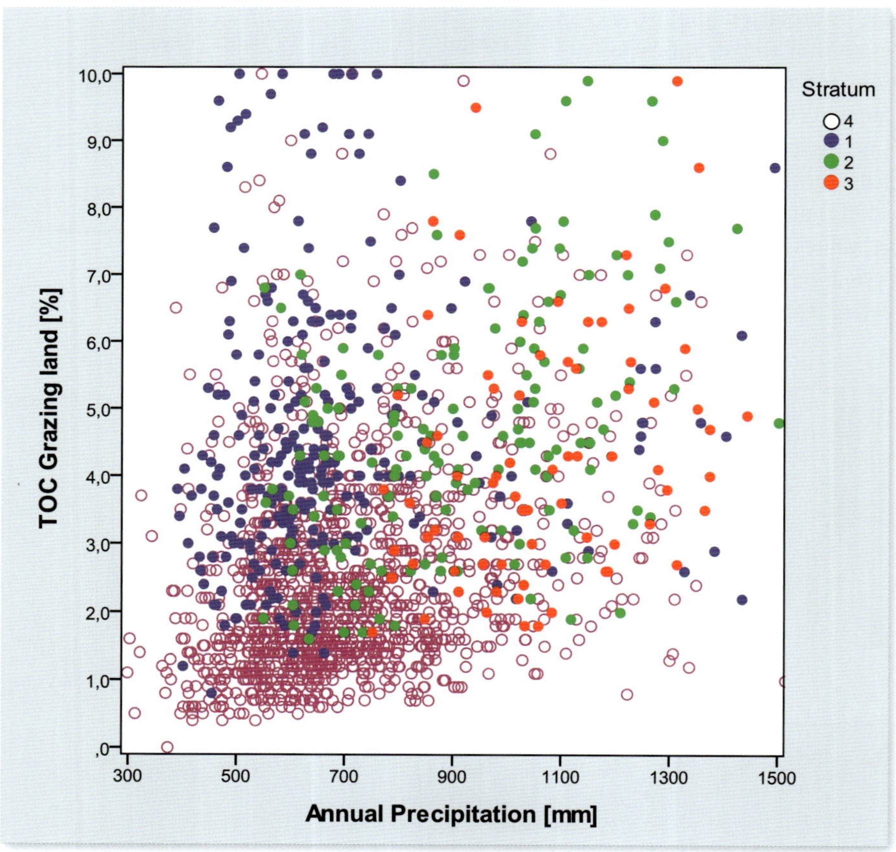

Fig. 6.7: Statistical distribution of TOC in grazing land topsoil according to annual averages of (a) precipitation and (b) temperature (Stratum 1: Boreal; Stratum 2: United Kingdom and Ireland; Stratum 3: High mountainous; Stratum 4: European continent).

Table 6.3: Relationship between LUCAS soil survey cropland types and soil texture.

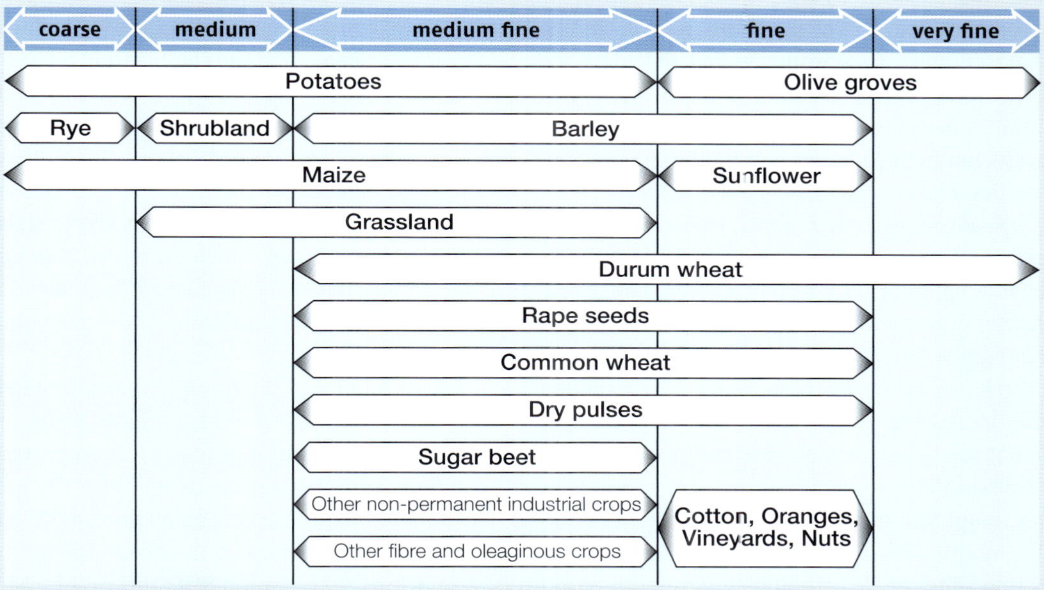

Texture class (European Soil Geographical Database 1 : 1,000,000)

coarse	18% < clay and >65% sand
medium	18% < clay <35% and ≥15% sand, or 18% < clay and 15% < sand <65%
medium fine	<35% clay and <15% sand
fine	35% < clay <60%
very fine	clay >60%
peat	no mineral texture

▸ 6.4 Discussion

The distribution of soil carbon in the GEMAS agricultural and grazing land soil samples clearly follows trends reported in the literature (e.g., Jones et al. 2004, 2005a; Lettens et al. 2004). Being faced with a lack of measured data across Europe, Jones et al. (2004, 2005a) have derived a first map using pedotransfer rules, which were applied to the European soil database. The map shows the distribution of the soil organic carbon concentration in the topsoil layer 0–30 cm. CORINE land cover data and annual accumulated temperature (AAT) were used as additional parameters. While measured carbon data are available for forest soil (Baritz et al. 2010; 2011), no such evaluations were possible up until now for European agricultural and grazing land soil. While this is now possible with GEMAS, additional results will become available soon through the LUCAS Soil project, which was recently conducted by the European Commission. A recent assessment of national soil carbon data revealed that comparable measured data are needed rather than estimates, in order to produce a reliable baseline for Europe (Panagos et al. 2013). The GEMAS database for the first time provides such a comparable data set with measured carbon values for agricultural and grazing land top soil according to REACH regulation specifications. Compared to GEMAS, it can be concluded that the existing estimates by Jones et al. (2004, 2005a) substantially overestimate (e.g., north-eastern Europe, Sweden) and underestimate (e.g., Norway) the real conditions. This observation is also made by Panagos et al. (2013). However, also the GEMAS data set is also facing some restrictions, especially for northern Europe. The variability of soil in northern Europe is so extremely high that prediction of regional soil organic carbon values based on a relative small number of samples seems difficult. This is especially true when considering inaccuracies from the evaluations. Such additional inaccuracies arise from the fact that data for some important parameters (soil type, relief position, climate) were not available at sample site level from fieldwork, but had to be derived from existing maps via point-polygon analysis.

Despite these restrictions, the statistical results of the regression analysis seem to provide reliable regional soil carbon values. Because this chapter only introduces the statistical possibilities that can be performed with the GEMAS data set, it can be said that the combination of parameters (predictors) and stratification have delivered regional regression models with coefficients of determination between 0.3 and 0.4. For example, it was found that regarding climatic data, the available monthly averages are quite sufficient for predicting soil organic carbon in a regression-based baseline map for Europe (e.g., maximum temperatures in May and August for agricultural soil; results not shown here).

It also has to be considered that parent material had to be aggregated into few groups in order to avoid too many substrata due to the relatively small amount of observations. For example, further improvements may be necessary for the analysis and stratification in stratum 4 in particular. However, the gradients for driving the current (very low) soil carbon values in ploughed agricultural (Ap) and grazing land topsoil seems to be dependent on macro-climatic large gradients across Europe. However, looking in more detail, relevant site factors, which are well-known from the literature, do not appear very clearly. This was investigated here for texture class, parent material and land use (crop type). Even if we individually just look at annual climatic values, the variability is enormous. It seems that the combination between stratification according to macro-climate combined with monthly climatic data and site factors, such as soil type, texture and slope position, allows for the plausible spatial prediction of TOC. This interpretation follows the results of the recent study by Goidts et al. (2009), which examined changes of soil organic matter in agricultural soil, clay content and climate (precipitation) as the dominant site factors.

The geochemical distribution maps for all elements and parameters were produced using the geostatistical kriging method for interpolation (Chapter 8 → Part A). The interpretation of the spatial element distribution based on geostatistics is difficult, because there is no knowledge about the relationship between TOC and biophysical environmental parameters. Regression analysis as implemented here can help to identify the driving forces for carbon storage at landscape-level. The spatial prediction, using high-resolution digital elevation models, also helps explicitly to delineate spatially high and low-carbon sites. Such differentiation is important for example in climate change research, but also soil protection policies in Europe (delineation of priority areas in order to better focus soil management against carbon loss; see also Kibblewhite 2012). However, it must be stressed that the planning of soil protection measures, based on soil monitoring, require evaluation targets, such as thresholds for natural, optimal and degraded carbon values in topsoil (Hüttl et al. 2007; Huber et al. 2008).

▸ 6.5 **Conclusions**

The spatial distribution of total organic carbon (TOC) derived from sampling and analysis of the GEMAS database was investigated. Even though the GEMAS sampling density is comparatively low, when compared to monitoring requirements (see also Saby et al. 2008), the spatial distribution of carbon in soil and its predictors in regression analysis shows plausible results. Despite the high quality of the database (homogenous sampling and analysis), a very large proportion of variability cannot be explained. To some extent this fact certainly results from the lack of information at the sampling site, such as soil type, management history, parent material, climate.

Regression-based spatial modelling seems to provide an efficient upscaling method, which only needs sufficient sampling points within the strata of investigation. Because the strata (e.g., macro-climate 'x' soil type, 'x' slope class) often extend over a large territory, and beyond country borders, only a limited number of samples is needed for the assessment of the spatial distribution pattern. However, it was also observed that in some strata, specific soil types appear as predictors (such as Histosol, Rendzic Letosol). In these cases, and because of the low coefficient of determination, the soil map units provide a much simpler upscaling method. This effect has been demonstrated by Baritz et al. (2006) comparing national and regional carbon inventories for forest soil with different sampling densities. It appears that with decreasing database densities, simple means for combinations of few predictors (e.g., soil type 'x' soil region) might be sufficient to develop regional assessments. However, this needs to be investigated further. With increasing database densities, meso-scale site factors including management (e.g., liming of forest soil; dominant tree species) appear in the prediction models. Despite these limitations, the GEMAS database for the first time provides information about the spatial distribution of carbon in agricultural soil for all of Europe, based on harmonised measurements, and according to REACH regulation specifications. It must be stressed that despite the different sampling depths, differences between agricultural and grazing land are pronounced, and that further evaluations of soil inventories for non-forest land must always distinguish between these land cover types (for the spatial assessment of carbon in forest soil, see Baritz at al. 2011). Further stratification according to crop type shows very limited trends; but this might be due to the lack of reliable management data. If the GEMAS data are used as a baseline for modelling changes of soil organic carbon, it has to be considered that the measured values for agricultural land (cropland) very likely do not represent steady state conditions due to management effects. This might be different, however, for the grazing land soil.

The GEMAS maps clearly show the effects of high elevation (mountain ranges), summer dryness (southern and eastern Europe), soil under oceanic influences (United Kingdom and Ireland), and hydromorphic conditions (Pleistocene lowlands). However, agricultural soil with very low organic carbon values can be found all over Europe, and not just in the southern parts. Thus, without profound ecosystem knowledge, it is difficult to judge whether agricultural soil is degraded after intensive management, or whether these values reflect an equilibrium, which can be managed sustainably with high yields and optimised fertiliser input. The meaning of optimal carbon values has been reviewed and discussed by Hüttl et al. (2007). They cite Domsch (1992) who found that the optimal soil biological activity correlates with organic matter values between 1.5 and 3.5%. For example, at 10% organic matter, decomposition would take 2 to 10 times longer than at the 'optimal' value range; below this range, decomposition is reduced up to one third. However, the main reason for the reduced decomposition under very low organic matter conditions is due to the low productivity (reduced amount of substrate for microbial growth). This rationale demonstrates that the evaluation of high or low values seems to correlate with environmental factors, but its interpretation for management is still difficult without thresholds for sustainable soil management. |◂

Rainer Baritz
Federal Institute for Geosciences and Natural Resources (BGR);
Hannover; Germany
E-mail: rainer.baritz@bgr.de

Vibeke Ernstsen
Geological Survey of Denmark and Greenland (GEUS);
Copenhagen, Denmark

Dietmar Zirlewagen
INTERRA Environmental Survey and Research;
Kenzingen, Germany

Chapter 7

Distribution of Selenium in European Agricultural and Grazing Land Soil

MICHAL POŇAVIČ & ANDREAS SCHEIB

▸ 7.1 Introduction

Selenium (Se) is a trace element naturally occurring in rocks, soil, water and air. In the surface environment, a primary control on the distribution of Se is the underlying geology. Other secondary considerations such as climate and human activities (e.g., coal burning) can also be important. Low-Se rocks (0.05 – 0.09 mg/kg Se) make up the majority of the Earth's crust and subsequently account for generally low levels in most types of soil (JOHNSON et al. 2009).

Selenium is essential to human and animal health at trace levels but harmful in excess. The intake range between dietary deficiency and excess is very narrow. The element plays a key role in a number of metabolic processes, including antioxidant systems, thyroid hormone metabolism, immune function and reproduction (JOHNSON et al. 2009). In China, for example, an endemic selenium-responsive cardiomyopathy (known as the Keshan disease) is associated with areas of extremely low Se concentrations in the environment (JOHNSON et al. 2000). Selenium toxicity on the other hand is at a global scale far less widespread, and can cause, for example, selenosis associated with a variety of symptoms including hair and nail loss (FORDYCE 2005). In most cases, food forms the main source of Se and the content in foodstuff is primarily linked to Se concentrations in the soil on which foods are being produced (JOHNSON et al. 2009). Therefore, soil is of great importance in determining the Se status of crops and livestock, and periodic monitoring of Se levels in soil and food is necessary (NAVARRO-ALARCON & CABRERA-VIQUE 2008).

This chapter presents Se concentrations in agricultural and grazing land soil across Europe with the aim to improve the knowledge and understanding of the spatial variability of Se-deficient and Se-toxic areas, as well as linking these to potential causes. Because of the close relationship between foodstuff and soil, information on soil-Se is particularly important to the agricultural community and European regulators to ensure livestock health and food production, and subsequently human health.

▸ 7.2 Selenium (Se)

This chapter gives a brief description of the chemistry and distribution of Se in the terrestrial environment. It focuses primarily on soil and the implications for human and animal health. For more detailed information refer to key publications by HAYGARTH (1994), PLANT et al. (2003), FORDYCE (2005), REILLY (2006), and JOHNSON et al. (2009) and references therein.

7.2.1 The chemistry of selenium

Selenium is a non-metal element, with atomic number 34 and an atomic mass of 78.96. It is grouped along with oxygen (O), sulphur (S), tellurium (Te) and polonium (Po), and between arsenic (As) and bromine (Br) in the Periodic Table. This location accounts for many of its biological similarities with S as well as As and its neighbour phosphorus (P). Selenium has six stable isotopes ^{74}Se, ^{76}Se, ^{77}Se, ^{78}Se, ^{80}Se, and ^{82}Se, which occur naturally with approximate abundances of 0.87 %, 9.02 %, 7.58 %, 23.5 %, 49.8 % and 9.19 %, respectively (HOFFMAN & KING 1997). Naturally occurring oxidation states of Se in elemental and chemical forms are selenide (Se^{2-}), elemental Se (Se^{0}), selenite (Se^{4+}) and selenate (Se^{6+}) (FORDYCE 2005).

The biologically interesting seleno-organo compounds are selenols having the general formula RSeH (hydrogen selenide, dimethylselenide and trimethylselenonium ions), selenides with the general formula RSeR (e.g., diphenylselenide), and diselenides, with the general formula RSeSeR (e.g., diphenyldiselenide). The reduced form of Se may also be found in the place of S in amino acids (e.g., selenomethionine and selenocysteine; Fig. 7.1). The other biologically interesting compounds are selenosugars, selenoxides with the general formula RSe(O)R, and selenyl chlorides with the general formula RSeCl.

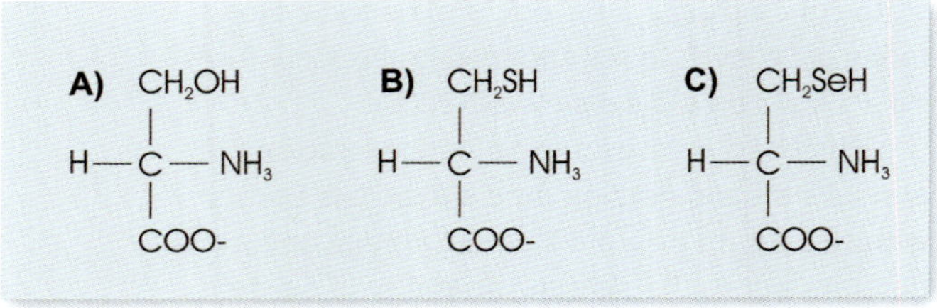

Fig. 7.1: Chemical structures of (A) serine, (B) cysteine, and (C) selenocysteine (after REILLY 2006).

Selenium is strongly chalcophile and generally substitutes for sulphur in sulphide minerals resulting in Se minerals, such as clausthalite (PbSe), tiemannite (HgSe), klockmanite (CuSe), and feroselite (FeSe) (FORDYCE 2005; JURKOVIČ et al. 2011). Elemental Se occurs rarely in rocks and soil, and common soil minerals include sodium selenite, sodium selenate, potassium selenite, and selenides of heavy metals (ALLOWAY 1995).

7.2.2 Selenium in the natural environment

The distribution and cycling of Se in the terrestrial environment is very complex. The primary sources of Se are rocks and minerals at the Earth's surface. RUDNICK & GAO (2003) have estimated a total Se concentration of 0.09 mg/kg for the upper continental crust (UCC), compared with a value of 0.05 mg/kg cited by FORDYCE (2005) for the Earth's crust. Table 7.1 lists Se concentrations in some common rock types. With the exception of phosphatic rocks and some types of mudstone, most common rock types, such as intrusive (and some extrusive) igneous rocks, sandstone and limestone rarely exceed the UCC concentration, which explains the overall low Se concentrations over much of the Earth's surface.

Table 7.1: Selenium (Se) concentrations in some common rock types (after FORDYCE 2005).

Rocks	Se [mg/kg]
Ultramafic	0.05
Mafic	0.05
Granitic	0.01 – 0.05
Volcanic	0.35
Limestone	0.03 – 0.08
Sandstone	< 0.05
Shale	0.05 – 0.06
Mudstone	0.1 – 1500
Phosphates	1.0 – 300

Figure 7.2 shows a schematic diagram of the global Se cycle. Input of Se into the terrestrial environment, particularly soil, is released primarily by weathering and leaching of the parent material. The marine system, however, constitutes the main natural pathway for the Se cycle.

Anthropogenic emission sources of atmospheric Se include coal and oil combustion, base metal smelting and refining industries, mining and milling operations (including copper, zinc, uranium, phosphate or lead) and end-product manufacturers (e.g., some semiconductor manufacturers; YAN et al. 2001; FORDYCE 2005). Natural atmospheric releases of Se result from volatilisation by plants and bacteria, and from volcanic activity. Dry and wet deposition also contributes to soil Se levels (Fig. 7.2).

The global mean of Se in soil is 0.4 mg/kg, and total Se concentrations in soil of < 0.6 mg/kg are generally considered indicative of deficiency for livestock production (FORDYCE et al. 2010). The concentrations and chemical forms of Se in soil, and subsequently its bioavailability to plants, are governed by various physical and chemical factors, including pH, chemical and mineralogical composition, adsorbing surface, organic matter (OM), microbial activity, climate and oxidation – reduction sta-

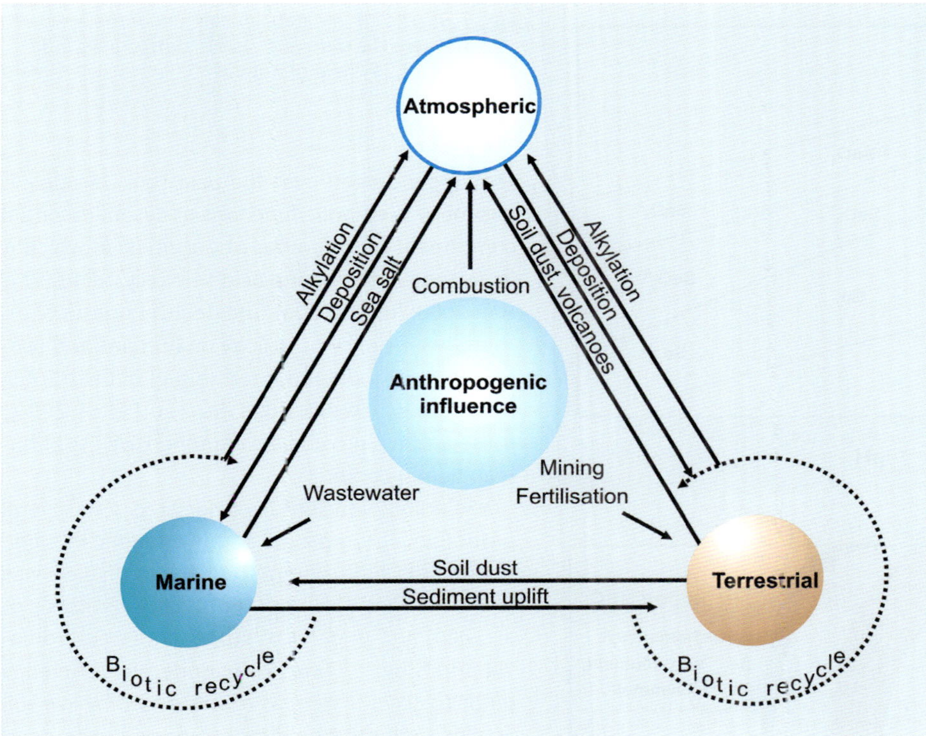

Fig. 7.2: Global selenium cycle showing the interference of anthropogenic activities (after HAYGARTH 1994).

tus. The oxidation stage of Se in soil, for example, is critical for its availability for plant uptake. Selenates (Se^{6+}) are more soluble and mobile in soil than selenites (Se^{4+}), which have a greater affinity to sorption. This means selenates are more bioavailable for crops (JOHNSON et al. 2009, FORDYCE et al. 2010). REILLY (2006) describes sodium selenate as one of the most mobile forms of Se due to its high solubility and inability to adsorb to soil particles. Therefore, oxidising and alkaline conditions that promote the formation of selenate improve the bioavailability of Se (FORDYCE 2005). The same author suggests the following decreasing order of bioavailability for the different forms of selenium found in soil: selenate (Se^{6+}) > selenomethionine > selenocysteine > selenite (Se^{4+}) > selenium (Se^{0}) > selenide (Se^{2-}).

Overall, high OM, Fe oxide and clay mineral content can inhibit the uptake of Se by plants and animals (FORDYCE et al. 2010). Elemental Se and inorganic Se-compounds such as sodium selenite can also be methylated by microorganisms and subsequently volatilised to the atmosphere (FORDYCE 2005). Methylation and hydride formation are important, and sulphur and iron compounds play an important role in cycling of Se (MAYLAND et al. 1991). Microbiological volatilisation of organic Se, particularly dimethyl selenide, is known to be an important factor in the loss of Se from some Se-rich soil and water (FORDYCE 2005).

Selenium in soil can also be affected by the presence of Fe, Mn and Al oxides. Under acid conditions, Se occurs mainly as Se^{4+} (Fig. 7.3), which strongly adsorbs to Fe oxides and may form highly insoluble Fe compounds such as ferric selenite ($Fe_2(OH)_4SeO_3$) or iron selenide (FeSe). In alkaline soil, most of the Se is likely to be present as Se^{6+} and therefore very weakly adsorbed (LANGNER 2000). LOFFREDO et al. (2011) show that elevated concentrations of Se are often associated with the clay fraction of sediments due to the abundance of free Fe or Al oxides. Both selenite $(SeO_3)^{2-}$ and selenate $(SeO_4)^{2-}$ species adsorb onto Fe oxy-hydroxides, but the affinity for Se^{6+} is generally smaller than for Se^{4+} (Fig. 7.4; HAYES et al. 1988; SU & SUAREZ 2000).

There is also a high degree of association between Se and OM (primarily humic substances) in aquatic and terrestrial environments (DHILLON & DHILLON 1999; COPPIN et al. 2006). However, the interaction between Se and OM is not yet

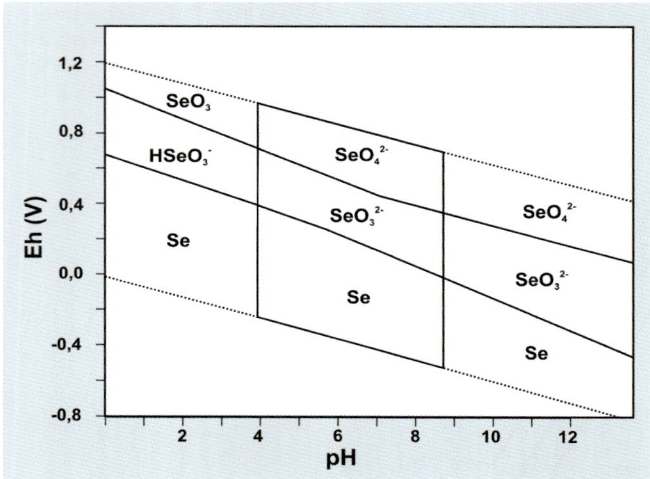

Fig. 7.3: The Eh–pH stability diagram of the Se–O–H system (from MAYLAND et al. 1991); dotted lines indicate the stability field of water.

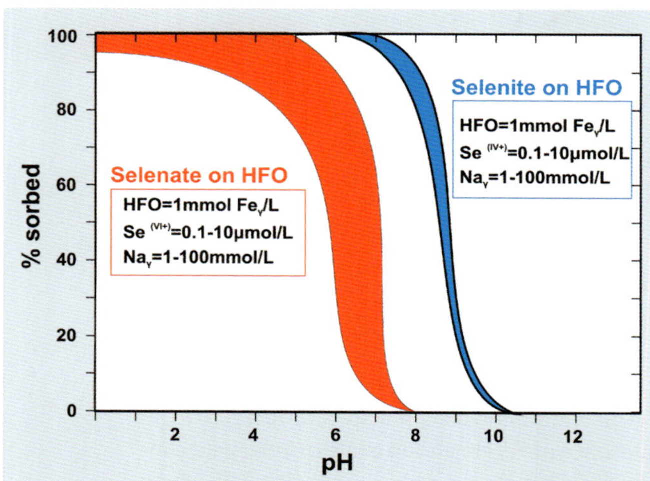

Fig. 7.4: Sorption of oxidised and reduced selenium (Se) species by hydrous ferric oxide (HFO) as a percentage sorbed against the pH; infilled areas show the adsorption of total Se concentrations (0.1 – 10 µmol/L) in selenate (red) and selenite (blue; modified after PLANT et al. 2003: Fig. 11, p. 52).

clearly understood. Recent hypotheses concerning the interaction between Se and OM are:

- Se forms complexes directly with OM (DHILLON & DHILLON 1999),
- Se is indirectly absorbed by OM–metal (Fe, Al, etc.) complexes to form a 'Se–OM–metal' ternary system (GUSTAFSSON & JOHNSSON 1994; COPPIN et al. 2006), and
- Se^{2-} produced by microbes or plants is incorporated into amino-, carboxyl- or hydroxyl-groups (ABRAMS et al. 1990; GUSTAFSSON & JOHNSSON 1994).

7.2.3 Health implications of Se deficiency and toxicity

Selenium is recognised as being essential in the nutrition of animals and humans through research conducted over the past four decades. It has many metabolic functions and is an essential component of a number of enzymes, in which it is present as the amino acid selenocysteine (RAYMAN 2000; REILLY 2006). The health impacts of either Se deficiency or toxicity on humans and livestock are well documented (e.g., FORDYCE 2005; REILLY 2006; RAYMAN 2008; JOHNSON et al. 2009); these authors provide a comprehensive list of human health conditions.

Selenium deficiency is associated with several pathological symptoms and is endemic in some regions of China, where the Keshan disease was first described (JOHNSON et al. 2000; FORDYCE et al. 2000) and in parts of Japan, Korea, Scandinavia and Siberia (TAN 1989). Endemic deficiency tends to occur in regions characterised by low soil Se. For example, the distribution of Keshan and Kashin-Beck disease in China relates to low soil Se, which leads to Se-deficient foodstuff, such as rice, maize, wheat and forage (TAN et al. 2002).

Deficiency also causes lower glutathione peroxidase enzyme activity and can lead to several other health conditions such as cardiovascular disease or myocardial infarction (SALONEN et al. 1982), arteriosclerosis (MOORE et al. 1984) and malignancies (SHAMBERGER 1970; SALONEN et al. 1984; MUECKE et al. 2009).

There is also a moderate to high health risk from Se toxicity. Chronic toxicity of Se in humans results in selenosis, a condition characterised by brittleness or loss of hair and nails, gastrointestinal problems, rashes, garlic breath odour, and nervous system abnormalities (JOHNSON et al. 2009).

The World Health Organization (WHO) recommends a minimum oral intake (MOI) of Se from 0.47 – 0.49 µg/kg/bw/day for adult males and 0.42 – 0.43 µg/kg/bw/day for adult females. The oral health critical value (HCV) for adult males and females is approximately 13 µg/kg/bw/day (values for England and Wales; ENVIRONMENT AGENCY 2009). Figure 7.5 displays the very narrow optimal intake range between dietary deficiency and excess in relation to body weight.

Additionally to Se concentrations in soil, an important factor that determines whether Se-related health problems affect humans and animals is the variable capacity of different plant species to accumulate Se (SPADONI et al. 2007). Based on their Se uptake, plant species can be divided into three groups: Se-accumulators, secondary Se-accumulators and non-accumulators. Crops and plants relating to the latter group rarely assimilate 100 mg/kg Se (dry weight; JACOBS 1989) and are generally associated with deficiencies in grazing animals and crops. The differences in the uptake between plant species and cultivars is genetically controlled and is due to various factors, including surface area of the root, root cation exchange capacity (CEC), root exudates and the rate of evapotranspiration (ALLOWAY 1995). Selenium has no proven function in plant nutrition (BROADLEY et al. 2006).

7.2.4 Selenium production and uses

Worldwide, there are no mines specifically extracting Se; instead, it is a by-product of other metal extraction and refining processes, such as of Pb and Cu, or recovered from sludge from sulphuric acid manufacturing industries (FORDYCE 2005).

Selenium is used in several industrial processes, of which use in electrical components, including low-voltage rectifiers, represents the most important applications. Selenium is also used in the glass industry as a decolourising agent and in the rubber industry as a vulcanising agent. Other uses are in insecticides, in photoelectric cells, in toning baths used in photography, and in xerography (LANGNER 2000). Selenium oxychloride has been used as a solvent, and selenium hexafluoride ($SeFe_6$) is used for electrical insulation. Medical uses include selenium sulphide in shampoos as an antidandruff agent. Previously, ^{75}Se-selenomethionine was used as a scanning agent for pancreas in roentgen diagnostics.

In agriculture, many pesticides, lime and phosphate fertiliser contain varied amounts of Se (FORDYCE 2005). Some countries, routinely add Se (as selenate or selenite) to their crop by foliar feeding or as an additive in artificial fertilisers. Finland is the only country in Europe to add sodium selenate to all fertilisers; a continuing practice since 1985 (ARO & ALFTHAN 1995; KESKINEN et al. 2011).

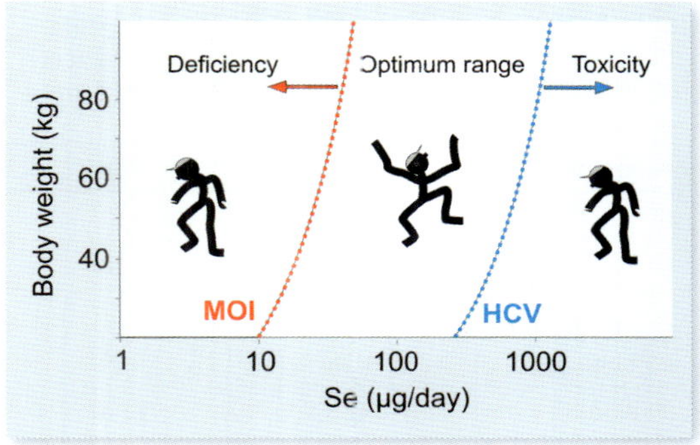

Fig. 7.5: Diagram showing the optimum oral selenium (Se) intake per day in relation to body weight; red dotted line indicates the minimal oral intake (MOI) and the blue dotted line the health critical value (HCV); MOI and HCV lines are based on values published by the WHO (2003) and the ENVIRONMENT AGENCY (2009).

▸ 7.3 Results

Summary statistics of the aqua regia extractable Se concentrations for 2108 agricultural (Ap) and 2023 grazing land (Gr) soil samples are listed in Table 7.2. Selenium concentrations in Ap soil range from < 0.05 to 3.84 mg/kg, with a median of 0.35 mg/kg. Out of the 2108 Ap samples, 2.2 % reported values below the lower limit of detection (LLD) of 0.1 mg/kg Se. Selenium concentrations in Gr soil range from < 0.05 to 6.84 mg/kg, with a median of 0.40 mg/kg, which is slightly higher than for Ap soil. Out of the 2024 Gr samples, 1.7 % returned values below the LLD. Data for both sample types are positively skewed and can be modelled as a lognormal distribution with some anomalous outliers (Section 11.50 → Part A).

Figures 7.6 and 7.7, displaying the spatial distribution of Se in European soil, show various areas of low and high Se contents in Ap and Gr soil samples, respectively.

Low soil Se concentrations correspond to areas with values below the 5th percentile of 0.10 and 0.11 mg/kg for Ap and Gr soil samples, respectively. Elevated Se concentrations correspond to areas with values above the 95th percentile of 0.89 and 1.06 mg/kg for Ap and Gr soil samples, respectively. Each of these anomalous areas is labelled with a number (Figs. 7.6, 7.7). The corresponding numbers are summarised in Tables 7.3 to 7.6 listing potential natural and anthropogenic sources or causes for the observed low or elevated concen-

Table 7.2: Summary statistics of aqua regia extractable selenium (Se) concentrations (mg/kg) in agricultural (Ap) and grazing land (Gr) soil samples of Europe.

Parameter	Ap	Gr
N	2108	2023
Minimum	< 0.05*	< 0.05*
10th percentile	0.16	0.17
25th percentile	0.24	0.28
Median	0.35	0.40
75th percentile	0.49	0.57
90th percentile	0.68	0.83
Maximum	3.84	6.84
Skewness	3.94	5.45

* Half the lower detection limit at 0.1 mg/kg Se

trations. In total, the maps indicate 22 and 20 Se-low, and 47 and 37 Se-high areas for Ap and Gr soil samples, respectively.

In general, the spatial distribution of low and elevated Se concentrations in Ap and Gr samples appear very similar. Both maps show that elevated Se concentrations are predominantly found in soil samples along the west coast of Ireland, the UK, Norway, Spain and France. Across the remaining continent the distribution of both Se-low and Se-high areas are very much localised and relate in most cases to individual sample localities.

However, geochemical maps (Figs. 7.6, 7.7) do indicate a spatial trend of generally low soil Se concentrations in both Ap and Gr soil samples across the East European Platform (Poland, Ukraine, Estonia, Lithuania), the Fennoscandian Shield (northern Sweden and Finland, and central and western parts of Norway), the Paris Basin and basins in central Spain, the Czech Republic and Hungary. Areas of elevated Se concentrations also occur in southern Finland and Sweden, as well as the Dinarides Mountains.

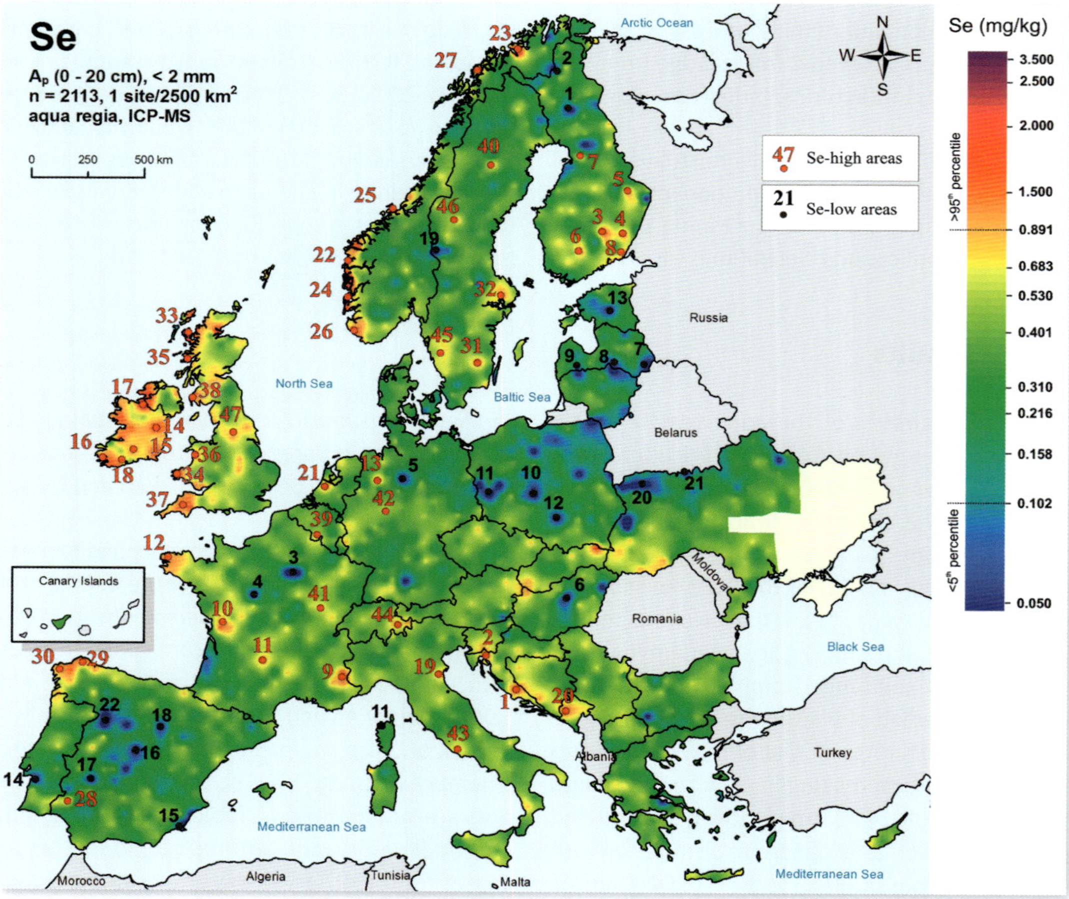

Fig. 7.6: Interpolated map of the spatial distribution of aqua regia extractable selenium (Se) concentrations in agricultural (Ap) soil of Europe.

Table 7.3: Explanation of potential sources and causes of Se-high anomalies (>1.05 mg/kg) in agricultural (Ap) soil; number labels correspond to Figure 7.6 (red colour numbers).

Label	Country / Region	Source; Cause
1, 2	**Croatia**	Geology; terra rossa soil
3 – 8	**Finland**	Geology; biogenic soil; Anthropogenic; enrichment due to Se-fertiliser
9	**France /** Lubéron area	Geology; Eocene–Oligocene sediments; Oligocene salt (brine) at depth
10	**France**	Geology; Armorican thrust (regional shear zone with hydrothermal alteration)
11	**France**	Geology; Quaternary volcanics and Pb – Zn mineralisation in the Cevenole belt
12	**France**	Geology; Armorican thrust; Anthropogenic; industrial pollution, military port and ship factory
13	**Germany /** North-west German lowland	Geology; high soil organic matter
14, 15, 18	**Ireland**	Geology; black shale of the Namurian period
16, 17	**Irish west coast**	Natural; sea spray and high soil organic matter
19	**Italy /** Po river	Geology; alluvial soil
20	**Montenegro /** Podgorica	Geology; terra rossa soil
21	**Netherlands /** South Holland Province	Natural; sea spray
22 – 27	**Norwegian coast**	Natural; sea spray and high soil organic matter
28	**Spain /** Iberian Pyrite Belt	Geology; site is very close to the world class Rio Tinto ore deposit (mostly VMS type) and porphyritic rock hosting a stockwork with Au – Ag – Cu and minor Bi mineralisation
29, 30	**Spain /** North-western part of Galicia	Geology; mineralisation, vein type Sn – W – As – Cu
31	**Sweden /** Småland	Geology; volcanic rocks of the Fennoscandian Shield
32	**Sweden /** Mälaren region	Geology; Se-rich marine clay
33 – 35, 38	**United Kingdom**	Natural; high soil organic matter
36, 37	**United Kingdom**	Geology; associated to sulphide mineralisation
39	**Belgium**	Anthropogenic; industrial pollution; coal burning
40	**Sweden /** Sorsele region	Geology; Cu – Zn – Pb – Ag – U – Au mineralisation
41	**France**	Geology; the Hercynian per-granitic mineralisation
42	**Central Germany**	Geology; high soil organic matter
43	**Italy /** Cicero area	Geology; alluvial sediments from carbonatic rocks and pyroclastics of Roman Magmatic Province
44	**Switzerland**	Geology; Permo-carboniferous volcanism in the Austroalpine area
45	**Sweden /** west coast	Geology; high soil organic matter
46	**Sweden /** Jämtland	Geology; mineralisation in Palaeozoic black shale
47	**UK**	Geology; also, possible anthropogenic input from coal burning in the Merseyside conurbation

Table 7.4: Explanation of potential sources and causes for Se-low anomalies (<0.09 mg/kg) in agricultural (Ap) soil; number labels correspond to Figure 7.6 (black colour numbers).

Label	Country / Region	Source; Cause
1, 2	**Finland**	Geology; clay soil, with low trace element contents
3	**France /** Paris Basin	Geology; sandstone of Fontainebleau (Pliocene – Quaternary) leached by a palaeo-alteration in tropical semi-arid climate
4	**France /** Paris Basin	Geology; palaeo-terraces of the Loire River, palaeo-alterite (ferralite)
5	**Germany**	Geology; Cretaceous sediments
6	**Hungary**	Geology; Pliocene sediments
7 – 9	**Latvia**	Geology; podsolic soil developed on glacial sandy sediments
10 – 12	**Central Poland**	Geology; Pliocene sediments
13	**Estonia**	Geology; podsolic soil developed on glacial sandy sediments
14	**Portugal**	Geology; Neogene sediments (sand, conglomerate, clays)
15	**Spain /** Murcia province	Geology; Tertiary (Neogene) volcanism and hydrothermal alteration processes
16	**Spain /** Madrid-Toledo provinces	Geology; Tajo Tertiary sedimentary basin (sand, clays, marls, limestone, gypsum, sodic sulphate)
17	**Spain /** Badajoz province	Geology; Neogene sediments (sand, conglomerate, clays, marls)
18	**Spain /** Burgos province	Geology; Neogene sediments (sand, conglomerate, clays, marls)
19	**Sweden /** Dalarna	Geology; quartzitic sandstone, depleted in most metals
20 – 21	**Ukraine**	Geology; podsolic soil developed on glacial sandy sediments
22	**Spain /** Zamora province	Geology; Neogene sediments (sand, conglomerate, clays, marls)

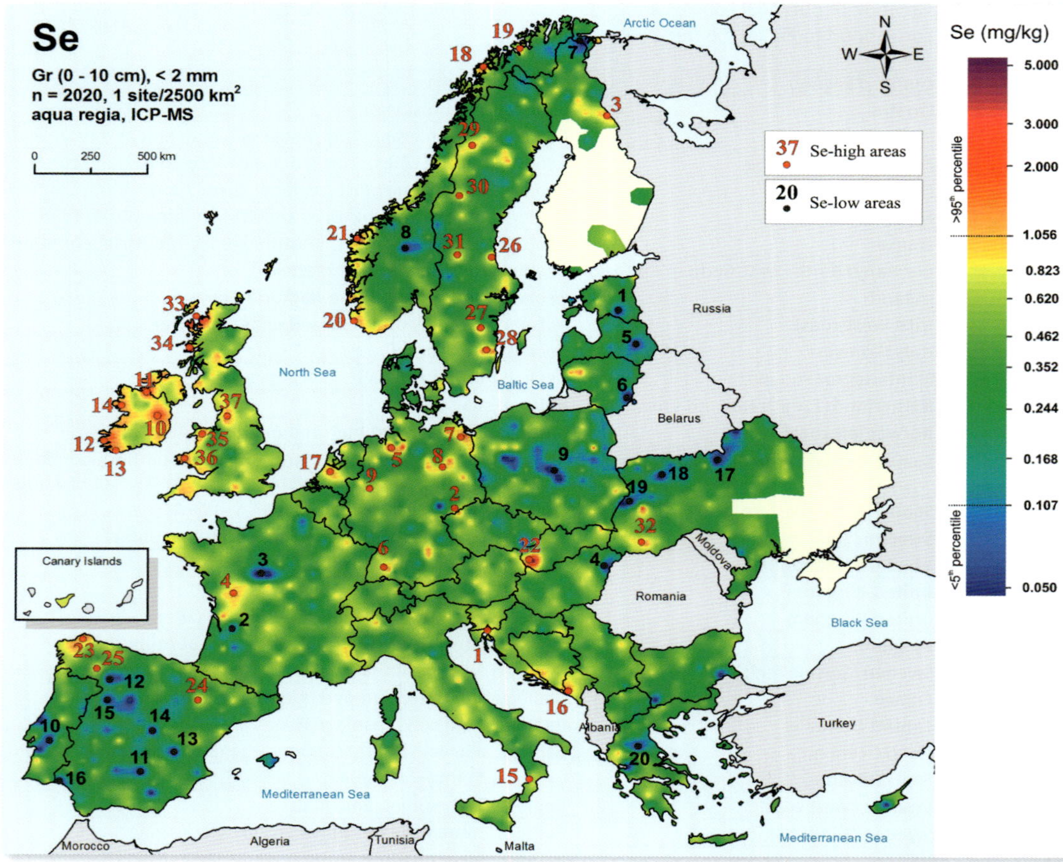

Fig. 7.7: Interpolated map of the spatial distribution of aqua regia extractable selenium (Se) concentrations in grazing land (Gr) soil of Europe.

▸ 7.4 Discussion

The statistical and spatial distribution of Se concentrations in Ap and Gr soil samples are fairly similar. Based on the medians, Se concentrations in Gr soil samples are approximately 13 % higher. Comparison of the 90th percentiles indicates approximately 17 % higher Se concentrations in Gr soil samples. This is likely the result of plant uptake and subsequently depletion of Se in Ap soil. The enrichment in the Gr soil samples is related to higher TOC (Chapter 6; see also Chapter 11, Section 11.15 → Part A), giving the grazing land soil a greater potential for sorption of Se to organic and mineral-organic compounds.

Median Se concentrations in Ap and Gr soil types are 0.35 and 0.40 mg/kg, respectively, and are thus higher than the UCC value of 0.09 mg/kg (Rudnick & Gao 2003), but very similar to the global soil mean of 0.4 mg/kg (Fordyce 2005). The difference to the UCC is likely a reflection of Se input to soil from atmospheric, marine and anthropogenic sources, as well as the accumulation of Se as a result of sorption to organic material, or the UCC estimate is too low (Chapter 11, Section 11.50 → Part A). The following sections discuss the spatial distribution and occurrences of Se-high and Se-low areas with respect to geological and anthropogenic sources and causes.

7.4.1 Se-high areas

Across Europe, Se-high areas correlate well with soil having elevated TOC values. This is particularly evident for coastal areas of countries, such as Ireland, UK and Norway, bordering the Atlantic Ocean and North Sea, where the organic matter content in soil is generally high. The widespread distribution of Se-high soil in these areas (e.g., Fig. 7.6, 22Ap – 27Ap, and Fig. 7.7, 18Gr – 21Gr) is the result of the adsorption of Se to organic

Table 7.5: Explanation of potential sources and causes of Se-high anomalies (>1.05 mg/kg) in grazing land (Gr) soil; number labels correspond to Figure 7.7 (red colour numbers).

Label	Country / Region	Source; Cause
1	**Croatia**	Geology; terra rossa soil
2	**Germany**	Geology; high soil organic matter; Anthropogenic; industrial pollution; coal burning in Czechia
3	**Finland /** Kuusamo	Geology; peat soil, high soil organic matter
4	**France**	Anthropogenic; Pb–Zn smelter near sample location (Mmelle mine) Geology; may correspond to Se enrichment in Eocene cover (lignite)
5–8	**Germany**	Geology; high soil organic matter
9	**Germany /** Ruhr region	Anthropogenic; industrial pollution; coal burning
10	**Ireland**	Geology; black shale of the Namurian period
11–14	**Irish west coast**	Natural; sea spray and high soil organic matter
15	**Italy /** Calabria	Geology; soil derived from metamorphic rocks of Calabrian Arc (Catanzaro graben)
16	**Montenegro /** Podgorica	Geology; terra rossa soil
17	**Netherlands /** South Holland Province	Natural; sea spray
18–21	**Norwegian coast**	Natural; sea spray and high soil organic matter
22	**Slovakia**	Natural; Se-bound to organic matter in soil
23	**Spain /** North-western part of Galicia	Geology; mineralisation associated with a Sn–W–As–Cu vein
24	**Spain /** Navarra and Zaragoza province	Geology; mineralisation associated with a Cu district
25	**Spain**	Geology; mineralisation associated with S–Ta–Nb of Penouta ore deposit
26	**Sweden /** Gävle	Geology; metasedimentary rocks and metavolcanics
27	**Sweden**	Geology; organic rich soil overlying ultramafic rock embedded in granite
28	**Sweden**	Geology; metavolcanic rocks and minor sulphide mineralisation
29	**Sweden /** Västerbotten	Geology; Zn–Cu–Pb mineralisation
30	**Sweden**	Geology; mineralisation in Palaeozoic black shale
31	**Sweden /** North-western Bergslagen	Geology; Zn–Pb–Cu–Ag mineralisation
32	**Ukraine /** Transcarpathia	Geology; associated with Alpine folding
33–36	**United Kingdom**	Geology; associated with sulphide mineralisation
37	**United Kingdom**	Anthropogenic; industrial pollution, e.g., coal burning

Table 7.6: Explanation of potential sources and causes for Se-low anomalies (<0.09 mg/kg) in grazing land (Gr) soil; number labels correspond to Figure 7.7 (black colour numbers).

Label	Country / Region	Source; Cause
1	**Estonia**	Geology; podsolic soil developed on glacial sandy sediments
2	**France /** Aquitaine	Geology; Palaeo-terraces of the Garonne river (sandy alluvium, very low in trace elements)
3	**France /** Paris Basin	Geology; palaeo-terraces of the Loire River, palaeo-alterite (ferralite)
4	**Hungary**	Geology; Pliocene sediments
5	**Latvia**	Geology; podsolic soil developed on glacial sandy sediments
6	**Lithuania**	Geology; podsolic soil developed on glacial sandy sediments
7	**Finland**	Geology; clays soil with low trace element contents
8	**Norway**	Geology; clays soil with low trace element contents
9	**Central Poland**	Geology; Pliocene sediments
10	**Portugal**	Geology; Neogene sediments (sand, conglomerate, clays)
11	**Spain /** Guadalquivir River Basin	Geology; sands, clays and marls
12, 15	**Spain /** Castilla-León Basin	Geology; Neogene sediments (sand, conglomerate, clays and marls)
13	**Spain /** Iberian Range	Geology; sands, clays and marls
14	**Spain /** Tajo basin	Geology; evaporitic sediments
16	**Spain**	Geology; Tertiary sands and clays
17–19	**Ukraine**	Geology; sod-podsolic soil developed on glacial sandy sediments
20	**Greece**	Geology; soil developed on Triassic crystalline mica-chlorite schist and gneiss with interbedded quartzite and marble

(humic) compounds and subsequent accumulation in the soil matrix (Låg & Steinnes 1978). Whilst sea water itself does not have any significant enrichment of Se (*ca.* 0.1 – 0.2 µg/L – Plant et al. 2003), the continuous supply of Se from sea spray, however, represents a potential source for elevating Se concentrations.

Away from coastal areas, Se-high areas are more localised and related to the geology as well as anthropogenic activities. In Ireland, for example, Se anomalies in Ap and Gr soil samples (14^{Ap}, 15^{Ap}, 18^{Ap}, and 10^{Gr}) may be traced to leaching of Se compounds from Se-rich parent materials and their subsequent deposition in low-lying areas, such as old glacial lake beds or river floodplains. The most important Se-bearing rock is the black shale of the Namurian (mid-Carboniferous) period. In Ireland, black shale occurs mainly in County Meath, north County Dublin, and in Counties Limerick and Tipperary (Finch 1971; McGrath 1996). Other sources of Se is the Calp limestone (impure and muddy), which occurs in mid to north Leinster (14^{Ap}). It is often interbedded with Namurian black shale. The highest Se concentrations occur in low-lying peaty swamps with high pH (Rogers et al. 1990).

Elevated levels of Se in UK Ap soil samples (anomalies 34^{Ap}, 35^{Ap}, 38^{Ap}) generally relate to the elevated levels of organic matter in soil. Additionally, areas indicated by 36^{Ap}, 37^{Ap}, 47^{Ap}, 35^{Gr} and 37^{Gr} in UK Ap soil are associated with sulphide mineralisation. Similarities between the chemistry of sulphur and selenium mean that selenium will substitute for sulphur in sulphide minerals (Fordyce 2005). Anomaly 47^{Ap} and 37^{Gr} may also be a result of anthropogenic contribution (e.g., coal burning) from the Merseyside conurbation.

In Southern Finland, elevated Se concentrations in Ap soil (3^{Ap} – 8^{Ap}) relate to binding in clay-rich biogenic soil. However, the source and subsequent Se enrichment in this soil is not natural and relates to the application of Se-supplemented fertilisers, which have been used nationwide since 1985 (Keskinen et al. 2011). Elevated Se concentrations in Gr soil near Kuusamo (3^{Gr}) are associated with peaty soil.

In southern Sweden, a broad Se anomaly in Småland Sweden (31^{Ap}, Växjö and Kalmar) seems to relate to occurrences of intermediate to mafic volcanic rocks of the Fennoscandian Shield. In the Mälaren region, the anomaly at 32^{Ap} might be related to the presence of marine clay, whilst slightly higher Se concentrations in clay- and organic-rich agricultural soil of south-west Sweden (45^{Ap}, River Göta valley) may also be due to the use of Se-supplemented fertilisers. In central Sweden, Se-high areas are more closely related to bedrock. For example, in the Sorsele region (40^{Ap}, Västerbotten) and central Jämtland (46^{Ap}, Östersund) the potential sources are Cu-Zn-Pb-Ag-U-Au mineralisation and Palaeozoic black shale, respectively. The latter are rich in organic matter and contain sulphides that also contain high concentrations of other metals, such as vanadium (V), uranium (U) and molybdenum (Mo) (Chapter 11, Sections 11.60, 11.59 and 11.36 → Part A).

Grazing land soil in Sweden indicates highest Se concentrations in areas where horses, sheep and cattle graze; a likely relationship to Se-supplements used in feed for livestock, especially breeding animals. Out of the six numbered anomalies (Fig. 7.7), five correspond to grassland grazed by horses, sheep and cattle. Nevertheless, a geological source cannot be excluded. For example, the soil sample at location 26^{Gr} (Gävle) is underlain by metasedimentary rocks and metavolcanics. Further south, elevated Se concentrations occur in organic-rich soil overlying ultramafic rocks embedded in granite (Fig. 7.7; 27^{Gr}), just 60 km east of the Se-bearing 'Skrickerum' mine; this area is characterised by widespread and strong base-metal geochemical anomalies in the host rocks, as well as the overlying till and plant materials (Chapter 11 → Part A). Elevated Se concentrations in Gr soil occurs between Vetlanda, Kalmar and Oskarshamn (28^{Gr}) and relate to occurrences of metavolcanic rocks and minor sulphide mineralisation.

Three anomalies (29^{Gr}, 30^{Gr}, 31^{Gr}), located near the Norwegian border are largely related to local mineralisation. Elevated Se concentrations at 29^{Gr} (Västerbotten) can be attributed to either (a) Caledonides and numerous Zn–Cu–Pb–Ag mineralisation in higher nappes (also on the Norwegian side) or (b) Zn–Cu–Pb mineralisation hosted by metavolcanic rocks in the Skellefte mining district. Soil anomalies at 30^{Gr} and 31^{Gr} are associated with Se-enriched black shale at the Caledonian front and to Zn–Pb–Cu–Ag mineralisation hosted by skarn, metavolcanic and metasedimentary rocks, respectively.

In Germany, most anomalous Se concentrations relate to high soil organic matter (13^{Ap} and 42^{Ap}; Fig. 7.6). Concerning anomaly 42^{Ap}, Se concentrations may also be influenced by Kupferschiefer (copper-shale), which outcrops on the eastern edge of the Harz Mountains. Anomalies in Gr soil (5^{Gr}, 6^{Gr}, 7^{Gr} and 8^{Gr}; Fig 7.7) also relate to elevated TOC in soil. The latter do not correspond to anomalies in Ap soil, which have generally low Se concentrations across most of northern Germany. This may indicate an additional anthropogenic source of Se through the application of Se-supplemented animal feed.

Additionally to the strong binding of Se in organic-rich soil, elevated Se concentrations found in soil at locations 2^{Gr} and 9^{Gr} (Fig. 7.7) may be related to emission from coal combustion in Czechia and the Ruhr region in Germany, respectively.

In Spain, the origin of anomalies at locations 29^{Ap}, 23^{Gr} (near Cabo Ortegal – El Ferrol) and 30^{Ap} (near La Coruña) is not clear, but may be due to the numerous Sn–W–As–Cu vein and disseminated mineralisation related to granite intrusions in the north-western part of Galicia. Elevated Se concentrations at location 28^{Ap} are in close proximity to Rio Tinto's world-class ore deposit within the Spanish part of the Iberian Pyrite Belt. It comprises several deposits, most of them of VMS type, and porphyritic rocks hosting Au–Ag–Cu and minor Bi mineralisation.

Anomaly 24^{Gr} (Navarra and Zaragoza provinces) is only found in Gr soil, and may be associated with the Cu district between Pamplona and Zaragoza trending WNW – ESE. This area contains many minor disseminations of Cu minerals in Middle Oligocene sandstone with carbonate cement.

In French Ap soil samples, elevated Se concentrations are geologically controlled and relate to Eocene – Oligocene cover with Oligocene salt deposits at depth (9^{Ap}; Lubéron area), the Armorican thrust zone and associated hydrothermal alteration (10^{Ap}) and Quaternary volcanics as well as the well-known Pb–Zn Cevenole belt (11^{Ap}). Anomalous Se concentrations at location 12^{Ap} (Fig. 7.6) are potentially affected by industrial pollution from a nearby military port and shipyards. The Se map for Gr soil (Fig. 7.7) only highlights one area (4^{Gr}) in western France with elevated Se concentrations. Whilst this anomaly is likely to be related to a Pb–Zn smelter (Mmelle mine) near the sample location, it may also be naturally enriched from Eocene lignite deposits.

Elevated Se concentrations in both soil types (1^{Ap}, 2^{Ap}, and 1^{Gr}) from Croatia correspond to the distribution of coastal terra rossa soil (Miko et al. 2000). Selenium enrichment in Ap and Gr soil from Montenegro (20^{Ap} and 16^{Gr}; near Podgorica and Orjen National Park, respectively) may also relate to coastal terra rossa soil.

Anomalous Se concentrations in an Ap soil sample from Belgium (39^{Ap}; Fig. 7.6) are most likely related to an anthropogenic source, for example, coal burning. Robberecht et al. (1983) found elevated Se concentrations amongst other trace elements (Pb, Cu, Zn, As), and particularly in the topsoil of nearby residential areas.

The highest Se concentration of 6.84 mg/kg in Gr soil is at location 22^{Gr} in the southern part of Slovakia, which corresponds to anomalous high TOC in soil.

The spatial distribution patterns of Se across Italy are explained by the biogeochemical studies of the behaviour of Se in agricultural soil by Spadoni et al. (2007) and are entirely geologically controlled. The highest Se concentrations (3.84 mg/kg) measured in Ap soil relates to location 19^{Ap} and can be attributed to an enrichment in alluvial soil of the River Po estuary. Elevated Se concentrations of the southern Latium (43^{Ap}; Cicero) area also are due to alluvial sediments, but derived from carbonatic and pyroclastic rocks of the Roman Magmatic Province. Figure 7.7 only indicates one area with elevated Se concentrations in Gr soil (15^{Gr}). In this area, soil is derived from metamorphic rocks of the Calabrian Arc (Catanzaro graben).

The only elevated Se concentrations determined in soil across the Alps is located in eastern Switzerland (44^{Ap}) and is most likely related to Permo-Carboniferous volcanism in the Austroalpine area.

7.4.2 Se-low areas

Total Se concentrations in soil < 0.6 mg/kg are generally considered indicative of deficiency for livestock production (Fordyce et al. 2010). Based on this value, over three quarters of all Ap and Gr soil samples (Table 7.2) would indicate a potential deficiency. Spatially, Se concentrations below the median values of 0.35 mg/kg (Ap) and 0.40 mg/kg (Gr) are evenly distributed across Europe (green and blue colours; Figs. 7.6, 7.7). However, some of the lowest Se concentrations are found in soil samples from the Iberian Peninsula and eastern Europe.

Across Spain, Se-low areas are linked to predominantly sedimentary rocks. Selenium-low areas indicated by Ap soil (Fig. 7.6) relate to the Tertiary Tajo sedimentary basin, comprising sand, clay, marl, limestone, gypsum and sodic sulphate (16^{Ap}; Madrid-Toledo provinces) and the great Neogene sedimentary basin of Castilla-León, which comprises conglomerate, sand, clay, marl and limestone that were deposited in a lacustrine environment (18^{Ap} Burgos province; 22^{Ap} Zamora province). The area of 17^{Ap} (Badajoz province; Extremadura) also relates to Neogene sediments, which represent the filling of a graben crossed by the Guadiana River. Low Se concentrations in Ap soil at location 15^{Ap} correspond to the La Unión-Cartagena mining district (Murcia province) and are probably not related to Tertiary sediments. In this area, late Tertiary (Neogene) volcanism gave rise to very widespread hydrothermal alteration and a large number of ore deposits of Pb–Zn–Ag–Sn mineralisation. Causes for Se-low areas in Gr soil at locations 12^{Gr}, 14^{Gr} and 15^{Gr} (Fig. 7.7) largely correspond to those of 16^{Ap}, 22^{Ap} and 18^{Ap}. In the Iberian Range (13^{Gr}; Sierra de Alcaraz), Se-low soil relates to clay filled grabens that appear between the Jurassic – Cretaceous limestone hills. At 11^{Gr}, low Se in soil corresponds to Neogene sand, clay and marl of the Guadalquivir River Basin. Area 16^{Gr} is situated south of the Iberian Pyrite Belt, and occurs on the Tertiary cover (sand, clay) overlying Palaeozoic rocks.

In the eastern part of Europe, all Se-low areas in the Ukraine (20^{Ap} and 21^{Ap}; 17^{Gr} to 19^{Gr}) coincide with sod-podsolic soil, which has developed on deep (5 – 10 m) predominantly Pleistocene glacial and aeolian deposits, such as cover sands and loess (Chapter 9). Figure 9.1 (in Chapter 9) shows that aeolian deposits cover most of Poland, north-east Germany and the Baltic States, which explains the abundance of Se-low areas (5^{Ap} and 7^{Ap} – 13^{Ap}; 1^{Gr}, 5^{Gr}, 6^{Gr} and 9^{Gr}).

The remaining areas of low Se concentrations (Figs. 7.6, 7.7) are affected by the local rather than regional geology. In central Scandinavia, for example, Se-low soil (19^{Ap} near Dalarna) is caused by sandy soil derived from weathering of quartzitic sandstone. The Se-low areas in Finland occur near Rovaniemi (1^{Ap}) and Enontekiö (2^{Ap}) and are related to clay soil.

In France, Se-low soil is found in the Paris Basin (3^{Ap}) and in the Loire Valley (4^{Ap} and 3^{Gr}). The former corresponds to Plio – Quaternary sandstone of Fontainebleau, which is totally leached by a palaeo-alteration in tropical semi-arid climate. The latter relates to Eocene palaeo-terraces and palaeo-alterite. Low Se concentrations in Gr soil near Aquitaine (2^{Gr}; Fig. 7.7) may be related to palaeo-terraces and sandy alluvium of the Garonne River.

7.4.3 Relation of Se to other trace elements and soil parameters

In an attempt to assess Se data against other trace elements and soil parameters from the GEMAS project survey, principal component analysis (PCA) was carried out with the aim to establish interrelations and associations. PCA is a multivariate statistical technique that has been widely used in regional geochemistry to identify key variables within a data set and to determine how variables inter-relate (Grunsky & Smee 1999; Cheng et al. 2006; Reimann et al. 2008). PCA calculates eigenvectors (or 'principal components') which are loadings that describe the relative significance of a component (e. g., trace element) and its geochemical variability within a data set (for more detailed explanations on PCA, see Chapter 9 → Part A).

Figure 7.8 shows a series of biplots, three for each Ap and Gr soil samples, comparing the first three principal components (PC). Principal component 1 represents 25.2 % and 26.1 % of the total variance in Ap and Gr soil, respectively, followed by PC2 and PC3 with 17.2 – 14.3 % and 12 – 13.1 %, respectively.

PC coefficients for Se are generally low and negative in case of PC1 (Gr) and PC2 (Ap and Gr).

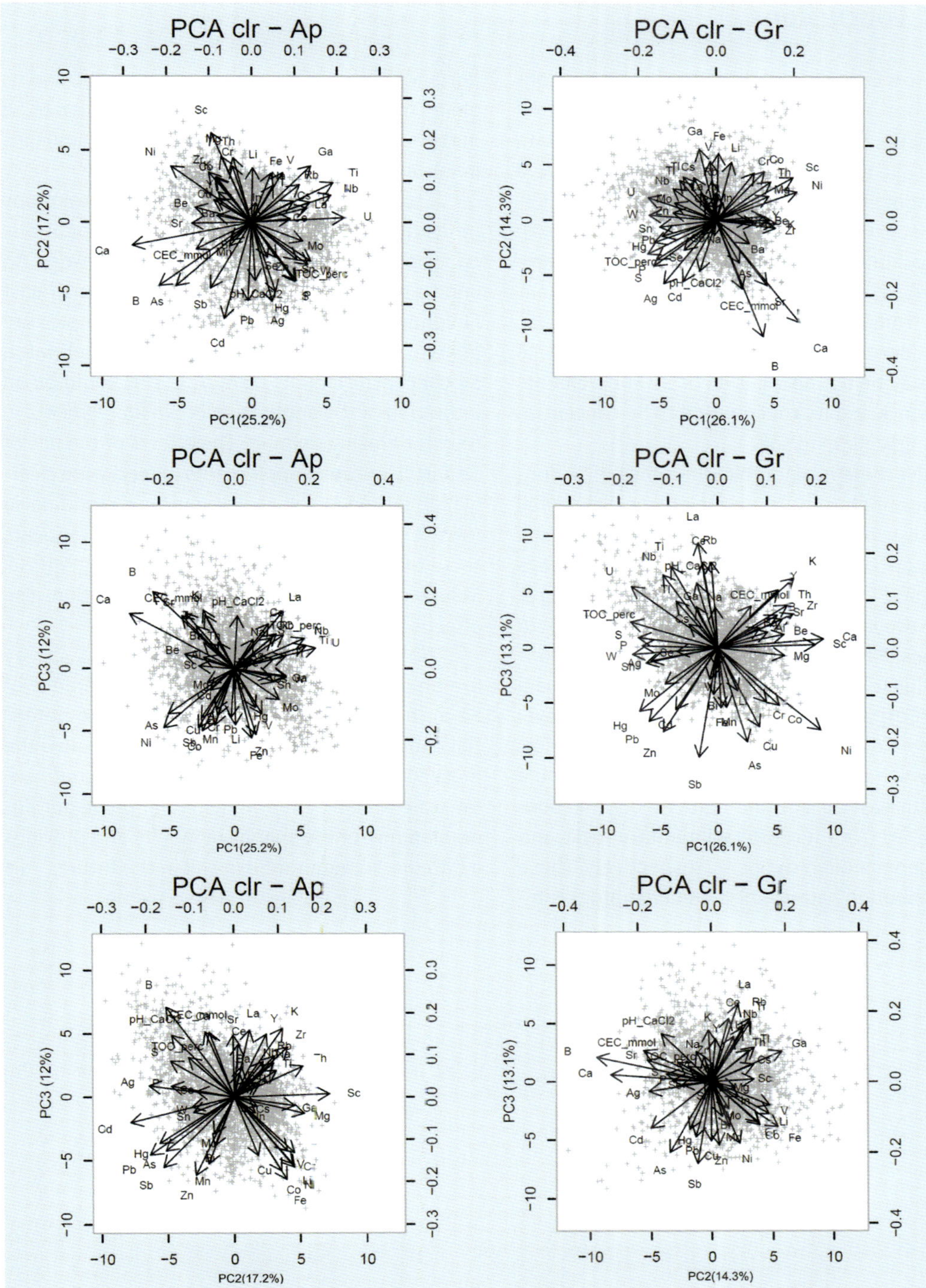

Fig. 7.8: Biplots displaying principal components (PC) 1, 2 and 3 of agricultural (Ap; left) and grazing land (Gr; right) soil data; percentage values in brackets of axis labels refer to the variance of each PC.

Overall, Se coefficients seem to be associated with those of total C, TOC, S and P in both Ap and Gr soil samples (Fig. 7.8). These results are clear evidence for selenium's chemical properties and biological interactions with S and P as well as its association with soil organic matter. Thus, the bulk of Se is bound to organic matter in both soil types.

Figure 7.8 also demonstrates a weak correlation between Se and Hg, Zn and Ag, which suggests a natural origin for some of the Se.

▸ 7.5 Conclusions

Across the European continent, median Se concentrations in Ap and Gr soil samples are 0.35 and 0.40 mg/kg, respectively. Evaluation of Se data shows that the two different land uses have a substantial effect on the Se concentrations in European soil. On average, Se concentrations in Ap are approximately 13 % lower than in Gr soil samples, a result of depletion in Ap soil through plant uptake from crops, as well as the higher organic carbon content in grazing land soil. An additional factor may be the application of Se-supplements for livestock and its contribution to Se concentrations in grazing land soil.

Whilst there is a notable difference in Se concentrations between the two soil types, the overall spatial distribution patterns of Se in Ap and Gr soil samples are broadly similar. The geochemical maps of high and low Se concentrations across Europe, present two large continental-scale trends for both soil types. Firstly, Se-high soil is most abundant along the coastal areas of western Europe as a result of organic-rich soil and sea spray. The second, large-scale pattern can be observed in Se-low areas, which appear across the Iberian Peninsula and north-east Europe and relate to organic-poor soil derived from Tertiary sedimentary rocks and Pleistocene aeolian deposits, respectively. These two distribution patterns also demonstrate the importance of the soil organic matter content, and its ability to enhance binding and accumulation of Se in the soil matrix. In fact, most high Se anomalies in Europe coincide with organic-rich soil.

Across the two soil types, geochemical maps indicate only two areas where the regional distribution of Se substantially differs between Ap and Gr soil. In north-east Germany and north-east Finland, Se concentrations in Ap soil samples are generally low whilst Gr soil samples have elevated Se concentrations > 1.05 mg/kg. This could be an indication of the application of Se-supplements in feed for grazing animals. The effects of the use of Se-fertilisers and supplements in feed were clearly highlighted in soil across southern Scandinavia (Sweden and Finland). Whilst agriculture appears to be the dominant source for anthropogenic Se, industrial processes, such as coal combustion, also affect soil in Germany and UK.

The Se data provided by the GEMAS survey are an invaluable source for the assessment of Se in the environment and research into potential deficiency and toxicity across Europe. It is the most comprehensive data set for Se at a continental-scale to date, and will help to improve our understanding of the spatial variability and provides a baseline to monitor Se in soil in the future. |◂

Michal Poňavič
Czech Geological Survey (CGS);
Praha, Czech Republic
E-mail: michal.ponavic@geology.cz

Andreas Scheib
Geological Survey of Western Australia (GSWA);
Perth, Western Australia

Chapter 8

Natural Radioactive Elements Uranium, Thorium and Potassium in European Agricultural and Grazing Land Soil

Domenico Cicchella, Stefano Albanese, Manfred Birke, Benedetto De Vivo, Walter De Vos, Enrico Dinelli, Annamaria Lima, Patrick O'Connor, Ignace Salpeteur & Timo Tarvainen

▸ 8.1 Introduction

Uranium (U), Thorium (Th) and Potassium (K) are the principal naturally occurring radioactive elements on the Earth. Uranium has three naturally occurring isotopes, ^{238}U, ^{235}U and ^{234}U, with an abundance at 99.284 %, 0.711 % and 0.005 % of its total mass, and half-life of 4.468 Ga, 701 Ma and 269,000 years, respectively (Magill et al. 2006).

Thorium-232 is the only naturally occurring Th isotope with a half-life of 14.05 Ga (Magill et al. 2006).

Potassium has three naturally occurring isotopes in nature. Two are stable (non-radioactive), ^{39}K and ^{41}K. Potassium-39 comprises most of naturally occurring potassium (about 93 % of its total mass), and ^{41}K accounts essentially for the remainder. Potassium-40 is a minor naturally occurring radioactive isotope of K (0.01119 % of total K), with a half-life of 1.28 Ga (Magill et al. 2006). It decays to ^{40}Ca by emitting a beta particle with no attendant gamma radiation (89 % of the time), and to the gas ^{40}Ar by electron capture with emission of an energetic gamma ray (11 % of the time). Potassium-40 is an important radionuclide in terms of the dose associated with naturally occurring radionuclides.

Daughter products of the U and Th decay-series, include radon (Rn) and radium (Ra), and their presence and relative abundance has implications for human and animal exposure-related health hazards. Therefore, concerns about the effect of these radioactive elements on the environment and human health have increased in the last few decades (Lima et al. 2005). Weathering of bedrock, the main reservoir of natural radionuclides, releases U, Th, and K to soil. Daughter products of the U and Th decay series also take part in this transfer. From soil, plants take up K, Ra, small amounts of U and extremely small quantities of Th (Morton et al. 2001). Radium, an important long-lived decay product of ^{238}U (the half-life of ^{226}Ra is 1600 years), is taken up by the plant, not due to being an isotope of an essential element, but because it is an alkaline earth metal chemically similar to Ca, which is essential. The uptake of U and Th by plants is usually small or negligible, since these radionuclides normally are relatively insoluble.

Taking into account that many rock types have high K content, the contribution of ^{40}K may be an important component of the total radioactivity to which humans are exposed. Plants utilise ^{40}K in the same manner as the essential major element, stable K. Many plant physiologists consider potassium second only to nitrogen in importance for plant growth (Chapter 1). It is a key nutrient of plant tolerance to stresses, such as cold/hot temperatures, drought, wear and pest problems. In plants, potassium plays a key role in the regulation of water use (osmoregulation), and acts as a catalyst for many of the enzymatic processes that are necessary for plant growth to take place.

▸ 8.2 Geochemical features

Uranium (U), Thorium (Th) and Potassium (K) are lithophile elements; they are strongly partitioned in the Earth's continental crust, and their average content varies as a function of rock type. Felsic rocks usually contain more K, Th and U than mafic rocks, and in sedimentary rocks, these elements are likely to be more concentrated in argillaceous lithologies, particularly black shale, than in sandstone and limestone (Table 8.1).

The lowest concentrations of U and Th are in ultramafic and mafic rocks, such as ophiolite. Uranium and Th are present in trace amounts in all crustal rocks, and they are highly concentrated in certain types of mineral deposits.

Uranium exists in the +4 and +6 oxidation states in most geological environments, whereas Th mostly occurs in the +4 oxidation state. Both elements are known to replace rare earth elements (REE) in some minerals, and to occur in trace amounts in apatite and sphene. Tetravalent thorium is readily soluble, and also quickly adsorbed or precipitated as hydrolysates. Consequently, the concentration of dissolved Th in natural water is quite low, and it is transported in the fluvial environment almost exclusively in particulate form (Andersson et al. 1995). The tendency of Th to be retained in the soil column can be used to infer the quantity of rock weathered to produce a given thickness of regolith (Rosholt 1982; Osmond & Ivanovich 1992). Tetravalent thorium forms solid solutions with Ce and Zr in resistate minerals, such as monazite and thorite, which become concentrated in heavy-mineral sand and placer deposits. In contrast, U, in the weathering environment, is much more soluble and mobile in oxic natural water.

According to Deneux-Mustin et al. (2003), the behaviour of U in soil is very different according to its oxidation state, i.e., hexavalent uranium (U^{6+}) in its uranyl cation (UO_2^{2+}) is the more mobile form, whereas U^{4+} is nearly immobile at room temperature. In neutral to acid conditions, U^{6+} is stabilised by argillic-humic complexes (Wood 1996). Under oxidising conditions, with low organic content, U behaves as most of the other cations (Ca, Mg, K, Na) and is leached. This is the case for tropical lateritic soil or brown Mediterranean soil. In the presence of dissolved CO_2, the solubility of U is increased by the formation of uranyl carbonate complexes. However, U tends to form strong associations, mainly with organic material and with dissolved phosphates. Some authors (Meunier 1992; Disnar & Sureau 1989) discuss the mechanism of concentration of uranium by organic matter, i.e., either direct or indirect reduction by H_2S-producing bacteria, and either by complexation or adsorption. According to Szalay (1964), quoted by Wood (1996), a geochemical enrichment factor of 100,000 (with respect to U in solution) was observed in humic peat-derived material. Under arid conditions, U forms several relatively stable compounds (e.g., oxides, carbonates, phosphates, vanadates, and arsenates).

Potassium is a major constituent of many rock-forming minerals, and is enriched in felsic relative to mafic igneous rocks. In sedimentary systems, K_2O may be fixed in authigenic illite or adsorbed onto the surfaces of other clay minerals (Table 8.1). Potassium is highly soluble in surface water under a wide range of pH and Eh conditions. However, sorption by clay minerals limits its mobility.

Table 8.1: Average radioelement contents in soil, water, sediment and rock in mg/kg, except where otherwise stated.

Natural material	U	Th	K
	[mg/kg]		
Soil [1]	2.7	9.4	19,700
Subsoil [2]	2.03	7.63	20,200
Topsoil [2]	2	7.24	19,200
	[µg/L]		
Water [2]	0.32	0.009	1600
Water [3]	0.5	< 0.001	2300
	[mg/kg]		
Stream sediment [2]	2	10	20,100
Floodplain sediment [2]	2	8	20,000
Upper Continental Crust [4,5]	2.6	10.5	23,244
Ultramafic [1,5,7]	0.001	0.003	5000
Basalt [1,6,7]	0.6	2.2	8000
Granodiorite [1,6,7]	3	10	33,000
Andesite [6]	2	-	-
Granite [1,6,7]	4.8	17	33,000
Shale [1,6,7]	4	12	27,000
Sandstone [1,6,7]	2.2	-	11,000
Limestone [1,6,7]	2	2	3000

Sources:
(1) Koljonen (1992) (2) de Vos & Tarvainen (2006) (3) Ivanov (1996) (4) Hu & Gao (2008)
(5) Rudnick & Gao (2003) (6) Taylor (1964, 1966) (7) Weaver & Tarney (1984)

▸ 8.3 Health effects

8.3.1 Uranium

Uranium is present in the human body at very low or trace concentrations, and is not known to be an essential element. Human intakes are constant through very small amounts of natural uranium in food and water, and even smaller amounts in air. The daily intake of uranium from food sources ranges from 0.6 to 1.0 pCi/day (0.9 – 1.5 µg/day) (ATSDR 2011). Uranium from soil is not taken up by plants, but rather is adsorbed onto the roots. Thus, the highest levels of uranium are found in root vegetables, primarily unwashed potatoes. People, living near uranium mills, mines, or other areas with elevated uranium in soil, may be exposed to higher levels of uranium from locally grown vegetables. Uranium levels in drinking water vary widely, with a mean population-weighted average of 0.8 pCi/L. Compared to the ingestion route, the intake of uranium via inhalation is small; intakes range from 0.0007 to 0.007 pCi/day (0.001 – 0.01 µg/day) (ATSDR 2011).

The following anthropogenic activities increase the potential for human exposure to uranium, i.e., mining, milling, and handling uranium, including depleted uranium products; processing uranium ore end products (uranium dioxide, uranium hexafluoride); producing nuclear energy and nuclear weapons; producing phosphate fertilisers from phosphate rocks that contain much higher than average levels of uranium, and improper disposal of wastes. Occupational exposure to airborne uranium ore dust occurs in uranium mines and mills and in processing plants.

The chemical action of all isotopes and isotopic mixtures of uranium are identical, regardless of the specific activity, because chemical action depends only on chemical properties. Thus, the chemical toxicities of natural, depleted, and enriched uranium are identical. Current evidence from animal studies suggests that the toxicity of uranium is mainly due to its chemical damage to kidney tubular cells, leading to nephritis. Other sensitive targets of toxicity include the respiratory tract (inhalation only), neurological system, reproductive system, and the developing organism

Since uranium is weakly radioactive, it has been assumed to be potentially carcinogenic at occupational levels by NIOSH (the *USA National Institute for Occupational Safety and Health*). The International Agency for Research on Cancer (IARC) has no classification for uranium. Although significant increases in the occurrence of respiratory tract cancer (predominantly lung cancer) have been found in numerous studies of uranium miners, radon progeny in the mines, and not the uranium itself, were clearly identified as the carcinogenic agents. Studies of workers at uranium milling or nuclear facilities, and residents living near uranium mining and milling facilities, have not found significant increases in cancer mortality associated with uranium exposure. Most studies of animals exposed to soluble or insoluble uranium compounds via inhalation or oral exposure did not report increases in the incidence of malignant tumours. However, increases in pulmonary neoplasms were observed in dogs exposed to uranium dioxide for 1 – 5 years, and rats exposed to uranium ore (containing 44 % uranium) for 65 weeks (ATSDR 2011). The investigators for the uranium dioxide study suggested that the increased incidence of lung tumours might have been the result of radiation exposure.

8.3.2 Thorium

Thorium is a relatively reactive, metallic radioactive element. Because thorium is a radioactive element, evaluation of adverse health effects, due to exposure to thorium, requires a slightly different approach than with chemicals. Radiation is a health risk, because radioactive elements can emit energetic particles or electromagnetic radiation that can damage cells. Very little data exist on health effects due to inhalation, oral, or dermal exposure of thorium in humans or animals. Hazards of thorium or its compounds may generally be of two types: radiological and chemical. The radiological hazards of thorium are connected essentially with the radioactivity of the isotopes in the decay chain from ^{232}Th to ^{208}Pb. Ordinary chemical toxicity of thorium or its compounds by means of introduction into the body is generally considered low

The existing health effects data suggest that thorium may pose a potential health threat to an exposed population. Some evidence of respiratory disease and increased incidence of pancreatic, lung, and haematopoietic cancers in humans was found following inhalation exposure (Archer et al. 1973; Polednak et al. 1983; Stehney et al. 1980). These effects were observed in thorium workers exposed to many toxic agents, so that the effects cannot be attributed directly to thorium exposure, but a causal effect cannot be discounted.

The majority of human thorium data deal with the effects of Thorotrast (colloidal thorium-232 dioxide) administered intravenously to patients as a radiological contrast medium. The literature suggests that the toxic effects of Thorotrast are due to the alpha radiation effects of thorium, and not to the chemical effects of thorium or of the colloid (BEIR IV 1988; Taylor et al. 1986). Cirrhosis of the liver, hepatic tumours, and blood dyscrasias were the most common effects of intravenously injected Thorotrast.

8.3.3 Potassium

Potassium is an important element for the proper function of all cells, tissues, and organs in the human body. It is also an electrolyte, a substance that conducts electricity in the body, along with sodium, chloride, calcium, and magnesium. Potassium is crucial to heart function, and plays a key role in skeletal and smooth muscle contraction, making it important for normal digestive and muscular function. It is involved in the storage of carbohydrates for use by muscles as fuel. Human muscle and nerve cells have specialised channels for moving potassium in and out of the cell. Sometimes potassium moves freely in and out, and occasionally a special energy-driven pump is required (Griffith 2001). When the movement of potassium is blocked, or when it is deficient in the diet, both muscles and nerves are unable to function normally, and a variety of muscular symptoms may occur including weakness, spasms, twitching and cramps. In cases of extreme hypokalemia (K deficiency), the muscles can go into a sustained involuntary state of contraction called tetany.

Having too much potassium in the blood is called hyperkalemia. Elevated blood levels of potassium can be toxic, and may cause an irregular heartbeat or even heart attack. Under most circumstances, the body maintains blood levels of potassium within a tight range, so it is not usually possible to produce symptoms of toxicity through intake of potassium-containing foods and/or supplements (Table 8.2). However, high intakes of potassium salts (potassium chloride and potassium bicarbonate) may cause nausea, vomiting, diarrhoea, and/or ulcers.

Keeping the right potassium balance in the body depends on the amount of sodium and magnesium in the blood. Too much sodium may increase the need for potassium. Diarrhoea, vomiting, excessive sweating, malnutrition, malabsorption syndromes (such as Crohn's disease) can also cause potassium deficiency, as well as use of a kind of heart medicine called loop diuretics.

In addition, the kidneys play an important role in eliminating excess potassium from the body, so older people have a greater risk of hyperkalemia, because their kidneys are less efficient at eliminating potassium as they age. Hence, they must limit their intake of potassium (Table 8.3).

▸ 8.4 Uses

8.4.1 Uranium

Before the advent of nuclear energy, uranium had very limited uses. It was only used in filaments for lamps. Compounds of uranium have been used in photography for toning, and in the leather and wood industries for stains and dyes. Uranium salts are mordants of silk or wool. In the past, uranium salts have been used to produce coloured ceramics and glasses. A 2000 year old sample of yellow glass found near Naples (Italy) contains uranium oxide. Additionally, uranium was used in dental porcelains for many years, but this practice has been discontinued.

By far the most important use of uranium lies in its application for nuclear (or atomic) energy. The ^{235}U isotope is used in nuclear reactors, as it is fissile and, therefore, capable of sustaining a nuclear chain reaction in the presence of energetic neutrons. A process known as enrichment is used to isolate the ^{235}U isotope from uranium ore for

Table 8.2: Potassium rich foods (USDA, National Nutrient Database for Standard Reference).

Food	K content per 100 gram serving [g]	percent Daily Value [% DV]
Sun dried tomato	3.4	98
Paprika	2.3	67
Whey powder	2.1	59
Chili powder	1.9	55
Dried apricot	1.9	53
Cocoa powder	1.5	44
Rice bran	1.5	42
Molasses	1.5	42
Prune	1	30
Pistachio	1	30

Food	K content per 100 gram serving [mg]	percent Daily Value [% DV]
Chestnut	960	28
Pumpkin	919	26
Raisin	825	24
Sunflower seed	850	24
Flax seed	813	23
Medjool date	696	20
Whelk	694	20
Dried fig	680	19
Watermelon seed	648	19
Salmon	636	18
Clams	628	18
Beans	561	16
Spinach (raw)	558	16
Baked potato (with skin)	535	15
Avocado	485	14
Sweet potato (with skin)	475	14
Brussels sprout (raw)	389	11
Banana	358	10
Coconut water (juice)	250	7
Yogurt (plain, low fat)	234	7
Orange juice	200	6
Palm heart	177	5

Table 8.3: Minimum daily intake of potassium from dietary sources (recommended Dietary Allowances 1989).

Age	Adequate K intake
0 – 5 months	500 mg
6 – 11 months	700 mg
1 year	1 g
2 – 5 years	1.4 g
6 – 9 years	1.6 g
10 – 18 years	2 g
> 18 years	2 g
Pregnant women	5.1 g
Lactating women	5.1 g

use as fuel in nuclear reactors. The enrichment process produces enriched uranium (uranium hexafluoride with enhanced ^{235}U concentration), and depleted uranium (uranium hexafluoride with reduced ^{235}U concentration). The enriched uranium is more radioactive, and the depleted uranium is less radioactive than natural uranium (Weigel 1983).

Uranium has been produced for use in the commercial nuclear power industry as low-enriched metal or ceramic UO_2 fuel pellets; smaller quantities of high-enriched fuel are produced for USA Navy ships and for weapons manufacture (ATSDR 2011 and references therein). One pound of completely fissioned uranium produces the same amount of energy as 1500 tonnes of coal. Depleted uranium is used in the manufacture of armour-piercing ammunition for the military, in inertial guidance devices and gyro compasses, as counterbalances for helicopter rotors, as counterweights for aircraft control surfaces, as radiation shielding material, and as X-ray targets.

8.4.2 Thorium

Thorium may be used as fuel in the generation of nuclear energy, and in the production of core-retention beds used in nuclear reactors to contain and possibly diffuse heat generated by accidental core meltdown. The non-energy uses include refractory applications, lamp mantles, aerospace alloys, welding electrodes nuclear weapon production, and other applications including ceramics and special lighting. Specific applications include production of investment moulds for casting high-temperature metals and alloys, crucibles and alloys of special shapes for use in high-temperature vacuum or oxidising furnaces. Other special applications include magnesium-thorium alloys for strategic aircraft, such as military jet fighters and bombers; mantles for incandescent lanterns, such as those used on camping trips; thoriated tungsten electrodes used to join stainless steel and other alloys, which require controlled weld applications; special lighting, such as airport runway lighting; computer memory components; photoconductive film, and target material for X-rays. Natural thorium is also used in ceramic tableware glaze and in flints for lighters.

Table 8.4: World Mine Production and Reserves of potash in thousand metric tonnes of K_2O (USGS 2010).

Producing country	Mine production		Reserves
	2008	2009	
United States	1100	840	90,000
Belarus	4970	3850	750,000
Brazil	471	500	300,000
Canada	10,500	6500	4,400,000
Chile	559	600	10,000
China	2750	2750	200,000
Germany	3280	2300	710,000
Israel	2300	2000	40,000
Jordan	1220	1100	40,000
Russia	6730	3600	1,800,000
Spain	435	400	20,000
Ukraine	11	10	25,000
United Kingdom	427	400	22,000
Other countries	–	–	50,000
World total (rounded)	35,000	25,000	8,500,000

8.4.3 Potassium

Potassium is an element widely used by humans (Table 8.4). The main use of potassium globally is in fertilisers. As potassium ions are a vital part of plant nutrition, crops and trees must be grown in soil with high concentrations of potassium ions. Many soil types have been depleted of available and slowly available K by decades or even centuries of continuous farming with crop removal exceeding inputs. In any given year, the major field crops grown in Europe, including hay and forage, often remove many times more K than is applied. Potash refers to a variety of K-bearing minerals that are used for fertilising in agriculture, horticulture and hydroponics. It includes potassium chloride (KCl, or muriate of potash (MOP)), potassium sulphate (K_2SO_4, or sulphate of potash (SOP)), potassium-magnesium sulphate ($K_2SO_4 \times MgSO_4$, or sulphate of potash magnesia), potassium nitrate (KNO_3, or saltpetre), and mixed sodium-potassium nitrate ($NaNO_3 + KNO_3$, or Chilean saltpetre). Potassium chlorate is also used for killing weeds. Another compound of potassium, potassium bisulphate, is used to preserve foods (except meat), wine and beer. It can also be used to bleach textiles and leather, and to coat glass with silver. Toughened glass (stronger than ordinary glass) is made using potassium nitrate. Potassium hydroxide, as strong base, is used in industries and science laboratories to neutralise acids and make potassium salts. It is also added to fat and oil to make soap. Potassium cyanide is used in gold mining.

▶ 8.5 Results and discussion

Summary statistics of the U, Th and K data for the samples collected in the GEMAS project are given in Table 8.5. For statistical computation, the element concentrations below the instrumental detection limit (IDL) have been assigned a value corresponding to 50 % of their detection limit. To produce the colour surface maps, kriging was used to convert the values from the irregularly distributed sampling sites to a regular grid. Class limits used for the colour maps are based on percentiles. In this section, the analytical data on U and Th concentrations, determined by ICP-MS, and on K measured by XRF analysis are discussed. Indeed, as tabulated in Table 8.5, most of the samples analysed by XRF show U and Th concentrations lower than the laboratory detection limit (LDL).

8.5.1 Uranium

The aqua regia extractable U concentrations over the survey area vary from < 0.1 to 23.55 mg/kg in agricultural soil (Ap) and from < 0.1 to 73.32 mg/kg in grazing land soil (Gr), with a median value of 0.77 and 0.74, respectively (Table 8.5).

Table 8.5: Summary statistics of agricultural soil (Ap samples, 0 – 20 cm, < 2 mm fraction) and grazing land soil (Gr samples, 0 – 10 cm, < 2 mm fraction).

	Agricultural soil (Ap)						Grazing land soil (Gr)					
Anal. method	Aqua regia (ICP-MS)			XRF			Aqua regia (ICP-MS)			XRF		
N	2218			2218			2127			2127		
Element	U	Th	K	U	Th	K	U	Th	K	U	Th	K
	[mg/kg]			[mg/kg]			[mg/kg]			[mg/kg]		
LDL	0.1	0.1	100	2	3	50	0.1	0.1	100	2	3	50
N < LDL [%]	0.33	0.71	0.33	63.5	24.3	0	0.3	1.19	0.79	77	14	0
	[mg/kg]			[mg/kg]			[mg/kg]			[mg/kg]		
Minimum	< 0.1	< 0.1	40	< 2	< 3	241	0.05	0.03	43	< 2	< 3	241
Q5	0.30	0.35	230	< 2	< 3	5710	0.25	0.24	198	< 2	< 3	4560
Q10	0.37	0.67	342	< 2	< 3	7620	0.34	0.45	295	< 2	< 3	6620
Q25	0.52	1.59	691	< 2	5	11,500	0.51	1.17	608	< 2	7	10,500
G-mean*	0.82	2.44	1154	1.64	6.40	14,502	0.80	2.02	1044	1.37	8.45	13,315
Median	0.77	2.89	1250	1	9	16,000	0.74	2.50	1126	< 2	11	14,900
Q75	1.20	4.62	2121	3	12	19,800	1.18	4.34	1965	< 2	14	19,100
Q90	2.17	6.29	3138	5	16	24,300	2.21	6.03	3116	4	18	23,820
Q95	3.36	7.94	3884	6	18	27,800	3.39	7.46	3948	5	20	27,960
Q98	5.25	10.46	5099	8	22	34,000	5.35	10.35	5392	8	25	34,088
Maximum	23.55	63.10	26,739	25	84	79,200	73.32	55.64	27,957	94	80	50,000
MAD	0.29	1.51	667	0	4	4200	0.29	1.52	623	0	4	4300

* Geometric mean
N Number of samples
LDL Laboratory Detection Limit
N < LDL percent of samples with concentration < LDL
Q Quantiles
MAD Median Absolute Deviation

Both distribution maps (Ap and Gr) show a very similar pattern (Figure 8.1).

Relatively low U values (< 0.51 mg/kg) occur in the glacial drift zone, extending from northern Ukraine to The Netherlands, where generally Quaternary soil of sandy affinity cover the underlying bedrock (PLANT et al. 2003a). Other sandy sedimentary basins show low background values for uranium, e.g., the London basin, the Garonne basin (south-west France) southern Spain and Portugal (Guadalquivir and Guadiana basins), and central Hungary (Danube basin).

Anomalous uranium areas (> 3.25 mg/kg) are found in the western Iberian peninsula related to the granitic masses that were intruded during the last stages of deformation of the Variscan orogeny (Fig. 8.2). The lithology of this area is comprised mostly of leucogranite, calc-alkaline granite that contains accessory uraninite and generally has a lesser amount of Th.

Other anomalous values occur in Italy over alkaline volcanic rocks of the Roman Neapolitan Province, in correspondence with mineralisation in metamorphic terrains of the eastern Alps, and in the north-eastern part of Sardinia (where they are probably associated with a monazite-bearing ore deposit, which may represent a likely source for both U and REE.

The high uranium values in north-eastern Greece, south Bulgaria and eastern FYROM are related to granitic intrusions of the Rhodope Massif of Late Cretaceous to Palaeogene age, which are known to host uranium mineralisation. The U anomalies in Croatia and along the Adriatic coast are probably associated with clay enrichment in terra soil on karst areas; in a circumneutral pH, this geochemical pattern is observed for many clay-related elements. The anomalous U concentrations in Slovenia are probably caused by agricultural influences (use of fertiliser).

Uranium enrichment also occurs in soil on the crystalline basement of the Massif Central (France), where several U deposits were exploited, and are related either to granitic cupola and

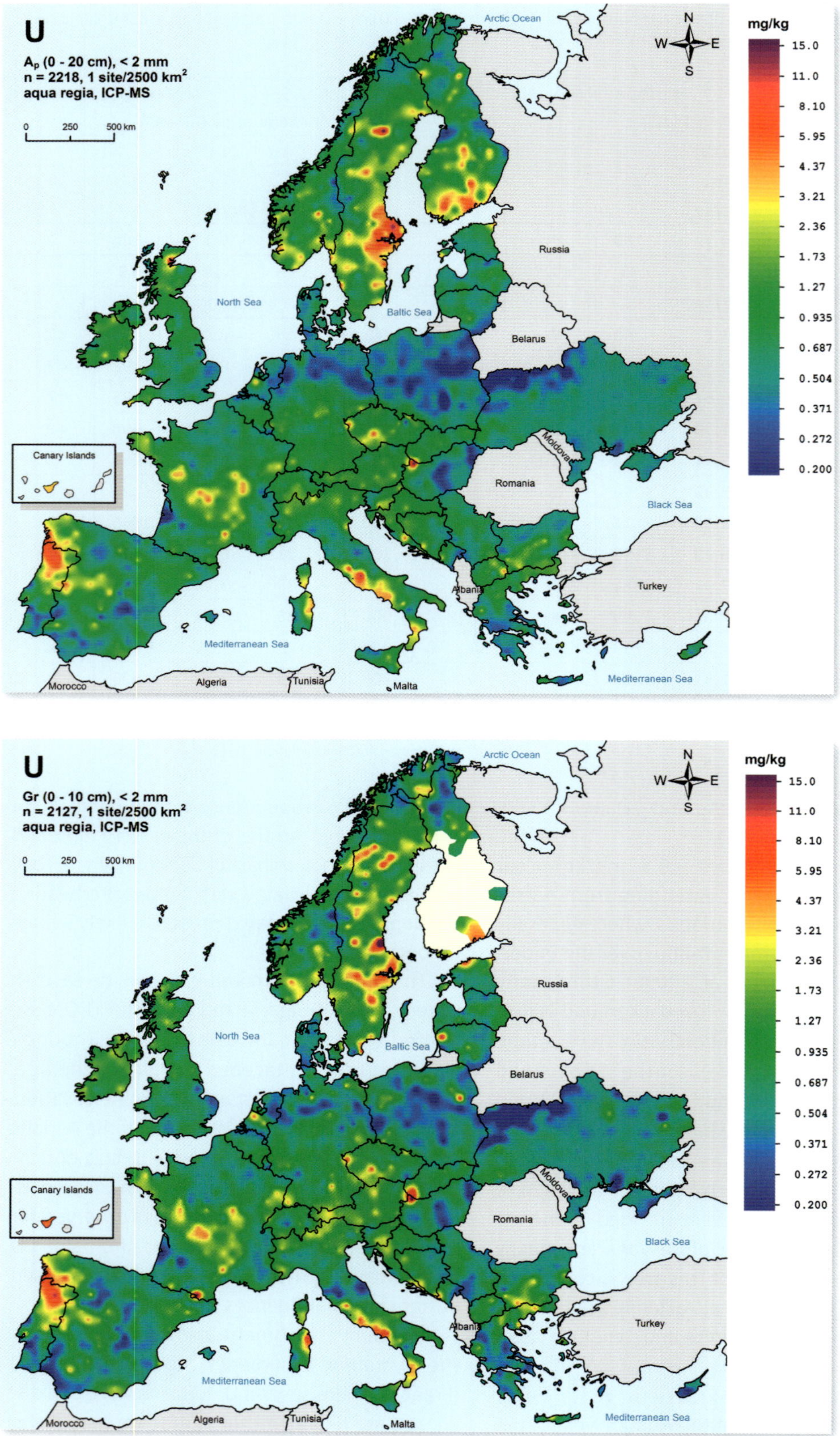

Fig. 8.1: Spatial distribution of aqua regia extractable uranium (U) in agricultural (top) and grazing land (bottom) soil of Europe.

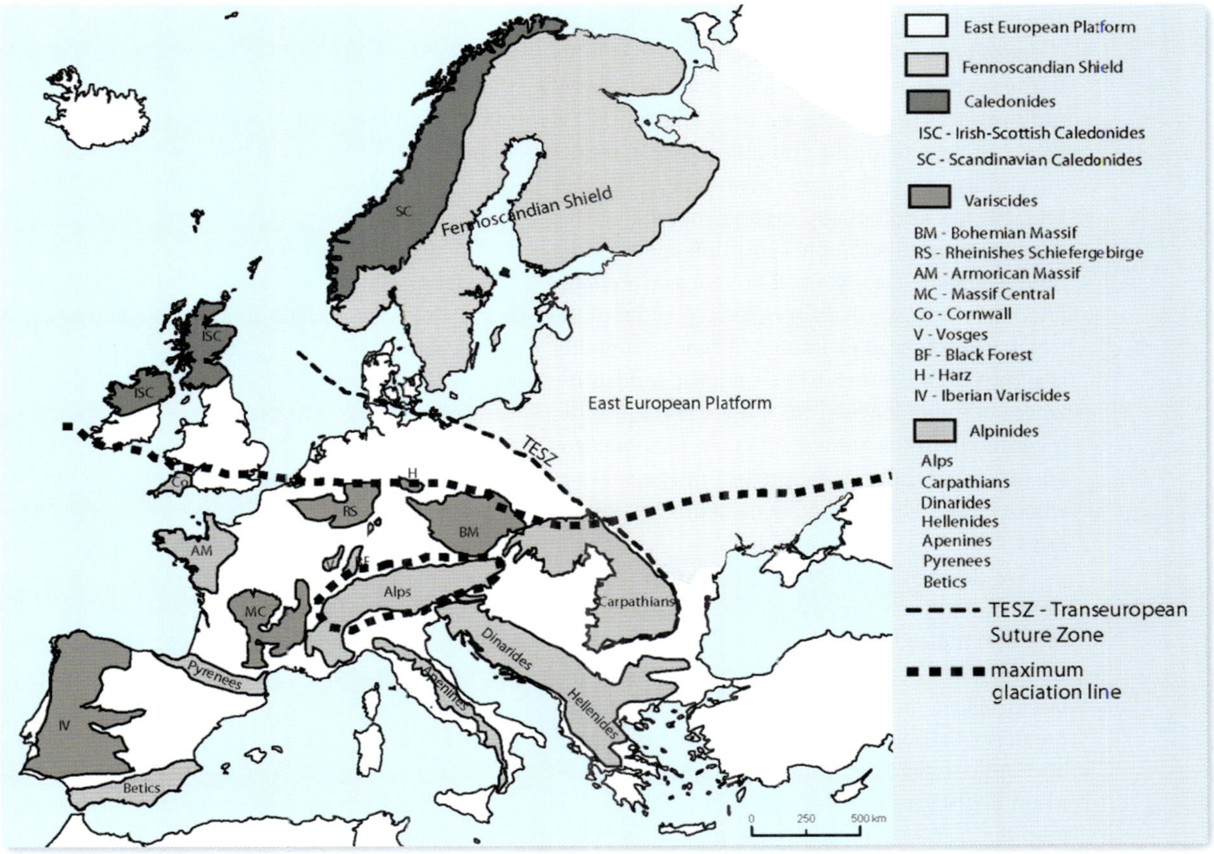

Fig. 8.2: Generalised geological terrane map of Europe. For a detailed geological map of Europe see Asch (2003) and also Chapter 2.

pervasive hydrothermal secondary alteration (episyenite type) or to vein type deposits. In the Vigan area (south-east Massif Central), karstic filling in Cambrian dolomite is rich in organic and phosphatic materials, and is strongly enriched in uranium and LREE. A slight enrichment (between 1.16 and 3.25 mg/kg) can be observed in soil developed on the crystalline rocks of Brittany, the Vosges granite at the border between France, Switzerland and Germany and the shale lithologies in the Variscan Ardennes-Rhenish Massif straddling the Belgian-German border. Anomalous U values (>2.12 mg/kg) are also found in the crystalline Bohemian Massif (Czech Republic), the Scottish Highlands (Caledonian granite) and the Variscan granite of south-west England.

Moderately elevated uranium concentrations (between 1.16 and 2.12 mg/kg) occur in Ap and Gr German soil samples over the crystalline rocks in the Ore Mountains, the southern Black Forest, the Fichtelgebirge, the Bavarian Forest (Bodenmais), and the Upper Palatinate. The granitic and metamorphic rocks are all enriched in the radio-elements U, Th and K, and the hydrothermal vein mineralisation in these districts is locally uranium-bearing (the Bi-Co-Ni-Cu-Ag-U high to medium temperature quartz vein association). The elevated concentrations in the areas of the Saar-Nahe depression and the Thuringian Forest are caused by uranium mineralisation in the Rotliegend sediments as well as in the volcanic rocks. The glacial moraines in the subalpine area in south Germany (loess mixed Tertiary deposits), and the area of Keuper sandstone south of Coburg, are also mapped by regionally slightly elevated uranium levels.

In northern Estonia, a black shale lithology (Dictyonema shale) causes a strong U anomaly. This formation of Cambrian-Ordovician age is particularly enriched in organic matter, pyrite, U, Mo, V and other elements (Petersell 1997). In this area, major phosphorite deposits are located under the Dictyonema shale, which are quite enriched in U, F, Sr and REE (Raudsep 1997).

The anomaly in the grazing land soil of Lithuania is **pedogenic**, and related to a geochemical redox barrier in peat land with shallow and changeable groundwater level. Hexavalent uranium (U^{6+}) is very mobile in oxidising conditions, but in reducing conditions, it changes its valence to U^{4+} and precipitates. Additionally, organic material at this site serves as an additional sorption agent. Some other elements (Se, Hg, Pb, P, Cu) are also accumulated here on a so-called geochemical redox-sorption barrier. A good example of uranium mobility is shown in the Permo-Triassic arkosic sandstone, where uranium leached from the palaeo-alterite, inherited from the granitic rocks (Permian palaeosurface), has been transported in the arkose, where it has been trapped in palaeoredox boundaries.

Sweden has some of the highest concentrations of U in soil in Europe, directly related to the fact that it also hosts some of Europe's most important U ore deposits. Some of the highest U values in soil are not, however, associated with known ore deposits, but with younger granite, pegmatite and radioactive rocks. In southern Sweden, uranium is enriched in the Mälaren region where marine clay is overlying younger granite and pegmatite. The northern part of the anomaly reflects deposits in felsic volcanics, younger granite and pegmatite. An example of known uranium mineralisation that shows up in the dispersion pattern is north of Stockholm, where uranium occurs in skarn mineralisation. In the east and south, uranium occurs in skarn mineralisation and alum shale. Going northwards, elevated concentrations are observed in association with known ore deposits, including Caledonian black-shale deposits and felsic volcanic rocks.

Elevated and anomalous U values are also associated with granite masses of the Precambrian Fennoscandian Shield, as in southern Norway, including the Oslo rift, and in southern Finland, which is characterised by rapakivi granite and other granitic rocks.

8.5.2 Thorium

The median of aqua regia extractable Th content is 2.89 mg/kg in agricultural soil and 2.5 mg/kg in grazing land soil; the range varies from < 0.1 to 63.1 mg/kg in agricultural soil and < 0.1 to 55.64 mg/kg in grazing land soil. Northern Europe, with some exceptions, shows Th contents below the respective median values on both distribution maps, with low values (< 1.64 mg/kg in agricultural soil and < 1.23 in grazing land soil) over the glacial drift-covered area extending from Ukraine to The Netherlands, Sweden, northern Ireland, and Scotland (Fig. 8.3). The southern margin of the last glaciation through The Netherlands, Germany, Poland and Ukraine is well marked on both maps, with low Th values to the north of it, where soil is developed on the coarse-grained sandy sediments of the last glaciation.

Similar to U, Th shows higher background values in soil developed on granitic rocks, and fine-grained sedimentary and metamorphic rocks derived from them. As a result, high Th values in soil appear in the Variscan crystalline basement of the Iberian Massif in Portugal and western Spain, in the alkaline magmatic province of Italy, the Massif Central and lower, but still anomalous contents in soil, over phosphatic chalk in the south-east Paris Basin in France, Slovenia, Croatia, Bosnia, Serbia, FYROM and Bulgaria. A weak Th arcuate anomaly stretches from Lorraine to the northern Massif Central and Vendée in France. It may be related to a Tertiary or Quaternary palaeo-shoreline enriched in heavy minerals (zircon, monazite, etc.). In southern Germany, the elevated Th concentrations mark the crystalline basement rocks (e.g., Thuringian Forest, granite of the eastern Ore Mountains, Bavarian Forest). Elevated Th values in north-eastern Greece are related to granitic intrusions and Caenozoic sediments, and on the Greek islands of Limnos, Lesvos and Patmos they are associated with felsic igneous intrusions.

In Fennoscandia, Th concentrations are highest in clay soil over areas with rapakivi granite and, in general, in areas where Proterozoic bedrock prevails. In Finland, the aqua regia extractable Th patterns show a good correlation with those of K (Fig. 8.4), but not with corresponding total XRF potassium distribution (Fig. 8.5). Aqua regia hardly extracts K from unweathered K-feldspar grains, while K-bearing biotite is leached in hot aqua regia. Clay soil in the Central Scandinavian Clay Belt, extending from southern Finland into the Stockholm region in Sweden, shows, apart from U, Th and K, elevated concentrations in several other

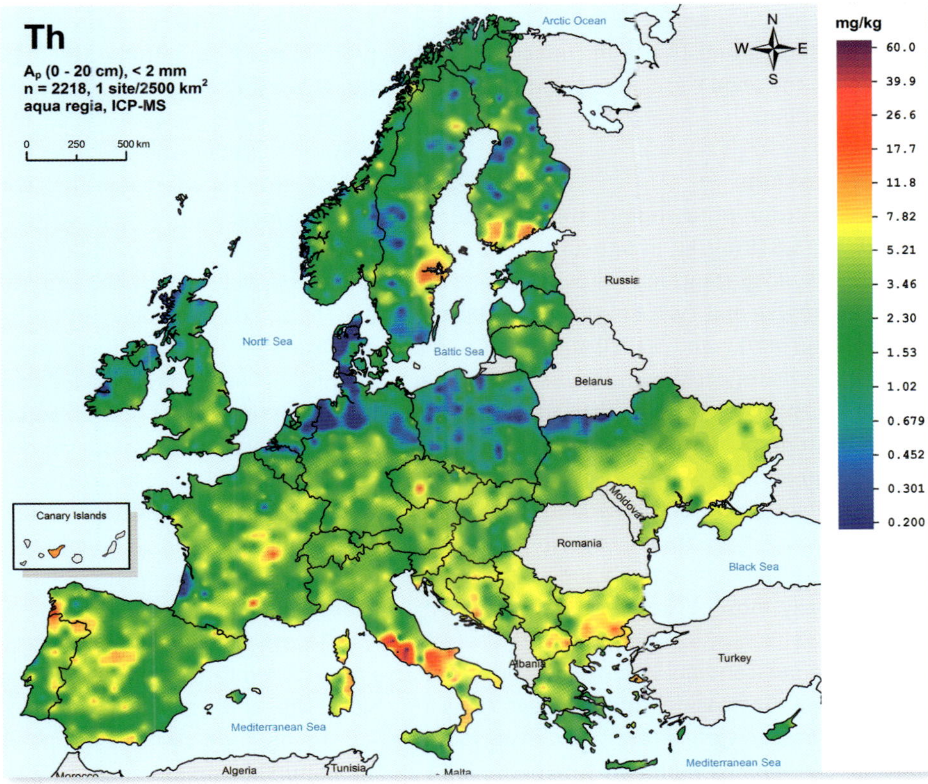

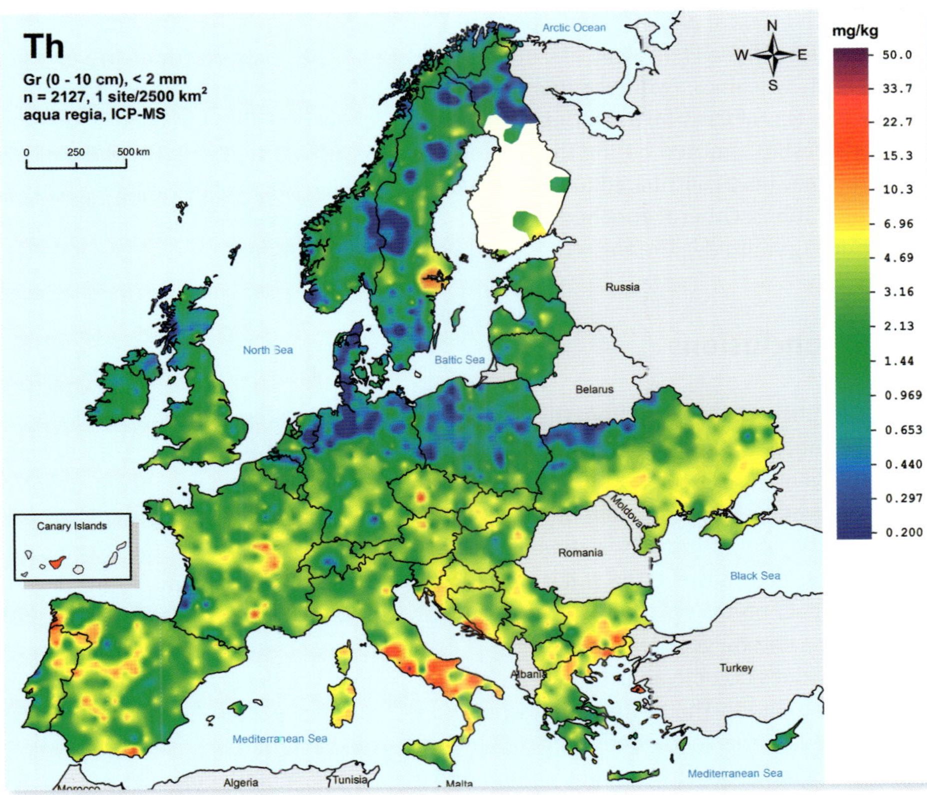

Fig. 8.3: Spatial distribution of aqua regia extractable thorium (Th) in agricultural (top) and grazing land (bottom) soil of Europe.

trace elements (Chapter 11 → Part A). Zircon is one of the most important Th bearing minerals, which can explain the good correlation between zirconium and thorium.

8.5.3 Potassium

The median total potassium content, measured by XRF, is 15,818 mg/kg in agricultural soil and 14,780 mg/kg in grazing land soil, with a range between 241 mg/kg and 79,221 mg/kg in Ap soil and between 240 mg/kg and 50,020 mg/kg in Gr soil. The K patterns reflect the underlying geological diversity of the areas to which they are related.

On both maps (Fig. 8.5), low K values occur over central and southern England, Ireland, the glacial drift area from The Netherlands to Ukraine, northern France, the Garonne basin in south-west France, the calcareous southern and eastern parts of Spain, Sicily, most of Greece, and Cyprus.

High K values (> 24,300 mg/kg) are observed over the Fennoscandian Shield, with concentrations generally increasing from Archaean rocks in the north-east towards Proterozoic rocks in the south-west. High K_2O values are also found in Estonia and Latvia (Devonian clay and basal Quaternary till rich in mica) and Lithuania, parts of the Bohemian Massif, the French Massif Central (granitoids, felsic volcanics, two-mica granite). In the core of the Massif Central, a K influx by rhyolitic phonolitic ashes, emitted during the Quaternary eruptions may have contributed to the anomalous K concentrations. Other K enriched soil areas are in the Alps (granitoids), the western Iberian Peninsula (Central System of the Iberian Massif with granitic and metamorphic lithologies), especially the crystalline areas of northern Portugal, Corsica, Sardinia and the west-Italian alkaline province (felsic and alkaline magmatic rocks).

In Germany, anomalous K values occur in areas with felsic to intermediate igneous and metamorphic rocks (Ore Mountains, Bavarian Forest). Anomalous values mark the Late Variscan granite masses and felsic volcanics of the Thuringian Forest. The K patterns at the northern edge of the Thuringian Forest may also be caused by outcrops of Zechstein and its saliniferous successions. Slightly elevated potassium values are distributed in regions of Palaeozoic clay and silt lithologies (e.g., Bergisch-Sauerland mountains, Westerwald, Eifel, Hunsrück), and carbonate rocks of the Middle Triassic (e.g., the West Hessian highlands, the Middle Franconian plates).

Elevated K values in southern Bulgaria and north-eastern Greece are related to granitic intrusions, leucocratic gneiss and Caenozoic sediments.

The higher K concentration values found in agricultural compared to grazing land soil (Table 8.2; Fig. 8.5) are most likely ascribed to the use of fertilisers.

The comparison of the two maps (Figs. 8.4, 8.5) clearly identifies areas where the aqua regia extraction does not convey the same information as XRF, such as in Scandinavia or the Iberian peninsula. This reflects different bedrock mineralogy, soil pH, grain-size and climatic factors (Dinelli et al., submitted).

▸ 8.6 Conclusions

The main conclusions arising from this study of U, Th and K concentrations in agricultural and grazing land soil of Europe can be summarised as follows. The overall U-Th-K distribution patterns are related to the compositional variation of the geological substrate on which the soil is developed. Human impact on the environment has not concealed these natural geochemical patterns.

Low U values occur in an extensive zone, from northern Ukraine to The Netherlands, where soil is developed on coarse-grained sandy sediments of the last glaciation. Other sandy sedimentary basins also show low U values. Anomalous uranium areas are observed in Variscan terranes of the western Iberian peninsula, in the alkaline magmatic province of Italy and also over the Caledonides of northern Scotland and southern Norway, Sweden and Finland. Another anomalous area of U is over the crystalline French Massif Central, where several U deposits were exploited. Other high U values also occur in relation to granitic rocks in western Sardinia, Brittany (France) and the eastern Alps; over the karst of Slovenia and Croatia; point anomalies in FYROM and Bulgaria,

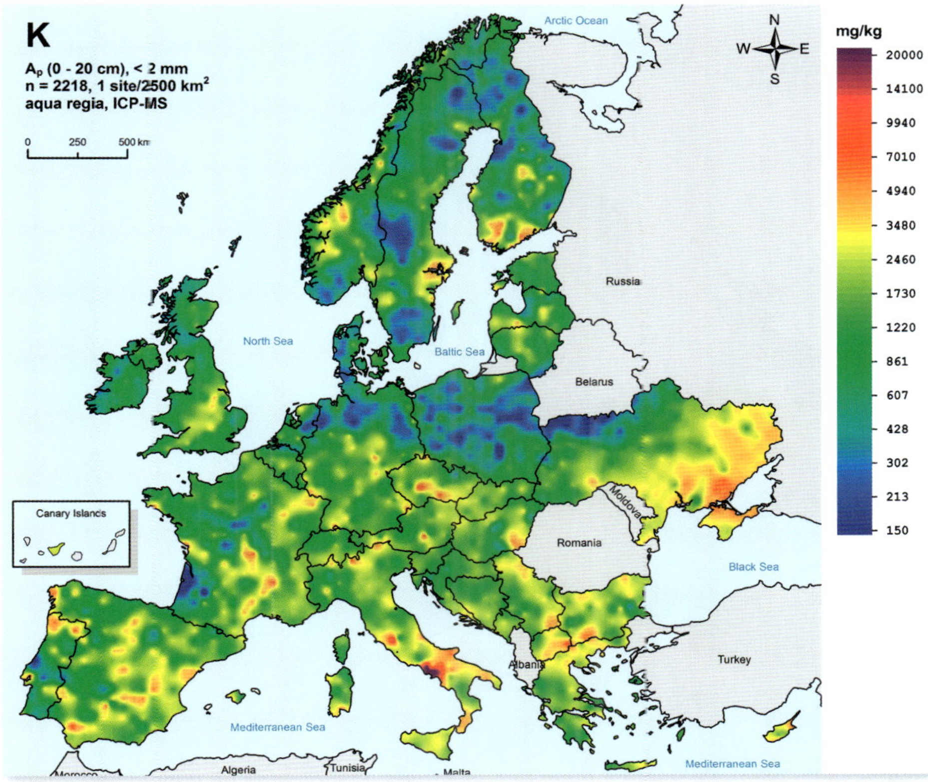

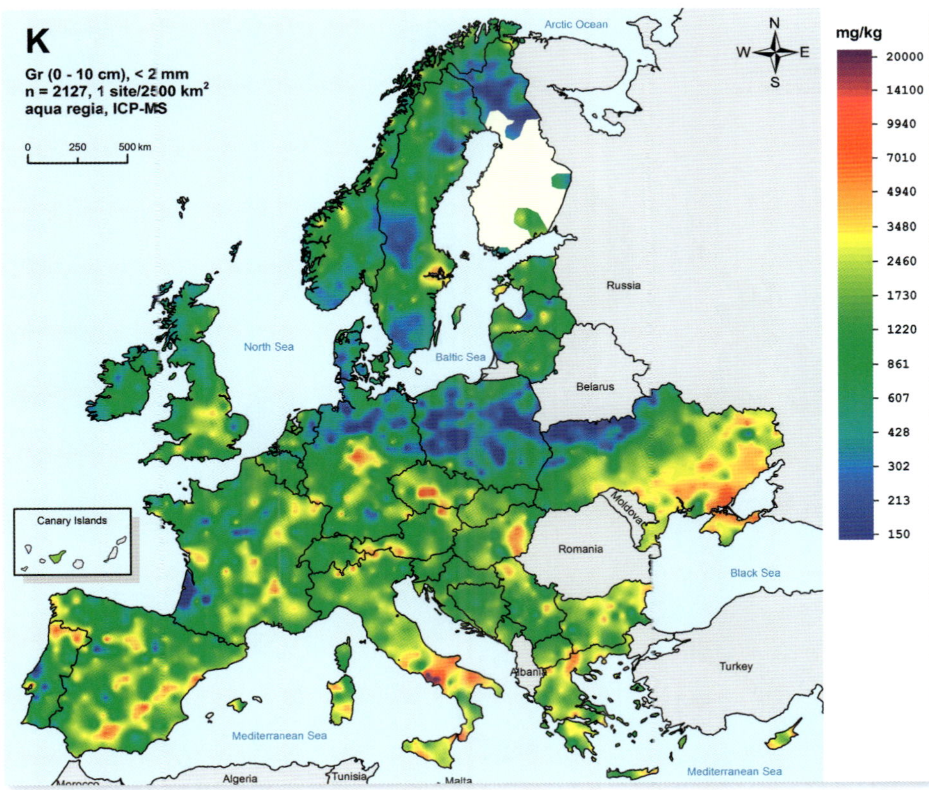

Fig. 8.4: Spatial distribution of aqua regia extractable potassium (K) in agricultural (top) and grazing land (bottom) soil of Europe.

Fig. 8.5: Spatial distribution of total potassium (K) in agricultural (top) and grazing land (bottom) soil of Europe.

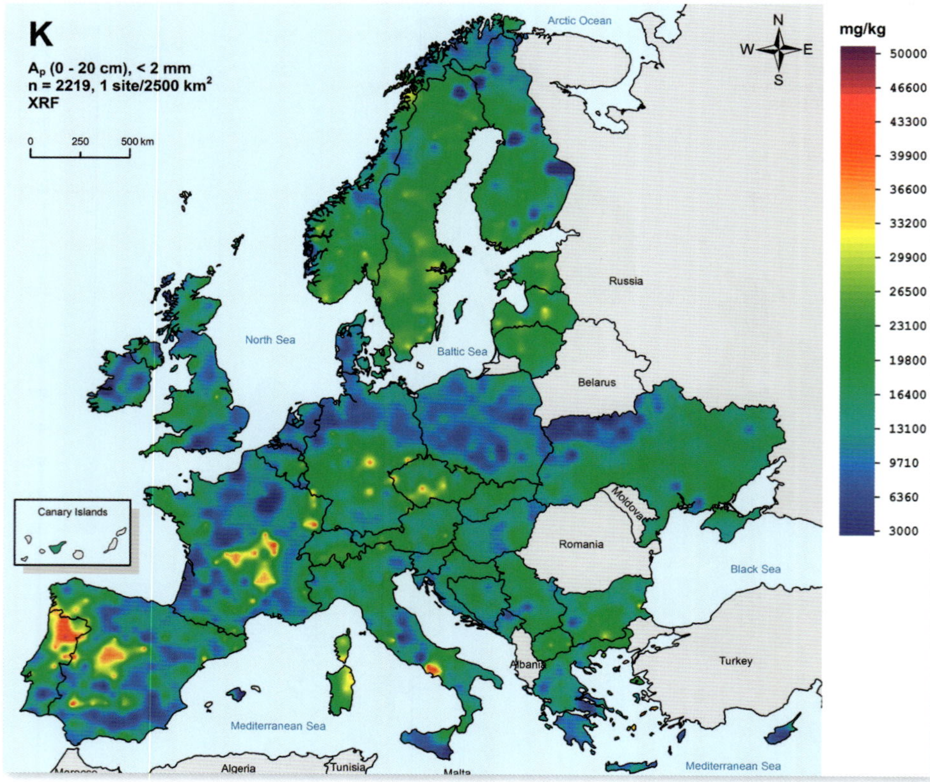

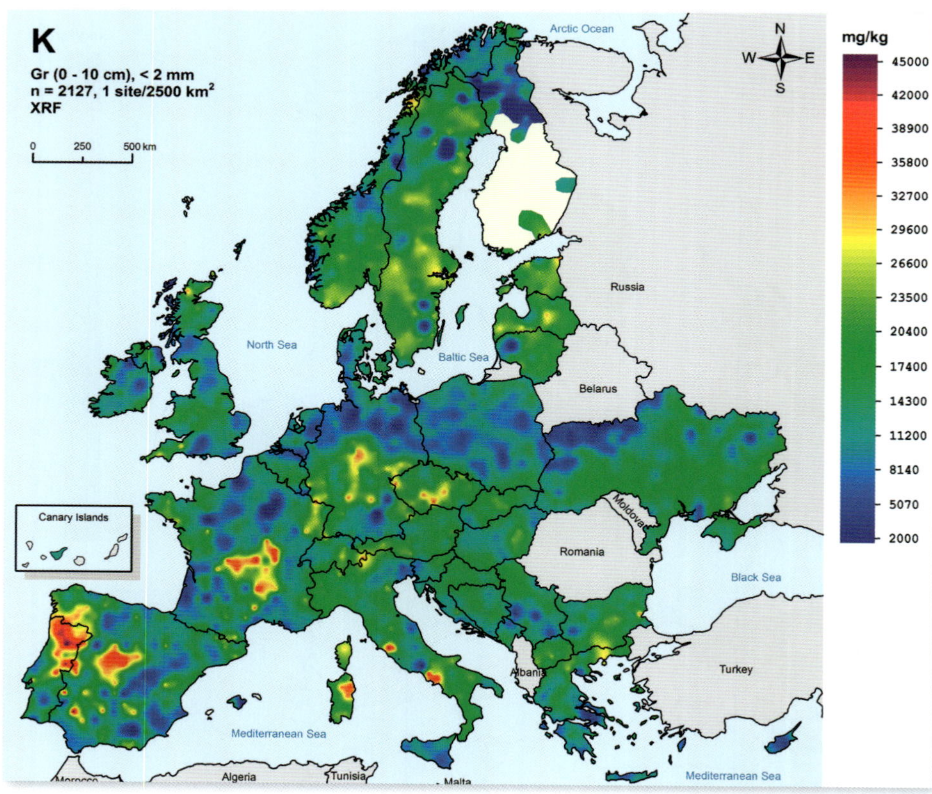

and in north-eastern Greece (related to granitic intrusions, U mineralisation and their weathering products).

The southern margin of the formerly glaciated area through The Netherlands, Germany, Poland and Ukraine is well marked on the Th maps, with low Th values to the north of it. High Th values in subsoil appear in soil over the Variscan crystalline basement of the Iberian Massif in northern Portugal and western Spain, in the alkaline magmatic province of Italy, the Massif Central in France, in Slovenia, Croatia, Bosnia, Serbia, FYROM and Bulgaria. Elevated Th values in north-eastern Greece are related to granitic intrusions.

Low K values occur in southern England, Ireland, the glacial drift area extending from The Netherlands to the Ukraine, in northern France, the Garonne basin in south-west France, in calcareous south-eastern Spain, Sicily, eastern Greece and Crete. High K values are found in soil of the Fennoscandian Shield, with concentrations increasing from Archaean rocks in the north-east towards Proterozoic rocks in the south-west. Potassium-anomalous areas are also found in Estonia and Latvia (Devonian clay and basal Quaternary till rich in mica), Lithuania, central Germany, the French Massif Central (two-mica granite is known to be enriched in K, Rb and Cs), the Alpine area, western Spain (Central System of the Iberian Massif with granitic and metamorphic lithologies), the crystalline areas of northern Portugal, the eastern Pyrenees, Corsica, Sardinia and the west-Italian alkaline province. Differences between uranium and thorium distribution in European soil seems to be inherited from differences in the weathering pattern related to climate. |◂

Domenico Cicchella
University of Sannio (UNISANNIO);
Benevento, Italy
E-mail: cidom@unisannio.it

Stefano Albanese
University of Napoli Federico II (UNINA);
Napoli, Italy

Manfred Birke
Federal Institute for Geosciences and
Natural Resources (BGR);
Hannover, Germany

Benedetto De Vivo
University of Napoli Federico II (UNINA);
Napoli, Italy

Walter De Vos
Geological Survey of Belgium (GSB);
Brussels, Belgium

Enrico Dinelli
University of Bologna (UNIBO);
Bologna, Italy

Annamaria Lima
University of Napoli Federico II (UNINA);
Napoli, Italy

Patrick O'Connor
Geological Survey of Ireland (GSI);
Dublin, Ireland

Ignace Salpeteur
Bureau de Recherches Géologiques
et Minières (BRGM);
Orléans, France

Timo Tarvainen
Geological Survey of Finland (GTK);
Espoo, Finland

Distribution of Aeolian Deposits in Europe and Their Influence on Soil Geochemistry

Andreas Scheib

▸ 9.1 Introduction

Aeolian (or wind-blown) deposits such as loess and loess derived sediments cover as much as 10 % of the World (Smalley et al. 2001), and approximately 20 % of the European land surface (Fink et al. 1977), and are thus a major soil parent material. Soil developed on loess is amongst the most fertile in the World (Catt 2001). However, they give rise to many geotechnical and stability problems, in particular structural collapse caused by loading and soaking; a process called hydroconsolidation (Rogers et al. 1994). The knowledge of their spatial distribution is, therefore, very important.

In Europe, loess as well as coversand deposits accumulated predominantly along the southern margin of the last glaciation (approximately 18 – 22 ka), which covered most of northern Europe, forming a discontinuous belt that extends from southern Britain as far as the Ukraine (Antoine et al. 2003; Haase et al. 2007; Zeeberg 1998).

This chapter briefly outlines the distribution and formation of aeolian deposits across Europe, and describes and examines the potential effects they had on the geochemistry of agricultural soil. In turn, soil geochemical data are also evaluated as a potential indicator for the distribution of loess and coversands.

▸ 9.2 Aeolian deposits in Europe

9.2.1 Formation of loess and coversands

The widespread loess and loess derived sediments were formed and deposited during the Quaternary under two main conditions: (a) periglacial, in mid-continental areas beyond the limits of major ice sheets, and (b) perimontane, along the margins of high mountain ranges (Pye 1995). Whilst some of the earlier loess dates from the Middle Pleistocene, most correspond to the Late Pleistocene Weichselian (Devensian) and earliest Holocene stages (Singhvi et al. 2001). Hence, most of the deposits can be found in the corridor confined by the extent of the last glacial maximum of the Weichselian glaciations (18 to 22 ka) as indicated on Figure 9.1. In places, however, vertical loess sequences extend back much further enabling reconstruction of detailed records of climate changes over the full glacial cycle corresponding to marine (oxygen) isotope stages (MIS) 2, 3, 4 and 5 (Frechen et al. 2003).

Pye (1995) defines loess as terrestrial clastic sediment, composed predominantly of silt size particles (30 – 70 %), and formed essentially by the accumulation of wind-blown dust. The formation requires: (a) sustained sources of silty material, (b) adequate wind energy to transport the latter, and (c) a suitable accumulation site (Pye 1995).

In Europe, the availability of large amounts of fine- to coarse-grained silt was the result of processes, such as glacial grinding and cold climate

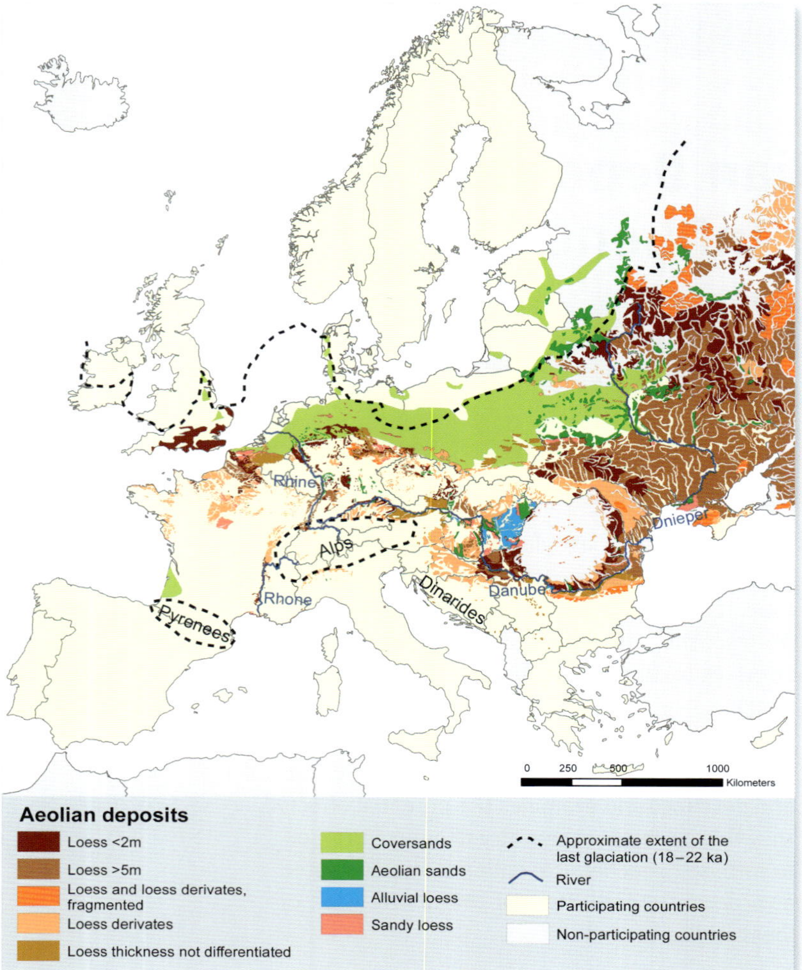

Fig. 9.1: Map of aeolian deposits in Europe with maximum extent of the last glaciations (Weichsel; 18 – 22 ka). Loess data for continental Europe redrawn from Haase et al. (2007); loess of southern Britain are based on Antoine et al. (2003). Generalised extent of coversand is based on Kasse (1997), Zeeberg (1998) and Bertrain et al. (2009).

weathering (Rogers et al. 1994), particularly over areas such as the Alps and Scandinavia. Large river systems, like the Danube, Rhine, Rhône or Garonne, were the main initial aid for carrying outwash material from glaciers, and were the origin of the bulk of the central European loess. The Danube, for example, carried and spread outwash material from glaciers in the Alps and the Carpathians across south-east Europe (Flint 1971). Sources of loess in Brittany and Normandy, north-west France and south-east England on the other hand were probably fluvial or marine sediments of the Channel palaeovalley and southern North Sea Basin (Bertrain et al. 2009; Antoine et al. 2003). Overall, the chief agents of distribution and transport of the silty material across central Europe were mostly westerly winds (Flint 1971), carrying the silt particles tens or hundreds of kilometres from these outwash plains. However, wind directions varied locally. For example, across southern Britain loess was deposited by north-easterly winds (Chorley et al. 1966; Catt et al. 1973; Antoine et al. 2003), and in the Pannonian Basin by north to north-westerly winds (Bruggle et al. 2008; Újvári et al. 2008).

Frechen et al. (2003) estimated that loess mass accumulation rates (MAR) in Europe were relatively high in a global context during the last glacial period (MIS 2). They state the lowest rates ($\approx 100 - 600\ g/m^2/yr$) for loess sites in France and Belgium, which are likely due to the maritime climatic conditions and lower supply rates. Higher rates corresponded with more continental environments of the plains in the east (Ukraine, Russia) and south-east (Czech Republic, Hungary). However, highest accumulation rates were estimated for sites on terraces along large river systems, such as the Rhine. Rates for the Rhine and eastern Europe were consistently between $\approx 800 - 3200\ g/m^2/yr$ (Frechen et al. 2003).

Similar to the deposition of loess, factors that favour coversand formation are the abundance of unconsolidated sediments, absence of topographic barriers, sparseness of vegetation cover, and the high ratio of wind energy to sand availability (Kasse 2002). Opposite to loess, however, the transport range of coversands is much shorter and more local, which is expressed by the proximity to the ice centres and glaciers of northern Europe (Fig. 9.1).

Dating of coversands in the Breckland area (Bateman 1995) and north-west Norfolk (Clarke et al. 2001) of eastern England, the Landes region in France (Bertrain et al. 2009) and Jutland in Denmark (Kolstrup et al. 1990), places the age of deposition of coversands during the cold stages of the Late Glacial (13 to 10 ka), with reactivation by local aeolian activity sometime into the early Holocene (Zeeberg 1998; Kolstrup et al. 1990; Kasse 2002). Kasse (2002) distinguished three phases of prominent aeolian sand deposition, and places the major period of coversand formation a little earlier, at the waning stage of the last glacial (Phase II) at *ca.* 18 – 14 ka.

9.2.2 Distribution of loess and coversands

Figure 9.1 shows the distribution of aeolian deposits in Europe. The map was compiled from various sources. Loess deposits of southern Britain were redrawn from ANTOINE et al. (2003) with remaining loess data supplied by DAGMAR HAASE of the Umweltforschungszentrum (UFZ) in Leipzig, Germany (HAASE et al. 2007). The latter are based on the review and compilation of loess data and information dating back to the early 1930s, when the first European loess map was published (GRAHMANN 1932). The spatial distribution of coversands is based on publications by KASSE (1997), ZEEBERG (1998), BERTRAIN et al. (2009) and HAASE et al. (2007).

The formation and accumulation processes, described above, produced widespread deposits of loess and loess derivates, as well as coversands during the Pleistocene to Holocene. Most aeolian deposits are mapped between the major ice margins of central and eastern Europe (Fig. 9.1). The most dominant spatial feature is an extensive, but discontinuous 'sheet' of loess that extends from Yorkshire (CATT et al. 1973) to East Anglia (CLARKE et al. 2001; BATEMAN 1995) and southern England (PARKS & RENDELL 1992; PERRIN et al. 1974), over France (ANTOINE et al. 2003), Belgium, The Netherlands (KASSE 1997; KOSTER 2009) and central Europe (Germany, Poland, Czech Republic) as far as the Ukraine and Russian plains in the east (FRECHEN et al. 2003; HAASE et al. 2007).

The European coversands (KASSE 1997; ZEEBERG 1998; BERTRAIN et al. 2009) and aeolian sands (HAASE et al. 2007) accumulated along the Weichselian ice margin (Fig. 9.1) in an almost continuous zone throughout the north-west and central European lowlands (ZEEBERG 1998; GIJSSEL 2006), extending from southern Britain to the Polish-Russian border and beyond (KOSTER 2009). This feature has been described as the European sand belt by KASSE (1997, 2002) and ZEEBERG (1998), and comprises Late Pleistocene to Holocene aeolian sand sheets and dunes. Figure 9.1 displays seven loess types and two aeolian sand units. Loess types are based on characteristics such as thickness, texture, and whether a primary loess deposit was reworked. Detailed information on the various loess facies are described by HAASE et al. (2007).

Generally, the largest aerea covered by loess, as well as the thickest accumulation of loess, can be found in Eastern Europe (e.g., the Ukraine) and the Russian plain reaching as far as the Black Sea and the Caucasian Mountains in the south. The loess belt across Central Europe is much more discontinuous, and varies in width from 10 to 200 km in the north-west to 50 to 100 km towards the northern fringe of the Carpathians. Fragmented loess units dominate in areas of altitudes between 400 and 600 m, for example, at the fringe of the Paris Basin, the Ardennes, the northern edge of the Ore Mountains (eastern Germany) or the margin of the Lower Danube Basin (Fig. 9.1).

Large areas of loess correspond to basins of south-eastern Europe, especially the Pannonian and Lower Danube basins (HAASE et al. 2007). Confined by the Carpathians in the east and the Dinarides to the west and south-west, the Pannonian Basin (Fig. 2.8 in Chapter 2) also contains significant areas of aeolian sands, alluvial loess and sandy loess.

Some of the thickest loess deposits can be found along the major rivers systems of the Rhine, Po, Rhône, Dnieper and Danube (Fig. 9.1). The Danube, for example, was very important for supplying glacial material from the Alps and the subsequent deposition of loess across the Pannonian and Lower Danube Basins.

▸ 9.3 Influence of aeolian deposits on soil geochemistry

There are various chapters in this book describing element distributions across Europe, which describe a spatial relationship with the extent of the last glaciation of northern Europe and subsequently associated aeolian deposits. The following two sections present results for high-field strength elements (HFSE) hafnium (Hf) and zirconium (Zr), and SiO_2, respectively, evaluating their distribution in relation to loess and coversand deposits..

9.3.1 Relationship of Hafnium (Hf) and Zirconium (Zr) concentrations to loess deposits

TAYLOR et al. (1983) established that elevated concentrations of Hf and Zr are characteristic for soil developed on aeolian deposits within the United States, China, Europe and New Zealand. More recently, BRUGGLE et al. (2008) and ÚJVÁRI et al. (2008) used geochemical data to characterise and provenance loess samples in south-eastern and eastern Europe. Their results also indicated that Hf and Zr concentrations were enriched in loess samples. STÜCKRAD et al. (2010) showed similar characteristics for loess in the Rhenish Slate Mountains north-west of Frankfurt, Germany. Loess samples from the south of England also gave high Zr concentrations (FALL 2003). A study by SCHEIB & LEE (2010) also identified very good correlations between elevated total concentrations of Hf (> 9.7 mg/kg) and Zr (> 369 mg/kg) in topsoil from East Anglia, with loess and coversand deposits of north-east Norfolk (CATT 1977, 1985) and the Breckland area (CLARKE et al. 2001).

Table 9.1 lists summary statistics for total and aqua regia (AR) extractable Hf and Zr concentrations in GEMAS soil samples. Total XRF data of Hf and Zr are normally distributed (skewness < 1), whilst AR data is strongly positively skewed. The median total concentrations of Hf and Zr, for both Ap and Gr soil, are significantly higher than those determined by AR, thus highlighting a very low extractability of < 1.7 %. This is an indication that total Hf and Zr concentrations correspond mainly to resistant host minerals, such as zircon. Detrital zircon is an accessory mineral also commonly found in the silt fraction of loess deposits.

Based on the potential relationship between Hf and Zr and loess deposits, Figure 9.2 displays a series of boxplots of total Hf and Zr concentrations grouped by the loess and coversand types shown in Figure 9.1. It shows that the highest medians (> 10 mg/kg Hf; > 300 mg/kg Zr) correspond to loess types 'loess no thickness', 'loess > 5m', 'loess < 2m' and 'loess derivates'. Data for samples collected over coversands correspond to a much weaker geochemical signal.

Figure 9.2 clearly highlights a relation between elevated Hf and Zr concentrations in soils and loess deposits. In an attempt to use Hf and Zr soil concentrations, as indicators for loess deposits at the European-scale, SCHEIB et al. (2014) applied non-linear discriminant analysis to total data (for Gr samples only) and derived threshold values of 10 mg/kg for Hf and 318 mg/kg for Zr. Figure 9.3 a – b displays the spatial distribution of areas that correspond to concentrations above the thresholds, which may indicate the presence of loessic material.

Above-threshold concentrations of Hf and Zr in northern Europe (Scandinavia and Scotland) relate to Neoproterozoic to Early Palaeozoic crystalline igneous intrusions and metamorphic rocks. The spatial distribution of soil concentrations above the thresholds beyond the glacial margins, indicate a strong correlation to mapped loess and loess derived sediments across central Europe (Fig. 9.1). Highest Hf and Zr concentrations relate

Table 9.1: Descriptive summary statistics for total (XRFS) and aqua regia (AR) extractable hafnium (Hf) and zirconium (Zr) concentrations (mg/kg) in GEMAS agricultural (Ap) and grazing land (Gr) soil.

Statistical parameters	XRFS data				AR** data			
	Hf		Zr		Hf		Zr	
	Ap	Gr	Ap	Gr	Ap	Gr	Ap	Gr
N	2108	2024	2108	2024	2108	2024	2108	2024
Minimum	< 5*	< 5*	4	4	< 0.02*	< 0.02*	< 0.1*	< 0.1*
25th Percentile	6	< 5*	192	178	0.02	< 0.02	0.7	0.7
Median	8	8	259	245	0.05	0.05	1.8	1.6
75th Percentile	11	12	346	324	0.10	0.11	3.7	3.7
90th Percentile	14	14	463	429	0.18	0.18	6.4	6.4
Maximum	28	24	963	871	2.97	6.73	165	376
Mean	8.6	8.7	281	263	0.08	0.08	3.3	3.4
Skewness	0.46	0.45	0.93	0.96	10.3	23.7	12.0	22.9

* values below the reported lower limit of detection (LLD)

** ICP-MS analysis following aqua regia extraction

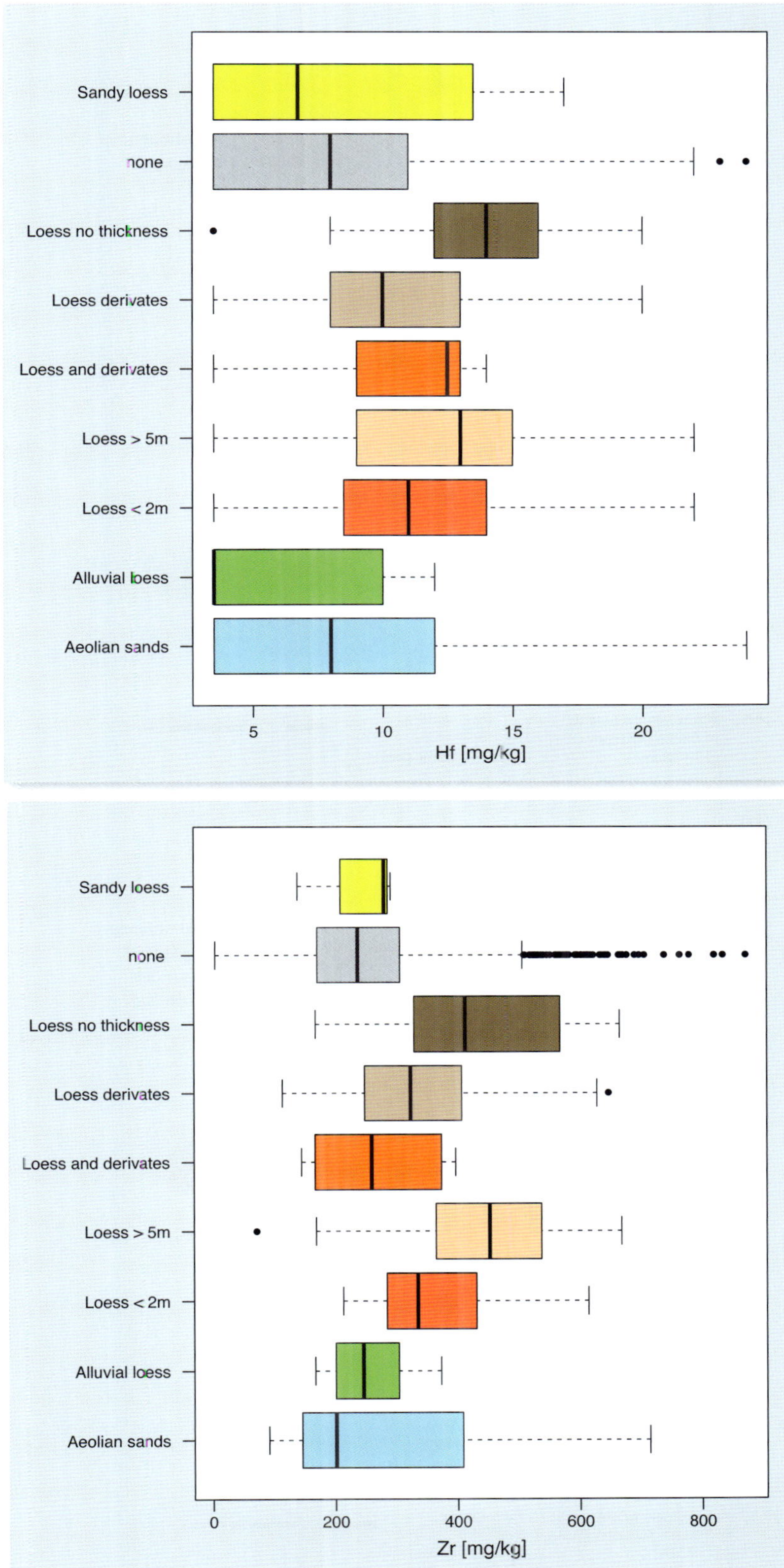

Fig. 9.2: Boxplots of total (XRFS) concentrations of hafnium (Hf) and zirconium (Zr) (mg/kg) in Gr soil samples grouped by aeolian deposits shown in Figure 9.1 (modified after Scheib et al. 2014);
Aeolian sands (N = 23),
Alluvial loess (8),
Coversand (175),
Loess < 2 m (38),
Loess > 5 m (70),
Loess and derivates (6),
Loess derivates (49),
Loess no thickness (20),
Sandy loess (4),
none (1628).

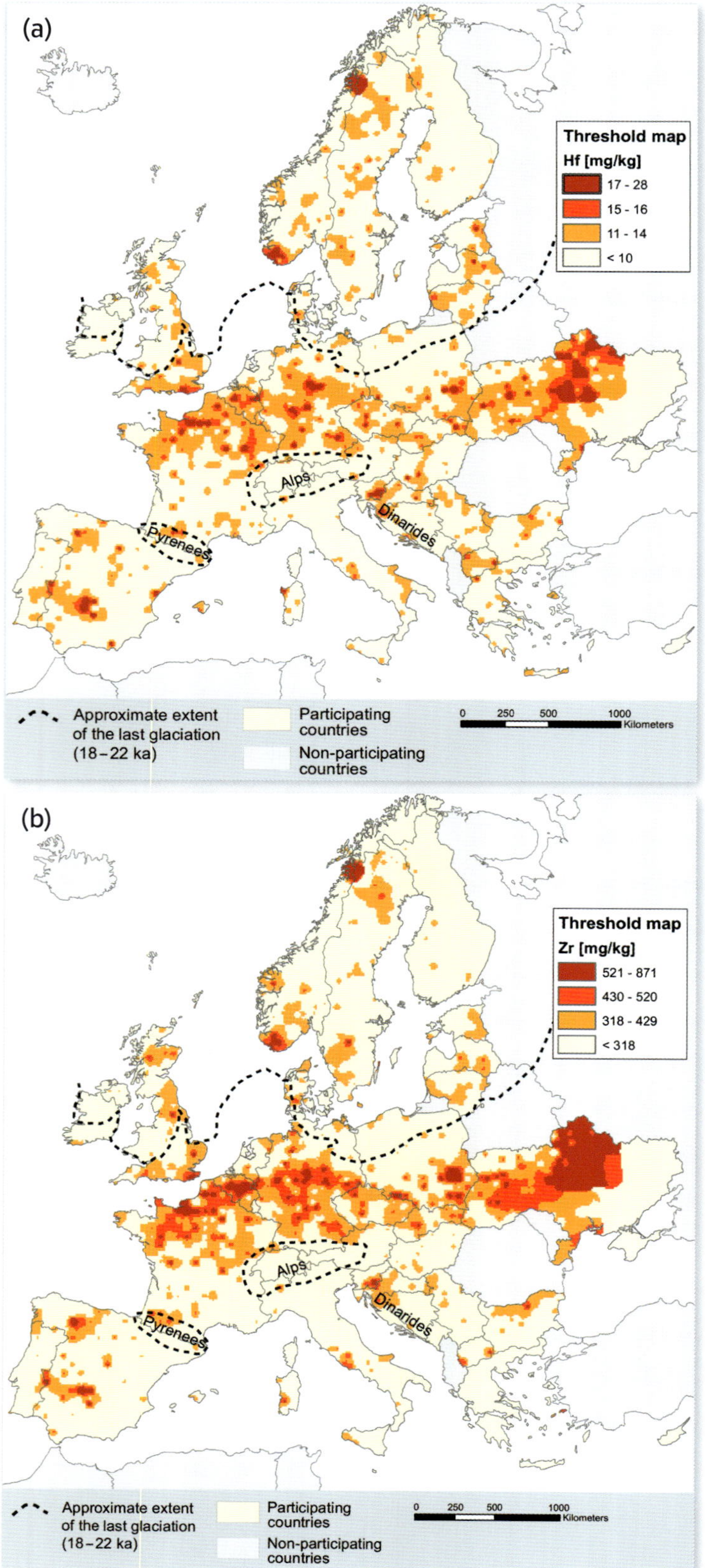

Fig. 9.3: Threshold maps of (a) hafnium (Hf) and (b) zirconium (Zr) indicating areas of total (XRFS) concentrations > 10 mg/kg and > 318 mg/kg, respectively; colour class intervals correspond to the 90th and 95th percentile; map is based on grazing land (Gr) soil data (modified after Scheib et al. 2014).

to soil samples from the Ukraine and northern France and Belgium.

Areas of soil concentrations above the thresholds in south central Italy and the northern Dinaric Mountains are not related to loess deposits; they rather reflect alkaline volcanic rocks (particularly tephra) and residual soil over karstified limestone, respectively (Scheib et al. 2014).

At the European scale, threshold maps of Figure 9.3a – b suggest a much more extensive and coherent loess belt across central and eastern Europe than indicated by the loess distribution map in Figure 9.1. Threshold maps indicate potential areas of loess deposits across the Paris Basin and Brittany, along the foothills of the Pyrenees and to the west of the Vosges in eastern France. Figure 9.3a – b also suggests loessic material within the large basins of central and northern Spain (e. g., Duero Basin).

With regard to coversands, above-threshold values of Hf and Zr correspond only to a few small areas in the Baltic States and Jutland of western Denmark, whilst the majority of coversands of the European Sand belt (Fig. 9.1) do not agree with either indicator element. This may relate to the high silicate contents and the lack, or dilution of finer, silty material containing resistate accessory minerals.

9.3.2 Relationship of SiO_2 concentrations to coversands

As the name suggests, coversands comprise largely sand and consequently contain large amounts of quartz grains and silicate minerals. Figure 9.1 displays the coversands as a broad and continuous belt across central and eastern Europe. This distribution, however, is very generalised and in reality likely to be more discontinuous and irregular. This section assesses the silicate contents in Ap soil samples in relation to coversands. This will allow to indicate areas of particularly high and low sand contents, and provide some refinement at the continental-scale.

The variation of SiO_2 content in Ap samples is enormous with values ranging from 2.5 to 96.5 wt% reflecting, for example, almost pure limestone to almost pure quartz sand end members (Reimann et al. 2012a). The median and average (mean) concentrations are 67.2 and 66.3 wt%, respectively, and the skewness of − 0.92 indicates a near-normal distribution of the data.

Figure 9.4 displays the spatial distribtuion of SiO_2 concentrations in Ap samples. Elevated SiO_2 concentrations > 77 wt% correspond strongly with the mapped extent of coversands of the European Sand Belt. Most of the highest concentrations (> 89.6 wt%) are located in the centre and in close proximity to the ice margins marking the locations of end moraines associated with the last glaciation (Reimann et al. 2012b). Within this area highest concentrations (> 89 wt%) correspond to soil from Poland, Jutland in western Denmark and north-east Germany. Figure 9.4 also indicates a consistent and steady decrease of SiO_2 concentrations in southerly and northerly directions away from the central part of the sand belt. Coversand deposits of the Gascony region on the south

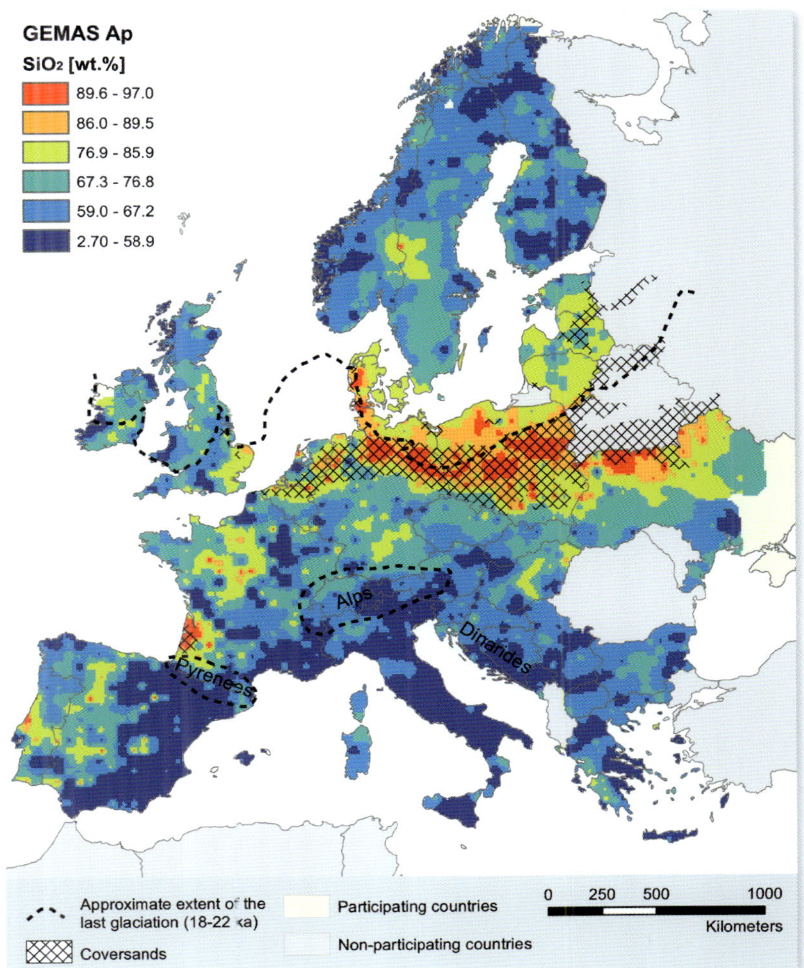

Fig. 9.4: Interpolated map of total (XRFS) SiO_2 concentrations (expressed in weight % oxides) for agricultural (Ap) soil; colour class intervals correspond to the 25th, 50th, 75th, 90th and 95th percentile.

Atlantic coast of France also correspond to enriched SiO_2 concentrations in Ap soil samples.

On the other hand, elevated SiO_2 concentrations in soil from the Paris Basin correspond to sandy material derived from the underlying Mesozoic and Tertiary sedimentary rocks. Figure 9.4 also shows that lowest SiO_2 concentrations are found in southern Europe and in areas underlain by, or containing, carbonates or gypsum (Reimann et al. 2012b).

▶ 9.4 Conclusions

The statistical and spatial distribution of total Hf and Zr as well as SiO_2 concentrations are strongly influenced and affected by the presence of loess and coversand deposits, which formed extensively under periglacial conditions during the Pleistocene across central and eastern Europe.

The GEMAS data set shows that soil developed over loess deposits are enriched in Hf and Zr, which are, based on their poor extractability, related to the presence of resistant accessory minerals, such as zircon. The application of Hf and Zr as indicator elements and threshold values of 10 mg/kg and 318 mg/kg, respectively, seems to be effective and the spatial distribution of areas above the latter thresholds suggests a much wider distribution of loess across central Europe than indicated by the map of aeolian deposits (Fig. 9.1). This discrepancy may be related to the fact that (a) some loess deposits were not mapped due to an insignificant thickness or (b) evidence of their structure was destroyed by tillage, whilst the geochemical signature of loess, i.e., resistant minerals, has been retained in topsoil.

The close relation between elevated SiO_2 concentrations and coversand deposits was predictable. However, the application of GEMAS SiO_2 data as a proxy for the sand content provides more detailed spatial information on the heterogeneity of the coversands at the European scale. For example, soil data suggest that the sand content within the European Sand Belt is highest along, and in close proximity to the ice margins of the last glaciation; a reflection of the much shorter aeolian transport of sandy material from the source to accumulation sites than for loess. |◀

Andreas Scheib
Geological Survey of Western Australia (GSWA); Perth, Western Australia
E-mail: andreas.scheib@dmp.wa.gov.au

Distribution of Boron, Chlorine and Fluorine in European Topsoil and Subsoil (based on samples from the FOREGS project)

Reijo Salminen, Manfred Birke, Clemens Reimann, Virgilija Gregorauskiené & Xueqiu Wang

▸ 10.1 Introduction

Boron (B), Chlorine (Cl) and Fluorine (F) are three elements that are rarely analysed for in whole rock or soil geochemical surveys. The reason is that they require element specific analytical techniques in order to obtain useful data. Good regional data for these three geochemically quite important elements are thus missing in the literature. In 1996, the Forum of European Geological Surveys (FOREGS) initiated a programme to collect a standardised and controlled geochemical database for Europe, with the aim of compiling the first geochemical atlas of Europe (Bølviken et al. 1990, 1993, 1996; Plant et al. 1996, 1997). In 1997, 26 European countries established a project under the auspices of the Forum of European Geological Surveys (FOREGS) to provide high quality environmental geochemical baseline data. The project followed recommendations set by the IGCP 259 (International Geochemical Mapping; 1988-1992) project (Darnley et al. 1995). The main aims of the project were:

- ▸ to establish uniform methods for sampling, analysis and data management,
- ▸ to collect and analyse a variety of geochemical materials throughout Europe,
- ▸ to prepare a standardised reference data set of geochemical baseline data,
- ▸ to carry out the European contribution to the IUGS/IAGC "*Global Geochemical Baselines*" programme.

The IGCP 259 was followed by the IGCP 360 "*Global Geochemical Baselines*" (1993 – 1998; Darnley & Garrett 1998), and the successor of this programme is the IUGS/IAGC Task Group on "*Global Geochemical Baselines*" continuing this work until its final completion (Reeder 2007; Smith et al. 2012). The goal of this programme is to establish a land-surface global geochemical reference network, providing multi-media, multi-element baseline data for a wide range of environmental and resource applications (see *www.globalgeochemicalbaselines.eu/* and IUGS/IAGC TGGGB 2013).

The sampling grid adopted was based on the recommendations of IGCP 259, and Europe was divided into 160 by 160 km GTN grid cells (Global Terrestrial Network; Darnley et al. 1995). Stream water, minerogenic stream sediment, minerogenic topsoil and subsoil, and the soil organic layer (humus) samples were collected from five randomly selected 100 km² catchment basins in each grid cell. Minerogenic floodplain sediment (the uppermost 25 cm) was sampled from five 1000 – 6000 km² catchment basins (which included the 100 km² catchment basin). Fieldwork was carried out according to the published field manual (Salminen, Tarvainen et al. 1998). Samples were analysed for more than 50 elements and other parameters (such as pH and grain size) (Sandström et al. 2005). Both total and aqua regia extractable concentrations were determined. Nine laboratories of European Geological Surveys carried out the analytical work.

Altogether, some 400 geochemical maps describing the distribution of elements throughout the whole of Europe were prepared. All results and field observations were organised in a common database and the maps were published as the Geochemical Atlas of Europe (Salminen et al. 2005, de Vos et al. 2006). The sampling sites were photographed and this photo archive is available via internet. Samples are archived for possible future reference at the British Geological Survey, UK. As a result, a large database of geochemical data, covering 26 European countries and containing chemical analyses of more than 50 elements and other parameters from different environmental media (soil, stream and floodplain sediment and stream water), is freely available together with a data description via internet.

At the time the FOREGS samples were analysed, it was not possible to obtain reliable determinations for total B, Cl and F concentrations. Ten years later, they were analysed from soil samples after compositing samples from each GTN cell. These results are presented and discussed in this chapter (note that for the GEMAS project, B was determined in an aqua regia extraction, but again no total concentrations were obtained).

Boron (B) is a light non-metallic trace element, which is widespread in the hydrosphere. It is more common in sedimentary rocks than in igneous rocks (Rudnick & Gao 2003). In endogenous processes, it is strongly enriched during the serpentinisation of ultramafic rocks, in the form of abyssal impurities in kimberlite, during skarn formation, in hydrothermal deposits, and in pegmatite. The exogenous accumulation of B is mainly connected with borates in sedimentary and volcanogenic sedimentary deposits, but elevated B concentrations are also common in saline loam and deep-sea clay. Combustible shale (oil shale) and phosphates are also characterised by elevated B concentrations. In the hydrosphere, high B concentrations are typical for sea water and especially for carbonate-rich and highly mineralised thermal water, but can also be due to contamination with organic waste.

Although B is one of the most mobile elements in soil, it is sorbed more strongly by soil than other anions such as Cl^- and NO_3^-. It may also be entrapped in the clay lattice by substitution for Al^{3+} and/or Si^{4+}, thus forming B-silicate compounds in soil. Adsorption on Fe and Al oxides is considered an important mechanism governing B solubility in soil. Adsorbed B on soil minerals can be easily leached, but irreversibility of B sorption has also been documented (Kabata-Pendias 2001). The presence of organic matter affects greatly the mobility and bio-accessibility of B, particularly in acid soil. Adsorption of B on soil humic acids increases with increasing pH up to a maximum near pH 9. Boron adsorption by oxy- and hydroxy-bonds by surface coatings, and by incorporation into interlayer or structural positions of aluminosilicate minerals are the likely dominant mechanisms in acid and neutral soil. Boron is accumulated in the uppermost A_0 horizon and in plants. In the mineral fraction of soil, hypergene borates are formed during weathering at carbonate-gypsum barriers (Ivanov 1996).

Boron compounds are used extensively as cleaning aids in the detergent industry and have a wide range of industrial uses, including as buffers for pH control (Waggott 1969). Boron pentahydrate is used in very large quantities in the manufacture of insulation fibreglass and sodium perborate bleach. Boron compounds are also extensively used in the manufacture of borosilicate glasses, in metallurgy as a flux and in the manufacture of steel alloys. Boric acid is used widely in the textile fibreglass, in cellulose insulation as a flame retardant and as a mild antiseptic. Amorphous B is used in pyrotechnic flares to provide a green colour, and as an igniter in rockets (Aggarwal 1999). Concentrations of boron are also likely to be high in sewage.

Although B is an essential element for plant growth and is often used in fertilisers. This often contributes to enhanced B levels in stream water in agricultural areas; B is highly toxic to plants in high concentrations.

Chlorine (Cl) is a widespread non-metallic element with strong biophile and hydrophile properties. It forms many different chlorides and hydrochlorides. In endogenous conditions Cl concentrations increase from ultramafic and mafic rocks to felsic and alkaline rocks (Rudnick & Gao 2003). Metamorphic and sedimentary (sandstone and limestone) rocks are characterised by low Cl contents, but in clay (especially modern deep-sea clay) and shale the Cl concentrations are rather high. Chlorine is a vitally essential element and its concentration in the biosphere is very high. Chlorine input to soil occurs mainly through rainfall, from fertiliser applications (KCl), irrigation water, sea spray, dust and air pollution. Chloride anions are not sorbed by solid soil particles, thus chloride in soil is very mobile and is found mainly in soil water. Concentrations of the Cl^- anion in water of different geological and climatic types vary over a wide range (Ivanov 1996). The factor controlling the amount of chlorine in soil is climate. Accumulation of chloride in soil is controlled by evaporation. In regions of hot climate and high groundwater level, Cl^- anions can accumulate on the soil surface via the soil salinisation process.

Chlorine is used in different branches of industry and its anthropogenic impact is high. The concentration of chloride in surface and groundwater can be influenced by human activities, such as the application of salt for de-icing of roads, leaching from landfill sites, discharges from sewage treatment works, and the intrusion of saline water following the pumping of fresh water from coastal aquifers. Chlorine is used in treating drinking water, swimming pools, hot tubs, etc., to kill bacteria and other micro-organisms. It is also a building block for polyvinyl chloride (PVC) and numerous other chemicals such as antifreeze, pesticides (DDT and chlordane), refrigerants, anti-knock compounds and industrial cleaners (carbon tetrachloride). As sodium hypochlorite, or bleach, it is widely used as a disinfectant and in bleaching.

Fluorine (F) is the most chemically active element in the halogen group. In nature, it occurs in oxidation state – 1 as fluoride. Fluorine is a strong oxidant and reacts with most metals. In igneous rocks, F concentrations increase rapidly from ultramafic to felsic and alkaline rock types. The highest F concentrations are found in ore deposits associated with alkaline granite, alkaline gabbro, nepheline-syenite and also with kimberlite and carbonatite. Very high F concentrations are typical of post-orogenic granite, such as rapakivi granite.

Fluorine is very active chemically and it reacts with nearly all elements. On the other hand, the low solubility of fluorite (CaF_2) is very important in the geochemistry of F. The tendency to form complex compounds, many of them easily soluble, favours the migration of F in complex form. Acid leaching removes fluorine from the weathering crust and soil. Biogenic accumulation and accumulation on calcium carbonates prevents it. As a result, eluvial soil in humid regions is low in F compared to the parent rock (Perelman 1977).

Fluorine, like Cl, is a cyclic element, which participates in the ocean-continent water cycle. Fluorine from atmospheric precipitation is adsorbed by soil where it is enriched. The soil sorption capacity depends on pH (6 – 7), and the amount of calcium and phosphorus. In sedimentary rocks, F contents do not vary much, with the exception of phosphorite, where F concentrations can be very high. Fluorine is a mainly hydrophilic and atmophilic element, and is enriched in sea water, groundwater, and also in some artesian basins and F-bearing rare-metal deposits.

Fluorine is an essential trace element for humans but it is toxic at higher concentrations. It is widely used in metallurgy (as fluorite), in aluminium smelting and brick making (Debackere & Delbeke 1978), and in glass, china and steel works (Kabata-Pendias 2001). Sewage, and the production and application of phosphate fertilisers and pesticides are other anthropogenic sources of F (Whalley 1976; Kabata-Pendias 2001).

▶ 10.2 Methods

10.2.1 Sampling – FOREGS project

Sampling was carried out in each European country by national teams following procedures set out in the Field Manual (Salminen, Tarvainen et al. 1998). Normally one team sampled all sites in each country during one field season between 1997 and 2001. Therefore, it can be stated that the FOREGS geochemical baseline mapping programme represents the end of the twentieth century state of the surficial environment in Europe.

The sampling grid was planned according to the recommendations of IGCP 259 on 160 by 160 km GTN grid cells. Residual or sedentary soil samples were collected from five randomly selected small, second order, drainage basins (< 100 km^2) in each GTN cell. The actual sampling site was at a suitable position above its alluvial plain and base of slope, where alluvium and colluvium are respectively deposited. Residual soil developed either directly on bedrock or on till was accepted. Residual soil from areas with agricultural activities was avoided, since such topsoil is usually affected by human activities. Colluvium or alluvium was not accepted as representing parent material.

Each residual soil sample was a composite of 3 to 5 sub-samples collected from pits located at a distance of 10 – 20 metres from each other. Two different depth related samples were taken at each site, i.e., (1) a topsoil sample from 0 – 25 cm (excluding material from the organic layer where present), and (2) a subsoil sample from a 25 cm thick section within a depth range of 50 – 200 cm (the C soil horizon).The residual soil sample represents the dominant sedentary soil type of the selected catchment basin. Minimum distance to roads was 10 m and to ditches 5 m. Living surface vegetation, fresh litter, large roots and rock fragments (stones) were removed by hand. Where the whole soil profile did not reach a depth of 75 cm, the lower sample was taken from a depth that can be undoubtedly identified as the BC- or C-horizon. The subsoil sample was always collected first, and then the topsoil sample in order to avoid mixing of the subsoil with topsoil material falling down the pit during sampling (Salminen, Tarvainen et al. 1998; Salminen et al. 2005).

The number of sampling sites was 790 for subsoil and 852 for topsoil. For analysis of B, Cl and F, the samples of each GTN cell were combined to form one composite sample. Thus, each composite sample was formed from 3 – 5 subsamples and the number of composite samples was 192 for subsoil and 198 for topsoil. The coordinates used for each composite sample represent the centre point of the 160 × 160 km grid cell. For this reason, some sampling site symbols are located in the sea on Figure 10.1.

10.2.2 Analysis

The composited soil samples were analysed for B, Cl and F in the laboratory of the Institute of Geophysical and Geochemical Exploration, Hebei, Peoples Republic of China.

The applied analytical methodologies were:

X-ray fluorescence (XRF): 4.0 g of sample were pelletised and the Cl concentration was determined by wavelength dispersive X-ray fluorescence spectrometry. GSS and GSD serial reference materials were used for calibration (Yao et al. 2011).

Emission spectroscopy (ES): 100 mg of sample was homogeneously mixed with 100 mg buffer by grinding for 5 min. The mixed powder was placed in a carbon electrode and the B concentration was determined with an emission spectrometer coupled with a photo plate photometer (Yao et al. 2011).

Ion selective electrode (ISE): The samples were first ignited in a Ni or graphite crucible at 400 °C temperature. After cooling, some drops of ethanol were added to wet the samples. Then the samples were decomposed by sodium peroxide fusion at 620 °C temperature for 15 minutes. The flux was digested in hot water. An aliquot of the clear solution was moved to a beaker and sodium citrate was added. The pH was adjusted to 6.8 – 7.1 with hydrochloric acid. Triethanol amine buffer solution was added and the solution moved to the beaker where the F concentration was measured using a fluorine selective ion electrode with a saturated calomel

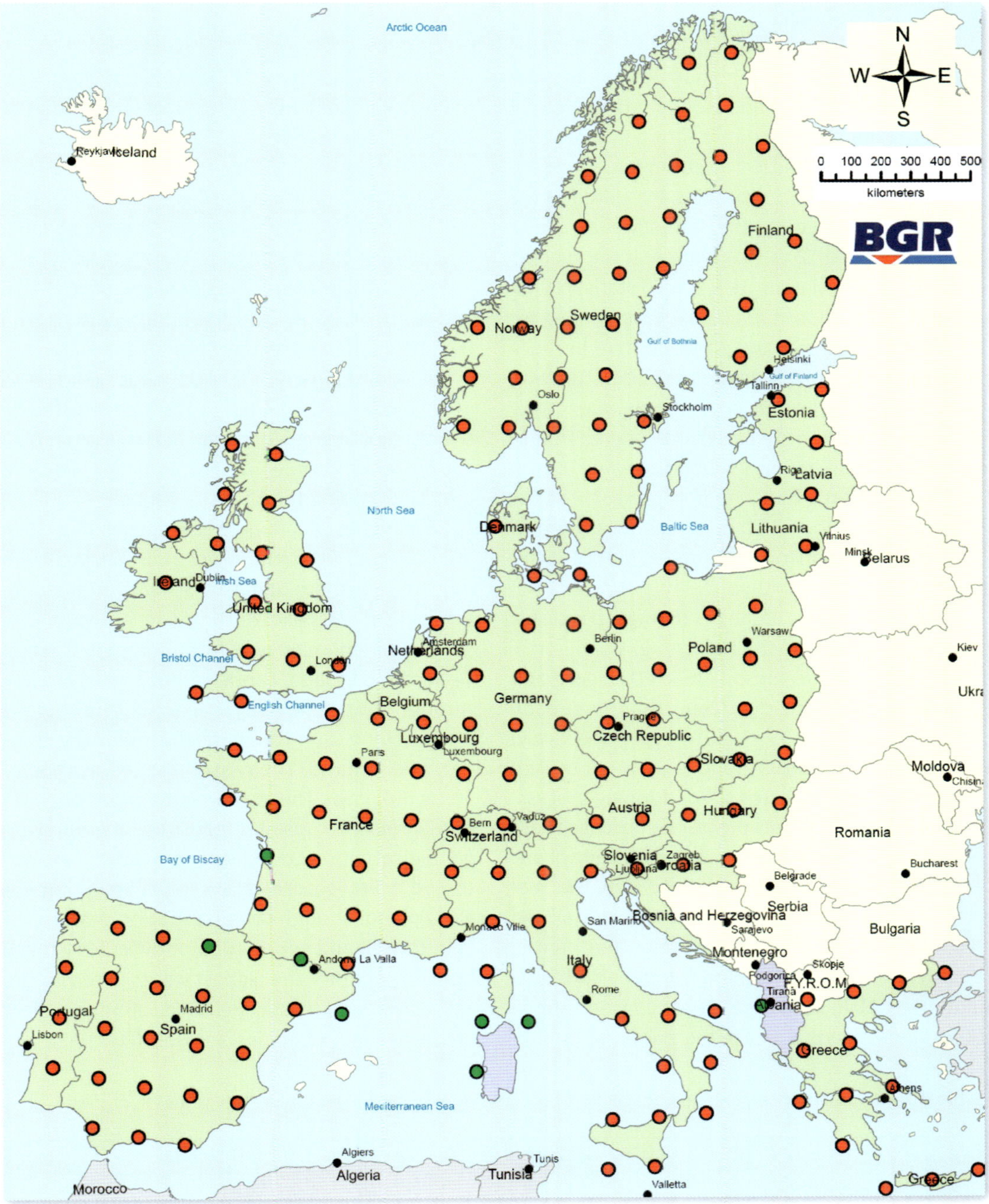

Fig. 10.1: Location of FOREGS composite soil sampling sites at the centre of the 160 × 160 km GTN cell (see text for explanation of sample sites falling in sea).

electrode as reference. For more details, see Li et al. (2005) and Zheng et al. (2005).

10.2.3 Map preparation

The data for map production were processed at BGR. The data were interpolated by kriging algorithms and plotted with ArcGIS. The kriging method was used to interpolate the few sampling sites to a high density grid for representation of the geochemical data on a map. A variogram analysis was made first to extract the parameters of the kriging function. The points of the resulting regular grid were classified according to the background and enrichment levels. Areas of the same class were combined and given the same colour on the map.

▶ 10.3 Results

Statistical parameters (minimum, maximum, median, arithmetic mean and standard deviation) of B, Cl and F concentrations are shown in Table 10.1. The regional distribution of each element, separately for topsoil and subsoil, is presented in the form of maps covering the sampled area of Europe in Figures 10.2 to 10.4. Note that the average and standard deviation are not valid statistical parameters for compositional data (Chapter 8 → Part A); nevertheless they are provided here because they have been much used in the literature and data on these three elements are very rare.

Generally, anomaly patterns observed in topsoil and subsoil are very much the same. Boron and Cl are slightly enriched in topsoil, whilst F is depleted from topsoil, but the distribution patterns between topsoil and subsoil do not differ too much from each other.

The range of B in subsoil (4–384 mg/kg) is clearly larger than in topsoil (3–164 mg/kg). For Cl and F ranges are larger in topsoil (36–1203 mg/kg and 70–2730 mg/kg, respectively) than in subsoil (31–719 mg/kg and 59–1830 mg/kg, respectively) (Table 10.1). These differences between topsoil and subsoil are connected to soil-forming processes and other exogenous processes, because these elements are important nutrients in the biosphere.

A characteristic feature of B and F are the negative anomalies (low concentrations) in subsoil and topsoil in the glaciated terrain of northern central Europe, i.e., Poland, northern Germany and Scotland (Figs. 10.2, 10.4). The lowest concentrations occur in the marginal area of the glacier (the southern limit of glaciation). This is a typical feature for many other elements in the Geochemical Atlas of Europe (Salminen et al. 2005). In this area, soil is formed from glacial sediments mainly derived from Precambrian bedrock of the Fennoscandian Shield, and this material was depleted in fine-grained material during glacial transport. High concentrations are observed in areas of residual soil of southern and central Europe. The distribution pattern of Cl (Fig. 10.3) differs essentially from B and F.

Most of the highest B and F values (Figs. 10.2, 10.4) appear to be related to the area affected by the Alpine orogeny. Other smaller areas with high values occur in the Pyrenees, Portugal, and the area surrounding the English Channel.

The highest Cl anomalies in subsoil (Fig. 10.3) are observed in northernmost Europe, southern Italy and southern Spain. Of specific interest are some linear anomalies cutting across Europe. One such linear anomaly runs from Denmark through Germany to Croatia, another from central Norway through central Sweden to Estonia, and there is also a strong anomaly in northern Norway, Sweden and Finland. The reason for these anomalies is unknown. These anomalies are stronger in subsoil than in topsoil, suggesting an underlying geological cause. Low Cl values in subsoil are observed in France, central Portugal, Spain, Greece and Poland. Though the distribution patterns of Cl in topsoil and subsoil do not differ very much from one another, the highest values in topsoil are concentrated in coastal areas indicating the effect of steady input of marine aerosols (Fig. 10.3).

Table 10.1: Statistical parameters of B, Cl and F concentrations in topsoil and subsoil; all values in mg/kg.

Statistical parameters	Topsoil			Subsoil		
	B	Cl	F	B	Cl	F
N	198	198	198	192	192	192
Minimum	3.33	35.6	69.5	4.31	30.7	58.7
Maximum	164	1203	2730	384	719	1830
Median	39.9	90.7	365	37.8	74.7	419
Average	40.2	114	410	42.6	98.6	464
StDev	26.6	106	285	37.9	79.5	253

N number of analysed samples **StDev** standard deviation

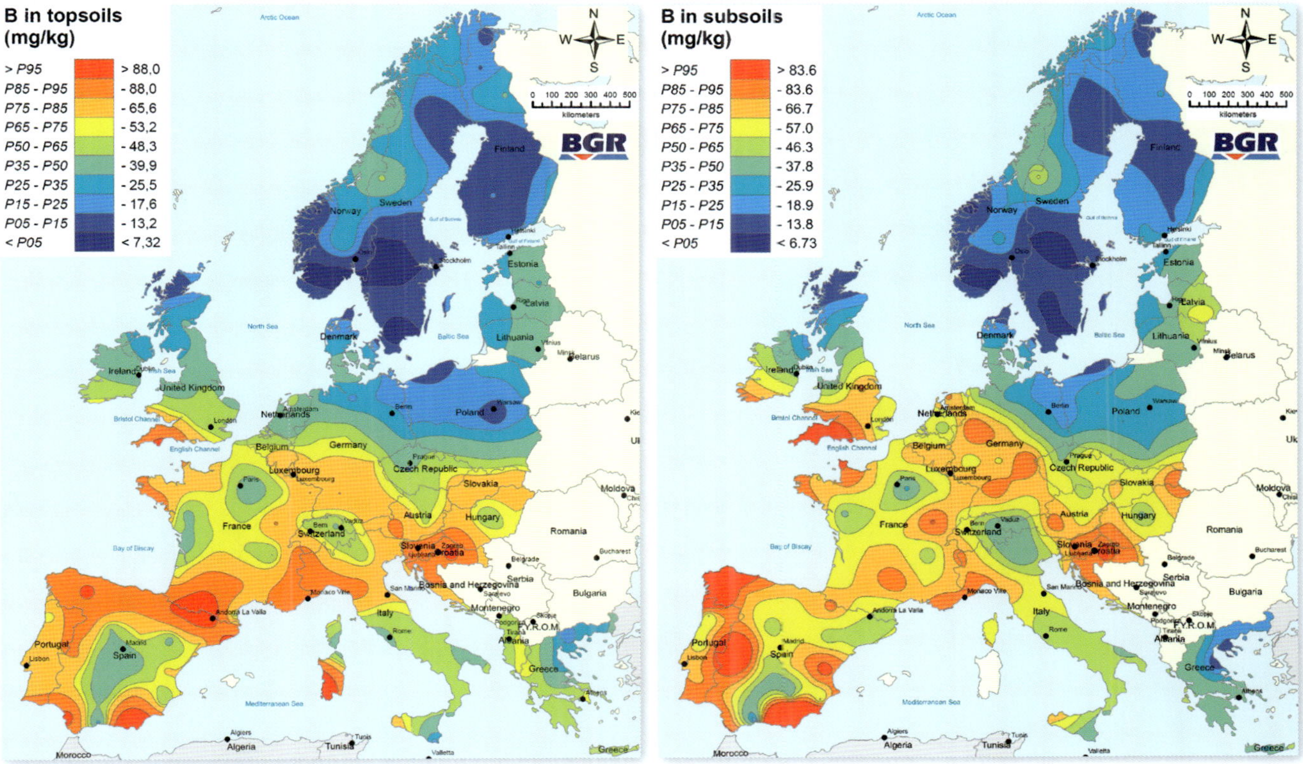

Fig. 10.2: Spatial distribution of boron (B) concentrations in topsoil and subsoil of Europe.

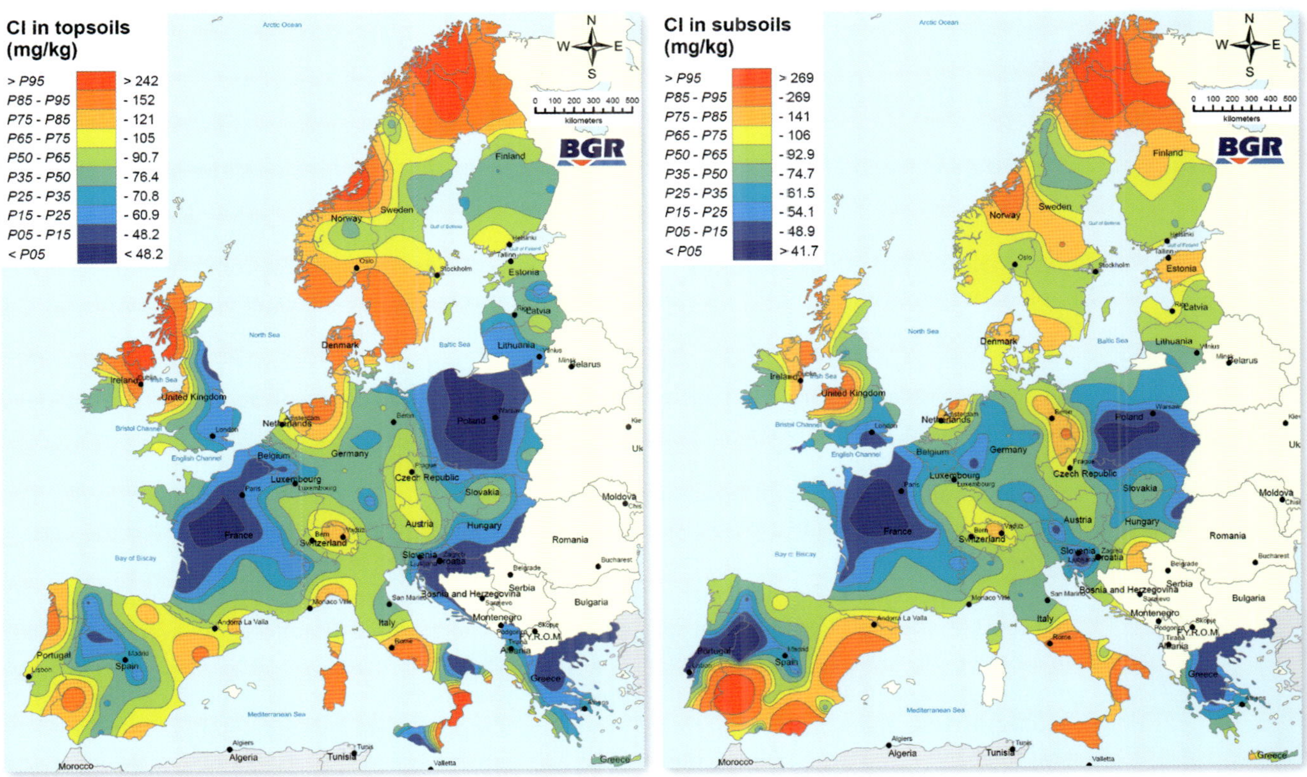

Fig. 10.3: Spatial distribution of chlorine (Cl) concentrations in topsoil and subsoil of Europe.

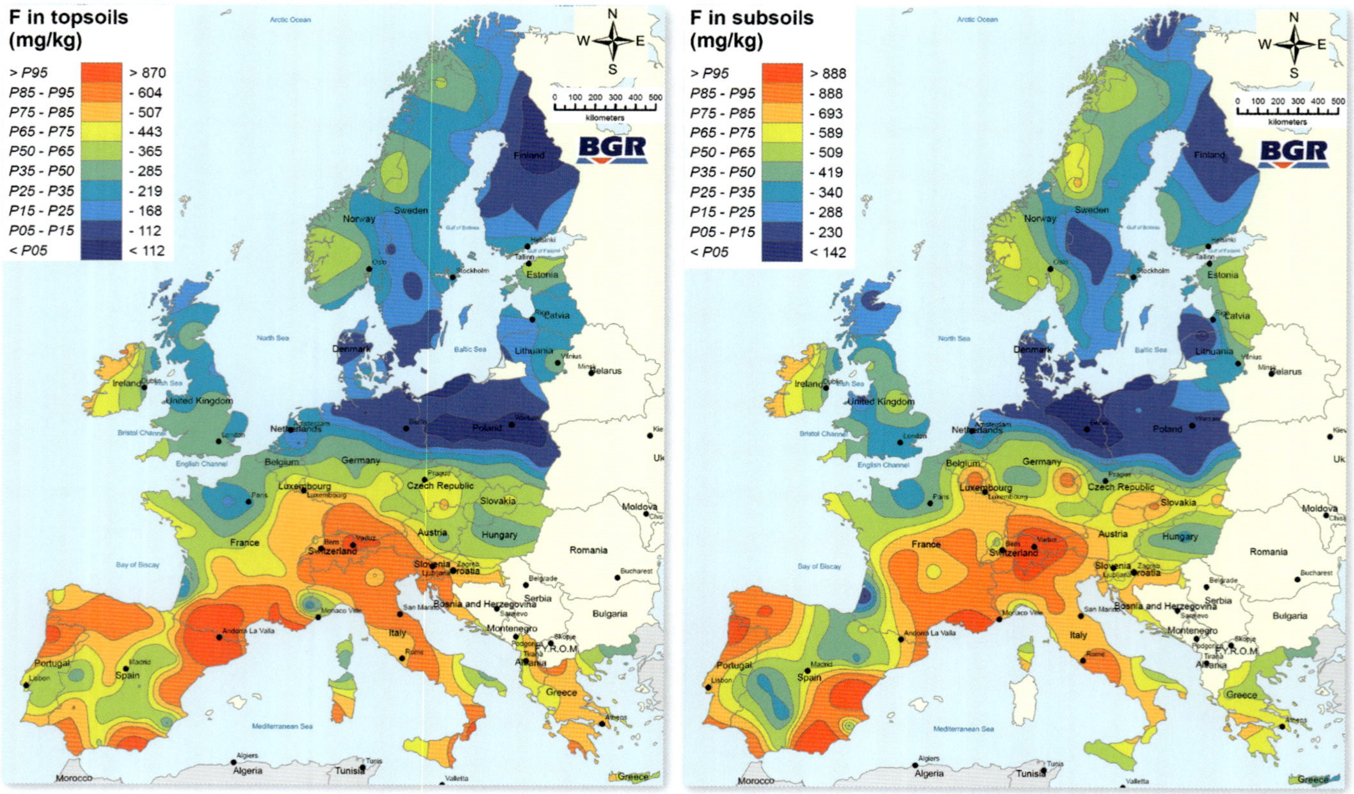

Fig. 10.4: Spatial distribution of fluorine (F) concentrations in topsoil and subsoil of Europe.

▸ 10.4 Discussion

Some data about B and its behaviour in surficial deposits are available in the literature (e. g., Waggot 1969; Ivanov 1996; Aggarwal 1999; Kabata-Pendias 2001; Table 10.2). The role of B as a plant nutrient is important. The availability of B in an aqua regia extraction is very low (probably less than 1 % of the total B in a soil sample is on average available in an aqua regia extraction). Data about total B concentrations in soil are rather rare due to analytical difficulties. Even fewer data are available for Cl and F concentrations in soil, although these are also important nutrients in the biosphere. Again, the main reason for this is that the appropriate analytical methodology at the concentration level prevailing in soil and biota are not easily available. Each of these elements requires an element specific analytical technique and these three elements are thus rather expensive to analyse. Thus, the FOREGS data presented here provide an important contribution to the understanding of the geochemistry of these elements in the Earth's crust.

The distribution and concentration levels of all studied elements can be connected to geological processes, either endogenous (differentiation of elements in magmas) or exogenous (glacial erosion and sedimentation). The atmospheric influence on the Cl concentrations in topsoil is clear. The steady input of marine aerosols has visibly increased the Cl abundances in the soil of coastal areas.

Minerogenic soil is derived from the underlying bedrock via different geological processes, which may also have mixed different materials to some extent. Therefore, the average element concentration of soil should reflect the average element concentration of bedrock. The area of Europe is certainly large enough so that the element concentrations can be compared to the estimated composition of the Upper Continental

Crust (UCC). The estimates for the B concentration in the UCC (Table 10.2) vary from 9.2 mg/kg (SHAW et al. 1967, 1976) to 28 mg/kg (GAO et al. 1998). Compared to these values the new FOREGS data presented here are considerably higher, 40 mg/kg for topsoil and 38 mg/kg for subsoil. Because the difference between subsoil and topsoil is so small, it can be assumed that the soil forming processes have not changed the concentration levels of B in the surficial deposits, despite the fact that B is considered to be one of the most mobile elements (PERELMAN 1977; KABATA-PENDIAS 2001). The countrywide data from Lithuania (GREGORAUSKIENĖ 2012) show that the soil-forming processes have depleted B to some extent from A- and E-horizons in loam-clay soil, but did not affect the concentrations in B- and C-horizons (30 mg/kg, 32 mg/kg, 43 mg/kg and 46 mg/kg, respectively). In sandy soil, similar depletion was not observed (22 mg/kg, 22 mg/kg, 27 mg/kg and 24 mg/kg, respectively). In the FOREGS project, samples were not collected according to soil horizons, but instead the uppermost 25 cm of the mineral soil was sampled, although care was taken not to mix soil from different distinct horizons. In this kind of sampling the influence of rather thin A and E horizons is negligible and the results represent the average of the whole topsoil

Table 10.2: Boron (B) concentrations in different constituents of the Earth's surface.

Constituents of Earth's surface	B abundance solid [mg/kg]	(min.–max.) [mg/kg]	water [µg/L]	References
Upper Continental Crust	9.2			SHAW et al. (1967, 1976)
	28			GAO et al. (1998)
	17			WEDEPOHL (1995)
	17			RUDNICK & GAO (2003)
	15			TAYLOR & MCLENNAN (1995)
Lithosphere	12			IVANOV (1996)
	17			WEDEPOHL (1995)
	15			TAYLOR & MCLENNAN (1995)
Ultramafic rocks	5			IVANOV (1996)
	3			KOLJONEN (1992)
Mafic rocks	5			IVANOV (1996)
	8			KOLJONEN (1992)
Intermediate rocks	15			IVANOV (1996)
Felsic rocks	15			IVANOV (1996)
	15			KOLJONEN (1992)
Clay, shale	50			IVANOV (1996)
	130			KOLJONEN (1992)
Sandstone	35			IVANOV (1996)
	35			KOLJONEN (1992)
Limestone	20			IVANOV (1996)
	20			KOLJONEN (1992)
Stream water			20	IVANOV (1996)
			10	KOLJONEN (1992)
Sea water			4400	IVANOV (1996)
			4440	LIDE (1966)
Podsol and sandy soil, Poland	9	(5–134)		KABATA-PENDIAS (2001)
Loamy and clay soil, Poland	15	(3–75)		KABATA-PENDIAS (2001)
Sandy soil on sandstone, USA	35	(< 20–100)		KABATA-PENDIAS (2001)
Agricultural soil, northern Europe	< 0.2	(< 0.2–8)*		REIMANN et al. (2003)
Sandy soil, Lithuania	22.2	(9–54)		KADŪNAS et al. (1999)
Loam–clay soil, Lithuania	32.7	(8–100)		KADŪNAS et al. (1999)
FOREGS topsoil	**40**	**(3–164)**		
FOREGS subsoil	**38**	**(4–385)**		

* Ammonium acetate extraction

Table 10.3: Chlorine (Cl) concentrations in different constituents of the Earth's surface.

Constituents of Earth's surface	Cl abundance solid [mg/kg]	(min.–max.) [mg/kg]	water [µg/L]	References
Upper Continental Crust	100			SHAW et al. (1967, 1976)
	142			GAO et al. (1998)
	640			WEDEPOHL (1995)
	370			RUDNICK & GAO (2003)
Lithosphere	180			IVANOV (1996)
Ultramafic rocks	100			IVANOV (1996)
Mafic rocks	60			IVANOV (1996)
Intermediate rocks	100			IVANOV (1996)
Felsic rocks	200			IVANOV (1996)
Clay, shale	800			IVANOV (1996)
Metamorphic rocks	200			IVANOV (1996)
Sandstone	11			IVANOV (1996)
Limestone	150			IVANOV (1996)
Stream water			7.8	IVANOV (1996)
Sea water			19,500	IVANOV (1996)
Agricultural soil water, Lithuania	17 – 33	(8 – 100)		BUIVYDAITĖ (2001)
Soil	100			FLOWERS (1988)
Soil, Russia	100			VINOGRADOV (1957)
Agricultural soil, topsoil, northern Europe	80	(<10 – 1130)		REIMANN et al. (2003)
Agricultural soil, subsoil, northern Europe	70	(<10 – 1310)		REIMANN et al. (2003)
FOREGS topsoil	**91**	**(36 – 1203)**		
FOREGS subsoil	**75**	**(31 – 719)**		

layer. There are many mechanisms of how B can be sorbed in soil, especially on clay particles or in clay lattices, and it will also be enriched in the organic material present in soil.

The estimates for the Cl concentration in the UCC (Table 10.3.) vary even more than those for B. SHAW et al. (1967, 1976) give a value of 100 mg/kg, WEDEPOHL (1995) suggests 640 mg/kg and the most recent estimate by RUDNICK & GAO (2003) is 370 mg/kg. With the exception of the oldest values provided by SHAW et al. (1967, 1976), these estimates are very high compared to the new FOREGS soil data, 91 mg/kg for topsoil and 75 mg/kg Cl for subsoil.

The estimates of F concentration in the UCC (Table 10.4) do not differ much from each other. The lowest value of 500 mg /kg is given by SHAW et al. (1967, 1976) and the highest of 611 mg/kg by WEDEPOHL (1995). Again, the new FOREGS data are lower than any of these estimates for the UCC, i.e., topsoil 365 mg/kg and subsoil 418 mg/kg F.

The area covered by the FOREGS project, 4.2 million km^2, is large enough to represent the whole UCC. It includes geological formations of all ages from Archaean to Present. It also includes all phases of sedimentation processes with their resulting deposits. The material of minerogenic subsoil is derived from bedrock. It is not much affected by recent soil-forming processes, because the samples were taken from the C-horizon. It should thus provide a rather good average for the chemical composition of the UCC. Considering the new FOREGS results for B, F and Cl, it appears that the B concentrations estimated for the UCC are underestimated, the Cl concentrations of the UCC appear to be seriously overestimated, and the F concentrations of the UCC are apparently slightly overestimated.

The ratio of median values between topsoil and subsoil (Table 10.5) shows that there is practically no difference in B concentrations (topsoil/

Table 10.4: Fluorine (F) concentrations in different compartments of the Earth's surface.

Constituents of Earth's surface	F abundance solid [mg/kg]	(min.–max.) [mg/kg]	water [µg/L]	References
Upper Continental Crust	500			SHAW et al. (1967, 1976)
	561			GAO et al. (1998)
	611			WEDEPOHL (1995)
	557			RUDNICK & GAO (2003)
Lithosphere	640			IVANOV (1996)
Ultramafic rock	100			IVANOV (1996)
Basic rock	400			IVANOV (1996)
Medium rock	500			IVANOV (1996)
Acidic rock	800			IVANOV (1996)
Clay, shale	600			IVANOV (1996)
Metamorphic rock	700			IVANOV (1996)
Sandstone	300			IVANOV (1996)
Limestone	300			IVANOV (1996)
Stream water			100	IVANOV (1996)
Sea water			1300	IVANOV (1996)
Soil	200			VINOGRADOV (1957)
Podsol and sandy soil, Poland	83	(20–150)		KABATA-PENDIAS (2001)
Loamy and clay soil, Poland	418	(250–750)		KABATA-PENDIAS (2001)
Sandy soils on sandstone, USA	205	(<10–1100)		KABATA-PENDIAS (2001)
Agricultural soil, topsoil, northern Europe	<200	(<200–1800)		REIMANN et al. (2003)
Agricultural soil, subsoil, northern Europe	<200	(<200–2220)		REIMANN et al. (2003)
Sandy soil, Russia	85			SAET et al. (1990)
Clay soil, Russia	650			SAET et al. (1990)
Humus layer of forest soil	0.3–0.12			SAET et al. (1990)
FOREGS topsoil	**365**	**(70–2730)**		
FOREGS subsoil	**418**	**(59–1830)**		

subsoil ratio 1.05), whilst Cl is clearly enriched in topsoil (ratio 1.21) and F is somewhat depleted from topsoil (ratio 0.87). For the vast majority of the elements analysed in the FOREGS project this ratio is between 0.9–1.1 (DE VOS et al. 2006a, b). Thus, B conforms well to the general situation; F shows an exceptional depletion in topsoil and Cl a very exceptional enrichment in topsoil. The main reason for the enrichment of Cl is the input via marine aerosols, because the highest concentrations are seen to be located at coastal areas. Despite the fact that Cl and B both are considered as marine elements (e.g., PERELMAN 1977), Cl is more mobile, and it is not sorbed in surficial deposits, as does B when it is adsorbed on clay and organic matter of topsoil. Another interesting feature is that the area of elevated F concentrations coincides with the high weathering index of soil (DE VOS et al. 2006a, b).

Soil type has a clear influence on the concentrations of B and F according to the grain size. Both elements are depleted from sandy soil and enriched in loamy and clayey soil. KABATA-PENDIAS (2001) reported from Poland 85 mg/kg for F in sandy soil and 418 mg/kg for F in clayey soil. From Russia, SAET et al. (1990) reported respectively 85 mg/kg and 650 mg/kg F. The difference of B concentration between sandy and clayey soil is much smaller. KABATA-PENDIAS (2001) reported from Poland 9 mg/kg and 15 mg/kg B, respectively, and KADŪNAS et al. (1999) reported 22 mg/kg and 33 mg/kg B, respectively, from Lithuania (Table 10.2). This is in accordance with the FOREGS data, which show lower concentrations of these elements in the glaciated terrain of northern Europe where sandy glacial sediments prevail.

Table 10.5: Ratio of median values of element concentration in topsoil and subsoil.

Element	Topsoil/Subsoil
B	1.05
Cl	1.21
F	0.87

▸ 10.5 Conclusions

New values for total concentrations of B, Cl and F in soil at the continental (European) scale are presented. The median values are 40/38, 91/74 and 365/418 mg/kg for B, Cl and F in topsoil and subsoil, respectively. Compared to published estimates of the concentrations of these rarely analysed elements in the Upper Continental Crust, the published UCC values are probably underestimated for B, slightly overestimated for F and seriously overestimated for Cl.

Boron and F originate predominantly from geological formations and the observed spatial variation depends only on the spatial variation in geology. Chlorine in topsoil shows a remarkable additional input from sea spray.

Boron does not exhibit any remarkable difference between topsoil and subsoil. Chlorine is enriched in topsoil, mainly because of the additional input via sea spray, while fluorine is depleted in topsoil. |◂

Reijo Salminen
Geological Survey of Finland (GTK);
Espoo, Finland
E-mail: reijo.salminen@seutuposti.fi

Manfred Birke
Federal Institute for Geosciences
and Natural Resources (BGR);
Hannover; Germany

Clemens Reimann
Geological Survey of Norway (NGU);
Trondheim, Norway

Virgilija Gregorauskiené
Lithuanian Geological Survey (LGT);
Vilnius, Lithuania

Xueqiu Wang
Institute for Geophysical
and Geochemical Exploration (IGGE);
Langfang, China

Part B

Chapters 11 – 13

Mobility and Risk Assessment of Metals

Chapter 11

Prediction of Metal and Metalloid Partitioning Coefficients (K_d) in Soil Using Mid-Infrared Diffuse Reflectance Spectroscopy

LES J. JANIK, SEAN FORRESTER, JASON K. KIRBY, MICHAEL J. MCLAUGHLIN, JOSÉ M. SORIANO-DISLA & CLEMENS REIMANN

▸ 11.1 Introduction

Solid-solution partitioning coefficients (K_d value) play a major role in the mobility and bioavailability of metals in soil (DEGRYSE et al. 2009). Soluble metals are assumed to be mobile and potentially bioavailable to soil organisms. An understanding of the solid-solution partitioning can reveal potential exposure pathways of metals in soil and which biota are likely to be exposed once a contaminant is released into the environment. This exposure pathway may also be important for agricultural land use scenarios to manage exposure of crops and the general population (through the food chain) to these elements. Partitioning coefficients are generally determined as the ratio of the element concentration of the solid phase to the total element concentration in solution (DEGRYSE et al. 2009).

The problem with determining K_d values for metals in large soil sets is the requirement for large sample numbers for replicated incremental additions of metals for each sample, thus increasing analysis times and costs. Thus, a more rapid surrogate method would be an advantage. Methods using easily and rapidly acquired data from diffuse reflectance mid-infrared (MIR) spectra (4000 – 400 /cm), together with partial least squares (PLS) regression, could reduce time and expense. The partition coefficient of metals and metalloids in soil may be predicted if soil properties are related to MIR absorption frequencies and intensities, and there is a correlation between K_d values and soil properties. The partitioning of metals in soil has been shown to be related to soil physicochemical properties, most notably soil pH and occurrence of amorphous iron oxides, aluminium oxides, clay, soil organic matter content, dissolved organic matter and particulate organic carbon (e.g., MCBRIDE 1989; SAUVÉ et al. 2000), attributes capable of prediction by infrared spectroscopy (JANIK et al. 1998).

The aim of this study was to show, for the first time, that the MIR diffuse reflectance infrared Fourier transform (DRIFT)-PLS method could be used to predict K_d values for the metallic cations silver (Ag^{+}), cobalt (Co^{2+}), copper (Cu^{2+}), manganese (Mn^{2+}), nickel (Ni^{2+}), lead (Pb^{2+}), tin (Sn^{4+}), and zinc (Zn^{2+}), anionic metal and metalloid oxyanions molybdenum (Mo^{6+} as MoO_4^{2-}), antimony (Sb^{5+} oxidised from Sb^{3+} as Sb_2O_3 which hydrolyses to $Sb(OH)_6^{-}$), selenium (Se^{6+} as SeO_4^{2-}), tellurium (Te^{6+} as TeO_4^{2-}) and vanadium (V^{5+} as VO_3^{-}), and uncharged boron (B^{3+} as $H_3BO_3^{0}$), and then to use these models to predict K_d values for the complete GEMAS set of soil samples. K_d values were determined by a single-point spike method using radioisotopes of the elements of interest or a single concentration of unlabelled element if no suitable isotope was available. K_d values were calculated from element concentrations or isotope activities as follows:

$$K_d\ (L/kg) = \frac{M_{solid\ phase}(mg\ or\ Bq/kg)}{M_{solution\ phase}(mg\ or\ Bq/L)}$$

▸ 11.2 Partial-least-squares models

Partition coefficient values were determined using a single point soluble metal or radioactive isotope spike (OECD 2000) on a spectrally-representative subset of ~500 GEMAS soil samples, and a DRIFT-PLS calibration model from only soil spectra (DRIFT) was developed to predict K_d values (more information on DRIFT spectroscopy and PLS methods can be found in Janik et al. (1998, 2009). Because of the relationships between soil pH and K_d (McBride 1989), a further model was tested, including DRIFT plus laboratory-determined pH values (DRIFT + pH). Soil pH data can be approximately derived, of course, by PLS prediction from soil spectra based on correlations with other soil properties, e.g., carbonate, organic acids and Al oxy-hydroxides (Janik et al. 1998). Unfortunately, such pH regressions are not particularly accurate over large geographical areas, since the soil components responsible for soil acidity can vary greatly from one soil type to another (Reeves et al. 2001).

The calibration models were trained by 'leave-one-out' cross-validation. However, this method tends to be optimistic in reporting modelling errors so that an independent test set was used to validate regression models properly. The quality of regression models can be ascertained from the residual predictive deviation (RPD = standard deviation/root mean square error), with values < 1.5 considered poor, 1.5 – 1.9 suggesting indicator quality, 1.9 – 3.0 indicating good quality, and > 3.0 of analytical quality (Janik et al. 2009).

11.2.1 Cations

The PLS models could be separated into two classes: (a) models that predicted experimental K_d values well (Class 1), and (b) models that predicted experimental K_d values poorly (Class 2) (Table 11.1). Inclusion of soil pH into the input variables together with DRIFT spectra improved the accuracy of the models for Class 1 metals. Class 2 metal regressions using DRIFT+pH models remained 'poor'; with regard to their ability to predict K_d values based on both R^2 and RPD values (Table 11.1). These regression models were considered as 'poor', suggesting that partitioning of these metals was not related either to pH or to other soil components that are MIR-active. The prediction performance of the DRIFT+pH models in the test validation soil samples was similar to that of the full calibration set (Table 11.1).

11.2.2 Oxyanions and boron

Similar to cationic metals, the DRIFT models could be separated in two classes: (a) models that predicted experimental K_d values well (Class 1), and

Table 11.1: Summary statistics; range and median (L/kg), class, coefficient of determination (R^2), root-mean-square-error (RMSE, L/kg) and residual predictive deviation (RPD) for the PLS cross-validation with DRIFT only and DRIFT + pH data, and test set validation using DRIFT + pH data.

Metal	Range [L/kg]	Median [L/kg]	Class	log-K_d (DRIFT) R^2	RMSE	RPD	log-K_d (DRIFT+ pH) R^2	RMSE	RPD	Test set validation (DRIFT+ pH) N Cal	N Val	R^2	RMSE	RPD
Zn	2 – 20,276	1737	1	0.78	0.47	2.1	0.93	0.27	3.7	292	206	0.90	0.33	3.0
Mn	1 – 14,288	1195	1	0.70	0.79	1.8	0.88	0.50	2.9	278	158	0.86	0.59	2.5
Co	3 – 15,739	2285	1	0.71	0.62	1.9	0.83	0.48	2.4	275	178	0.81	0.56	2.1
Ni	4 – 3925	549	1	0.72	0.35	1.9	0.87	0.24	2.8	289	205	0.83	0.30	2.2
Pb	10 – 339,624	10,939	1	0.70	0.48	1.9	0.84	0.35	2.6	284	197	0.85	0.40	2.2
Cu	23 – 8589	1643	2	0.40	0.30	1.3	0.46	0.28	1.4	255	195	0.43	0.30	1.3
Sn	60 – 22,079	2500	2	0.32	0.47	1.2	0.32	0.47	1.2	144	171	0.32	0.43	1.3
Ag	159 – 4655	2623	2	0.33	0.24	1.2	0.35	0.23	1.2	171	111	0.31	0.24	1.2

N number of samples used in the models, divided into Cal (calibration) and Val (validation) for the test-set validation

(b) models that predicted poorly (Class 2) (Table 11.2). The derived RPD values ranged from 1.6 – 1.9., suggesting indicator quality predictions for all oxyanions and B, except for Se ('poor'; RPD = 1.3). A significant improvement in accuracy was found for Mo, Te and Sb (RPD now promoted to 'good'), when DRIFT was considered together with pH in agreement with the role of pH controlling the K_d of these elements. On the other hand, only a slight improvement was observed for B, and no improvement for Se and V. As it was the case of the cation models, the derived R^2 and RPD values for the DRIFT+pH and the 'Test' set were similar (Table 11.2).

▸ 11.3 Prediction of K_d values from DRIFT + pH models

The DRIFT + pH models were applied to the prediction of K_d values for Class 1 metals in 4313 unknown soil samples, and partitioning maps for agricultural and grazing land soil across Europe were developed. An example of the spatial distribution of predicted K_d values is provided in Figures 11.1 and 11.2 for Ni and Mo, respectively. These maps are representative of elements of which K_d values were highly positively (Co, Mn, Pb, Te and Zn) and negatively (Sb) related to soil pH. A good degree of agreement can be observed between maps of agricultural and grazing land soil.

11.3.1 Cations

The distribution of predicted K_d values for the GEMAS soil samples was well represented by the DRIFT + pH PLS models (data no shown). The accuracy of K_d value prediction of 'unknown' soil samples can also be expressed as empirical 'deviation' values.

According to this, there were few unknown samples for the different metals with high deviation values (i. e., the unknown soil samples are likely to be predicted with similar accuracy to the calibration models). Thus, the prediction of future 'unknown' samples using the models developed for Class 1 metals is reliable.

Figure 11.1 clearly shows the lowest strength of Ni partitioning in northern Europe (especially the Caledonides and Baltic Shield) and a low variability. The rest of Europe exhibits a higher variability for Ni partitioning, with the highest values generally found in southern and eastern Europe. Exceptions to high values in southern Europe are granite areas in the north-west of the Iberian Peninsula and the Massif Central in France. These spatial trends are associated with the geographical patterns of soil pH induced by climate (principally rainfall) and soil parent material.

11.3.2 Oxyanions and boron

The distributions of predicted K_d values were found to be broadly similar to that of the calibration samples, except for Mo and V that had higher predicted values, and Se with lower predicted values. This suggested that the distribution of predicted K_d values for Sb and Te were well represented by the DRIFT + pH models, but Mo, V and Se could have been improved with additional samples in the calibration models.

Table 11.2: Summary statistics; range and median (L/kg), class, coefficient of determination (R^2), root-mean-square-error (RMSE, L/kg) and residual predictive deviation (RPD) for the PLS cross-validation with DRIFT only and DRIFT + pH data, and test set validation using DRIFT + pH data.

Metal	Range	Median	Class	log-K_d (DRIFT)			log-K_d (DRIFT+ pH)			Test set validation (DRIFT+ pH)				
	[L/kg]			R^2	RMSE	RPD	R^2	RMSE	RPD	N Cal	N Val	R^2	RMSE	RPD
Te	0.32 – 2433	193	1	0.72	0.52	1.9	0.79	0.45	2.2	260	186	0.78	0.45	2.1
Mo	0.70 – 7078	41.7	1	0.63	0.48	1.7	0.75	0.38	2.1	297	212	0.67	0.45	1.8
Sb	0.51 – 5494	67.9	1	0.64	0.31	1.7	0.74	0.27	2.0	295	212	0.72	0.27	2.0
V	0.35 – 11,507	596	1	0.61	0.39	1.6	0.62	0.39	1.6	290	212	0.64	0.41	1.7
B	0.38 – 51.88	2.15	1	0.66	0.19	1.7	0.68	0.18	1.8	300	215	0.65	0.18	1.6
Se	0.58 – 6339	2.20	2	0.43	0.56	1.3	0.43	0.56	1.3	296	213	0.39	0.59	1.3

N number of samples used in the models, divided into Cal (calibration) and Val (validation) for the test-set validation

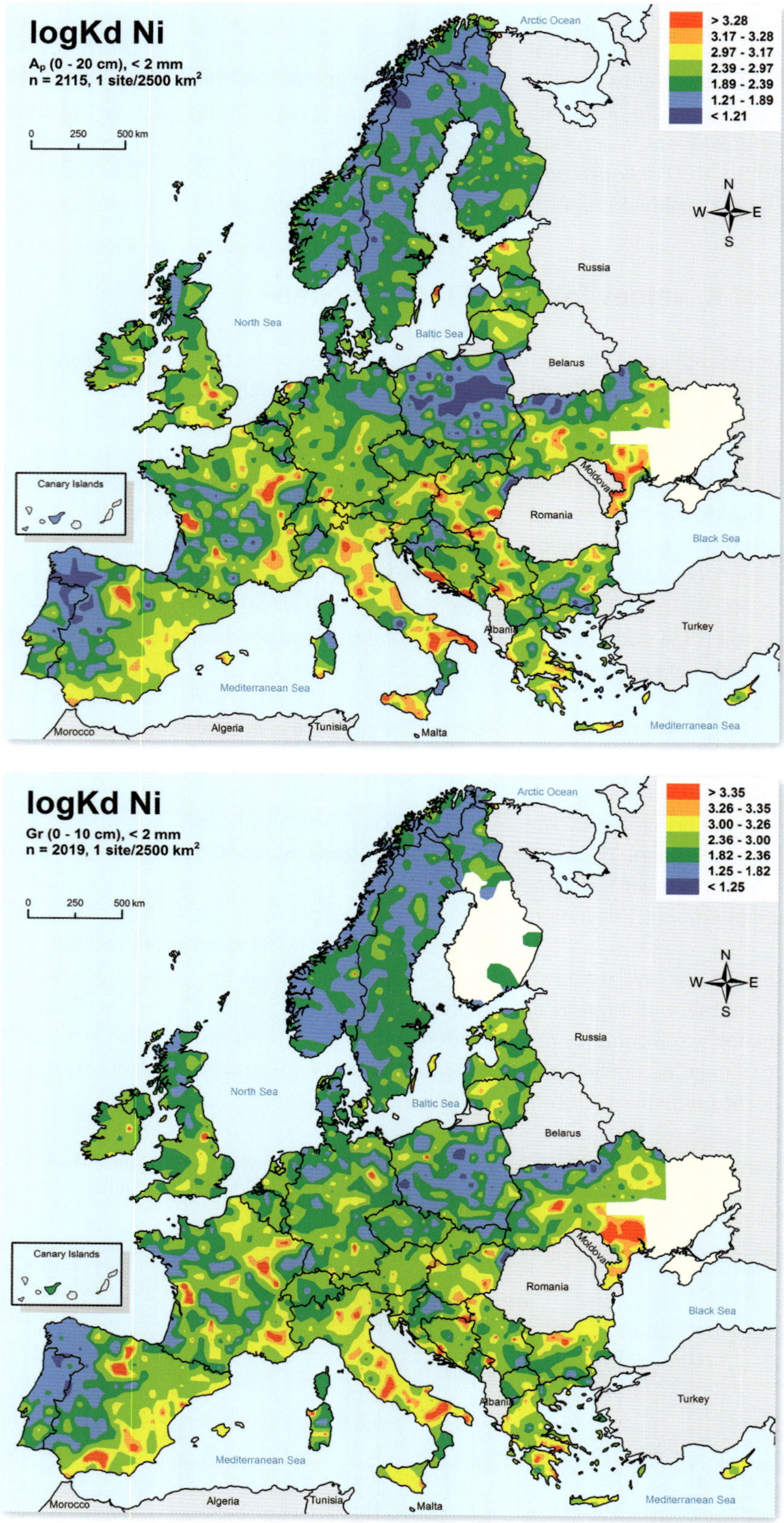

Fig. 11.1: Maps of soil-solution partitioning values (logarithm10 transformed, L/kg) for nickel (Ni) (K_d values) in agricultural (Ap) and grazing land (Gr) soil across Europe.

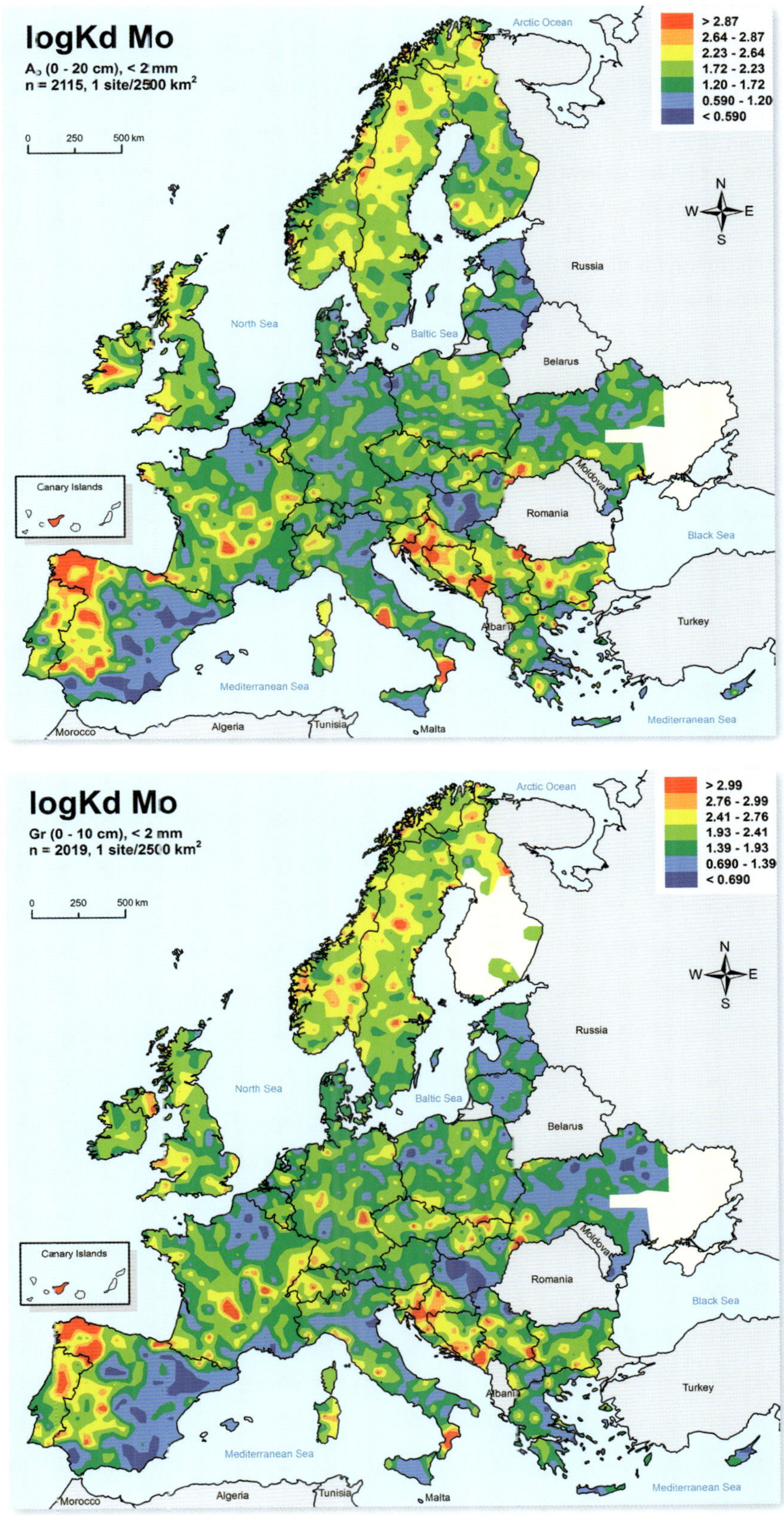

Fig. 11.2: Maps of soil-solution partitioning values (logarithm 10 transformed, L/kg) for molybdenum (Mo) (K_d values) in agricultural (Ap) and grazing land (Gr) soil across Europe.

With regard to the 'deviation' values for the predicted samples almost all samples in the GEMAS predicted sample set had low 'deviation' values. Deviations for B were, however, slightly lower than the calibration experimental values, while Mo and Sb were slightly higher.

The geographical distribution of predicted K_d values for Mo (Fig. 11.2) is somewhat opposite to that found for Ni, due to the opposite relationship of K_d with soil pH values. However, a higher variability is observed for Mo, especially in southern and eastern Europe with the lowest values found for eastern Spain (carbonates). These low K_d values contrast with the very high values found in western Spain and Portugal, central and southern Italy and the Dinarides. The variability is lower in the UK and Scandinavia, where moderate to high K_d values are observed. The spatial trends are also associated with the geographical patterns of soil pH and the fact that Mo partitioning in soil is related to properties controlling anion sorption.

▸ 11.4 Conclusions

Depending on the accuracy of K_d predictions needed and the metal or metalloid considered, DRIFT-PLS analysis can provide good predictions for a number of metal and metalloids and is extremely cheap and inexpensive. Furthermore, the infrared method offers the further possibility of predicting K_d values in-situ in the field using hand-held spectrometers. |◂

Les J. Janik
CSIRO Land and Water;
Adelaide, South Australia
E-mail: les.janik@csiro.au

Sean Forrester
CSIRO Land and Water;
Adelaide, South Australia

Jason K. Kirby
CSIRO Land and Water;
Adelaide, South Australia

Michael J. McLaughlin
CSIRO Land and Water;
Adelaide, South Australia

José M. Soriano-Disla
CSIRO Land and Water;
Adelaide, South Australia

Clemens Reimann
Geological Survey of Norway (NGU);
Trondheim, Norway

Use of Monitoring Data for Risk Assessment of Metals in Soil under the European REACH Regulation

KOEN OORTS & ILSE SCHOETERS

▸ 12.1 Introduction

The European REACH Regulation (Registration, Evaluation, Authorisation and Restriction of Chemicals (EC No 1907/2006) specifies that industry must prove that it can produce and use its substances safely for humans and organisms living in the environment. Therefore, risks due to the exposure to a substance during production and use have to be reliably assessed. This must be done at the local, regional and continental scales (Fig. 12.1). The local or 'site-specific' scale considers the risks in the vicinity of a point source, e.g., an industrial plant producing or using the substance. Risks at regional scale (country, state, region) are assessed by taking into account all releases from point and diffuse sources in the region. The third spatial scale – the continental scale – takes into account all sources of exposure at the European scale. Exposure in the various environmental compartments (water, sediment, soil, air) is assessed using modelled and, where available, measured data. Finally, for each compartment, risks are assessed at different scales by comparing the typical and reasonable worst-case exposure values to a derived safe limit value for the substance (effect thresholds).

Metals are naturally occurring substances and as such, they can always be found at natural background concentrations, which vary in space due to natural geological, biogeochemical and other processes. Because of historical and current anthropogenic input from diffuse (e.g., fossil fuel combustion, traffic, agricultural inputs) and local (from industry) sources, the direct measurement of real natural background concentrations (i.e., the natural metal concentration in the environment that reflects the situation before any human activity disturbed the natural equilibrium) is not feasible in most parts of the world (ELBERLING et al. 2010; REIMANN & GARRETT 2005). Hence, metal concentrations in the environment are the result of the natural background, historical contamination and current emissions associated with the production and use of metals and metal compounds. These anthropogenic inputs differ with land use and the proximity of point sources.

Due to the variations of natural background concentrations and anthropogenic inputs, large differences in metal levels can be observed between different locations. Soil is thus not characterised by one single background concentration, but instead is represented by a background con-

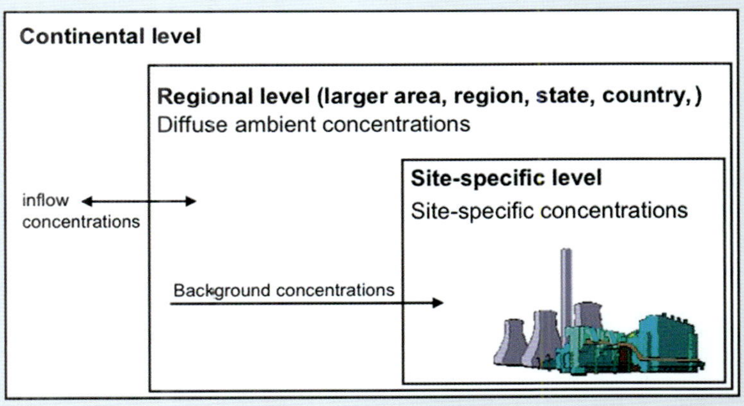

Fig. 12.1: The relationships between exposure assessments at different spatial scales (from DEFRA et al. 2007).

centration range for metals. This spatial variation in background levels in soil may have a significant impact on the outcome of the risk assessment process through, e.g., the direct variation in exposure to metals in soil, but also through potential adaptation of terrestrial organisms to high or low metal concentrations in their native environment. Therefore, the spatial variation in background levels should be taken into account in a sound risk assessment for metals in soil to prevent that, for example, risks are identified at natural background levels.

Environmental exposure concentrations of substances can be assessed using modelled or measured data. Modelling the fate and behaviour of estimated metal emissions in the environment has the advantage that the contribution of specific sources can be assessed. However, there is often a high degree of uncertainty associated with modelled data, because the natural background concentrations for natural elements are not taken into account. Measured data of sufficient quality, on the other hand, provide a more realistic reflection of the current environmental exposures and include all possible contributions. Therefore, when sufficient representative and reliable measured data are available, monitoring data are generally preferred over model calculations for exposure assessment of natural elements.

Apart from the exposure concentrations for natural elements in soil, the toxicity of metals also strongly varies among soil types. Metal toxicity to terrestrial organisms (plants, invertebrates and micro-organisms) is typically studied by exposing these organisms to an uncontaminated soil recently treated with increasing doses of a soluble metal salt (e.g., $CuCl_2$, Na_2MoO_4). From the resulting dose-response relationship, toxicity thresholds (NOEC, No Observed Effect Concentrations, i.e., highest dose at which no significant inhibitory effect is observed, or EC10 values, i.e., effective concentration yielding 10 % inhibition in response) can be calculated for the effect of that specific metal on the endpoint studied (e.g., plant growth, invertebrate reproduction, microbial respiration). Metal toxicity thresholds for the same endpoint can vary up to two orders of magnitude among different freshly spiked soil (e.g., Li et al. 2009; Oorts et al. 2006; van Gestel et al. 2011). Generally, it is indeed accepted that the total metal concentration in soil is a poor predictor for toxicity to the environment. However, there is not yet a generally accepted method for measurement of the bioavailable fraction of metals in soil (Menzies et al. 2007; Zhao et al. 2006), and data for (pseudo-) total metal concentrations, based on concentrated acid digestions, are most commonly available and used.

During the last decade, a lot of research has been performed on bioavailability and toxicity of metals in soil. Physico-chemical soil properties (e.g., pH, organic carbon and clay content, cation exchange capacity) can explain a significant part of the variation in metal toxicity thresholds for plants, invertebrates and micro-organisms based on (pseudo-) total metal concentrations in soil and empirical regression models are derived, covering a wide range of European soil types (Smolders et al. 2009). Such regression models can be used to normalise effects data towards specific conditions of a site of interest and, hence, allow for the derivation of site-specific threshold concentrations (Predicted No Effect Concentration, PNEC), expressed as total metal concentration. This threshold value can be directly compared with the measured or modelled total exposure concentration to assess the potential risks at the site of interest.

In a standard risk assessment for REACH, worst-case approaches are typically used, but unfortunately these do not account for the large variation in both exposure and effect concentrations across soil types. Exposure and effect concentrations for metals typically overlap (Fig. 12.2),

Fig. 12.2: Probability distribution of copper (Cu) exposure concentrations in agricultural soil (GEMAS Ap data) and selected effect concentrations (NOEC or EC10) without bioavailability correction.

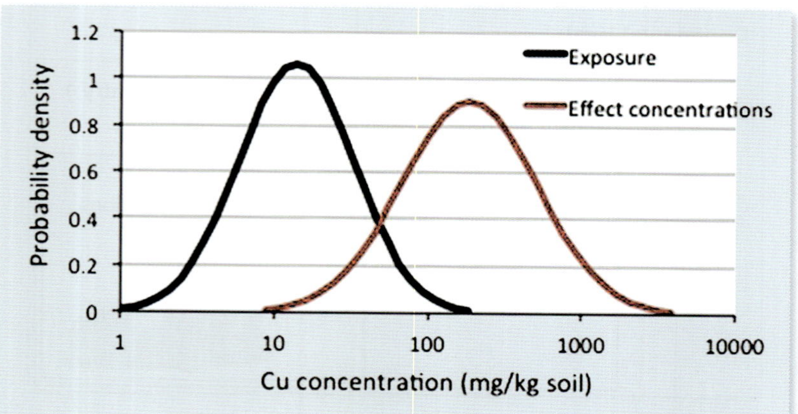

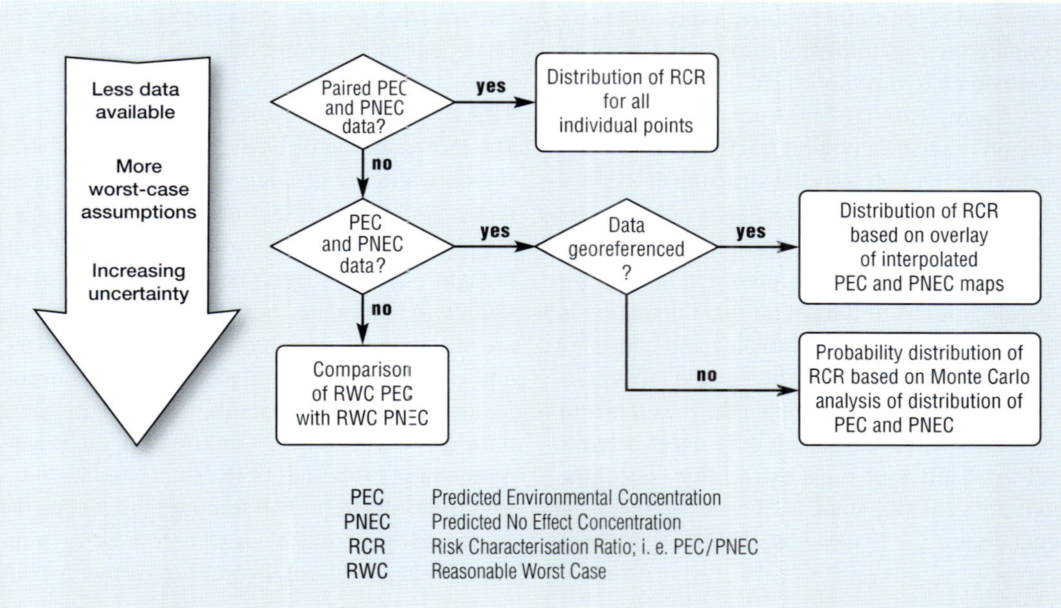

Fig. 12.3: Data availability and uncertainty for regional risk assessment for metals in soil.

and worst-case assumptions for both parameters (i.e., high exposure and low effect concentrations) result in over-conservative conclusions. A sound risk assessment for metals in soil must preferentially take into account the spatial variation of both exposure and effects concentrations. Data availability for soil metal concentrations and soil properties influencing the toxicity of metals in soil differs largely across different countries or regions in Europe, resulting in varying degrees of (worst-case) assumptions on exposure and effects of metals (Fig. 12.3). The use of different sampling protocols and analytical methods further complicated comparison of results. Therefore, there was a strong need for a harmonised European monitoring data set for metal exposure and soil properties known to affect metal bioavailability. The GEMAS project provices such a harmonised and geo-referenced, high quality data set for accurate assessment of metal exposure and effects concentrations in agricultural and grazing land soil across Europe.

▸ 12.2 Materials and methods

12.2.1 Exposure assessment

The GEMAS project provides high quality and comparable exposure data for elements in agricultural and grazing land at the European scale. The measured element concentrations from the GEMAS project can be considered as ambient background concentrations, i.e., the sum of the natural background of an element with diffuse anthropogenic input in the past or present (due to, e.g., agricultural inputs, combustion of fossil fuels, vehicle traffic, etc.). The influence of point sources (e.g., from local industrial activities) is avoided. The consistent land use and sampling depth (0–20 cm and 0–10 cm for agricultural (arable) and grazing land soil, respectively) ensure a comparable level of exposure for all samples within the same land use.

Aqua-regia (3 parts HCl + 1 part HNO_3) soluble metal concentrations are selected as basis for exposure in metal risk-assessments. The aqua regia digestion method is harmonised as an International Standard (ISO 11466: 1995) and is applied in most European Union (EU) countries. Although not a measure for the real total metal concentrations, data on metal concentrations, based on aqua regia, or comparable concentrated acid destructions, are, therefore, more commonly available in national or international monitoring programmes for metal concentrations in topsoil compared to real total metal concentrations determined by Hydrofluoric acid (HF) extraction or X-ray fluorescence (XRF). The aqua regia digestion method releases all metal fractions, except the fraction built into the crystal structure of the minerals in soil. This mineral fraction is not expected to be released over a reasonable time span under conditions normally encountered in nature. The fraction not covered by aqua regia digestion is, therefore, judged not bioavailable and hence not critical for a risk assessment. For the most common cationic metals, with regard to environmental effects (Cu, Ni, Zn, Cd, Pb, Co), differences between real-total and aqua regia extractable concentrations are generally small (>75% recovery in aqua regia digest; see REIMANN et al. 2011) and therefore the latter are often considered as pseudo-total or near-total concentrations.

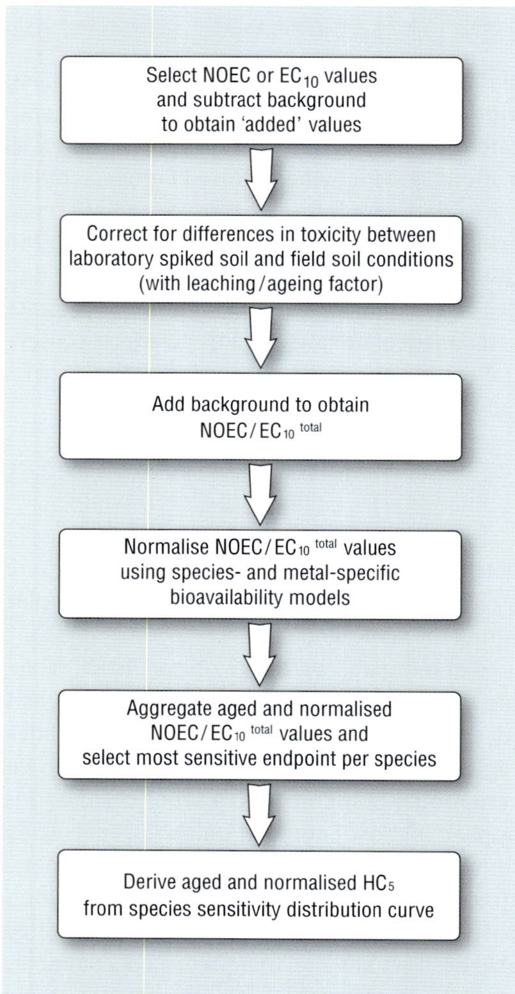

Fig. 12.4: Flow chart for the implementation of bioavailability factors into soil threshold concentrations for direct effects of elements to terrestrial organisms (plants, invertebrates and micro-organisms). The NOEC is the highest no-observed-effect concentration, the EC10 the concentration eliciting 10% adverse effect.

12.2.2 Effects assessment

The incorporation of bioavailability models into the derivation of ecological soil standards (Predicted No Effect Concentrations, PNEC) for metals in the European risk assessments is shown in Figure 12.4, and described by SMOLDERS et al. (2009). In summary, these effects assessments are based on reliable chronic toxicity data (NOEC and EC10) from standard ecotoxicity tests in soil freshly amended with soluble metal salts. Only data with sufficient information on the physico-chemical properties of the soil tested were selected. Results from such standard tests in freshly spiked soil generally overestimate toxicity in realistic field conditions. In order to correct for difference in bioavailability and toxicity of metals between such freshly amended soil, and realistic field conditions due to slow fixation reactions (ageing), as well as the absence of excess ions after spiking with a soluble salt (leaching effect) (e.g., LOCK et al. 2006; OORTS et al. 2006a; WENDLING et al. 2008; VAN GESTEL et al. 2012), all toxicity data, based on added metal concentrations, were multiplied with a so-called leaching-ageing (L/A) factor (Table 12.1). This L/A factor is only applied on the added dose, as the bioavailability of the background metal concentration in soil is not expected to further change with time. After adding the background concentration, the total 'aged' NOEC and EC10 values were normalised for differences in metal availability between the tested soil and the soil to which the PNEC is calculated, using the slope of the regressions between toxicity and selected soil properties. The input parameters required are dependent upon the metal under consideration (Table 12.1), and are generally readily available soil parameters likely to be determined in routine soil analyses (eCEC, pH, organic carbon and clay content). The effective CEC (eCEC) was generally the parameter that best explained the variation in toxicity among soil samples for the cationic metals (Cu, Co, Ni and Zn), while pH and clay content were best correlated with toxicity thresholds for Mo, which is present in soil as the oxyanion molybdate. Based on these normalised values, a species mean threshold was calculated

Table 12.1: Bioavailability corrections for metal toxicity to soil organisms.

Element	Leaching-ageing factor	Soil properties for data normalisation	Assessment factor
Co	1.1 – 3.5 (increasing as a function of pH)	eCEC	2
Cu	2	eCEC, % organic carbon, % clay and pH	1
Mo	2	pH and % clay	1
Ni	1 – 3 (increasing as a function of pH)	eCEC	2
Zn	3	eCEC, pH and background Zn	1

eCEC effective cation exchange capacity = CEC at prevailing soil pH

for each plant, invertebrate and microbial process covered. A critical concentration, protecting 95% of all species or processes, was derived as the 5th percentile of the cumulative distribution (HC_5) of species mean thresholds (species sensitivity distribution, Fig. 12.5). This value was further divided by an assessment factor, taking into account the level of residual uncertainty on the effects assessment.

It is important to stress that all methods for analysis of soil properties (pH, organic carbon, clay and eCEC) in the GEMAS project were consistent with methods used for the derivation of metal bioavailability models in the EU metal risk assessments for these metals. Therefore, data from the GEMAS project can be used directly for calculation of site-specific PNEC values.

12.2.3 Risk characterisation

In a risk characterisation, exposure concentrations (PEC: Predicted Environmental Concentration) are compared with effect concentrations (PNEC). The GEMAS results for aqua-regia extractions can be directly compared with the site-specific PNEC values, calculated based on the soil properties measured for the GEMAS sample sites. Each site-specific PNEC value was compared with the corresponding measured aqua-regia extractable metal concentration for the same site, and the risk characterisation ratio (RCR) for this site was calculated as PEC/PNEC. This approach results in distributions of RCR values for Europe and each individual country.

Distributions of PNEC and RCR values for the EU and individual countries are calculated non-parametrically for data sets with more than 25 data points. For countries with smaller sample size, parametric distributions were fitted to the data, and used for determination of the 10th, 50th (median) and 90th percentiles of the PNEC and RCR data. Because sites were sampled in a regular grid over Europe and there is no bias due to spatial heterogeneity of sampling density, it is appropriate to base the distributions of PNEC and RCR values on the measured observations, and no interpolation is needed to derive area-based distributions.

Analogous to the derivation of a generic or typical predicted environmental concentration (PEC), based on monitoring data (REACH Guidance on information requirements and chemical safety assessment, Chapter R.16.4, ECHA 2010), a generic or typical regional RCR was calculated as the median value of the 90th percentiles of RCR values for the individual countries.

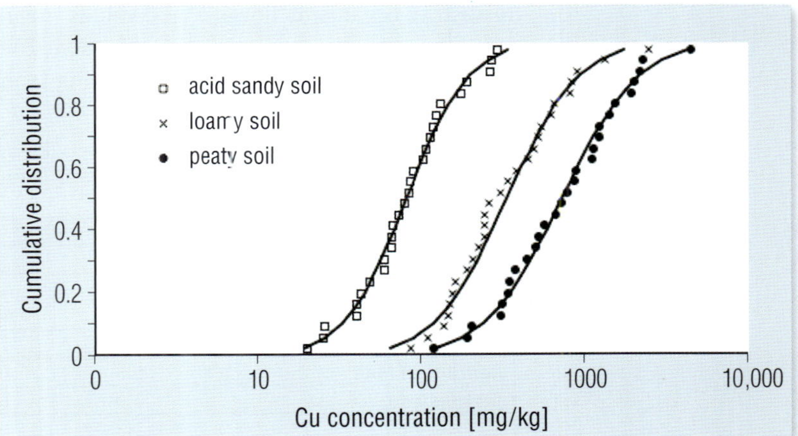

Fig. 12.5: Species sensitivity distributions (SSD) for the effects of copper (Cu) on plants, soil invertebrates and micro-organisms, normalised for bioavailability in three different soil types (**sandy soil**: pH 4.8, 2.8% organic matter, 7% clay and eCEC: 2.4 $cmol_c$/kg; **loamy soil**: pH 7.5, 2.2% organic matter, 26% clay and eCEC: 20 $cmol_c$/kg; **peaty soil**: pH 4.7, 40% organic matter, 24% clay and eCEC: 35 $cmol_c$/kg).

▸ 12.3 Results

Only results for effect concentrations (PNEC) and risk characterisation (RCR) will be reported here. For a discussion of the measured exposure concentrations, we refer to the element-specific chapters of this book (Section 11.23 for Cu and Section 11.36 for Mo → Part A). Generally, it can be stated that natural processes (i.e., geology) control the spatial distribution patterns of metals at this scale, and there is no evidence of significant effect of diffuse emissions on the regional distribution pattern. The spatial distribution of metal concentrations in agricultural and grazing land soil is almost identical.

12.3.1 Effect concentrations

Predicted no effect concentrations (PNEC) corrected for differences in metal bioavailability among soil samples typically vary 10–100 fold across European soil types (Fig. 12.6; Table 12.2). The distribution of site-specific PNEC values is generally related to the distribution in the soil properties used for normalisation (Section 11.5 for discussion of pH; Section 11.3 for clay content; Section 11.15 for organic carbon content and Section 11.1 for eCEC → Part A). For Cu, the PNEC map largely reflects the distribution in eCEC across European soil. The more acid and sandy soil in Scandinavia and north-eastern Europe has a low eCEC and, therefore, low PNEC values for Cu toxicity to terrestrial organisms (lower PNEC values mean higher toxicity in soil, i.e., more sensitive soil). The Cu PNEC map for agricultural and grazing land soil generally shows the same regional distribution patterns, but PNEC values for grazing land are generally higher than PNEC values in agricultural soil. This is directly explained by the higher eCEC and organic carbon content of the grazing land soil. Because organic carbon is accumulated in the topsoil, and the organic carbon content decreases with depth in soil, this difference in organic carbon content between agricultural and grazing land soil can be attributed to the different sampling depth (0–20 cm and 0–10 cm for agricultural and grazing land soil, respectively). Moreover, agricultural soil is generally ploughed and, therefore, the organic carbon is 'diluted' in the field over the upper 20 cm of the soil profile. Because organic carbon is a major contributor to the eCEC of soil, the difference in organic carbon content between agricultural and grazing land soil is also reflected in the eCEC, and hence in the PNEC values.

There is no strong general trend in Mo PNEC values across Europe. Predicted no effect concentrations for Mo are normalised based on pH (lower PNEC values, i.e., higher toxicity, in soil with high pH) and clay content of the soil (higher PNEC values, i.e., lower toxicity, in soil with higher clay content). The lowest PNEC values for Mo are found in eastern and southern Spain, with its predominantly calcareous lithology, and are attributed to the high pH values in these soil types (Section 11.5 → Part A). As Mo occurs as the oxyanion molybdate (MoO_4^{2-}), its bioavailability is positively

Table 12.2: Predicted No Effect Concentrations (PNEC) and Risk Characterisation Ratios (RCR) for copper (Cu) and molybdenum (Mo) in agricultural and grazing land soil sampled across Europe.

Land use	Minimum	10th percentile	Median	90th percentile	Maximum
			PNEC Cu		
Agricultural land	11.8	42.5	84.4	124	216
Grazing land	13.0	49.2	93.1	137	220
			PNEC Mo		
Agricultural land	0.6	10.7	34.1	168	458
Grazing land	0.9	10.2	37.5	179	407
			RCR Cu		
Agricultural land	0.01	0.08	0.17	0.43	4.03
Grazing land	0.01	0.06	0.15	0.38	2.85
			RCR Mo		
Agricultural land	0.0002	0.002	0.012	0.058	1.06
Grazing land	0.0002	0.002	0.011	0.061	1.41

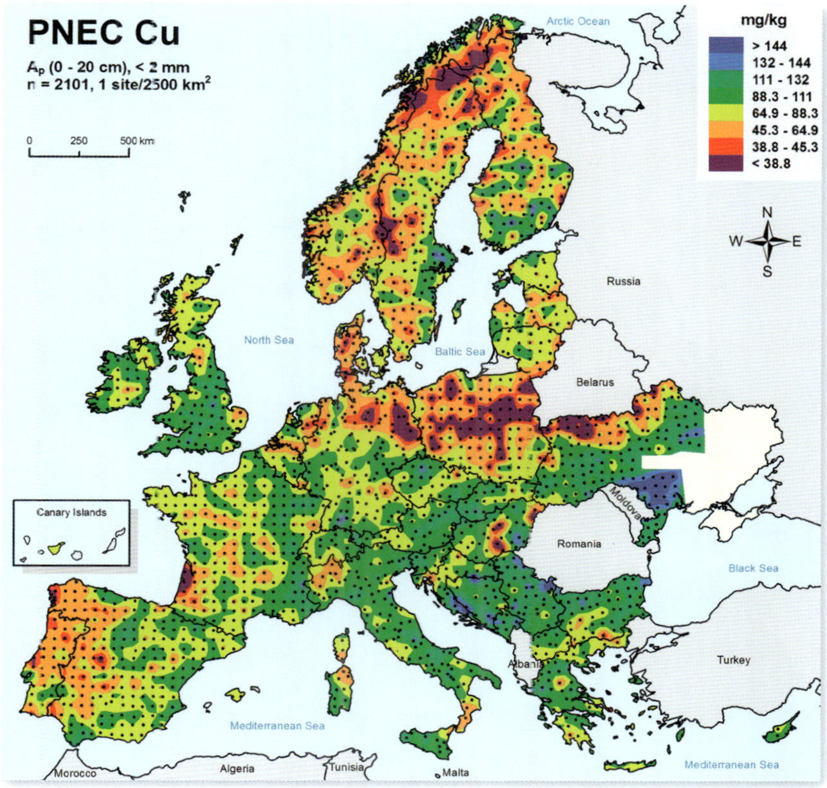

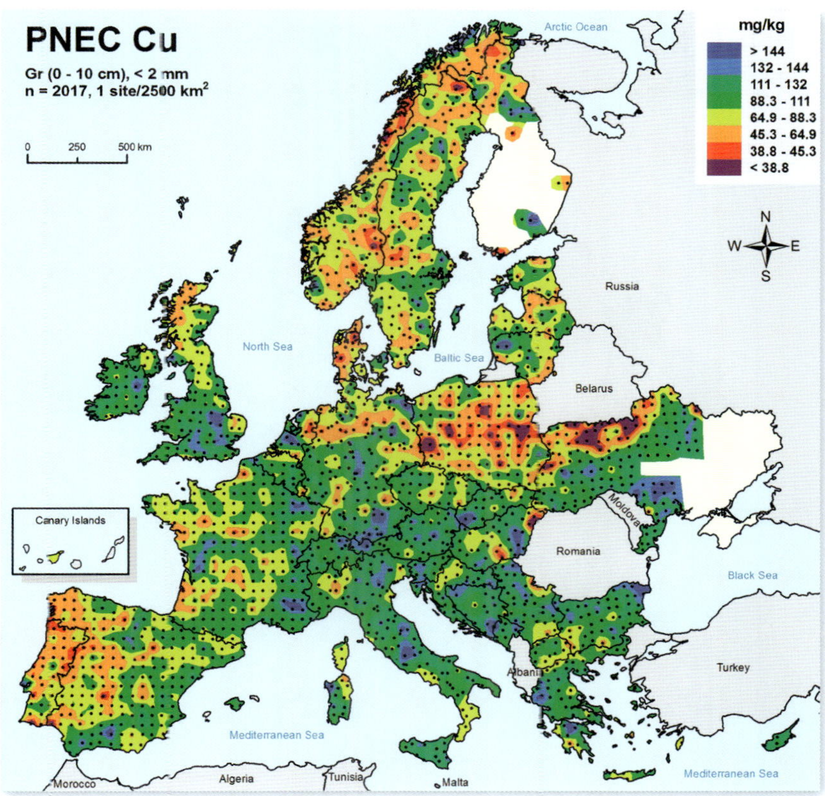

Fig. 12.6: Predicted No Effect Concentrations (PNEC) for copper (Cu) and molybdenum (Mo) in agricultural and grazing land soil in Europe. Dots represent sampling sites. The PNEC values at locations between sample sites were interpolated using the kriging technique. Colour classes are based on 5th, 10th, 25th, 50th, 75th, 90th and 95th percentiles.

Fig. 12.6: continued

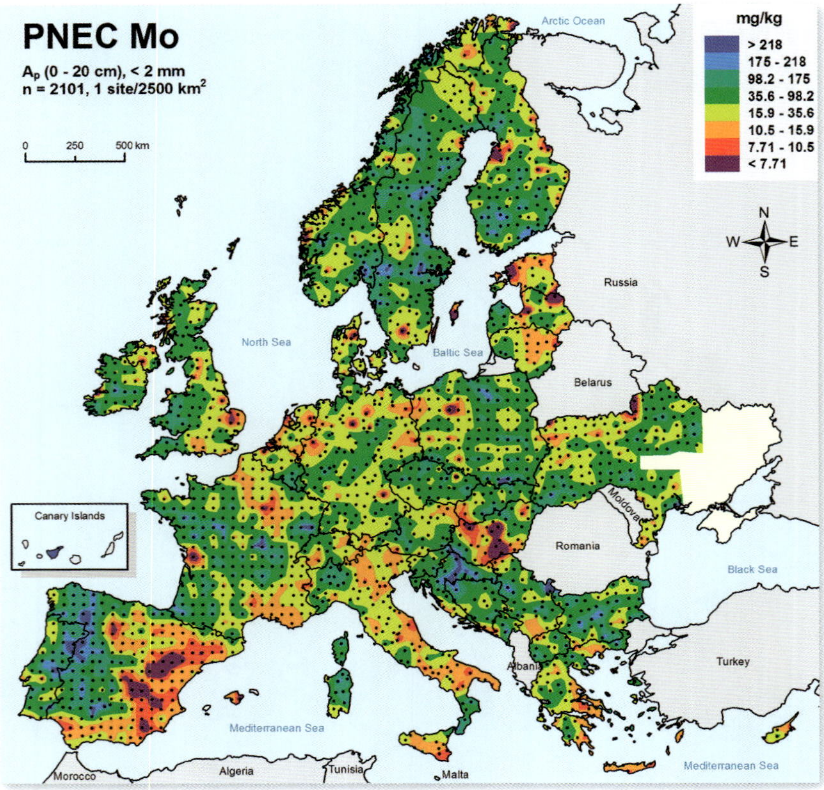

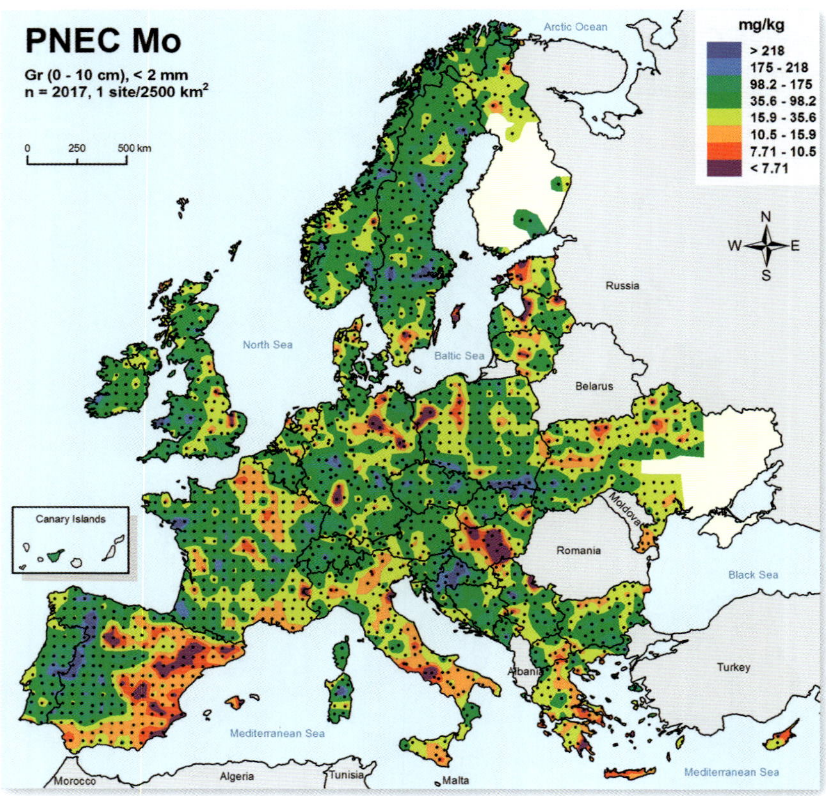

related to pH, i.e., higher bioavailability and toxicity, and hence lower PNEC values, in high pH soil due to lower sorption. There are no strong differences in distribution of pH and clay content between agricultural and grazing land soil and, therefore, distribution of PNEC values for Mo in both land uses is similar.

12.3.2 Risk ratios

Risk ratios for various metals vary 100 – 1000 fold among the sites sampled (Fig. 12.7; Table 12.2). For both Cu and Mo, there is no regional risk observed in agricultural and grazing land soil.

Direct risks of Cu for terrestrial organisms (i.e., soil Cu concentration ≥ PNEC or RCR ≥ 1) are only predicted for few, isolated sites: 1.5 % and 1.2 % of sites for agricultural and grazing land soil, respectively. Southern Europe (Italy, Greece) has relatively large RCR values compared to the rest of Europe. This is mainly due to the relatively high natural background concentrations for Cu in these regions (Section 11.23 → Part A). The high RCR for several sites in France and Greece can also be attributed to increased Cu content due to local application of Cu fungicides in vineyards and orchards. The RCR values for grazing land are generally slightly lower compared to agricultural soil, due to the higher PNEC values for Cu in grazing land soil. The general pattern across Europe is, however, the same for both land uses. The variation in Cu risk ratios is much better correlated with the exposure concentrations ($R^2 = 0.76$ and 0.80 for agricultural and grazing land soil, respectively) than with the variation in PNEC values ($R^2 < 0.03$ for both agricultural and grazing land soil).

Risk ratios for Mo are generally lower than for Cu, and out of > 2000 sites for agricultural and grazing land soil in Europe a risk is only predicted for 1 and 2 sites, respectively. The general regional distribution of RCR values is very similar between agricultural and grazing land soil. The highest RCR values are generally found in Spain, south France and Italy. In contrast to the situation for Cu, the variation in RCR values for Mo is better explained by variation in PNEC values ($R^2 = 0.56$ and 0.57 for agricultural and grazing land soil, respectively) than by the variation in exposure concentrations ($R^2 = 0.28$ and 0.31 for agricultural and grazing land soil, respectively).

▸ 12.4 Discussion

As mentioned in the introduction, correction for differences in bioavailability and toxicity is crucial for some metals in order to obtain realistic results from a risk assessment. This is illustrated by comparison of the outcome of a risk assessment, based on generic aged PNEC values, with the results after full normalisation of the effects data (correction for leaching-ageing and normalisation for variation in bioavailability due to differences in soil properties, Table 12.1; Fig. 12.4). Generic aged PNEC values are calculated, based on the same effects data (NOEC and EC10 values), but only corrected for ageing and leaching effects (L/A factor), and without normalisation for variation in bioavailability due to differences in soil properties. Except for Zn, larger risks are predicted, based on the generic PNEC values compared to the normalised, site-specific PNEC values (Table 12.3), with the largest difference observed for Co and Ni. The value of the generic PNEC is largely determined by the results for the most sensitive end point tested in the most sensitive soil and, therefore, the outcome of such a generic risk assessment without bioavailability correction strongly depends on the choice of the soil types used for ecotoxicity testing. The results for Co and Ni clearly show the over-conservative outcome of such an approach, because all sites sampled are under normal agricultural production, and no significant toxicity for terrestrial organisms is expected.

For those metals with existing normalisation models, the availability of good quality data for both exposure (metal concentrations) and those soil properties, affecting the bioavailability of metals in soil, is critical for the outcome of the risk assessment. The GEMAS data set provides such harmonised data for a representative range of agricultural and grazing land soil types across Europe.

Fig. 12.7: Risk Characterisation Ratios (RCR) for copper (Cu) and molybdenum (Mo) in agricultural and grazing land soil in Europe. Dots represent sampling sites. The RCR values at locations between sample sites were interpolated using the kriging technique. Colour classes are based on 5th, 10th, 25th, 50th, 75th, 90th and 95th percentiles. Areas with RCR values greater than one (> 1) are indicated by a red colour.

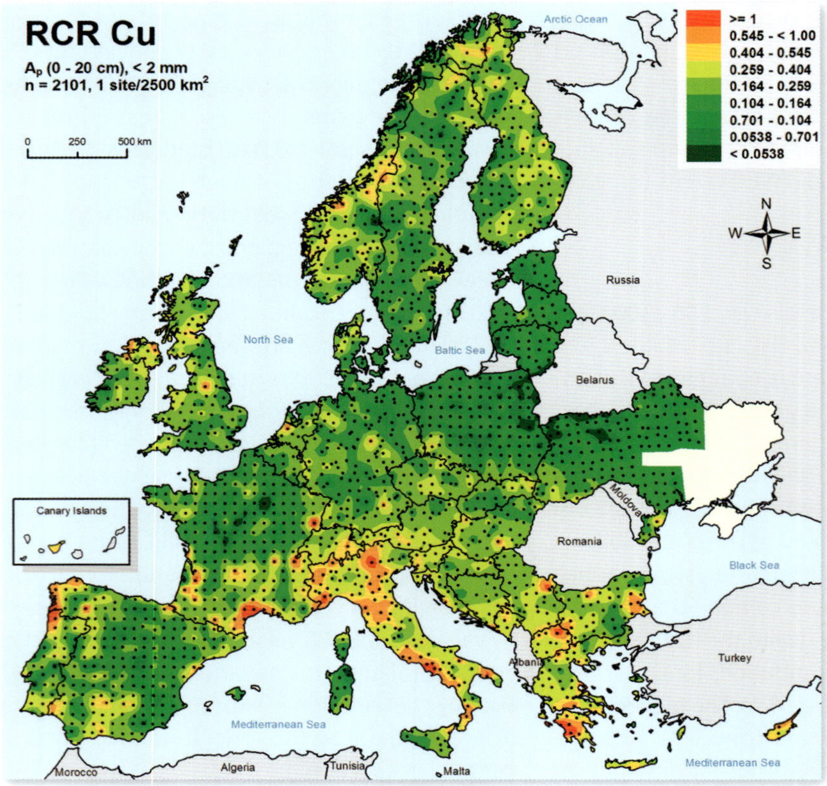

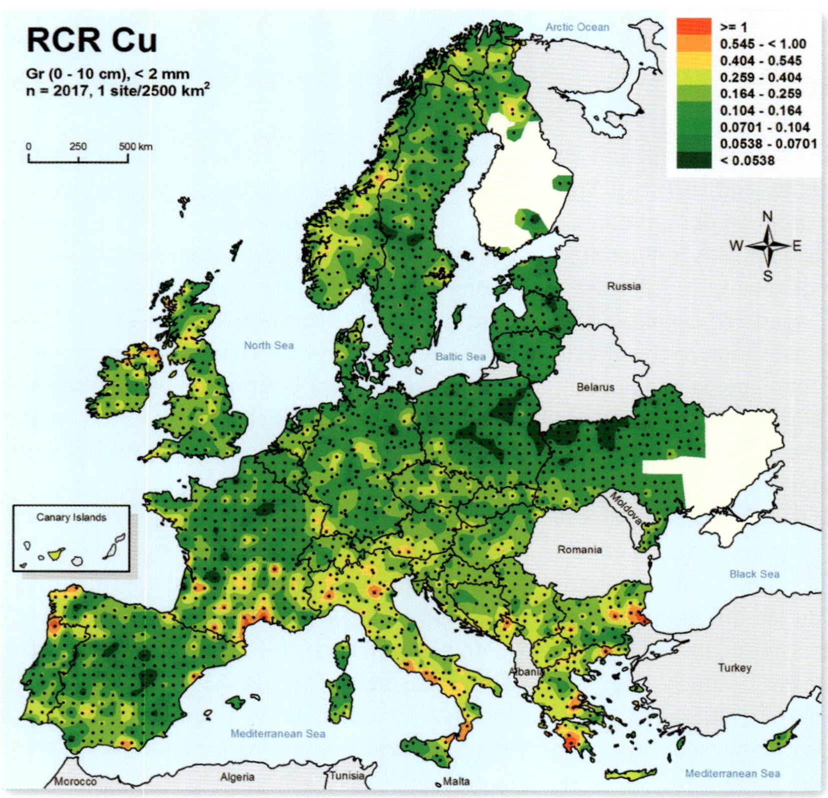

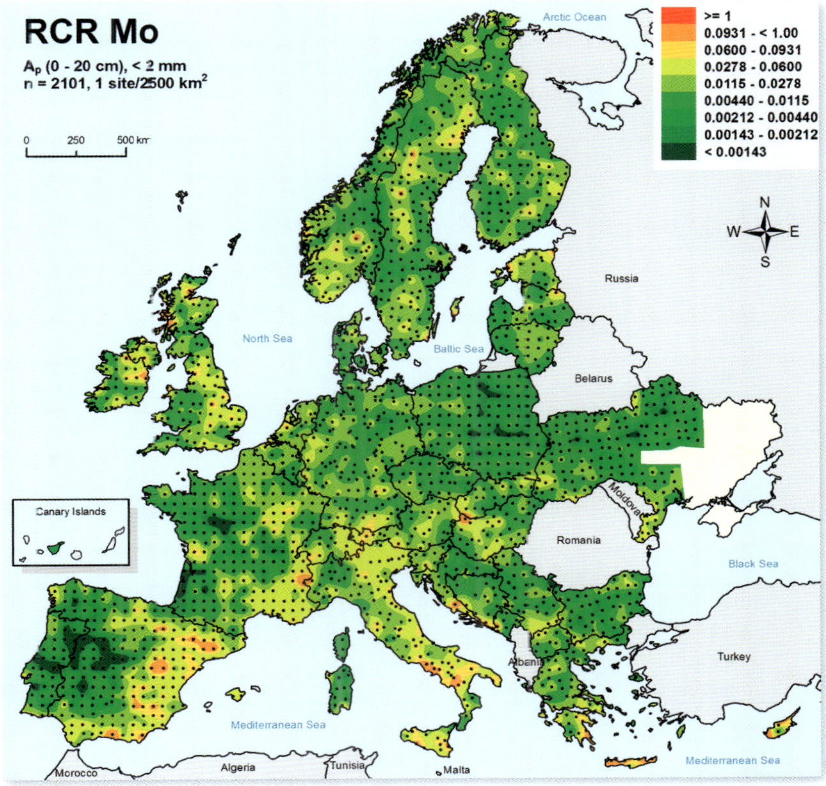

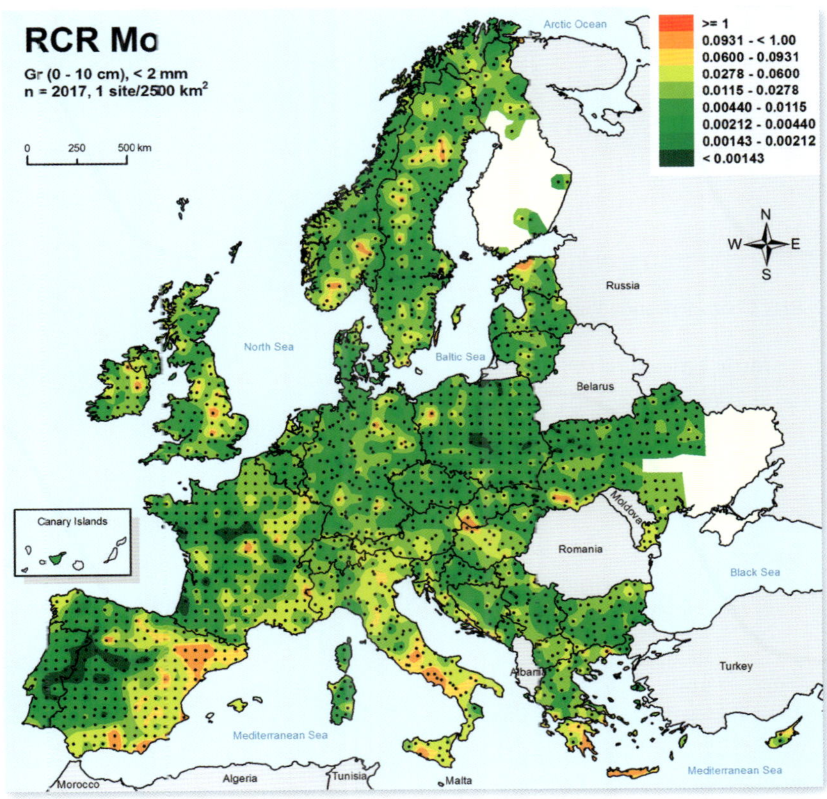

Fig. 12.7: continued

Table 12.3: Risks calculated in agricultural land soil from GEMAS project, based on generic aged PNEC values (only corrected for ageing and leaching effects, no normalisation for variation in bioavailability due to differences in soil properties), and aged PNEC values normalised for variation in bioavailability due to differences in soil properties (see Table 12.1, Fig. 12.4).

Metal	PNEC aged		Sites with risks predicted [1]	
	generic	normalised [2]	no normalisation	with normalisation
	[mg/kg]		[%]	
Co	4.5	13 – 72	68.9	2.5
Cu	58	43 – 124	3.7	1.5
Mo	9.8	11 – 168	0.2	0.1
Ni	15	18 – 80	49.3	10.3
Zn	132	71 – 276	1.6	2.1

[1] i.e., RCR ≥1 [2] 10th – 90th percentile for normalised PNEC values in agricultural land soil

Table 12.4: The 90th percentiles of observed RCR values for copper (Cu) and molybdenum (Mo) in European soil.

Land use		Europe	Country-specific
Cu	in agricultural land	0.43	(Lithuania) 0.12 – 0.91 (FYROM)
	in grazing land	0.38	(Estonia and Lithuania) 0.11 – 0.82 (Montenegro)
Mo	in agricultural land	0.06	(Poland) 0.015 – 0.36 (Cyprus)
	in grazing land	0.06	(Switzerland) 0.003 – 0.15 (Cyprus)

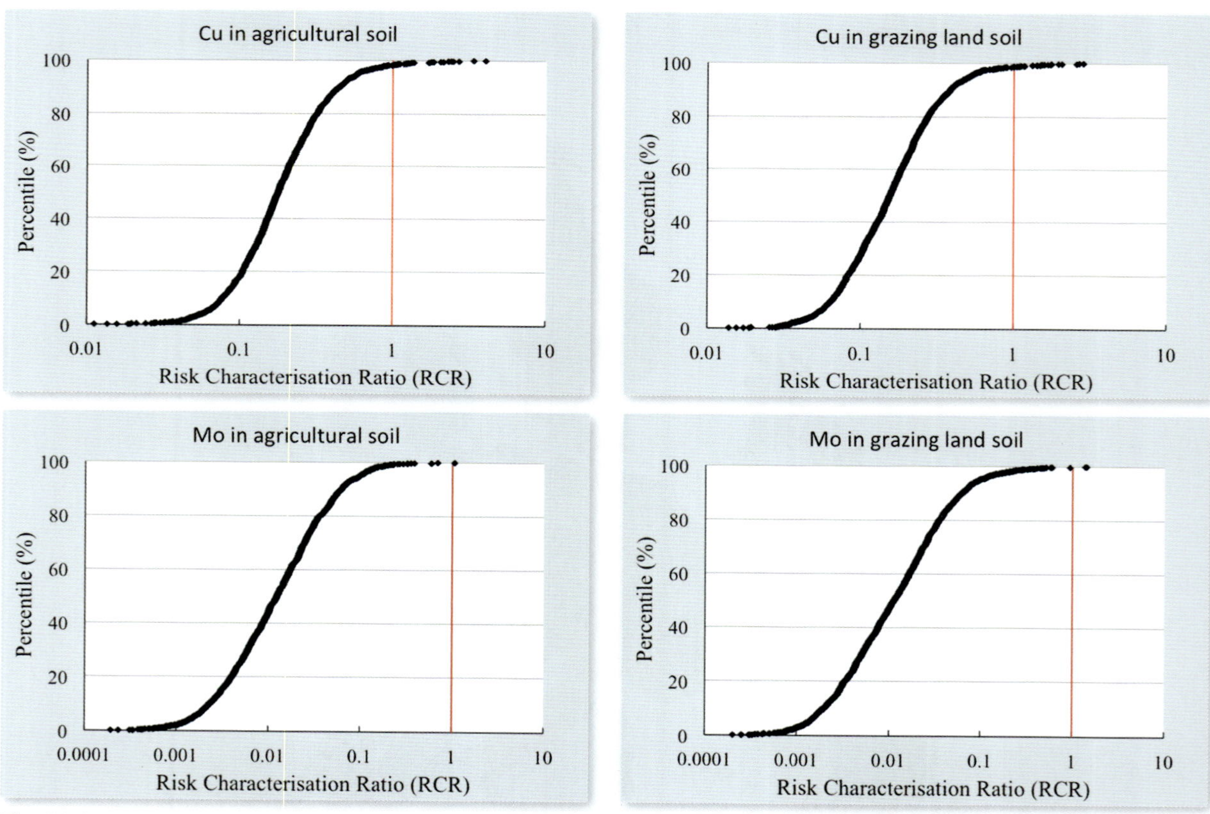

Fig. 12.8: Cumulative frequency distribution of observed site-specific RCR values for copper (Cu) and molybdenum (Mo) in agricultural and grazing land soil across Europe.

The distributions of site-specific RCR values across European soil, show that precicted risks for direct effects to terrestr al organisms are small (i.e., >90 % of all RCR values are <1, Fig. 12.8). For all individual countries covered, the 90th percentiles of RCR values are also well below 1, and for most of the countries, even the maximum observed RCR values are below 1 (Table 12.4). Therefore, it was concluded that there is no regional risk of ambient Cu and Mo concentrations towards terrestrial organisms.

Conclusions based on the distribution of site-specific RCR values avoid the need for (worst case) assumptions on both PEC and PNEC side for a certain region and, therefore, limits the uncertainty and over-conservatism in a regional risk assessment (Table 12.5). This approach is especially important for risk assessment of naturally occurring substances. Metal bioavailability (effects) and background concentrations (exposure) are often significantly correlated (Fig. 12.9). This correlation can be illustrated by the observation that metal background concentrations are generally higher, and their bioavailability lower, in clayey soil compared to sandy soil. Because both parameters are correlated, exposure and effects data should not be treated independently to derive generic reasonable worst-case (RWC) scenarios for risk assessment, as is done by default through comparison of reasonable worst-case PNEC with reasonable worst-case PEC values. Therefore, whenever possible, regional and local environmental risk assessments for metals must be based on comparison of site-specific exposure and effects data.

As mentioned in the introduction, the regional background concentrations must also be taken into account for assessments of risk due to local emissions. The predicted environmental concentration at the local scale (PEC_{local}) is the sum of local concentration in soil due to local emissions (C_{local}) and a reasonable worst-case regional background ($PEC_{RWC,\,regional}$). A generic local risk assessment is based on the comparison (risk characterisation ratio, RCR) of the PEC_{local} with a generic (reasonable worst-case) PNEC:

$$\text{Generic } RCR_{local} = \frac{PEC_{local}}{PNEC_{generic}} = \frac{C_{local} + PEC_{RWC,regional}}{PNEC_{generic}}$$

Table 12.5: Reasonable worst-case RCR values for generic regional risk from copper (Cu) and molybdenum (Mo) in European soil.

Land use		Based on distribution site-specific RCR values [1]	Based on RWC assumptions for both PEC and PNEC [2]
Cu	in agricultural land	0.37	0.90 (= 38.1/42.5)
	in grazing land	0.34	0.74 (= 36.5/49.2)
Mo	in agricultural land	0.05	0.12 (= 1.2/10.7)
	in grazing land	0.05	0.13 (= 1.3/10.2)

[1] RCR_{RWC} = reasonable worst-case (RWC) median of all country-specific 90th percentiles
[2] RCR_{RWC} = ratio of PEC_{RWC} (90th percentile of all data for this parameter) and $PNEC_{RWC}$ (10th percentile of all data for this parameter)

This approach entails again worst-case assumptions for both regional exposure and for effects concentrations. This equation can be rewritten, however, as:

$$\text{Generic } RCR_{local} = \frac{C_{local}}{PNEC_{generic}} + RCR_{RWC,\,regional}$$

By summing risk ratios for the local and regional contribution, instead of comparing the PEC_{local} with a $PNEC_{generic}$, the over-conservatism on the regional contribution can be avoided.

A further advantage of the GEMAS data set, is the fact that a generic realistic worst-case (RWC) PNEC can be calculated, based on the distribution of a representative set of site-specific PNEC values across Europe. Such an approach, again avoids worst-case assumptions on the individual soil properties selected for normalisation of the PNEC values and, therefore, increases the credibility of the selected $PNEC_{RWC}$. A $PNEC_{RWC}$, based on worst-case assumptions for each individual soil property, relies on the hypothesis that every combination of

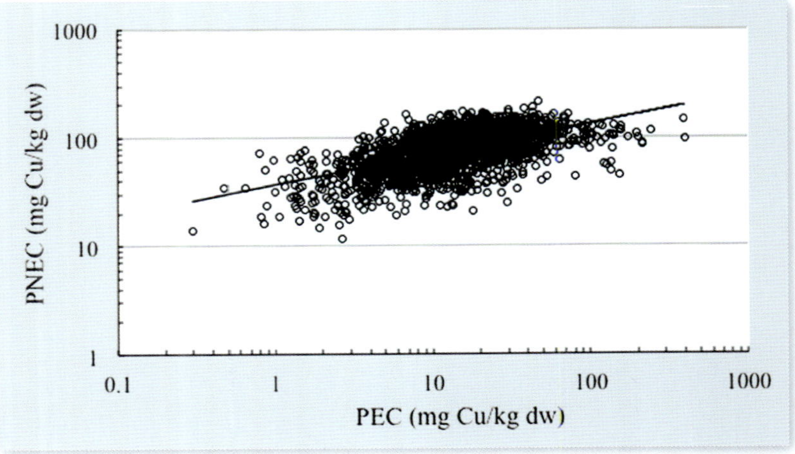

Fig. 12.9: Correlation between exposure concentrations (PEC, Predicted Environmental Concentration) and Predicted No Effect Concentrations (PNEC) for copper (Cu) in agricultural soil.

soil properties is equally randomly possible, and this assumption ignores any correlation among these soil properties in real soil. For example, the toxicity of Cu to the various soil organisms is normalised, based on eCEC, pH, organic carbon content and clay content of the soil. The $PNEC_{RWC}$ value for Cu in agricultural soil, based on reasonable worst-case assumptions (10th or 90th percentiles) for the individual soil properties, is 28.6 mg Cu/kg, which is significantly lower than the 10th percentile of all individual PNEC values (42.5 mg Cu/kg). Similarly, the $PNEC_{RWC}$ for Mo in agricultural soil, based on reasonable worst-case assumptions for individual soil properties (10th percentile of clay content and 90th percentile of pH), is as low as 1.0 mg Mo/kg, which is a factor 10 lower than the 10th percentile of all individual PNEC values for Mo in agricultural soil (10.7 mg Mo/kg).

► 12.5 Conclusions

Harmonised monitoring data for both soil metal concentrations, and general soil properties across Europe, are critical for a realistic assessment of risk of metals towards terrestrial organisms (plants, invertebrates and microbial processes). Such data provide a strong basis for taking into account the spatial variability of both exposure (metal concentrations) and effect concentrations (considering bioavailability through variation in soil properties) in a risk assessment for metals in soil, avoiding the need for (worst-case) assumptions in both respects and, thereby, increasing the transparency and reliability of the regional risk characterisation for Cu and Mo in soil. Harmonised data, with respect to land use, sampling and analytical methods, facilitate a consistent approach at the European scale, enabling direct comparison of results for various regions. |◄

Acknowledgements

This research was funded by the European Copper Institute (ECI) and the International Molybdenum Association (IMOA).

Koen Oorts
ARCHE Consulting;
Gent, Belgium
E-mail: koen.oorts@arche-consulting.be

Ilse Schoeters
Rio Tinto;
London, United Kingdom

Chapter 13

Mobile Metal Ion Analysis of European Agricultural Soil

ALAN MANN, CLEMENS REIMANN, PATRICE DE CARITAT & NICHOLAS TURNER

▸ 13.1 Introduction

Partial extraction techniques have long been favoured for agricultural soil analysis, because they provide to some extent a better indication of 'bioavailability' compared to total analyses. In the main, simple extraction solutions have been used for single nutrient analysis (e.g., COLWELL 1967, 1967a), or for a limited number of nutrients (LINDSAY & NORVELL 1978). Sequential partial extractions have often been used to establish to which phase of a soil particular elements are related (e.g., HALL 1998). Their general problem has been that they often work well (or can be optimised) for one or very few elements, but not for wide multi-element analysis. In this guise, partial extractions have also been used to a greater or lesser extent over a number of years in mineral exploration (e.g., HALL 1998). With respect to mineral exploration, a number of theories have been presented as to the relationship of the soil anomaly to mineralisation (e.g., HAMILTON 1998; MANN et al. 2005). Whilst there is no clear consensus as to an overall or universal mechanism for vertical transport of elements from a weathering ore body through to surface soil, i.e., from phreatic to vadose zone, three factors are evident: (1) oxidation of protolith (ore) to produce ions is necessary; (2) water is a preferred if not essential carrier; and (3) capillary rise (assisted by evaporation and evapotranspiration) is a very important process within soil that determines where the zone of maximum element accumulation (adsorption) occurs, and thus where sampling is optimised. The clear advantage of partial over stronger extractions is that they can provide an increased anomaly to background ratio; their disadvantage is that a number of samples will report values below the lower level of detection (LLD) for some elements. The development of Inductively Coupled Plasma Mass Spectrometry (ICP-MS), with its very low levels of detection for a wide range of elements has to some extent alleviated this situation; partial extractions are often a preferred carrier due to their lower levels of interference from major elements (Fe, S , Al).

The Mobile Metal Ion (MMI®) partial extraction technique is based on having a number of strong ligands in a neutral to alkaline buffered solution to enable ions, which are weakly adsorbed onto soil particle surfaces, to be extracted and retained in solution for subsequent analysis in an ICP-MS instrument. By having a large sample charge (50 g), and a large sample charge to solution ratio (50 g : 50 mL), good sensitivity and reasonably good repeatability is achieved for a majority of elements. Initially, the MMI® technique focussed on assisting mineral exploration to locate 'commodity' elements of economic interest in Australia; extension to glacial environments in Canada and other terrains soon followed (e.g., MANN et al. 1998). It was later found (MANN 2010) that expression of certain elements in soil could be related to parent lithology, even when surface exposure was not evident, i.e., that concealed lithology could be 'inferred' from MMI® soil analysis. This has important implications, not just for lithological interpretation; some of the electrochemical and mineralisation focussed models for anomaly development (HAMILTON 1998) are likely to be at best limited in their application. In addition, it suggests that there is probably no limit to which elements can or should be involved in re-distribution in the regolith at low concentrations via the aqueous phase.

In the agricultural sciences, numerous techniques, including partial extractions, have been used to estimate the degree to which nutrients are available to plants and animals (ISKANDER & KIRKHAM 2001). A number of factors, including the large number of elements, the range of soil types and the array of plant and animal species, make this a formidable task for partial digestion and extraction techniques. At best, bioavailability when estimated by partial extraction is operationally defined, which means no two extractants will provide identical estimates. In the case of MMI®, the choice of organic and inorganic ligands and their concentrations have been tailored to 'mirror' those available in soil. The degree to which this is successful was investigated for 11 nutrients on 1186 catchment outlet sediments in Australia (MANN et al. 2012). Correlation with other, simpler estimates of bioavailability was discussed. An estimate for the percentage bioavailability was obtained by comparing average extraction rates for MMI® with total content. In these sediment samples, average 'bioavailability %' was found to range widely for different elements.

The MMI® technique (MANN 2010) differs from most other partial extractions in that a number of strong ligands allow for a wide range of elements including alkali metals, alkaline earths, transition elements, base metals, actinides and precious metals to be simultaneously extracted from soil and held in one solution. Over 50 elements can be routinely analysed after application of 50 mL of the MMI®-M (M for multi-element) extraction solution to 50 g of soil. The reason for inclusion of MMI® derived data into the GEMAS project is succinctly addressed in an early publication of some preliminary GEMAS data (REIMANN et al. 2011). After making the case for obtaining a good regional geochemical database for a wide range of elements in Ap and Gr soil samples, this report asks the question *"what is the 'bioavailability' of metals and other chemical elements in soil?"* The need to establish (where possible) natural background and to differentiate industrial impact from geogenic background is formulated. With Europe's long and varied history in the discovery, exploitation and use of many elements, particularly metals, the potential for anthropogenic contributions to soil geochemistry here is high. This could include surface re-distribution around mining sites, dispersion during transport and storage, contamination during and after processing, and in the case of fertilisers, widespread application. Partial extraction of soil, using an extractant with good multi-element capability, has a good chance of achieving the aims of assessing bioavailability and determining relative importance of geogenic and industrial impacts, at least for a number of elements. In January 2012, aliquots of the Ap series of soil samples from the GEMAS project were forwarded to SGS Toronto Canada for MMI® extraction and ICP-MS analysis. The results of this analysis are the focus of discussion in this chapter.

▸ 13.2 Analysis and quality control

Extracts were prepared using standard MMI® protocol; 50 g of sample was mixed with 50 mL of MMI®-M solution and shaken for 30 minutes. They were allowed to stand overnight before centrifuging for ten minutes at 8500 rpm, using a Sorvall Evolution RC (ThermoElemental Corporation). Extracts were analysed, using a Perkin Elmer NexION 300D ICP-MS fitted with a Universal Cell, by measuring them against calibration standards prepared in MMI®-M matrix solution. Chromium, V and Se were measured in reaction mode using ammonia reaction gas to remove matrix interferences. Sulphur was measured by a Perkin Elmer Optima 4300DV ICP-OES using matrix matched standards. Laboratory internal quality control (QC) criteria included randomised duplicate analysis every 13 samples, plus blank and reference materials inserted every 46 analyses. The acceptance criteria were based on a limiting repeatability of 20 %, and a lower level of detection of 2.5 x the reporting limit for each analyte.

Laboratory independent, external QC for the GEMAS project started already during the sampling campaign: a field duplicate was collected at

Table 13.1: Results of an unbalanced ANOVA (%) for the MMI® extraction, GEMAS Ap-samples. All variables were log-transformed prior to the calculation. Left hand side ANOVA results in alphabetical order according to element, and right hand side ordered according to increasing 'geochemical (natural)' variation for an easy identification of 'problematic' elements (less than 80 % natural variation). **Notation**: natural (geochemical); site (sampling); analytical; natVar (natural variance) (from Reimann et al. 2012: Table 6, p. 26–28).

	GEMAS Ap samples			
	median	natural	site	analytical
Element	[mg/kg]	[%]		
Ag	0.014	97.4	0.0	2.6
Al	61	97.4	0.0	2.6
As	0.02	92.5	2.2	5.2
Au	0.0003	94.2	0.0	5.8
Ba	1.09	96.5	0.0	3.5
Bi	0.0005	96.8	0.0	3.2
Ca	595	96.3	3.2	0.6
Cd	0.054	97.6	1.4	1.0
Ce	0.135	93.9	1.6	4.4
Co	0.1815	95.7	3.2	1.1
Cr	0.065	97.0	0.0	3.0
Cs	0.0007	96.3	1.1	2.6
Cu	1.34	94.0	4.7	1.3
Dy	0.035	95.5	0.0	4.5
Er	0.0202	96.6	1.0	2.4
Eu	0.0075	93.3	2.2	4.5
Fe	31	97.7	0.6	1.8
Ga	0.002	94.3	0.0	5.7
Gd	0.036	94.2	0.0	5.8
Hg	0.0005	87.5	5.2	7.4
In	0.00025	92.3	0.0	7.7
K	64.2	90.7	8.9	0.4
La	0.049	96.9	0.0	3.1
Li	0.0025	97.0	0.6	2.5
Mg	52	97.1	2.5	0.4
Mn	19.6	95.1	3.9	1.0
Mo	0.015	95.2	0.0	4.8
Nb	0.0009	96.5	0.0	3.5
Nd	0.1015	95.0	1.3	3.7
Ni	0.652	98.3	1.2	0.5
P	4	98.3	0.7	1.0
Pb	0.32	92.9	3.6	3.5
Pd	0.0005	*	*	*
Pr	0.02	96.3	2.2	1.5
Pt	0.0005	*	*	*
Rb	0.069	95.4	4.0	0.6
S	24.2	94.4	0.0	5.6
Sb	0.002	94.6	2.9	2.5
Sc	0.016	93.2	3.0	3.8
Se	0.01	75.4	6.6	18.1
Sm	0.029	93.7	2.3	4.1
Sn	0.0005	93.4	0.0	6.6
Sr	1.67	97.6	1.7	0.7
Ta	0.0005	99.3	0.7	0.0
Tb	0.005	93.9	0.0	6.1
Te	0.005	8.2	54.8	37.0
Th	0.0221	93.1	0.4	6.5
Ti	0.038	95.2	0.0	4.8
Tl	0.0006	93.6	5.3	1.1
U	0.084	98.5	1.1	0.4
V	0.027	96.1	0.1	3.8
W	0.001	97.0	0.5	2.6
Y	0.1835	95.6	0.0	4.4
Yb	0.016	96.9	2.1	1.0
Zn	0.76	97.1	1.9	1.0
Zr	0.051	97.4	0.0	2.6

Element	nat Var (ordered) [%]
Pd	*
Pt	*
Te	8.2
Se	75.4
Hg	87.5
K	90.7
In	92.3
As	92.5
Pb	92.9
Th	93.1
Sc	93.2
Eu	93.3
Sn	93.4
Tl	93.6
Sm	93.7
Ce	93.9
Tb	93.9
Cu	94.0
Au	94.2
Gd	94.2
Ga	94.3
S	94.4
Sb	94.6
Nd	95.0
Mn	95.1
Mo	95.2
Ti	95.2
Rb	95.4
Dy	95.5
Y	95.6
Co	95.7
V	96.1
Ca	96.3
Cs	96.3
Pr	96.3
Ba	96.5
Nb	96.5
Er	96.6
Bi	96.8
La	96.9
Yb	96.9
Cr	97.0
Li	97.0
W	97.0
Mg	97.1
Zn	97.1
Ag	97.4
Al	97.4
Zr	97.4
Cd	97.6
Sr	97.6
Fe	97.7
Ni	98.3
P	98.3
U	98.5
Ta	99.3

* too many values near DL

each 20th sample site (Reimann et al. 2009, 2011, 2012 and Chapter 6 → Part A). During sample preparation, an analytical replicate sample was prepared of each field duplicate. Finally, all samples were randomised and standard samples were also hidden between the real samples so that the laboratory could recognise neither the standards nor the analytical duplicates. When the analytical results were received standard, field duplicate and analytical replicate results were retrieved from the data set and X-charts and Thompson and Howarth plots (Thompson & Howarth 1978) were prepared (Reimann et al. 2012a). Precision was calculated based on the field duplicates, and an unbalanced analysis of variance (ANOVA) was carried out to estimate the suitability of the data for the construction of geochemical maps. Precision calculated according to two different methods, practical detection limits and the results of the ANOVA are summarised in Table 13.1. The QC results suggest that with the exception of Pd, Pt and Te all elements can be mapped.

▶ 13.3 Results

13.3.1 Lithology related elements

Composition of the geological substrate (parent material) potentially has an important bearing on the geochemical composition of a soil (Jenny 1941). This is evident in partial extractions of soil samples, where important thresholds for diagnostic elements, such as Ce, Ni and Ca after MMI® extraction, have been established on the Archaean Yilgarn Block of Western Australia over a number of rock types (Mann 2010). 'Classic soil samples' over authenticated rock types were analysed for a wide range of elements. The antithetic relationship between MMI® Rare Earth Elements (REEs; exemplified by Ce) and Ni was immediately noted; high concentrations of other elements, such as Ca, Fe and Ti, were also recognised to be related to particular minerals. Soil samples overlying common rock types (felsic, mafic, and ultramafic), were found to lie along either of the two axes on a Ce versus Ni diagram (MANN 2010). Granitoids were found to have high Ce (Ce > 1 mg/kg) and little Ni; mafic and ultramafic rock types were found to have varyingly high levels of Ni and little or no Ce. A level of 2 mg/kg Ni was found to be indicative of ultramafic (vs. mafic) rocks, and Ni > 8 mg/kg to be indicative of adcumulate ultramafic, often comprising olivine and sometimes NiS. This phenomenon is related to the relative size of the ions; those with large ionic radii, such as Ce, the other lanthanides, and the univalent elements, such as K, Cs, Rb, and Ag, are precluded from crystallisation in mafic and ultramafic magmas, and are concentrated or differentiated into late stage crustal material, i.e., granite. Soil over sedimentary rocks (e.g., feldspathic sandstone, banded iron formation), derived from granite, also have high Ce and REEs contents. Soil over carbonate sedimentary rocks (dolomite or calcite) commonly attained levels in excess of 800 mg/kg Ca in the MMI® extraction solution. Levels of MMI® Fe > 20 mg/kg and Ti > 0.5 mg/kg together were found to be indicative of weathering magnetite; weathering ilmenite is also likely to release both Fe and Ti (ANAND & GILKES 1984).

For the European agricultural (Ap) soil samples, investigation of the element distribution maps for Ce (Fig. 13.1) and Ni (Fig. 13.2) shows wide ranges for these elements, and values similar in magnitude to those observed in Western Australia (see also EDA maps on attached DVD → Part A).

Application of the aforementioned thresholds for elements Ce, Ni and Ca, to the Ap soil samples results in a simple interpretive 'Inferred Lithology' map (Fig. 13.3).

The inferred lithology map shows firstly, that the distribution of MMI® Ce > 1 mg/kg in soil in the Scandinavian Shield, the north-west of Scotland, parts of France and north-west of Spain and Portugal is suggestive of areas with known plutonic rocks of granitoid origin. This is in general accord with most geological maps for Europe. Second-

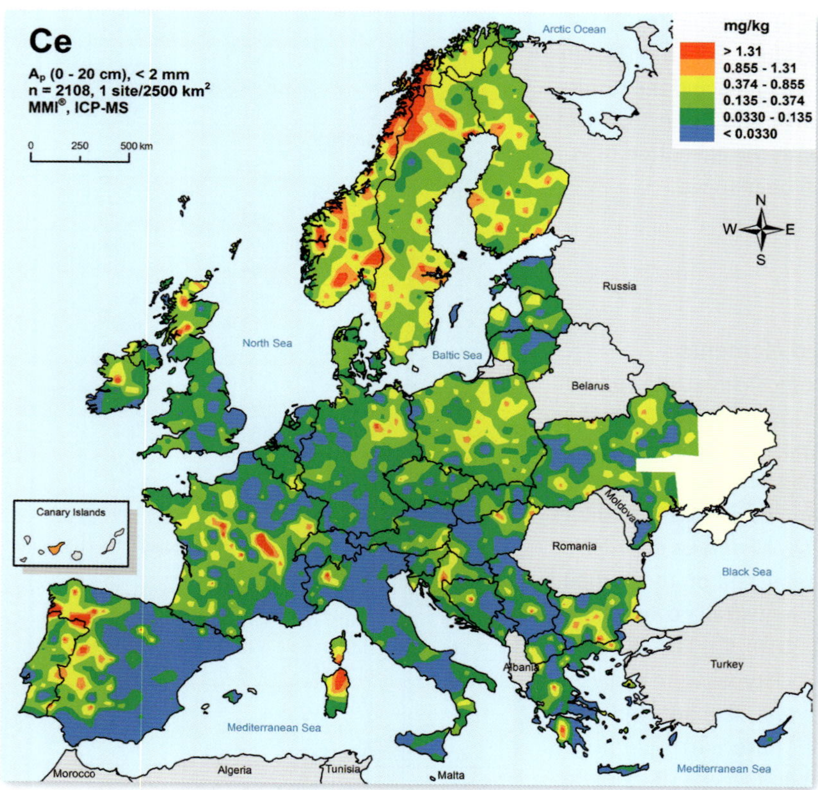

Fig. 13.1: Map of cerium (Ce) distribution in Ap soil samples.

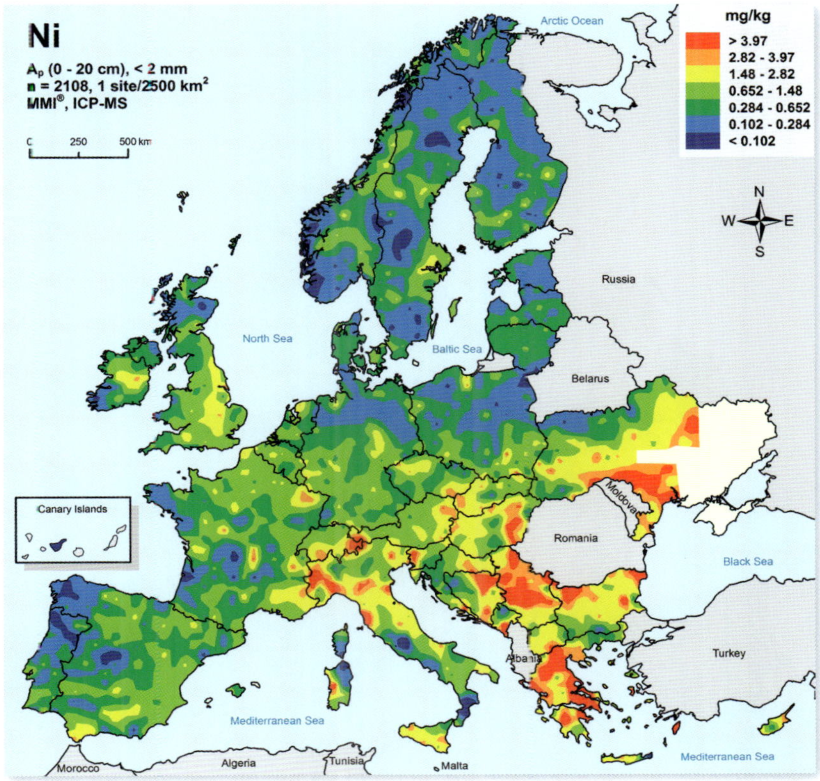

Fig. 13.2: Map of nickel (Ni) distribution in Ap soil samples.

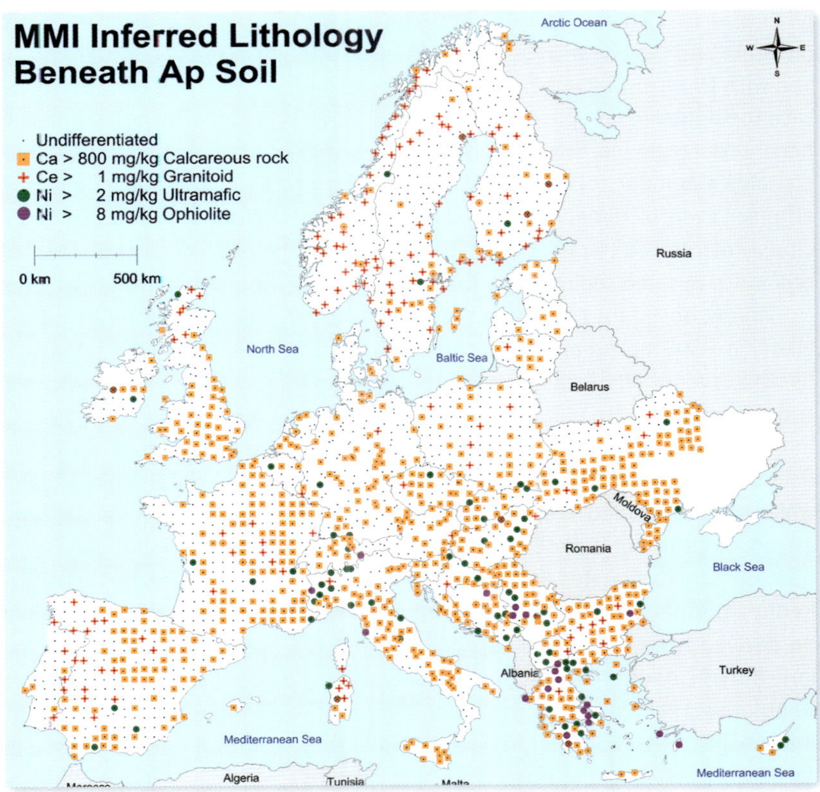

Fig. 13.3: Inferred lithology from Ap soil geochemistry. Significance thresholds after MMI® extraction and analysis are: for calcareous rock Ca > 800 mg/kg, for granitoid Ce > 1 mg/kg, for ultramafic Ni > 2 mg/kg and for ophiolite Ni > 8 mg/kg.

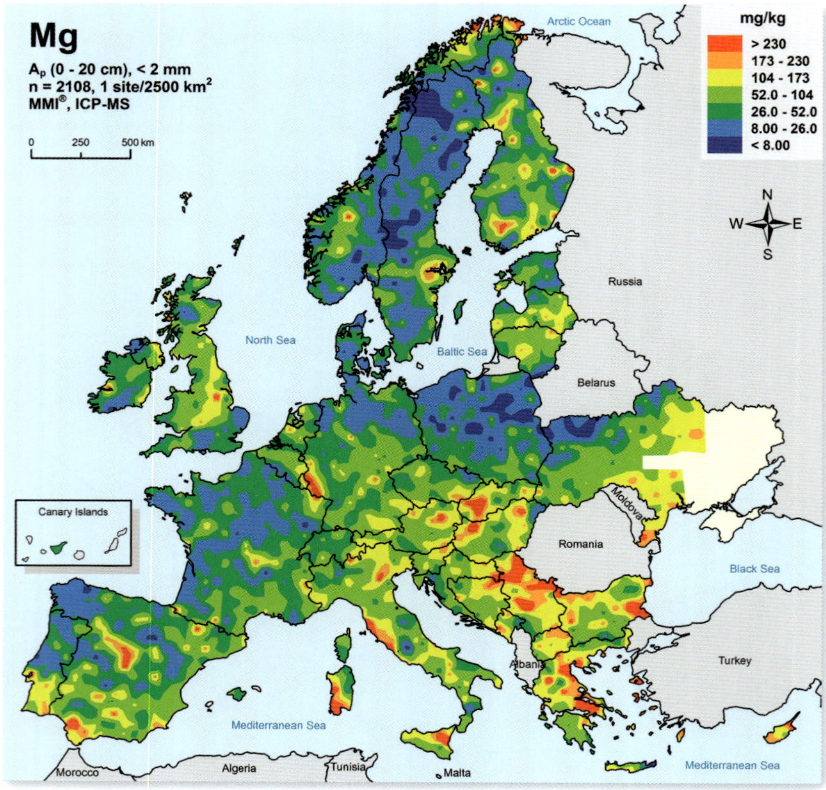

Fig. 13.4: Map of magnesium (Mg) distribution in Ap soil samples.

ly, Figure 13.3 shows Ap soil in northern Italy, Slovenia and Greece with very high levels of Ni. Soil with Ni > 8 mg/kg appears to coincide closely with the ophiolite complex of northern Italy, Bosnia and Herzegovina, Slovenia and Greece, which contains magmatic rocks of mafic and ultramafic affinity. Levels of Ni between 2 and 8 mg/kg, previously attributed to ultramafic rocks, appear to occur also in some sedimentary rocks here.

Thirdly, on Figure 13.3 those soil samples with MMI® Ca > 800 mg/kg are highlighted. In Australia, this level of MMI® Ca has been linked to soil containing carbonate material (Mann et al. 2012). In the present case, the possibility that Ca is reflective of naturally occurring gypsum, or gypsum, which has been added to ameliorate the soil, must also be kept in mind. Carbonate can be derived from primary carbonates (limestone including dolomite), but may also in arid and semi-arid areas be derived from secondary carbonate, particularly if mafic or ultramafic rock types are actively weathering. High Ca is most prevalent in southern Europe, but it is also apparent that sedimentary carbonates are contributing to high Ca soil values in the eastern UK, Sweden, and parts of France, Germany and Ukraine. In the present case, some limestone terrains, particularly in northern and central Europe appear not to be marked by this level of Ca.

Other alkaline earths, such as Mg, Sr and Ba, show broad similarities to Ca in their distribution patterns over Europe, although there are some notable differences. Comparison of the MMI® Mg map with the previous Ca distribution is illustrative (see EDA map on attached DVD → Part A).

Soil samples with Ca > 800 mg/kg have been highlighted in Fig. 13.3, the inferred lithology map. In general, high Ca occurs in Ap soil samples across southern Europe. The map for MMI® Mg (Fig. 13.4) also shows generally higher Mg values in southern European soil, with some very marked Mg 'anomalies' in Greece, Serbia, Bosnia and Herzegovina and Cyprus in particular; some of these are likely to be related to the ultramafic rocks of ophiolite complexes, which contain high concentrations of Mg as well as Ni. Magnesium is readily incorporated into silicate lattices along with Fe and Ni, whilst Sr and Ba, being much larger, are not. Highest values for Sr and Ba occur in southern and south-eastern European soil, which are related to calcareous and calc-alkaline igneous lithologies.

It might be expected that Ni, Mg and Cr should all highlight ultramafic terrain; to some extent this is true, but as the Cr map (Fig. 13.5) shows, there are also some subtle differences, and one has to consider sedimentary rocks with detritus from the ultramafic rocks, and base metal mineralisation (see EDA map on attached DVD → Part A).

Both Cr and Mg delineate the greenstone belt of northern Finland more clearly than does Ni. Nickel and Mg highlight the ultramafic rocks of the ophiolite belts (northern Italy, Greece, Serbia and Bosnia and Herzegovina) to a better extent than Cr. Whilst, ultramafic parent rock types are not all equally endowed in Ni, Cr and Mg, there may be an additional factor in play in this case. Chromium is much less 'mobile' under normal weathering conditions than either Mg or Ni; Mg is less 'mobile' than Ni.

The distribution of Ce has been discussed previously (Fig. 13.1) and its ability to be diagnostic of granitoid terrain displayed on Figure 13.3. As a group, the lanthanides or REEs show a closer correlation to one another than any other group of elements in the Periodic Table. This is due to their very similar outer atomic structure, and only a very small 'lanthanide contraction' separates the light from heavy lanthanides. The erbium (Er) map is here presented for comparison (Fig. 13.6).

The distribution patterns of MMI® Ce and Er are very similar. All lanthanides have their highest concentrations in north-west Iberian peninsula, Scotland, the Scandinavian Shield and the Massif Central in France, and lowest values in southern Europe, particularly where the geology is dominated by ophiolite complexes – mafic and ultramafic rocks from which elements with large ionic radii are excluded. Scandium and Y are, strictly speaking, not lanthanides but have similar electronic structure, a dominant trivalent oxidation state and large ionic radii. The Sc map is shown in Figure 13.7 (see EDA map on DVD → Part A).

Scandium and Y also have very similar, but not identical distribution patterns in Ap soil samples, after MMI® extraction and analysis to the lanthanide elements. Highest values of Sc occur

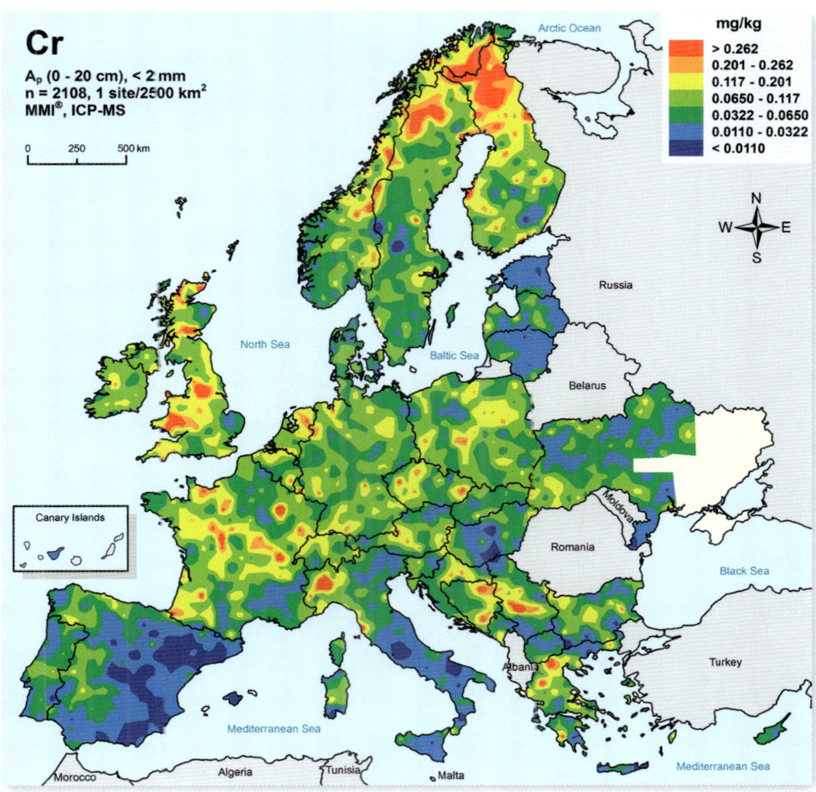

Fig. 13.5: Map of chromium (Cr) distribution in Ap soil samples.

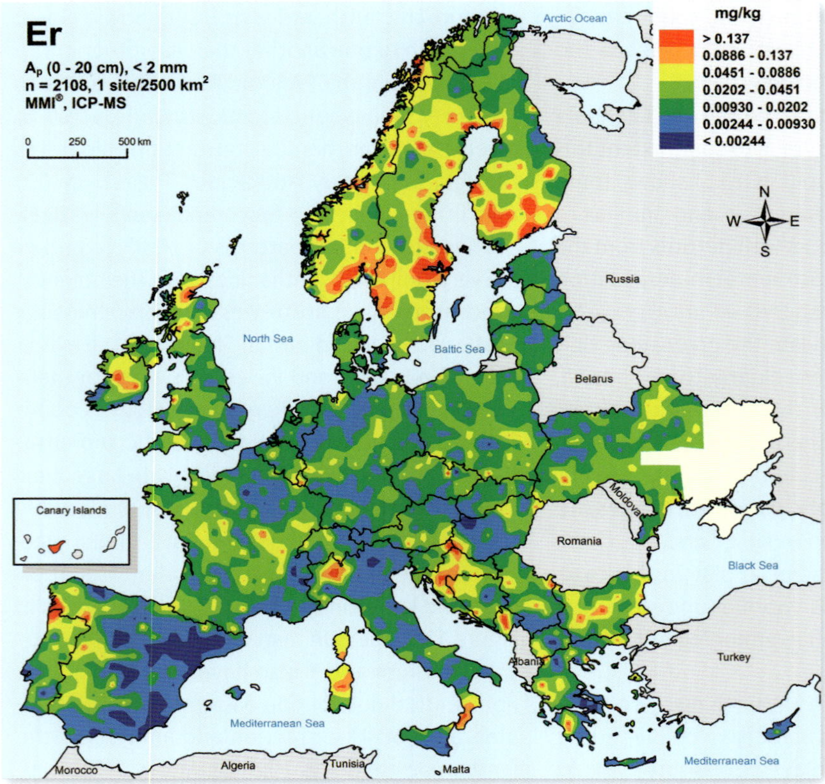

Fig. 13.6: Map of erbium (Er) distribution in Ap soil samples.

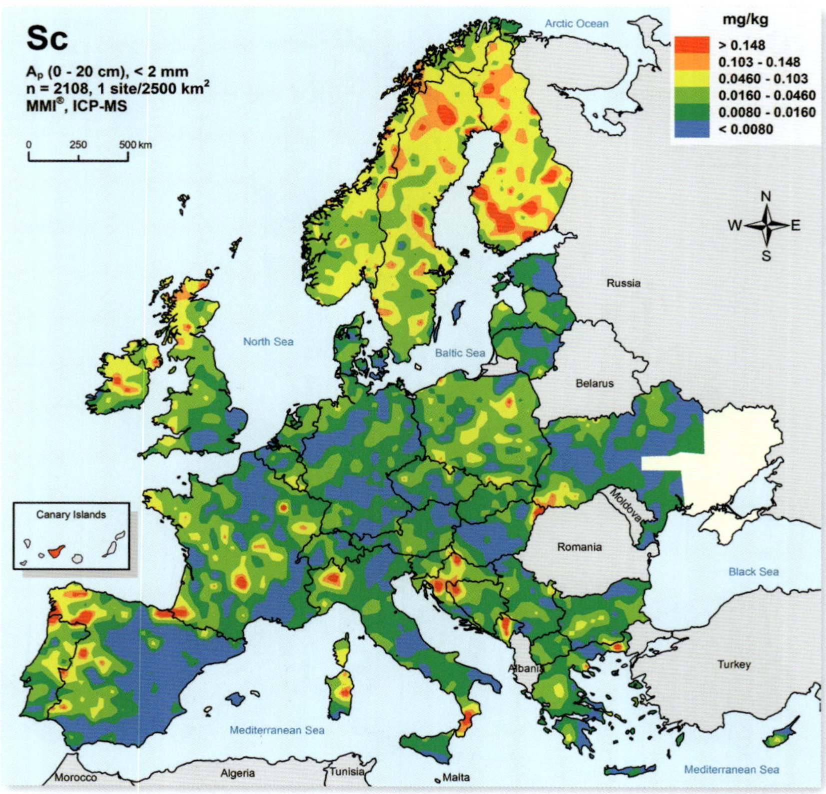

Fig. 13.7: Map of scandium (Sc) distribution in Ap soil samples.

in Scandinavia, particularly Finland, and northwest Iberian peninsula.

13.3.2 Elements influenced by climate and weathering

Rainfall in Europe ranges from in excess of 1000 mm per year on the western coasts of Ireland, the United Kingdom, Norway, Croatia, and parts of France and Italy to less than 500 mm per year in eastern Spain, Poland, eastern Germany, Ukraine and northern Finland. On the other hand, annual evaporation varies little from east to west, but markedly from north to south from 200 mm per year in Norway, Sweden and Finland, to greater than 600 mm per year in Spain, Italy and Greece. Even this most cursory climatic summary suggests a wide range, particularly from north to south, of an important parameter – net rainfall, i.e., precipitation minus evaporation. This is an important (derived) parameter with respect to soil geochemistry, as it controls the potential for annual flushing of the profile versus accumulation of secondary weathered products. Not surprisingly, there are a number of elements that show a marked trend in soil values from north to south, after MMI® extraction and analysis. Elements that have higher concentrations in northern European soil include Al, Fe, Ti, Nb, Ga, P and Zn. On the other hand, higher concentrations in southern European soil occur for Ca, Co, Ni, Au, Li, Cu and Mn. Not all distribution differences can necessarily be attributed to weathering; Co is probably a case in point, where parent rocks in southern Europe probably contain a greater concentration of Co. However, some elements, such as Fe, Al, Ca and Ti, occur in a wide range of rock types, represented in both northern and southern Europe, and weathering has to be considered as a factor, in addition to lithology, to explain their marked regional MMI® distributions.

Figure 13.8 shows a widespread distribution of mobile Fe throughout not just the Scandinavian Shield, but also the west coast of Scotland, Ireland, Spain, Portugal, The Netherlands and Poland. There is a pronounced gradient in general values from northern to southern Europe. In Australia, it has been shown (MANN et al. 2012) that the distribution of mobile or bioavailable Fe is not

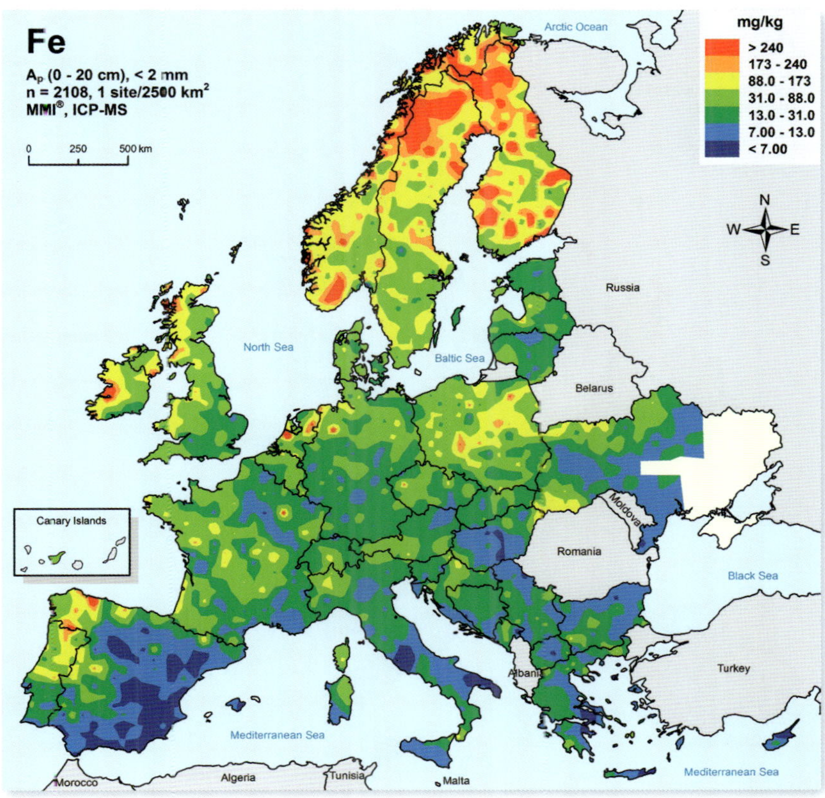

Fig. 13.8: Map of iron (Fe) distribution in Ap soil samples.

just related to mineralisation; the large haematite deposits of the Pilbara in Western Australia show a very subdued MMI® signal. Rather it is the high rainfall areas (where lateritisation is believed to be still occurring) of the south-western, eastern and north-eastern Australian coasts where highest Fe values are present in the catchment outlet sediments (MANN et al. 2012). In a similar manner, the Fe distribution in Europe appears to have some relationship to high rainfall; highest precipitation is in Scandinavia and the west coasts of Ireland, the UK, Spain, Portugal, Belgium, The Netherlands, Slovenia, Croatia, Bosnia and Herzegovina (see EDA map on attached DVD for sample site details → Part A).

Iron is mobile in the Fe(II) oxidation state, and relatively immobile in the Fe(III) oxidation state in the pH range 4 – 6 (POURBAIX 1963). The presence of organic matter may also influence the solubilisation of Fe; it was shown to be a factor in Tasmania, and is more likely to occur in northern Scandinavia, Scotland and Ireland than in southern Europe. The paucity of carbonate in the northern European soil is another factor that favours the mobility of Fe.

Northern and western European soil samples have high values of MMI® Al; those in the south and east have much lower values in general. Granitoid substrate of the Scandinavian Shield, Scotland and north-western Iberian peninsula is no doubt an important source of Al, but there are a number of areas in southern Europe where Al is available in parent material. Low soil pH originating from felsic bedrock lithologies, with relatively abundant organic matter and high precipitation in northern Europe contribute to the mobility of Al (and Fe). Conversely, high pH and Ca content, which arise from the evaporation-dominated weathering pattern, mitigate against Al mobility. Gallium, which is from the same group of elements in the Periodic Table, shows a similar distribution pattern.

Figure 13.10 shows (with one exception) highest values for Ti in the soil samples of the Scandinavian Shield, Scotland and parts of Poland (see EDA map on DVD → Part A); Ti frequently occurs as ilmenite and/or titaniferous magnetite in both felsic and alkaline rock types. The one exception is the highest value for MMI® Ti in Ap soil samples,

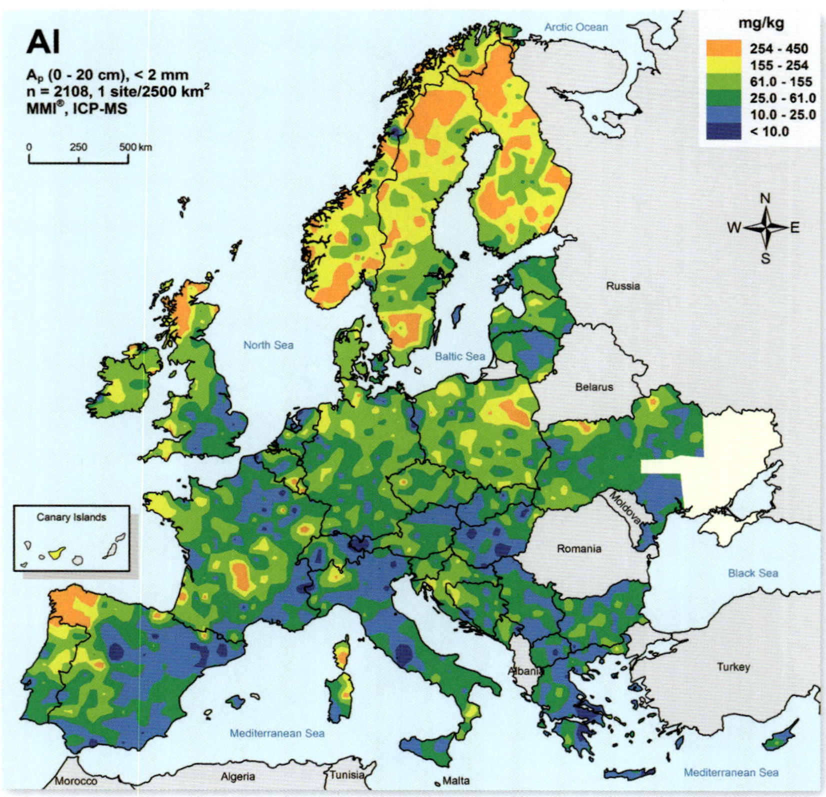

Fig. 13.9: Map of aluminium (Al) distribution in Ap soil samples.

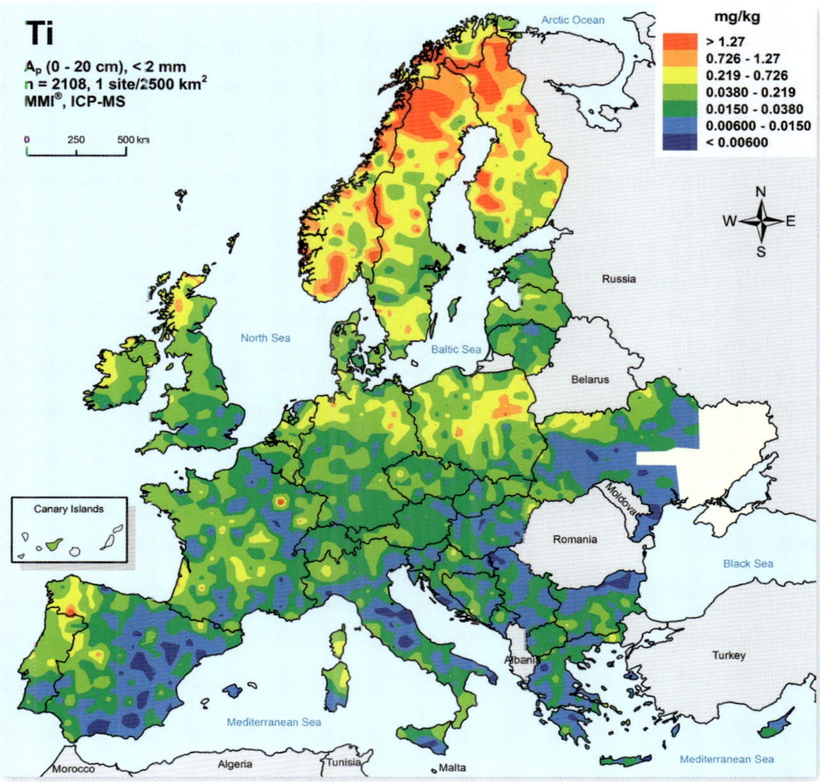

Fig. 13.10: Map of titanium (Ti) distribution in Ap soil samples.

which occurs on the outskirts of the town of Saint Dizier in the Champagne district of north-eastern France (see EDA map on DVD → Part A). This value (13.2 mg/kg Ti), is very high by MMI® standards, and accompanied by high MMI® Fe, which may be related to the nearby Bettancourt la Ferrie oolitic iron deposit (siderite, goethite, limonite, haematite). Nevertheless, it is not clear whether this should be attributed to the presence of weathering magnetite and/or ilmenite from a natural source, the oolitic iron mineralisation, or Ti from an anthropogenic source. Saint Dizier had a large airbase, and was bombed heavily in World War II. It constitutes a single point anomaly; surrounding values are all low.

13.3.3 Elements related to mineralisation

Mineralisation, as characterised by the above average concentration of an element in the Earth's upper continental crust, often has limited spatial dimensions, and is usually a term applied to commodity elements (metals). In this particular survey involving agricultural soil, the sample sites have not been optimised to mining districts or terrain necessarily favourable to mineralisation, and due to the wide spacing of sample locations, relatively few are likely to overlie directly mineralisation. In addition, the mineralisation must be actively weathering to produce ions that can migrate through the regolith and adhere to soil particles to respond to MMI® analysis. Despite these factors, a number of elements in this GEMAS survey appear to have their highest MMI® values in the vicinity of known mineralisation containing that element (see Chapter 3; and also mineral deposit maps on attached DVD → Part A).

The highest value for MMI® Fe is just 60 km east of Kiruna and 70 km north-west of Malmberget, both large underground Fe ore mining centres in northern Sweden. Whilst there is a climatic (weathering) component to Fe distribution in Ap soil samples (see above), it is pertinent in this case to attribute the extremely high Fe (and Ti) concentration in soil (2200 and 0.896 mg/kg, respectively) to the possibility of nearby mineralisation. The dominant mineral source of Fe here is magnetite, and weathering magnetite is known to produce MMI® signatures for both Fe and Ti.

Gold (Fig. 13.11) provides another case in point. The highest value for MMI® Au is 0.0237 mg/kg, from a field in north-eastern Portugal (see also EDA map on attached DVD → Part A). It is 8.6 km south-east of Mirandela, a Au mining area since Roman times. The same soil sample coincidentally has the highest value of all Ap samples for As of 0.530 mg/kg, probably indicating the association of Au with arsenopyrite. It is, of course, possible that the high Au (and As) value has arisen from surface re-distribution, as a result of mining, and this is a very difficult issue to resolve from the current data set alone. The second highest value for Au, 0.0144 mg/kg, arises from an Ap soil sample close to a known Au mining district in southern Serbia. It is interesting that the next two highest MMI® Au values (0.0121 and 0.0114 mg/kg Au, respectively) are within 20 km of Paris and London respectively. Whilst there is Au mineralisation in the Scandinavian Shield, the MMI® anomalies in this region are not as pronounced. The reasons for this are not yet fully understood, but it was noted above that Au is one of the elements with a significant negative correlation with respect to latitude, suggesting some climatic (weathering) influence on Au dispersion.

The maximum value of Sb in soil is 0.071 mg/kg from a soil sample in the Tyrol region of Austria. This is an area where Ag mines have stibnite as an accessory mineral. The anomalous Sb soil sample referred to has an anomalous level of Ag of 0.11 mg/kg. Other examples, where high values occur close to known mineralisation sources, are evident for Ag (1.34 mg/kg) close to the Rio Tinto mining district in southern Spain, several soil samples with high Cd close to the Zn mining centre of Navan (in Ireland), Hg (0.053 mg/kg) and S (2620 mg/kg) in the volcanic tuff zone near Cerveteri on the west coast of Italy, and Pb with its highest value in a south Pennine Pb mining district in the United Kingdom.

13.3.4 Elements suggesting anthropogenic influence

In the main, anthropogenic factors arise from the mining, smelting and processing of ores (principally metals), and the distribution of fertilisers

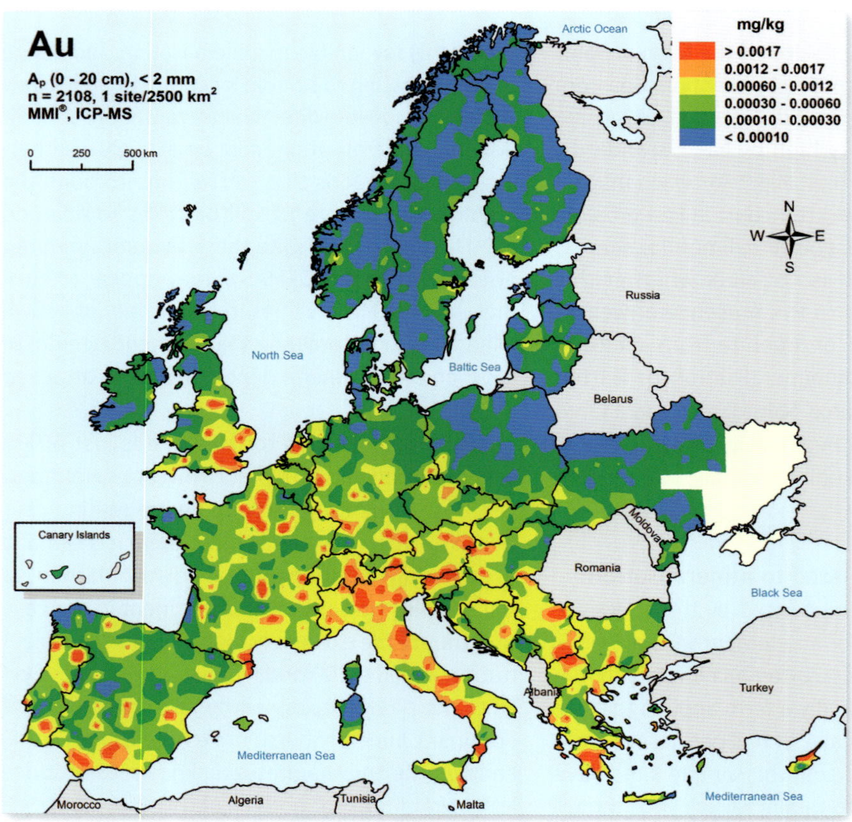

Fig. 13.11: Map of gold (Au) distribution in Ap soil samples.

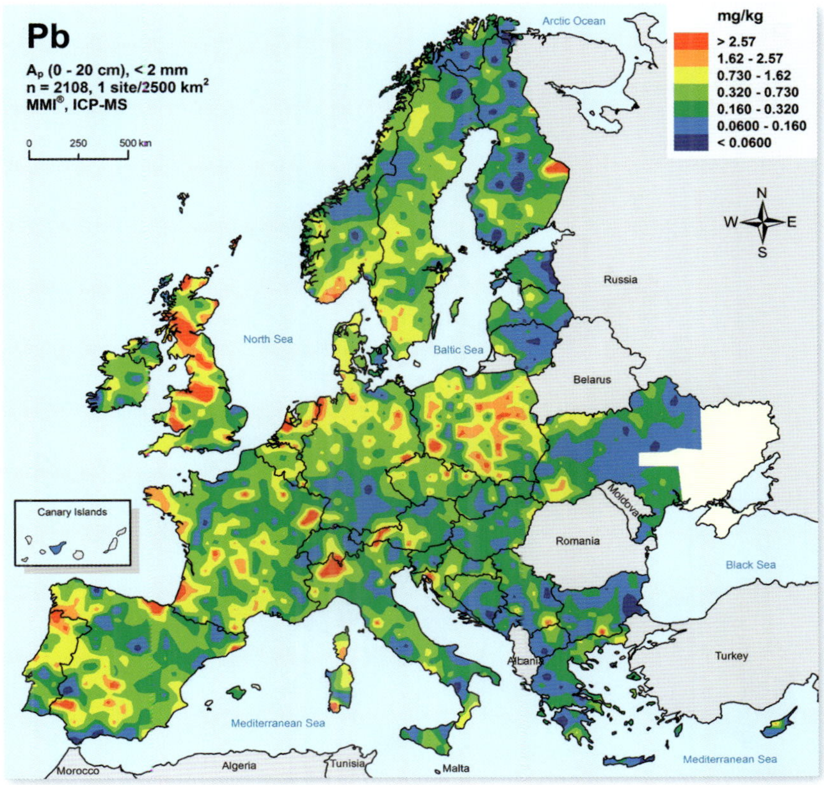

Fig. 13.12: Map of lead (Pb) distribution in Ap soil samples.

(including sewage sludge) and fungicides. The highest value of Pb in this survey of 58.3 mg/kg occurs west of Chesterfield in Derbyshire in the United Kingdom (see Fig. 13.12; and also EDA map on attached DVD → Part A). Not only is this a Pb mining district, it is an area where Pb has been smelted and processed over many centuries. Further, a number of Pb roofing companies operate in the area today. This level of Pb in the soil, which is one of the highest ever recorded by the MMI® technique, is hard to reconcile with just subsurface weathering mineralisation, given the district's long mining and smelting history.

The second highest Pb value of 18.9 mg/kg Pb is from a field 25 km south of Amsterdam and 20 km west of Utrecht, and unlikely to have mineralisation as a direct source (see EDA Pb map on attached DVD → Part A). Contamination from motorways is also unlikely in this case; the sample site is some distance from the nearest roadway. It is possible this site represents a repository for sewage sludge soil amendment.

Figure 13.13 shows an interesting distribution pattern for P (see also EDA map on attached DVD → Part A). Apatite is a known source of P in rocks of the Scandinavian Shield, which has one operating phosphate mine, at Siilinjärvi in central eastern Finland. High values in eastern Finland and northern Sweden could also possibly be attributed to the presence of apatite. It is also of interest that the P distribution delineates the extent of the last glaciation event.

However, the majority of the widespread above average values of P in Ireland, The Netherlands, Germany and Poland are more than likely to be a result of phosphate fertiliser distribution (see also EDA map on DVD → Part A). Cadmium, which is an impurity in many phosphate fertilisers (Syers et al. 1986, IFA 2012), has a similar but not identical distribution pattern. High values for Cd are displaced to the south of those for P, and may correspond more closely with fertiliser application; the influence of glaciation on exposure of apatite in the north has already been noted. Much of southern Europe (with two or three exceptions) appears to have low MMI® P values in Ap soil samples. This could be due to climatic influence as there is abundant Ca in this soil, and Ca phos-

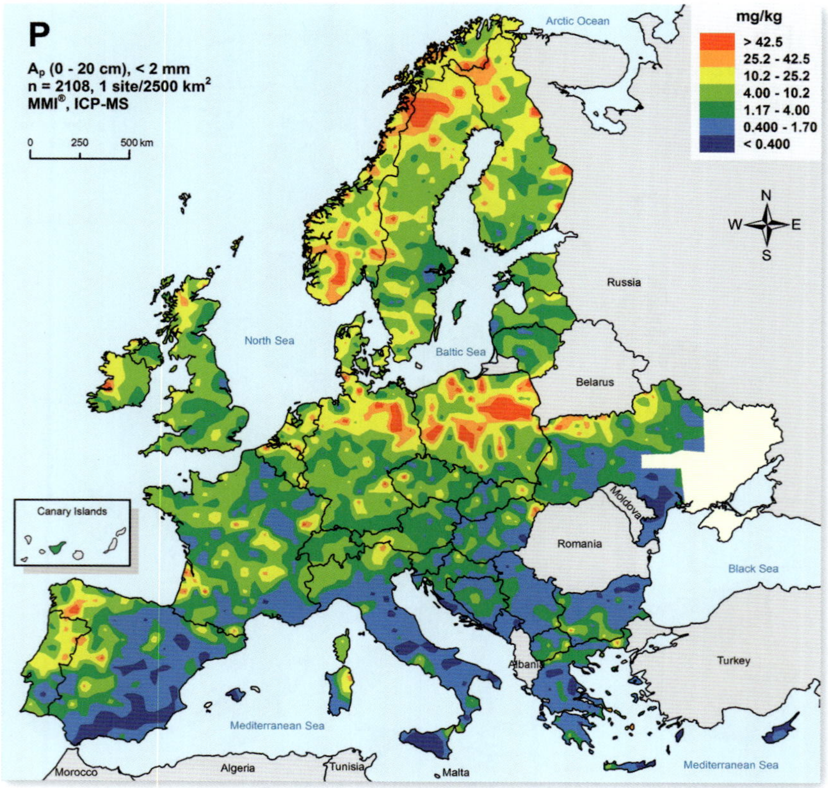

Fig. 13.13: Map of phosphorus (P) distribution in Ap soil samples.

phate is relatively insoluble, but economic factors related to the type of farming and rates of fertiliser application could also be a reason. The most significant exception is a single point anomaly of MMI® P in excess of 50 mg/kg from a location in north-eastern Greece. However, there are a number of other elements in high MMI® concentrations in this particular soil sample and the most likely source is fertilisers; this is a tobacco growing area.

Many high values of MMI® Cu appear to be related not to mineralisation, but rather to agricultural pursuits. The highest value of MMI® Cu (62.1 mg/kg) is 20 km south of Bordeaux in a vineyard soil, as shown in Figure 13.14 (see also EDA map on DVD → Part A).

'Bordeaux mixture' is the name given to a combination of copper sulphate and lime, which is used as a spray to control mildew, particularly in grape vines. It is almost certainly the reason for the extremely high value of MMI® Cu at this location. The second highest value (58.8 mg/kg) of MMI® Cu in Ap soil is from a field in Galicia, north-western Spain. This has been a Cu mining district since the Bronze Age, and accordingly the high soil value is more likely to be attributable to mining activities, rather than to spraying with Cu fungicide. The third highest Cu value of 58.6 mg/kg, and still 'extremely high' by normal MMI® standards, is from a vineyard soil 20 km north-east of Montpellier in south-eastern France, whilst the fourth highest Cu value is from a vineyard in the Alsace region of north-eastern France. The soil value of MMI® Cu for the latter is 56.7 mg/kg Cu. A value of 51.9 mg/kg Cu is registered from the Beaujolais wine growing district, some 25 km north-west of Lyon in central France, and the value of 46.9 mg/kg Cu from a soil in the recently expanded Languedoc Roussillon wine growing district of southern France, indicates that high values of MMI® Cu are widespread throughout most wine growing districts of France. It is surprising, therefore, to find that Cu values are comparatively low in the vineyards of Spain; instead, they appear to have a number of high values of S in soil. Sulphur is also used as a fungicide, particularly in vineyards (Fig. 13.15; see also EDA map on DVD → Part A).

As well as a fungicide, S is an essential nutrient. Figure 13.15 shows highest concentrations

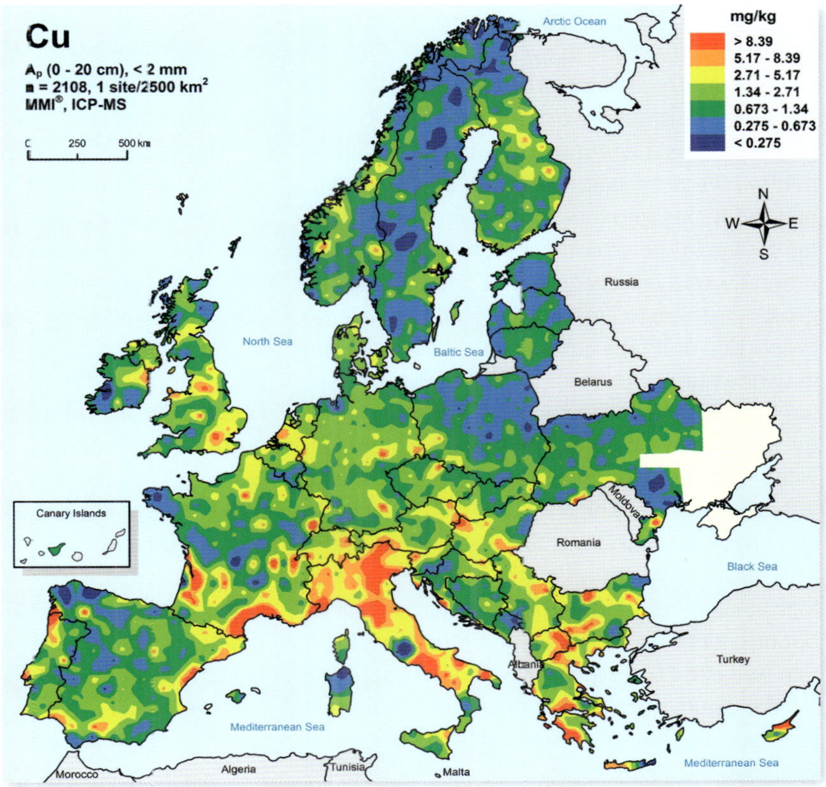

Fig. 13.14: Map of copper (Cu) distribution in Ap soil samples.

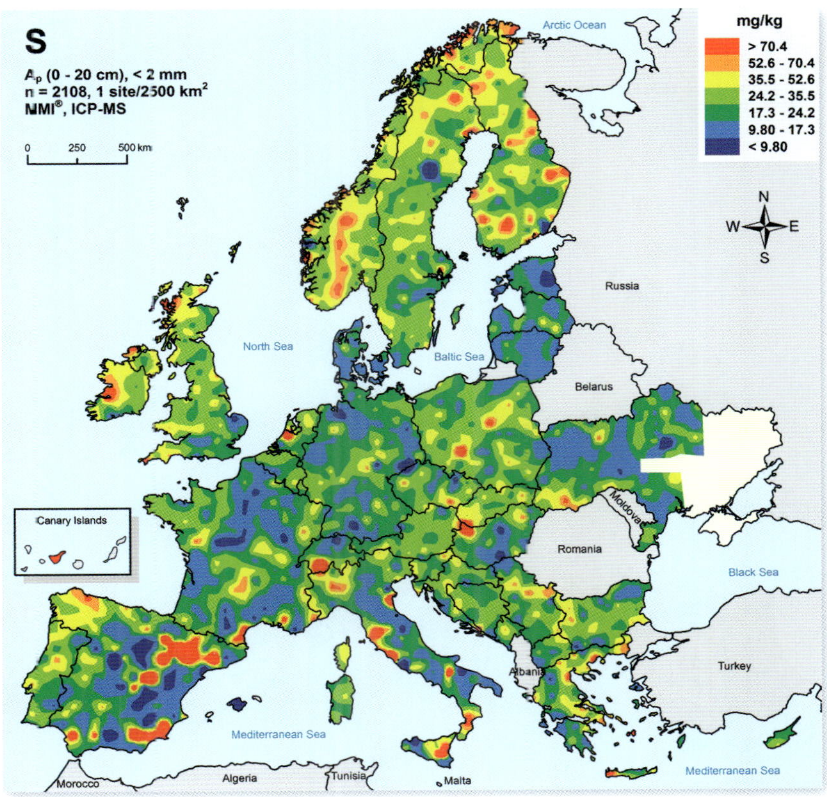

Fig. 13.15: Map of sulphur (S) distribution in Ap soil samples.

of S in Spain and Italy. Spain has the largest area under viticulture (spread over a number of zones) of any country in the world, and is the world's third largest producer of wine after France and Italy. The causal agent responsible for powdery mildew, *Unicula necator*, is favoured by warm humid conditions, and therefore is common in Spain (and Italy) in the growing season. Elemental S is normally sprayed every 10 – 14 days during the growing season to limit powdery mildew; a large proportion of it falls onto, and resides in the soil in various forms. Much of the S observed in Figure 13.15 for the Italian peninsula is also probably derived from viticulture. However, the highest S value (2620 mg/kg), which is on the west coast of Italy, has coincident high Hg and is possibly related to the alkaline volcanic tuff present in the area.

Figure 13.16 shows a number of isolated K high values in soil within the Scandinavian Shield, but also a number of low values in this same terrain (see also EDA map on attached DVD → Part A).

Potassium is also a major nutrient and used in fertiliser; the Ap soil samples of much of central and southern Europe show above average values of MMI® K. Potassium is very soluble and is associated with land (soil) affected by salinity; this is likely to be relevant in some semi-arid areas of southern Europe.

Figure 13.17 shows a cluster of high Zn values in The Netherlands, several on the Scandinavian Shield, and isolated high values in Scotland, northern Spain, south-western France, northern Italy, and northern and southern Poland (see also EDA map on attached DVD → Part A).

Navan is the centre of Irish Zn mining operations; there is no MMI® Zn anomaly in this vicinity, but there occurs a cluster of soil samples with high Cd. Cadmium is common in sphalerite, but is also, due to its concentration and retention in the seafood chain, present in many phosphate fertilisers. Zinc also is an essential nutrient and often added to soil as a trace element supplement. The highest value of MMI® Zn (36.2 mg/kg) is from a soil on a farm on the Isle of Mull off the coast of western Scotland (see EDA map on DVD → Part A).

The two actinides included in the MMI® analysis, Th and U, show similar but not identical distribution patterns. Both are clearly more abundant

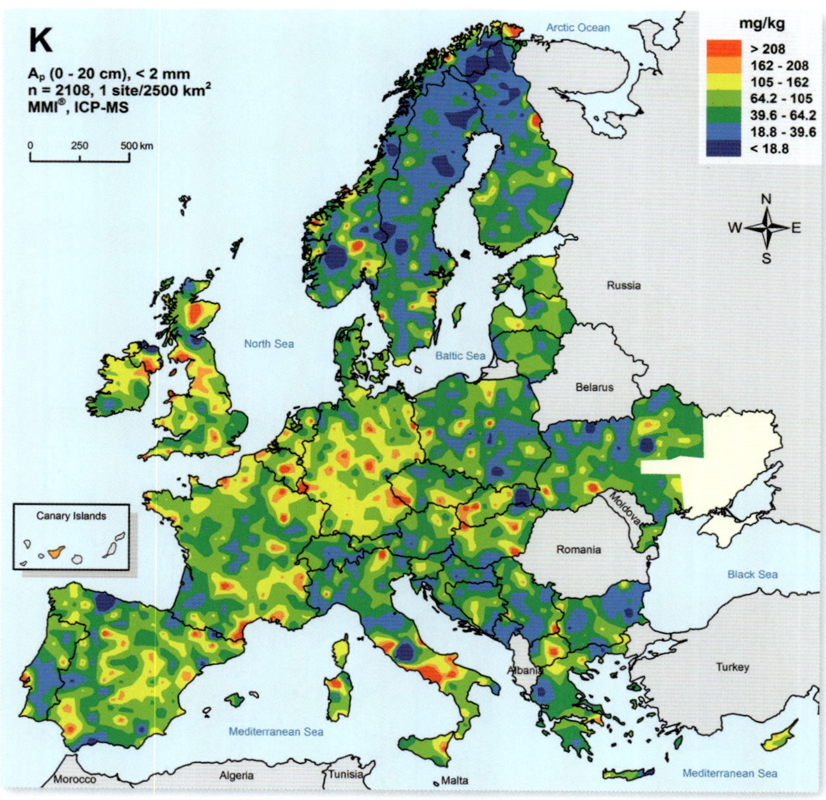

Fig. 13.16: Map of potassium (K) distribution in Ap soil samples.

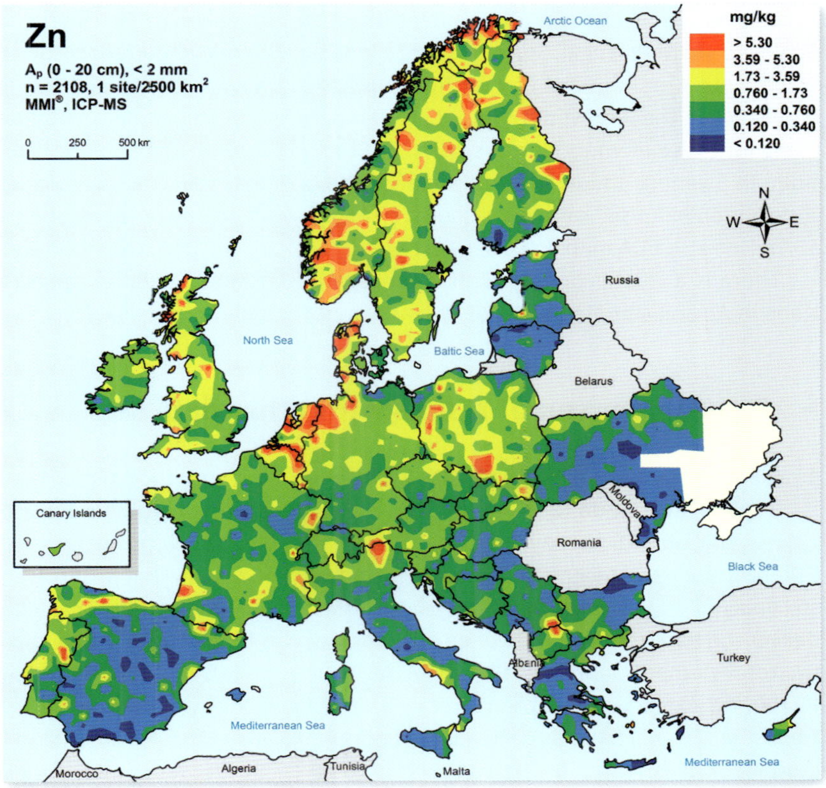

Fig. 13.17: Map of zinc (Zn) distribution in Ap soil samples.

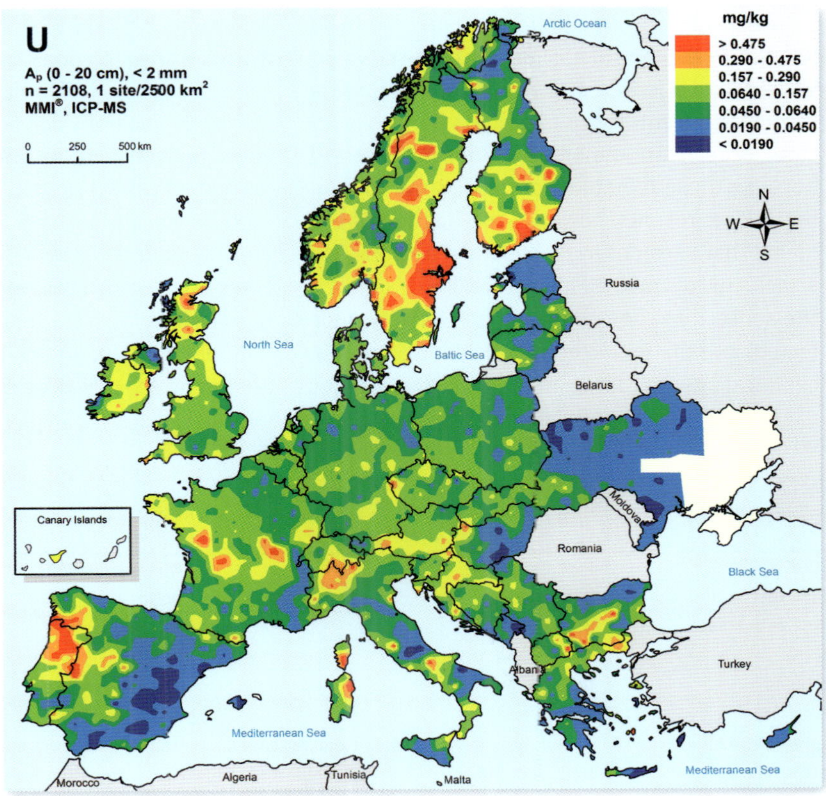

Fig. 13.18: Map of uranium (U) distribution in Ap soil samples.

in soil over granitoid rocks (Galicia in north-west Spain, northern Portugal, Scotland, the Scandinavian Shield, Corsica, northern Sardinia, Bulgaria and FYROM). The largest difference between patterns for these elements occurs north of Stockholm in Sweden, and in western Finland. The U distribution in Ap soil samples is shown on Figure 13.18 (see also EDA map on DVD → Part A).

There is a cluster of very high U concentrations in Ap soil samples north of Stockholm not matched by Th. The chemistries of U and Th differ in one important respect; whereas both elements have a stable, relatively immobile oxidation state (IV), only U has an additional very mobile (VI) oxidation state. Higher concentrations of U in soil may reflect either localised higher concentrations in granite or segregation after U has been oxidised to the U(VI) oxidation state, migrated and concentrated. The latter can occur naturally or anthropogenically assisted.

▸ 13.4 Comparison of results with other data sets

13.4.1 Comparison of MMI® results with XRF and Aqua Regia from the GEMAS survey

Different analytical techniques are used in routine soil analysis. X-ray fluorescence analysis (XRF) will deliver the total concentration of an element, i.e., the total element pool in a soil sample. An Aqua Regia (AR) extraction is more correctly referred to as a strong partial digestion, since it aims to dissolve at least some of the soil phases to release their element content. It is often described as 'pseudo-total' in the literature. Followed by an ICP-AES or -MS finish, it is most often used in environmental sciences, and will only deliver partial element concentrations. Elements bound in silicates or resistate minerals will not be detected, because these minerals do not dissolve in aqua regia. It represents, however, the fraction of elements that becomes available under a strong acid attack. By contrast, MMI® is a weak extraction in the sense that the soil matrix is relatively unaffected; ions are extracted and held in solution by chemical agents that complex with the analytes. Each method has its advantages and disadvantages as shown in Table 13.2.

Not surprisingly, each method produces different analytical results when applied to the GEMAS Ap soil samples. For Aqua Regia, 'extractability' (AR/XRF x 100) was shown to be close to 100 % only for Cu (115 %) (Reimann et al. 2011). Extractability was between 61 % and 83 % for 10 elements including Pb, Zn, Ni and P, and less than 21 % for 11 elements including W, Ti, Sb and Bi. The remaining 14 elements had extraction efficiencies between 22 % and 55 %. It is interesting that elements with extraction efficiencies less than 1 % include Na (a relatively soluble ionic species) and Zr (which is generally considered derived from resistate minerals). In soil sciences, it is an old dream to find an extraction that relates to 'plant uptake', and because plants will only be able to take up loosely bound elements, e.g., from ion exchange sites on clay surfaces, very weak extractions, like ammonium acetate (usually buffered to a certain pH), EDTA or a $CaCl_2$ solution, are often used. Generally, these weak extractions have problems,

Table 13.2: Comparison of features of the XRF, Aqua Regia digestion and MMI® extraction methods.

Feature	Analysis Technique		
	XRF	Aqua Regia	MMI®
Sample	Fusion disc	Partial digestion	Weak extraction
Sample preparation	Milled to < 63 µm	Pulverised	Nil
Sample	Mixed with borate	1 – 5 g	50 g
Detection method	XRF	ICP-AES and ICP-MS	ICP-MS
Analytical range	mg / kg to %	µg / kg to mg / kg	µg / kg to mg / kg
Proxy for	Total assay	Pseudo total	Bioavailability
Advantages	Approximates totals	Bound, non-resistate; wide element range	Labile / mobile fraction; wide element range
Disadvantages	Limited sensitivity	Limited as proxy	Elements close to LDL

unless strong complexing agents are employed; analytical quality is usually very difficult to keep constant when working with weak extractions, and simple extractions often work well only for some elements. The correlation observed, between element concentrations in soil for strong digestions and some weak extractions and concentrations in plants, is often far from convincing (MANN et al. 2012). The strengths of the multiple complexing agents in MMI® extraction solution have deliberately been tailored to extract (for nutrients at least) those concentrations thought to be available to plants from soil pore water. By intent, MMI® is a weak extraction with respect to effect on substrate, but 'strong' in complexing ability and is able to hold many analytes in solution through the analytical process. Comparison with XRF results has previously been carried out (MANN et al. 2012) for some nutrients, and with Aqua Regia digestion for Cu and Zn in Australian soil. In Table 13.3 the results of comparing median values from MMI® extraction versus XRF (i.e., MMI®/XRF x 100) and Aqua Regia digestion (i.e., MMI®/AR x 100) are presented for a number of elements.

Comparison of median values of MMI® and XRF results gives a good approximation of percentage bioavailability for a number of elements. Unfortunately, the detection limits for XRF (and in some cases also MMI®) do not allow for this to be carried out for Ag, Au, Bi, Cd, Hg, In, Li, Pd, Pt, S, Se, Ta, Te, and Tl. The highest extractability for MMI® versus XRF is for Cu (10.3 %). Ten elements (Cu, Ca, U, Ni, Mn, Co, Sr, Pb, Mo and Zn) have extractabilities greater than 1 %. Nine elements (Rb, Sb, W, V, Zr, Ga, Cs, Nb and Ti) have extractabilities less than 0.1 % relative to XRF. The remaining 13 elements have extractabilities in the range 0.1 % to 1 % relative to XRF. These low extraction percentages are designed to 'mimic bioavailability' and provide a measure of active, labile or mobile elemental forms during weathering.

Since the Aqua Regia results do not closely approximate 'total' assay concentrations for most elements, extractabilities versus Aqua Regia are not an accurate representation of overall percentage bioavailability. They are, however, interesting, and also show the (intentional) low extraction efficiency of the MMI® extraction. For the elements that could not be compared using XRF, comparison with Aqua Regia shows Ag, Au and Cd to have highest extraction efficiencies in MMI® relative to Aqua Regia of 37 %, 33 % and 30 %, respectively. For S the MMI® extraction efficiency is 11.7 %, for Se 2.82 % and for Tl 0.5 % relative to Aqua Regia.

13.4.2 Comparison of MMI® results from the GEMAS survey with NGSA

The National Geochemical Survey of Australia (NGSA; CARITAT & COOPER 2011a) pioneered the application of MMI® analysis to low-density continental geochemical surveys (CARITAT et al. 2011, 2012; MANN et al. 2012). Results have been interpreted to reflect lithology, land use, mineralisation and bioavailability, much like the GEMAS results discussed here. A brief comparison of the NGSA and GEMAS results follows.

In order to contrast the MMI® results from the NGSA and GEMAS surveys, it is imperative to demonstrate first that the data are consistent with one another. This is crucial when integrating data sets from diverse sources (REIMANN et al. 2012a), and particularly important for partial digestion techniques, which can be prone to larger errors, e.g., analytical errors, compared to more robust total analyses. The NGSA Top Outlet Sediments (0 – 10 cm) coarse (< 2 mm) samples (Tc) were digested and analysed at the SGS Laboratory in Perth, Western Australia, in early 2010, whereas the GEMAS Agricultural (0 – 20 cm, also < 2 mm) samples (Ap) were digested and analysed by the SGS Laboratory in Toronto, Ontario, in early 2012. Both surveys repeatedly analysed (N = 22 for NGSA; 32 for GEMAS) a laboratory standard (MMI®-M SRM18) (CARITAT & COOPER 2011a; REIMANN et al. 2012a). The averages of these repeat analyses by both projects are within 25 % RSD of one another for Sr (0 % RSD), Ba, Mo, Rb, P, Se, Cr, Mg (6 %), Au, Ni, Pd (12 %), V, Pt, K, Ca (15 %), Co, Ag, Al, Eu (20 %), Cu, Pr, Dy, Ti and As (25 %), elements which we thus consider to be comparable between the two studies. They are within 50 % RSD for Mn (27 %), Sm, Er, Nd, La, Zn, Th, U (30 %), Gd, Cd, Ce, Y (34 %), Pb (44 %) and Fe (50 %), which we consider acceptable for the technique. Comparability of MMI® results for Zr, Hg and Cs is not warranted, as the averages of the repeat standard analyses by both projects are over 50 % RSD of one another (54, 60 and 114 %, respectively).

Table 13.3: Extractability of the elements analysed by the MMI® method in relation to Aqua Regia extraction (near total), and XRF (true total concentrations). Left hand side in alphabetical order according to element names; right hand side ordered according to decreasing extractability of the MMI® method (from Reimann et al. 2012: Table 9, p. 30–31).

Element	GEMAS Ap samples				
		Aqua Regia extraction		XRF determinations	
	MMI® Q50	Aqua Regia Q50	Extraction	XRF Q50	Extraction
	[mg/kg]		[%]	[mg/kg]	[%]
Ag	0.014	0.038	36.8	-	-
Al	62	10,993	0.564	55,069	0.113
As	0.02	5.48	0.365	7.00	0.286
Au	0.0003	0.0009	33.3	-	-
Ba	1.09	62	1.76	384	0.284
Bi	<0.001	0.17	-	1.50	-
Ca	600	3034	19.8	8494	7.06
Cd	0.054	0.18	30.0	-	-
Ce	0.135	28.4	0.475	58.0	0.233
Co	0.182	7.5	2.43	9.00	2.02
Cr	0.065	20	0.325	62.0	0.105
Cs	0.0007	1.1	0.064	5.00	0.014
Cu	1.34	14.5	9.24	13.0	10.3
Fe	31	17,200	0.180	24,550	0.126
Ga	0.002	3.43	0.058	12.0	0.017
Hg	<0.001	0.03	-	-	-
In	<0.0005	0.018	-	-	-
K	64.2	1250	5.14	15,819	0.406
La	0.049	14.3	0.343	23.0	0.213
Li	0.0025	11.4	-	-	-
Mg	52	2860	1.82	5488	0.948
Mn	19.65	445	4.42	604	3.253
Mo	0.015	0.42	3.57	1.00	1.500
Nb	0.0009	0.48	0.188	13.0	0.007
Ni	0.652	14.7	4.44	20.0	3.26
P	4.1	653	0.628	786	0.522
Pb	0.32	15.8	2.03	21.0	1.52
Pd	<0.001	<0.001	-	-	-
Pt	<0.001	<0.001	-	-	-
Rb	0.069	14.3	0.483	73.0	0.095
S	24.2	207	11.7	-	-
Sb	0.002	0.234	0.855	2.50	0.080
Sc	0.016	2.15	0.744	8.00	0.200
Se	0.01	0.354	2.82	-	-
Sn	<0.001	0.723	-	2.00	-
Sr	1.675	18.1	9.25	101	1.66
Ta	<0.001	<0.005	-	2.50	-
Te	<0.01	<0.02	-	-	-
Th	0.0221	2.89	0.765	9.00	0.246
Ti	0.038	86	0.044	3621	0.001
Tl	0.0006	0.12	0.500	-	-
U	0.084	0.77	10.9	1.50	5.60
V	0.027	25.4	0.106	68.0	0.040
W	0.001	0.073	1.37	2.50	0.040
Y	0.184	6.69	2.75	27.0	0.681
Zn	0.76	45	1.69	60.5	1.26
Zr	0.051	1.79	2.85	259	0.020

Ordered GEMAS Ap samples according to Extractiablity of the MMI® method results			
Aqua Regia		XRF	
Element	Extraction [%]	Element	Extraction [%]
Ag	37	Cu	10.3
Au	33	Ca	7.06
Cd	30	U	5.60
Ca	20	Ni	3.26
S	12	Mn	3.25
U	11	Co	2.02
Sr	9.3	Sr	1.66
Cu	9.2	Pb	1.52
K	5.1	Mo	1.50
Ni	4.4	Zn	1.26
Mn	4.4	Mg	0.948
Mo	3.6	Y	0.681
Zr	2.8	P	0.522
Se	2.8	K	0.406
Y	2.8	As	0.286
Co	2.4	Ba	0.284
Pb	2.0	Th	0.246
Mg	1.8	Ce	0.233
Ba	1.8	La	0.213
Zn	1.7	Sc	0.200
W	1.4	Fe	0.126
Sb	0.85	Al	0.113
Th	0.76	Cr	0.105
Sc	0.74	Rb	0.095
P	0.63	Sb	0.080
Al	0.56	W	0.040
Tl	0.50	V	0.040
Rb	0.48	Zr	0.020
Ce	0.48	Ga	0.017
As	0.36	Cs	0.014
La	0.34	Nb	0.007
Cr	0.33	Ti	0.001
Nb	0.19		
Fe	0.18		
V	0.11		
Cs	0.064		
Ga	0.058		
Ti	0.044		

A number of elements (Bi, Ga, Li, Nb, Sb, Sc, Sn, Ta, Tb, Te, Tl, W and Yb; in alphabetical order) were found to be present too close to, or below, the LLD (by one or both studies) in the MMI®-M SRM18 laboratory standard; mutual consistency is thus difficult to demonstrate. Finally, two elements, In and S, were reported in the GEMAS survey but not in the NGSA survey. We will thus limit the subsequent discussion to the first two series of 38 elements above (Sr to Fe).

Table 13.4 gives a statistical summary of the MMI® results in GEMAS compared to NGSA. It shows that GEMAS has the higher median values for most listed elements. This imbalance can be explained by the generally fresher (less weathered) rocks in Europe (Reimann et al. 2012a, b). Thus, the parent materials in Europe provide, on the whole, a higher concentration of ions available in exchangeable sites on soil particles compared to Australia, where soil leaching is generally more intense due to relative tectonic stability and protracted weathering (Reimann et al. 2012a). Of the elements with a higher median value in Europe compared to Australia, Cd and Pb are the two most extreme, probably reflecting the potential for surficial anthropogenic contamination. Given the spatial distributions of MMI® Cd and Pb in Europe discussed above (e.g., Fig. 13.12), it is *likely* that a component of anthropogenic influence, including (a) the production and smelting of Pb-Zn sulphide ores in places over centuries (Tylecote 1992); (b) the formerly widespread use of leaded products such as petrol (Smith & Flegal 1995); (c) the application of sewage sludge to agricultural land by some European countries (Fuge 2005), and (d) the use of phosphate fertilisers containing Cd impurities (Syers et al. 1986; IFA 2012), is required to explain these higher median values over Europe. The few exceptions for which the NGSA MMI® medians are higher are K, Mg, Rb and Sr. These alkali and alkaline earth cations are normally fairly mobile in soil environments, and it appears at this stage that the slightly (K) to much (Mg) higher median values in the Australian compared to the European soil could be related to landscape position. Indeed, floodplain environments, sampled across Australia, sit low in the landscape, and are sites of preferential accumulation of elements mobilised higher up in the catchment, as exemplified by the common occurrence of calcrete in valleys in many parts of arid Australia. The European survey's sampling strategy was not based on landscape position, but rather on land use and as such sampled a variety of landscape positions.

Figure 13.19 shows a selection of cumulative frequency diagrams for both GEMAS and NGSA MMI® data.

This figure illustrates not only the above point about the generally higher median values in Europe compared to Australia, but also shows different groups in both populations (e.g., breaks in slope, changes in curvature), the different behaviours of both populations (e.g., cross-overs), and detection limit issues for some elements. Remarkably similar statistical distributions for Europe and Australia are obvious for Ag, Au, Ba, Ca, Ce, Co, Cr, Dy, Er, Eu, Gd, K, La, Nd, Ni, Pr, Sm, Sr, U (above the 25th percentile), V and Y. This similarity for such a long list of elements was unexpected, given the diverse sampling media, climate, depth of weathering profiles, land use practices, geology, etc. exhibited by both continents. In part, this may be due to the fact that MMI® analysis represents only the first, early stages of weathering (Mann et al. 1998), and although these continents have varying degrees of 'flushing' versus accumulation of weathering products, production of ions from initial weathering of lithology may be of a similar rate for many elements. It follows that MMI® results are less dependent on depth (and to some extent length of time) of weathering. Skeletal soil on, for example, a glaciated surface can produce MMI® values as high as those from a weathering front in a lithologically equivalent deep profile. Ion production occurs at the weathering interface, wherever it may be located, at rates dependent mostly on the presence of chemical factors, such as water and oxygen.

The elements for which a large discrepancy is noted between Europe and Australia are Al (systematically higher in Europe), As (ditto, except one very high value in Australia that surpasses the European maximum), and Cd (an order of magnitude higher concentrations in Europe up to the 75th percentile, after which both cumulative frequency curves converge to a similar maximum). For Cu, although the curves are close up to the 75th percentile, the European population continues

Table 13.4: Summary statistics of MMI® analysis of GEMAS Ap (N=2109) and NGSA Tc (N=1191) samples for 39 elements deemed mutually consistent between both studies (see text). LLD and ULD are Lower Limit of Detection and Upper Limit of Detection, respectively. Highest median values are highlighted in boldface.

Element	GEMAS project Ap samples (N=2109)							NGSA project Tc samples (N=1191)						
	LLD	ULD	min.	25%	median	75%	max.	LLD	ULD	min.	25%	median	75%	max.
Ag	0.001	N/A	<0.001	0.006	**0.014**	0.025	1.34	#	#	<0.001	0.002	0.005	0.011	0.302
Al	1	300	<1	25	**61**	155	>300	#	N/A	<1	7	12	26	453
As	0.01	N/A	<0.01	<0.01	**0.02**	0.03	0.53	#	#	<0.01	<0.01	<0.01	0.01	1.98
Au	0.0001	N/A	<0.0001	0.0001	**0.0003**	6E-04	0.024	#	#	<0.0001	<0.0001	0.0002	0.0005	0.02
Ba	0.01	N/A	<0.01	0.57	**1.09**	1.95	51.6	#	#	<0.01	0.4	0.87	1.74	15.3
Ca	10	N/A	<10	320	**595**	902.5	2490	#	#	<10	220	390	560	3680
Cd	0.001	N/A	<0.001	0.032	**0.054**	0.085	2.81	#	#	<0.001	0.002	0.005	0.009	1.75
Ce	0.005	N/A	<0.005	0.033	**0.135**	0.373	23.1	#	#	<0.005	0.019	0.101	0.429	12.1
Co	0.005	N/A	<0.005	0.077	**0.182**	0.434	8.94	#	#	<0.005	0.086	0.178	0.328	2.58
Cr	0.001	N/A	<0.001	0.033	**0.065**	0.117	1.3	#	#	<0.001	0.01	0.024	0.057	2.55
Cu	0.01	N/A	<0.01	0.68	**1.34**	2.71	62.1	#	#	0.01	0.47	0.96	1.59	30.5
Dy	0.001	N/A	<0.001	0.015	**0.035**	0.075	2.17	#	#	<0.001	0.009	0.023	0.079	1.49
Er	0.0005	N/A	<0.0005	0.009	**0.0202**	0.045	1.23	#	#	<0.0005	0.0043	0.011	0.038	2.23
Eu	0.0005	N/A	<0.0005	0.003	**0.0075**	0.016	0.208	#	#	<0.0005	0.0016	0.007	0.021	0.359
Fe	1	N/A	<1	13	**31**	88	2200	#	#	<1	2	4	12.5	555
Gd	0.001	N/A	<0.001	0.015	**0.036**	0.081	2.21	#	#	<0.001	0.008	0.029	0.096	1.61
K	0.1	N/A	<0.1	39.6	64.2	105	749	#	#	1.2	44.9	**82.2**	134	1040
La	0.001	N/A	<0.001	0.006	**0.049**	0.145	12.1	#	#	<0.001	0.005	0.027	0.123	3.31
Mg	1	500	<1	26	52	103.3	>500	#	N/A	2	70	**140**	260.5	2920
Mn	0.01	N/A	<0.01	11.4	**19.6**	31.2	180.2	0.005	#	0.079	2.24	3.99	8.14	59.7
Mo	0.005	N/A	<0.005	0.008	**0.015**	0.027	0.529	#	#	<0.005	<0.005	0.007	0.012	0.276
Nd	0.001	N/A	<0.001	0.021	**0.102**	0.263	11.2	#	#	<0.001	0.012	0.092	0.343	6.87
Ni	0.005	N/A	<0.005	0.286	**0.652**	1.48	30.1	#	#	0.006	0.181	0.411	0.912	16.1
P	0.1	N/A	<0.1	1.7	**4**	10.2	188	#	#	<0.1	0.6	1.2	2.5	32
Pb	0.01	N/A	<0.01	0.16	**0.32**	0.73	58.3	#	#	<0.01	0.02	0.04	0.11	5.89
Pd	0.001	N/A	<0.001	<0.001	<0.001	<0.001	0.002	#	#	<0.001	<0.001	<0.001	<0.001	0.012
Pr	0.001	N/A	<0.001	0.003	**0.02**	0.054	2.98	#	#	<0.001	0.002	0.016	0.063	1.25
Pt	0.001	N/A	<0.001	<0.001	<0.001	<0.001	<0.001	#	#	<0.001	<0.001	<0.001	<0.001	<0.001
Rb	0.005	N/A	<0.005	0.039	0.069	0.131	2.85	0.00025	#	0.003	0.07	**0.107**	0.182	2.63
Se	0.01	N/A	<0.01	<0.01	**0.01**	0.02	0.17	0.002	#	<0.002	0.002	0.007	0.015	1.28
Sm	0.001	N/A	<0.001	0.009	**0.029**	0.068	2.02	#	#	<0.001	0.004	0.025	0.083	1.34
Sr	0.01	N/A	<0.01	0.81	1.67	3.11	90.2	#	#	0.05	1.34	**2.4**	4.82	35.6
Te	0.01	N/A	<0.01	<0.01	<0.01	<0.01	0.03	#	#	<0.01	<0.01	<0.01	<0.01	0.02
Th	0.0005	N/A	<0.0005	0.009	**0.0221**	0.057	2.84	#	#	<0.0005	0.0026	0.021	0.104	3.52
Ti	0.003	N/A	<0.003	0.015	**0.038**	0.219	13.2	#	#	<0.003	0.01	0.019	0.04	26.5
U	0.001	N/A	<0.001	0.045	**0.084**	0.157	5.32	#	#	0.001	0.027	0.063	0.129	3.49
V	0.001	N/A	<0.001	0.013	**0.027**	0.057	12.9	#	#	0.002	0.014	0.025	0.048	2.61
Y	0.005	N/A	<0.005	0.082	**0.184**	0.387	16.2	#	#	<0.005	0.036	0.098	0.337	4.16
Zn	0.02	N/A	<0.02	0.34	**0.76**	1.73	36.2	#	#	<0.02	0.06	0.12	0.31	19.5

N/A not applicable

same LLD or ULD as for GEMAS project survey

to increase linearly; whilst the Australian population curves upward then flattens for the top two percentiles. Iron is similar to Cd, with a greater difference at low concentration and convergence of distributions above the 90th percentile. Magnesium provides a rare case where the Australian population is systematically above the European results, and the distributions are sub-parallel. Conversely, Mn and Mo represent almost the opposite situation to Mg; there is a significant LLD problem for Mo. For P, despite the LLD issues, both populations are close at the 10th percentile level, but progressively diverge linearly at higher percentiles, whereas Pb is similar to Cd and Fe with a wider gap at low concentrations narrowing at higher percentiles. Australia has higher Rb concentrations at the lower percentiles; the distributions cross at the 90th percentile, with Europe showing the greatest maximum. For Se (despite LLD and quantisation issues), values are similar from the 25th percentile until Australia displays much higher concentrations above the 90th percentile. Thorium is opposite to Rb, i.e., Europe has the higher concentrations below the 50th percentile, and Australia above; Europe's Rb distribution is quasi-linear, whilst Australia's is curving in an 'S' shape, probably reflecting multiple populations. In the case of Ti, both populations start at largely the same level up to the 25th percentile, after which European values diverge to higher values up to the 98th percentile, then converge again toward the Australian distribution. For Zn, Europe is systematically higher and has a linear distribution, whereas Australia's lower values first diverge, then are sub-parallel to the European distribution between the 25th and 75th percentiles, then converge at higher levels.

In MANN et al. (2012), the *'bioavailability %' of eight essential nutrients (Ca, Cu, Fe, K, Mg, Mn, P and Zn) and three micronutrients (Mo, Ni and Se) was operationally defined, as the average proportion of the MMI® extractable concentration relative to the total concentration of the elements.* The latter was determined by XRF analysis or (for Cu, Mo, Ni and Zn) by ICP-MS analysis of a total extraction using hot HF and HNO_3 acid. For Se, the 'total' concentration was determined by Aqua Regia digestion followed by ICP-MS analysis. This was done on 1186 catchment outlet sediments covering much of Australia.

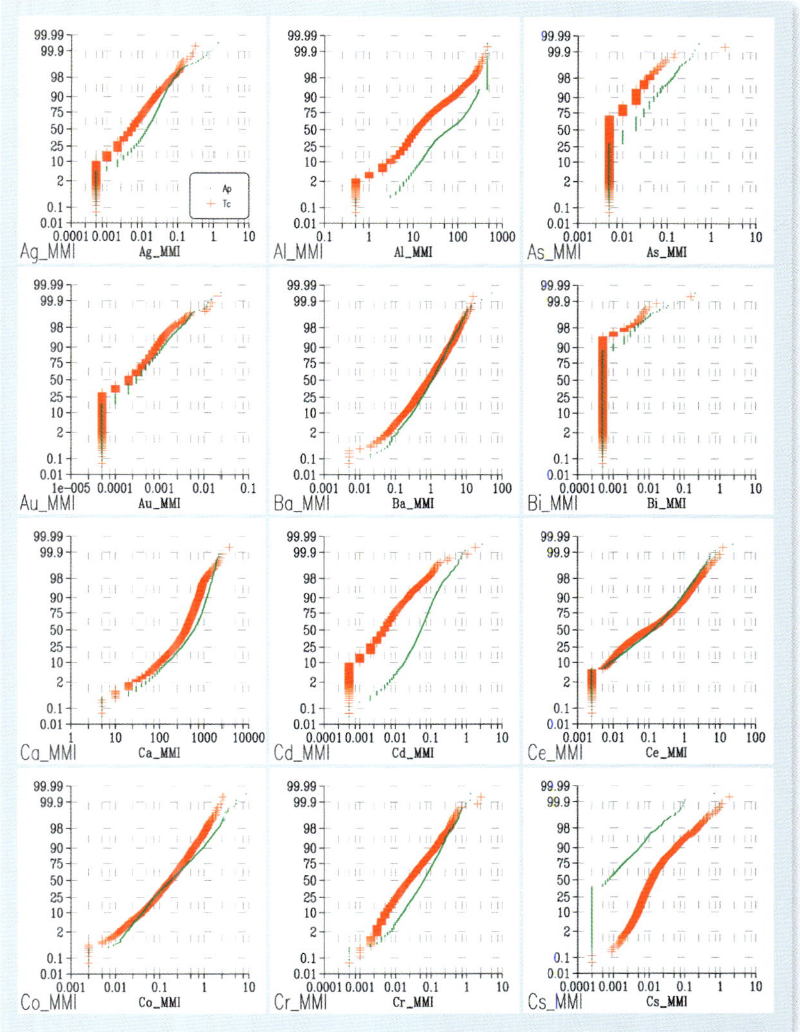

Fig. 13.19: Cumulative frequency distributions of MMI® results from Europe (green dots) and Australia (red crosses) shown on log-probability scales. In this space, a straight diagonal line represents a log-normal population. Breaks in the distributions (slope changes, curvatures) can be interpreted as reflecting major changes in the controls on the concentrations.

Thus defined, average 'bioavailability %' was found to range from 31 % for Se, to 15 % for Ca, to 8 % for Cu and Mg to less than 1 % for P and Fe.

Since not all elements are reported as true total concentrations (by XRF) in the GEMAS data set, we will instead here compare MMI® concentrations to Aqua Regia concentrations, this ratio being termed 'extractability %' here. The median values of this ratio are shown for both GEMAS and NGSA in Table 13.5. It can be seen that the extractability % is similar on both continents for many elements. If we use the same criterion of similarity as described above (< 50 % RSD between the values), then the following elements are deemed to have similar extractability % on both continents: Ni (4 % RSD), K (5 %), Ce, Ca, U,

Table 13.5: Median values of MMI®/Aqua Regia analysis ratios for GEMAS Ap and NGSA Tc samples for all elements deemed mutually consistent between both studies (see text) and for which both surveys have MMI® and Aqua Regia data. N represents the number of values used to determine the median value (ratios with either MMI® or Aqua Regia or both result(s) <LLD or >ULD were excluded from this computation) and can be compared to the total number of samples (GEMAS Ap = 2109 and NGSA Tc = 1191). Highest median values are highlighted in boldface.

Element	GEMAS		NGSA	
	median ratio MMI/AR [%]	N	median ratio MMI/AR [%]	N
Ag	**49**	1993	36	351
Al	**0.5**	1985	0.12	1155
As	0.44	1493	**0.71**	334
Au	38	1549	**46**	698
Ba	**1.9**	2106	1.6	1186
Ca	16	2096	**17**	1138
Cd	**32**	2100	14	898
Ce	**0.57**	1977	0.52	1115
Co	3	2100	**3.4**	1180
Cr	**0.32**	2100	0.095	1187
Cu	**11**	2107	9.5	1186
Fe	**0.2**	2105	0.033	1052
K	**5.5**	2101	5.1	1170
La	**0.46**	1895	0.28	1120
Mg	2	2088	**9.1**	1152
Mn	**4.5**	2107	2.1	1189
Mo	4.4	1844	**5**	711
Ni	**5.2**	2106	4.9	1183
Pb	**2.1**	2103	0.75	1048
Rb	0.53	2099	**1**	1187
Se	3.8	1328	**15**	903
Sr	8.4	2104	**12**	1169
Th	**0.84**	2080	0.66	1140
U	11	2104	**12**	1139
V	**0.11**	2106	0.08	1184
Y	**3.6**	2091	2.1	1164
Zn	**1.8**	2105	0.6	1130

Co, Mo, Cu (11 %), Ba, Au (15 %), Th, Ag (21 %), V (25 %), Sr, As (33 %), La (35 %), Y, Rb (41 %). This list contains macronutrients (Ca, K), other alkalis and alkaline earths (Ba, Rb, Sr), metals, including some noble and some that are micronutrients (Ag, Au, Cu, Co, Mo, Ni, V), REEs and other elements usually considered to be 'immobile' or less mobile (Ce, La, Y, Th), the metalloid As and the actinide U. The mobility of some of these elements in the environment is controlled by redox (e.g., As, Mo, U).

The fact that such a wide variety of elements has similar extractability % in European and Australian soil (despite the range of lithology, climate, topography, land use, vegetation, etc. covered) to some extent follows previous observations (Mann et al. 1998) that concentrations of commodity elements in soil above mineralisation were observed to be similar, following MMI® extraction in a variety of climatic and geographical situations, namely Scotland, Australia, and Venezuela. A three-stage model was presented in this paper suggesting that the concentrations of elements extracted from a soil by weak extractants, such as MMI®, are 'recent or active weathering products', which occur as intermediates ('unbound' or adsorbed species) in soil formation, between release of ions from weathering substrate, and their final incorporation as 'bound' material in soil phases after extended pedogenesis. That this intermediate form of adsorbed and/or weakly bound material is relatively robust to different weathering conditions is evident within Australia (e.g., from desert to high rainfall environments) (Mann et al. 1997). To some extent, this is also observed in Europe in the present study for many (but not all) elements. Possibly, there is a 'self-compensating' effect relating to rainfall – high annual rainfall promotes high rates of ion release from weathering substrate, but also promotes rapid flushing of ions or rapid conversion during pedogenesis of adsorbed and amorphous-held material to more inert forms, accessible only to stronger digestions. Conversely, the immediate weathering products of infrequent ionic release of ions in semi-desert and desert areas from episodic rainfall events appear to be to some extent 'protected' from solution dispersion or further rapid pedogenesis by low soil moisture. It is also interesting, that in both Australia and Europe it is often those elements usually associated with 'extremes' of weathering, Fe, Al, and Mn, which have relatively insoluble sesquioxides, and the soluble alkali and alkali-earth elements, which have marked distribution differences on both continents, and which appear to be related to climate.

Elements that have a very different extractability % on the two continents include Mn (52 % RSD), Cd, Pb (68 %), Zn, Cr (77 %), Se (85 %), Al, Mg (90 %) and Fe (101 %). All of these, excluding Se and Mg, have a significantly greater extractability % in Europe than in Australia (Table 13.5). The high mobility of Al, Mn and Fe in Europe probably relates to low pH situations (see Discussion below). Cadmium and Pb were discussed above, and their greater bioavailability (median MMI® concentration) and extractability % in Europe probably relates to historical non-point source (dispersed) contamination (e.g., by fertilisers, leaded petrol). The greater extractability % of Mg and Se in Australia possibly reflects the low-lying landscape position where the NGSA samples were taken, which would generally speaking be more a site of accumulation than leaching of mobile elements. The median 'total' Se (Aqua Regia extraction) in NGSA samples is 0.06 mg/kg, which is at the lower end of the range of world soil values (0.01 to 2 mg/kg; FORDYCE 2005), but the extractability % is relatively high at 15 % compared to the reported 0.3 – 7 % soluble range (JACOBS 1989). Floodplain environments sampled in the NGSA project would be consistent with the oxidising and alkaline conditions, Fe-oxyhydroxide poor and organic matter poor conditions that favour Se mobility (FORDYCE 2005).

▶ 13.5 Discussion and conclusions

Soil pH is a very important parameter controlling the composition and the mobility of elements in a soil. Figure 13.20 shows the pH in European Ap soil, as measured in the GEMAS project, classified into three percentile-based categories (each with approximately 700 samples) defining high, medium (or normal) and low pH values.

Low pH (pH < 5.23) is predominant in northern Europe, Portugal, and western Spain. These areas have been identified in broad terms as having granitoid lithology, but mention has also been made above of the possible correlation with high rainfall. The map for 'mobile' Fe was presented in Figure 13.8, and the apparently similar behaviour

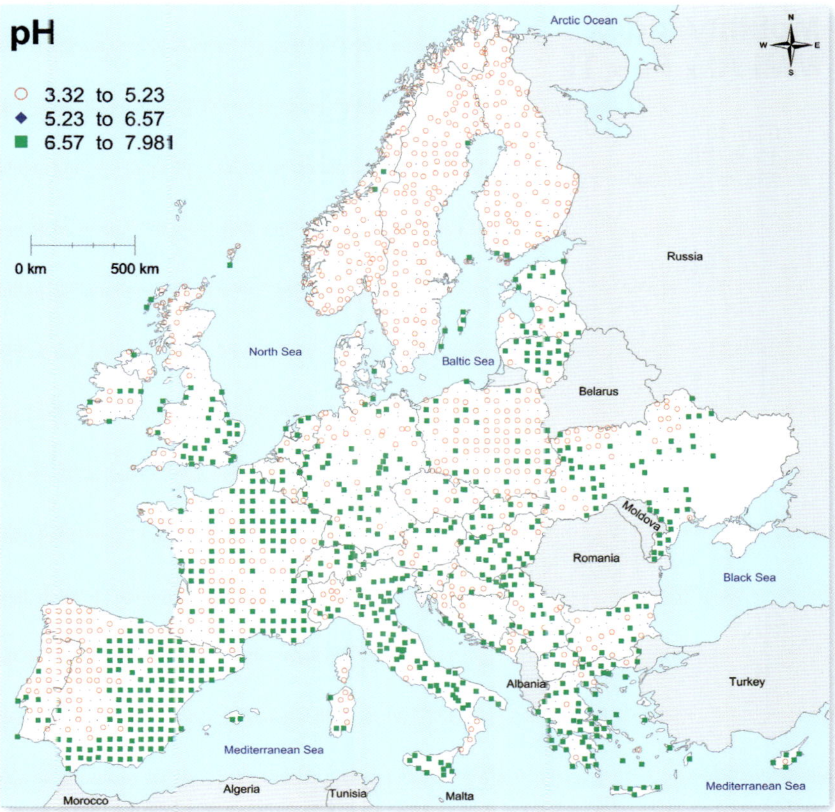

Fig. 13.20: Values of pH in Ap soil, placed into three categories based on equal numbers (approx. 700 samples in each).

of Al and Ti in response to high rainfall and weathering noted. A multiplicative mobility index, highlighting the first stages of weathering and combining the MMI® values of Al, Fe and Ti has been constructed; a multiplicative index is appropriate when all elements need to be equally represented. It tends to zero if any one element is missing or is in low concentration; this weathering index is also presented (Fig. 13.21) in three percentile-based categories, high, medium and low.

Comparison of Figures 13.20 and 13.21 is compelling; there is a high degree of correlation between low pH and where 'mobile' forms of these three elements is greatest. There is potentially a causative link. As mentioned previously, Fe is ubiquitous in most rock types, and if anything is normally higher in mafic and ultramafic rocks than granitoids; it is the weathering rate of it, which is the important variable. Weathering of Fe produces acid by the reaction:

$$2Fe^{2+} + 3H_2O + 0.5O_2 \rightarrow 2FeOOH + 4H^+$$

This reaction, termed 'ferrolysis,' has been shown to result in landscape pH changes in Australia. The lowering of pH is sufficient to not only promote the 'mobilisation' of Al, but many other elements as well (Mann 1983), including Fe itself. One of the most important considerations of pH of soil is related to the availability of nutrients to plants, i.e., there is a 'window' between pH 6 and pH 7 where the majority of nutrients are in their most available form (e.g., Handreck & Black 1994).

Ten of the most important nutrients have been measured in this study; whilst N, B and Cl are not amenable to MMI® extraction and ICP-MS analysis; bioavailable forms for P, K, Ca, Mg, Fe, Mn, Mo, Cu, Zn, and S are, and will be discussed. The nutrient list may be further subdivided into macronutrients (P, K, Ca, Mg, S), which end up in plant tissue in the percentage range, and micronutrients (Cu, Fe, Mn, Mo, Zn) that end up in plant tissue in the mg/kg range.

Multiplicative indices for the macro- and micro-nutrients are presented separately; again, a multiplicative index is appropriate where all nutrients in the index are considered important to plant growth; indeed, growth is often restricted

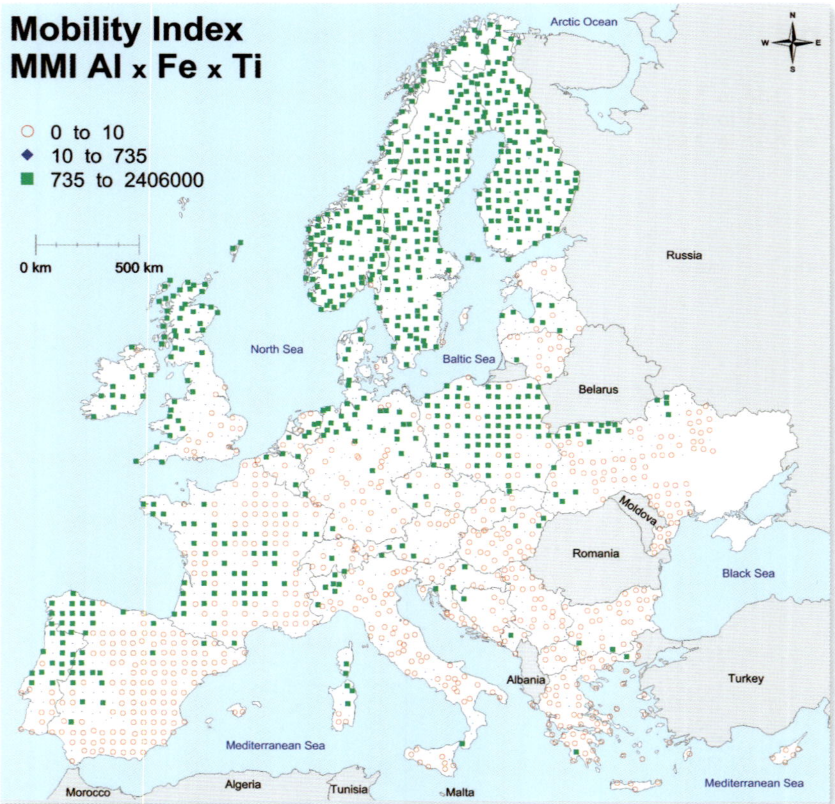

Fig. 13.21: Mobility index based on a multiplicative index of aluminium (Al), iron (Fe) and titanium (Ti) after MMI® extraction and analysis. Three categories are shown based on equal numbers (approx. 700) in each category.

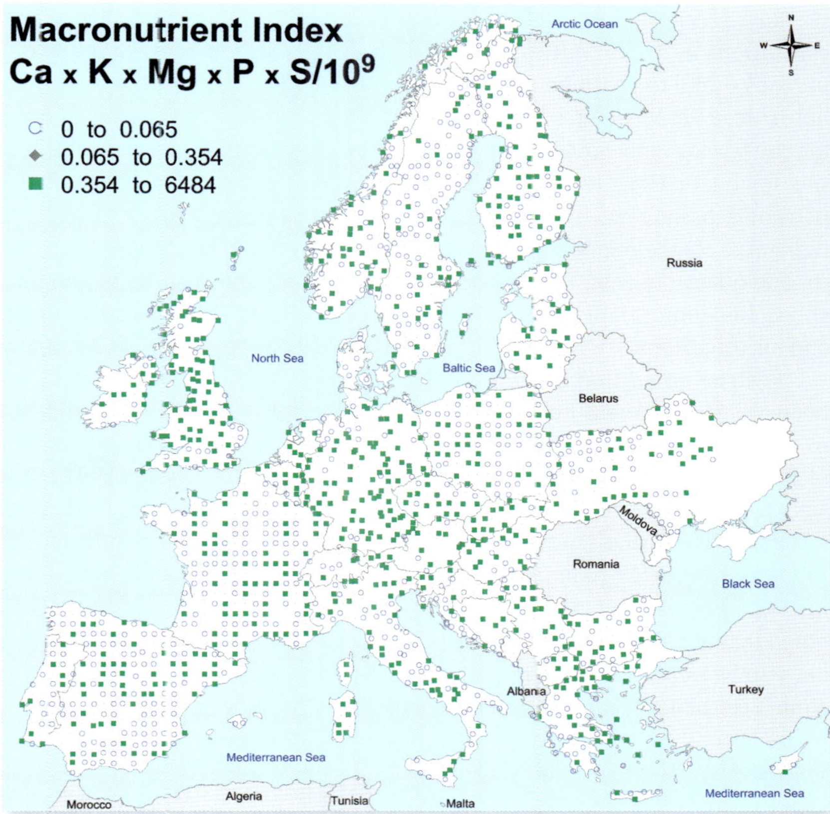

Fig. 13.22: Macronutrient index based on a multiplicative index of calcium (Ca), potassium (K), magnesium (Mg), phosphorus (P) and sulphur (S) after MMI® extraction and analysis. Three categories with approximately equal numbers (700) in each category are shown.

by the nutrient in shortest supply. Three percentile-based categories for each index (low, medium, high) are again represented in Figures 13.22 and 13.23.

Given the different land use types, fertiliser application rates and relatively wide spacing of Ap samples in the GEMAS project, it might be expected that coherent patterns might be rare. Very different nutrient levels exhibited by neighbouring samples would result in a uniformly 'patchy' distribution map for macronutrients. Whilst this latter feature is certainly evident, the map also shows some areas with coherent clusters of high values. These are evident for central Europe (UK, Belgium, The Netherlands and Germany), and may well be a result of a blend of natural inherent lithological attributes (Ca, Mg, K) of the soil and application of adequate levels of fertilisers (P) and the use of horticultural sprays (S). In several areas of Norway, Sweden, Denmark, eastern Poland, central France and eastern Spain and Italy, there are also zones with coherent areas of low macronutrients in soil (relative to the central interval). In southern Europe, this is more likely to be as a result of low P and/or K, since these areas have previously been shown to have high Ca, Mg and sometimes S.

The distribution map for micronutrients (Fig. 13.23) also contains coherent 'areas' with higher or lower than normal levels. In central Norway and Sweden, Ap soil samples appear to have abundant micronutrients as opposed to macronutrients. Central European countries (U.K., Germany, The Netherlands) have 'above normal' levels of both. Large areas of eastern Spain, Latvia, Ukraine and Lithuania exhibit 'below normal' levels of micronutrients. In some of these cases, the high carbonate content of the soil may have an effect on availability of micronutrients; the relative insolubility of Cu and Zn in the presence of carbonate is well known (Mann & Deutscher 1977, 1980). With respect to Fe and Mn, it should be pointed out that, in general, the concentrations of these micronutrients, released as a result of lithological content and weathering, far exceeds plant requirements at many locations. The same is also true of Cu in instances where Bordeaux mixture is used as a spray.

Soil scientists often classify soil nutrient concentrations as 'Low', 'Sufficient' and 'High'. It is perhaps useful to conclude this discussion of soil nutrients in European Ap soil by similarly classifying the levels of individual nutrients measured in this soil by the MMI® technique. The appropriate percentile-based classes were calculated for each nutrient as for the pH, weathering index, and macro- and micro-nutrient indices above. Table 13.6 shows the appropriate levels for the ten macro- and micronutrients.

It will be noted that the term 'sufficient' has been replaced by the word 'normal' in Table 13.6. As yet, there is no direct experimental correlation of MMI® values with what plants require; the results presented here are simply a summary of the results of MMI® analysis of the Ap soil samples. For this reason, inverted commas have been retained for 'low', 'normal' and 'high'. It has already been noted that observed values for some nutrients are likely to be well in excess of requirements; this is almost certainly true in the latter two categories for Fe, Mn and Cu.

This survey has been conducted over a large area; it covers over 50 degrees of longitude and 42 degrees of latitude. Accordingly, there is a wide range of climatic zones, and weathering conditions. The effects of this on element distribution patterns have been observed for some elements. The generally high abundance, 'resistate' elements Fe, Al and Ti have been shown to have greater 'mobility' in northern Europe, where rainfall is greater and evaporation a minimum. Conversely, Ca has been shown to be retained (and measured) at higher concentrations in the soil profile, where evaporation is dominant over precipitation. For commodity elements, such as Au, Cu, Co and Ni, where this study also suggests some variation in 'mobility' from north to south, it may be suggested by some that this is an impediment to the use of a partial extraction solution for mineral exploration. On a regional basis, this may be so – raw data that define an 'anomalous value' in the north is more significant (for these elements) than the same 'anomalous value' in the south. However, mineral exploration, using soil geochemistry, is

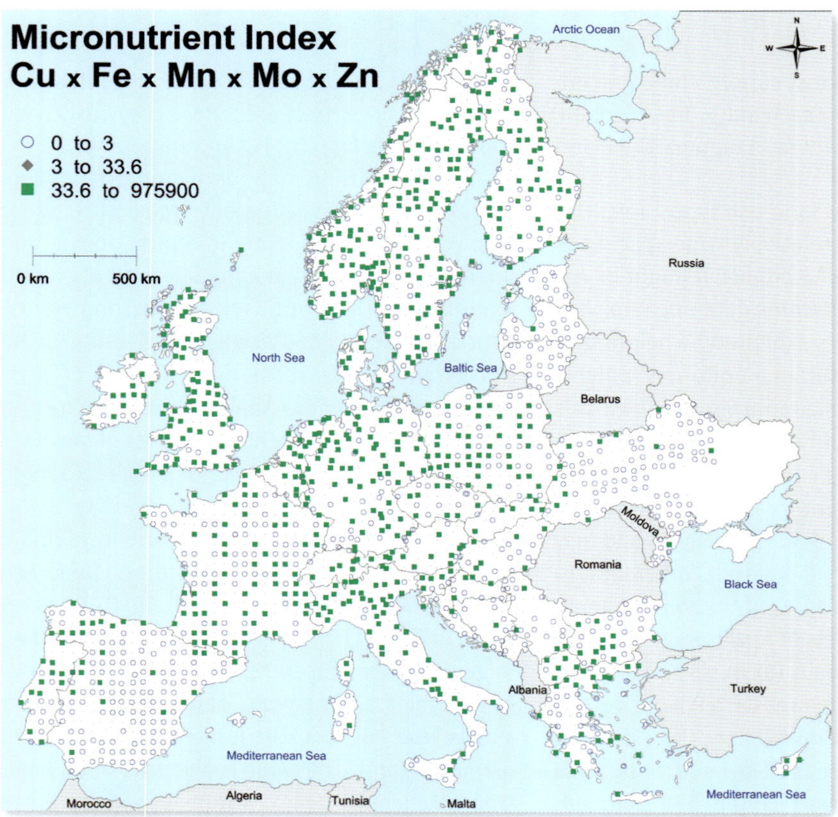

Fig. 13.23: Micronutrient index based on a multiplicative index of copper (Cu), iron (Fe), manganese (Mn), molybdenum (Mo) and zinc (Zn) after MMI® extraction and analysis. Three categories with approximately equal numbers (700) in each category are shown.

not normally carried out on such a regional scale. On a district, or prospect basis 'anomalies' are normally defined by comparison with 'local' background. The advantage of applying a method, which provides information on 'bioavailability' to such a regional data set is that regional geochemical patterns likely to have been affected by human activity or are likely to affect human activity are generated. Whilst many of the 'soil anomalies' generated by the application of the MMI® extractant to Ap soil samples have been attributed to lithology and/or mineralisation, a great number have been seen to be potentially related to anthropogenic activity. That anthropogenic influence has been evident in various forms; individual examples of mining activities, horticultural treatments and fertiliser distribution have been documented. To this end, it appears that those aims of the GEMAS study may have been met.

Table 13.6: Concentrations of nutrients after MMI® extraction and analysis for European Ap soil in three ranges, 'low', 'normal' and 'high'.

Nutrient	'low' range	'normal' range	'high' range
		[mg/kg]	
Ca	5 – 410	410 – 820	820 – 2490
Cu	0.005 – 0.870	0.870 – 2.08	2.08 – 62.1
Fe	0.5 – 18	18 – 63	63 – 2200
K	0.05 – 46.7	46.7 – 88.6	88.6 – 749
Mg	0.5 – 34	34 – 82	82 – 750
Mn	0.005 – 14.2	14.2 – 26.8	26.8 – 180
Mo	0.0025 – 0.010	0.010 – 0.021	0.021 – 0.529
P	0.05 – 2.3	2.3 – 7.3	7.3 – 188
S	0.25 – 19.5	19.5 – 31	31 – 2620
Zn	0.010 – 0.450	0.450 – 1.31	1.31 – 36.2

Acknowledgements

The NGSA project was part of the Australian Government's Onshore Energy Security Programme 2006–2011, from which funding support is gratefully acknowledged. NGSA was lead and managed by Geoscience Australia and carried out in collaboration with the geological surveys of every State and the Northern Territory under National Geoscience Agreements. |◂

Alan Mann
MMI Soil Services;
Fremantle, Western Australia
E-mail: awmann@iinet.net.au

Clemens Reimann
Geological Survey of Norway (NGU);
Trondheim, Norway

Patrice de Caritat
Geoscience Australia;
Canberra, Australia

Nicholas Turner
SGS Minerals Services;
Toronto, Canada

Part B

Chapters 14 – 15

Regional Interpretation of GEMAS Data

Chapter 14

Elemental Patterns in Agricultural and Grazing Land Soil in Norway, Finland and Sweden: What Have We Learned from Continental-Scale Mapping?

Anna Ladenberger, Jo Uhlbäck, Madelen Andersson, Clemens Reimann, Timo Tarvainen, George Morris, Martiya Sadeghi, Mikael Eklund & Peter Filzmoser

▸ 14.1 Introduction

The GEMAS project generated a valuable data set displaying baseline levels of elements for agricultural and grazing land soil, both on a European and a regional scale.

The Geological Surveys of Norden have been documenting the natural geochemical background of chemical elements in a variety of sample materials for more than 50 years, but most of the existing data sets vary between one another making it impossible to construct harmonised transnational geochemical reference maps. In fact, harmonised geochemical data on agricultural soil do, however, exist for ten countries in north-eastern Europe (Reimann et al. 2003), but data on grazing land soil are completely absent. Therefore, for Norway, Sweden and Finland, the new data set constitutes a unique opportunity to make complex regional comparisons and to unify baseline definitions. In Scandinavia, this is the first such attempt to build a common soil sample archive, which can be used by different authorities and research groups.

Norway, Sweden and Finland are the neighbouring countries in northern Europe that, apart from Iceland and Russia, occupy the northernmost latitudes of the European continent. They have the most populated regions in the world that are located north of the polar circle, with active farming and agriculture at latitudes higher than 65° N. On a European scale, they share many common features, such as geology, climate, and vegetation, but on the regional scale, these features are responsible for the variation in soil chemistry patterns. Here, we compare the element distribution patterns of these three countries, relate them to the European data set and discuss the importance of the results for mineral exploration, agriculture and environmental studies.

▸ 14.2 Topography and climate

Norway, Sweden and Finland are located on the Fennoscandian Shield, stretching over a geographical range that covers varying topography, precipitation and climate – important factors that affect the weathering conditions and development of soil regions in the area. Lowland topography is characteristic around the Gulf of Bothnia and the Baltic Sea, the Oslo Rift valley, and locally along the Norwegian coast. The north-eastern part of Finland and the central part of Sweden comprise mid to high altitudes of *ca.* 200–800 m.a.s.l., while the highest terrains (1000–2000 m.a.s.l.) are found in the Caledonian mountain range of the Scandinavian peninsula and in southern Norway. While the Norwegian topography is mostly a result of intensive glacial erosion, being characterised by extreme differences in elevation between fjords and mountain tops, the influence of glacial deposition and postglacial history are important in Sweden and Finland where they are responsible for the formation of extensive till deposits and land forms, such as eskers and drumlins. In Sweden, the landscape varies from high mountains in the west to lowlands in the southern and eastern parts of the country. Typical feature of the Finnish landscape is the rather flat terrain with numerous lakes and islands. The postglacial rebound is most prominent in the Gulf of Bothnia, where the uplift of *ca.* 1 cm/year continues to this day.

The Caledonian mountain range serves as a natural barrier for precipitation, leading to higher annual precipitation in Norway than in Sweden and Finland. Due to the high latitude and alpine topography, the vegetation of the highlands also starts at much lower elevations (at *ca.* 800 m.a.s.l. in the south and at *ca.* 300 m.a.s.l. in the north) compared to continental Europe. The climate differs among the three countries, from subarctic conditions in the northern part of Norway, cold climate in most of Sweden and parts of central Norway and Finland to a mild and humid climate in the south of the Scandinavian peninsula. The climate of western Scandinavia is strongly influenced by the warm Gulf Stream, while the eastern part and Finland is dominated by a continental climate.

▸ 14.3 **Geology**

14.3.1 Bedrock

The bedrock of Norway, Sweden and Finland comprises some of the oldest rocks in Europe (Fig. 14.1). Most of the area is underlain by crystalline basement of the Fennoscandian Shield (the north-eastern part of the East European Craton), which comprises Archaean to Proterozoic rocks formed during several orogenies (e.g., Svecokarelian between 2.4 and 1.75 Ga, Gothian between *ca.* 1.7 to 1.55 Ga, Telemarkian at *ca.* 1.5 Ga, and Sveconorwegian at *ca.* 1.0 Ga). A number of metamorphic events have produced large metamorphic belts (e.g., Transscandinavian Igneous Belt), with rocks ranging from low-grade greenschist to eclogite, which are often migmatised. Large granitoid massifs and metamorphic terrains are cut by mafic intrusions and dyke swarms of different ages.

The oldest (Archaean) rocks (*ca.* 3.5–2.5 Ga tonalite-trondhjemite-granodiorite gneiss, paragneiss, greenstone belts, granulite complexes and migmatitic amphibolite) occur in eastern and north-eastern Finland, and in northern Sweden and Norway (Hölttä et al. 2008). Palaeoproterozoic rocks form most of the Fennoscandian Shield (e.g., northern and central Sweden, central and southern Finland, and southern Norway), and comprise a large variety of types, from large granitoid intrusions to metasedimentary and metavolcanic complexes that host some of the most important ore deposits in the region (e.g., Bergslagen and Skellefte in Sweden, Outokumpu in Finland). Mesoproterozoic rocks comprise smaller intrusions of rapakivi granitoids, dolerite dykes, flood basalt and volcanic and sedimentary successions, such as the Jotnian sandstone. Late Mesoproterozoic rocks (*ca.* 1.1–0.9 Ga) dominate in southern Sweden and south-western Norway, comprising mostly igneous rocks deformed and metamorphosed under amphibolite to granulite-eclogite facies conditions into gneiss and migmatite (Bingen et al. 2008; Lahtinen et al. 2008). Neoproterozoic rocks are predominantly represented by sedimentary successions of marine and fluvial sandstone, conglomerate and carbonates, dolerite dyke swarms and tillite occurring within the Caledonian Orogen (Nystuen et al. 2008). Outside the Caledonides, Neoproterozoic rocks occur in southern Sweden by Lake Vättern (sandstone, shale, dolomite and conglomerate), in central Sweden (Alnö carbonatite complex) and in Norway (dolerite dykes and the Fen carbonatite) (Nystuen et al. 2008).

In the western part of Scandinavia, a mountain chain covers the older basement rocks. The mountains were formed during the Caledonian orogeny in Early – Mid Palaeozoic (Ordovician to Devonian) and are mainly composed of Neoproterozoic to Palaeozoic (mostly Silurian) marine to fluvial sedimentary and volcanoclastic successions, which were deposited at the continental margin of the Palaeozoic Iapetus Ocean (Chapter 2). During the orogeny these allochthonous units were transported as a set of nappes for at least 300 km westwards onto the Fennoscandian platform. The metamorphic grades within the Caledonian nappes vary from low-grade greenschist facies to granulite-eclogite facies (GEE et al. 2008).

The youngest rocks (younger than 250 Ma) are found in the Permian Oslo Rift (volcanic and plutonic intrusions) in southern Norway, and in the Mesozoic sedimentary cover occurring in southern Sweden and (locally) offshore Norway. During the Tertiary period most of the sedimentation occurred offshore and Tertiary deposits are known only locally in Fennoscandia, e.g., in southernmost Sweden (Scania) and in northern Finland (the Akanvaara Clay) (RASMUSSEN et al. 2008). These deposits comprise deep marine clay and turbiditic siltstone, deltaic deposits, limestone, chalk and ash layers, originating from the volcanic centres located in the North Atlantic (RASMUSSEN et al. 2008).

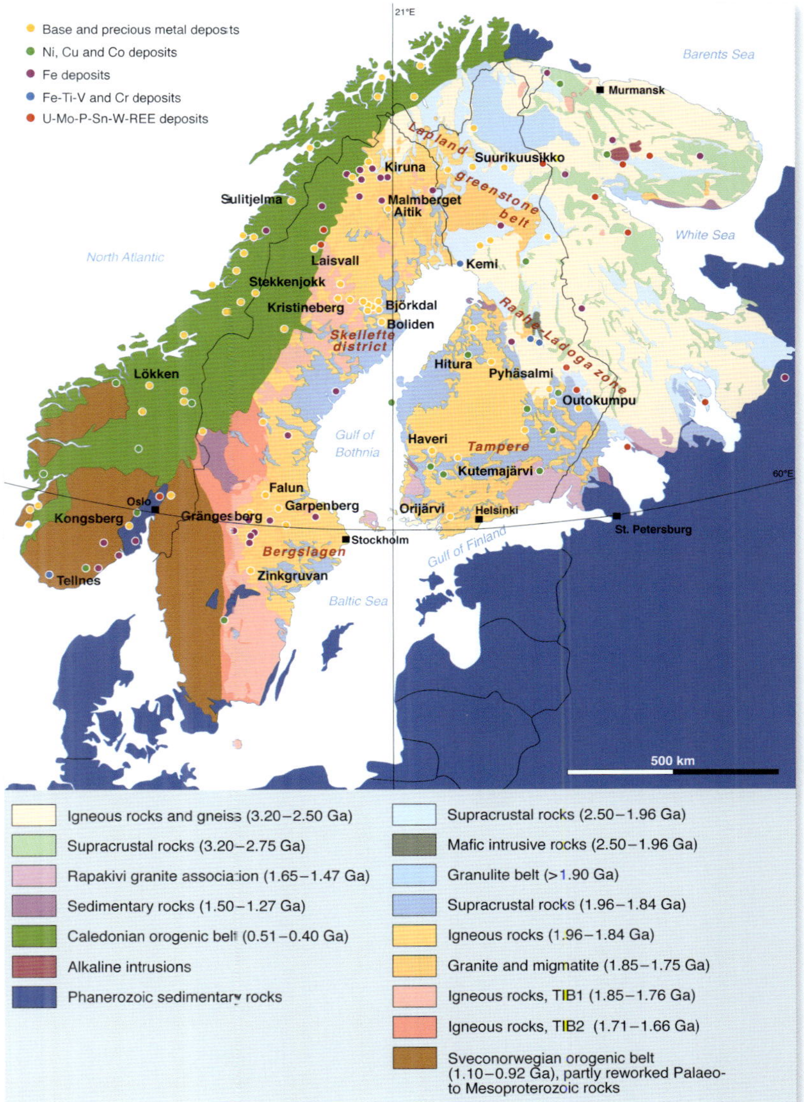

Fig. 14.1: Simplified geological map with major ore deposits in Fennoscandia (based on LAHTINEN et al. (2005) and SGU digital lithotectonic map).

14.3.2 Superficial Quaternary deposits and soil types

Surficial deposits (till, gravel, soil) in Norway, Sweden and Finland are largely the result of the latest Quaternary history with its periods of glaciation and deglaciation. Most of the youngest cover formed during the last (Weichselian) glaciation and deglaciation. At its maximum, the Weichselian ice covered Finland and the entire Scandinavian peninsula (WOHLFARTH et al. 2008). Glaciers generally spread from the Caledonian mountain range. These were thick ice sheets, transporting loosened bedrock fragments, weathered soil and pre-Quaternary debris. U-shaped valleys, fjords and lakes were formed by the erosive power of the glaciers. Melting of the ice cap and resulting isostatic rebound have a pronounced influence, not only on the topography, but also on the nature of the Quaternary cover. In the melting phase, sub-glacial streams transported sand and gravel, which were deposited to form eskers. The topography in the pristine landscape was dominated by barren bedrock or a cover composed of sand, gravel, till or marine clay. In the postglacial phases vegetation spread, clay was deposited and peat land developed. The last deglaciation (9 – 7 ka BP) was rapid, and isostatic rebound continues today with a rate of up to 10 mm/year. Erosion processes were very intense during that time with spectacular results, such as the development of Norwegian deep fjords and deposition of thick glacigenic deposits (up to 4 km thick), which occur along the western Scandinavian continental margin (WOHLFARTH et al. 2008).

Glacial melt water and the retreating sea (due to land-uplift) have also influenced the uppermost cover, leaving behind thick layers of freshwater sediment, sand and gravel, and both postglacial and marine clay. All these features can be recognised in the soil characteristics (composition, texture and geochemistry). Marine clay, for example, covers extensive areas from the Mälaren region in central Sweden (Fig. 14.2) and continues eastwards into southern Finland, the eastern coast of the Baltic Sea and, locally, inland. The marine influence on soil can be recognised up to an altitude of 290 m.a.s.l. in Sweden, *ca.* 210 m.a.s.l. in north-western Finland, and up to *ca.* 200 m.a.s.l. in Norway (e.g., near Trondheim). The presence of salty brines in the groundwater systems and their influence on the surficial deposits can also be locally important.

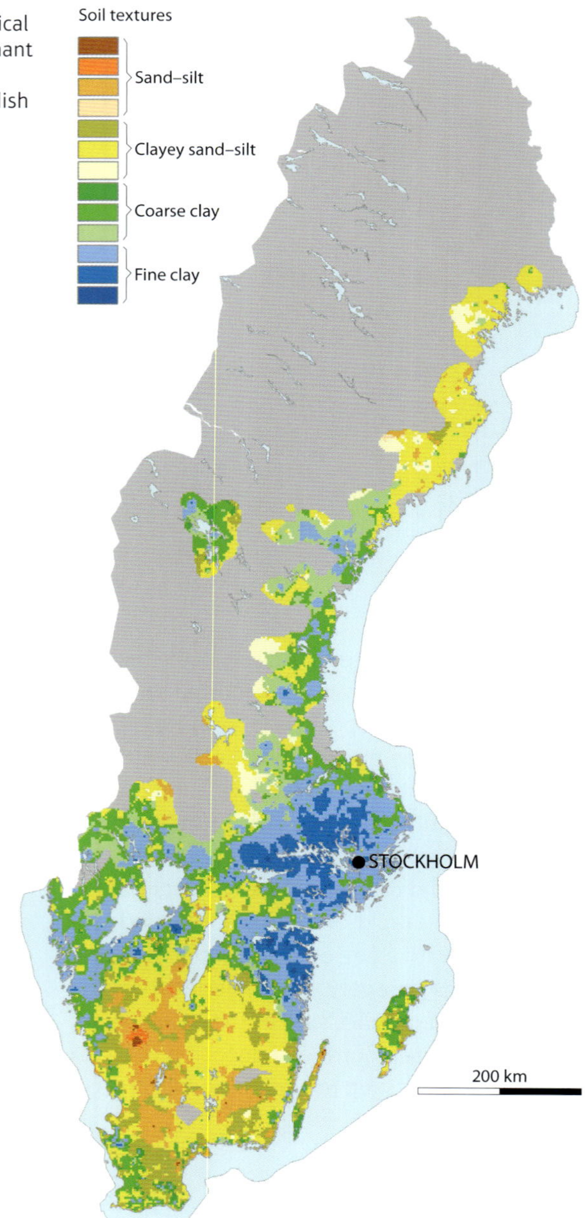

Fig. 14.2: Geographical variation in the dominant mineral soil texture according to the Swedish classification system (n = 4 954) (ERIKSSON et al. 2010).

Most soil in Norway, Sweden and Finland is developed on glacial deposits from the last glaciations and on postglacial clay-rich marine deposits below the highest coastline (REIMANN et al. 2003). Norway, Finland and Sweden have a rather thin soil layer compared with the rest of Europe covered with glacial debris, mainly composed of sandy till, being especially common above the highest coastline. In central Sweden and the southern part of Finland, heavy glacial or postglacial clay is common (Fig. 14.2). Peat lands frequently occur, especially in the northern regions. Soil types, classified according to HARTWICH (1998), are podsol (the most common), leptosol, cambisol, histosol with local occurrences of gleysol and luvisol in southern Sweden (Chapter 1). In the northern areas, nutrient poor acid podsol dominates, however, peat soil is also common. In central Sweden and southern Finland, soil developed on clay-rich postglacial deposits is characterised by good fertility. At many places around the Bay of Bothnia in Sweden and Finland, the problem of acid sulphate soil occurs as a result of isostatic rebound and frequent drainage activities (SOHLENIUS & ÖBORN 2004; TOIVONEN & ÖSTERHOLM 2011).

14.3.3 Mineral occurrences

There is a long tradition of mining and ore processing in Norway, Sweden and Finland. The mineral resources of the three countries are among the richest in Europe and are of great economic interest. Most of the ore deposits are defined within metallogenic provinces or districts (EILU 2012), and they represent many different types, e.g., volcanogenic massive sulphide deposits (VMS with Cu, Zn, Pb, Au, Ag), orogenic gold deposits (Au), layered intrusions (Ni, PGE, Ti±V), intrusive hosted copper-gold deposits, apatite-bearing iron oxide deposits, chromite- and anorthosite-hosted titanium deposits, iron oxide-copper-gold (IOCG) deposits, shale-hosted nickel-zinc-copper deposits and various types of uranium deposits (WEIHED et al. 2008; EILU 2012).

The largest deposits of Fe, Cu and Ni in Europe occur within the Palaeoproterozoic units of the Fennoscandian (Baltic) Shield (Fig. 14.1). Many of these deposits are situated in a composite sulphide-rich ore belt stretching south-east to north-west through central Finland into Sweden, and include the Outokumpu region in the south-east, Pyhäsalmi in the centre, and the Swedish Skellefte district in the north-west. The largest VMS ore district in Sweden, the Skellefte district, includes the Boliden polymetallic deposit, which was once the richest Au-As deposit in Europe and also produced Cu, Ag, Pb and Zn. Geologically, these ore bodies are related to quartz porphyry or other granitoid intrusions cutting volcanic units of dacitic to rhyolitic composition. Garpenberg and Zinkgruvan in the Swedish Bergslagen, the Viscaria mine in north-western Sweden and the Pahtavaara Cu-Zn mine in Finland are sometimes classified as VMS deposits. The largest Cu-Au mine in Scandinavia, the Aitik mine in northern Sweden, occurs within Svecofennian volcanoclastic rocks and is classified as a porphyry deposit.

Orogenic gold deposits of principally epithermal origin occur within Archaean and Proterozoic units (Weihed et al. 2008) with the largest gold deposits at Boliden (Sweden), Cu-Au ores in Bidjovagge (northern Norway), Suurikuusikko, Pahtavaara and Saattopora (northern Finland). In Sweden, the so-called Gold Line has been recognised in Västerbotten County with two recent mines at Svartliden and Blaiken. The Gold Line is a south-east trending Au anomaly detected by the State Mining Property Commission (NSG) during a till geochemical survey in the late 1980s.

Iron ore deposits in Fennoscandia are among the biggest iron ore deposits in the world with the main mines located in northern Sweden (Kiruna and Malmberget). The iron ore deposits belong to several genetic types: apatite-bearing iron oxide ores (Kiruna mine, Grängesberg mine), the Archaean-type banded-iron formations (BIFs, e.g., the Dannemora ore in Bergslagen in central Sweden), and skarn type iron ores. The largest iron-ilmenite deposits occur in south-west Norway (the Tellnes deposit). Ni-Cu±PGE deposits are located in several different settings within the Fennoscandian Shield, e.g., in Finland (Kotalahti, Hitura and Vammala) and in Sweden (e.g., Lainejaur). Other types of metal occurrences include Cr-, Be-, Li-, Mo-, Nb-, REE-, Sc-, Sn-, Ta-, W-, Zr- and U-bearing deposits. The Kemi chromite mine in Finland is one of the largest chromium deposits in Europe and the ore is hosted in mafic – ultramafic intrusive rocks. The biggest Mo deposits occur in southern Norway (in Kvinesdal, Telemark and Oslo Rift). Carbonatite hosted Nb-Fe-P-REE deposits are known in southern Norway (the Fen Complex). Uranium occurs as several mineralisation types, both in vein-type deposits within Precambrian rocks, younger granite and pegmatite (primarily in Finland), and in Palaeozoic shale overlying the Precambrian basement, e.g., Cambrian alum shale in Sweden (Weihed et al. 2008). Ni-Cu-S and Cr deposits, sometimes with PGE and talc, hosted by mafic to ultramafic rocks, occur in both Norway (Bruvann, Råna) and Sweden (Kukkola, Notträsk, Kläppsjö).

The Caledonian mountains host many important mineralisation types, e.g., Cu-Zn-Pb sulphide deposits (e.g., Bleikvassli, Mofjellet, Lökken, Röros, Sulitelma in Norway and Stekenjokk-Levi, Bjelke and Fröå in Sweden). Iron ores also occur within the Caledonian nappes, but only the Ørtfjell iron deposit in Norway is currently being mined. Hydrothermal gold deposits are Bindal and Mofjellet in Norway. In northern Norway, polymetallic tungsten skarn deposits have been described (Grenne et al. 1999; Sandstad et al. 2012; Hallberg et al. 2012). Silver mineralisation (mined from 18th century) occurs at many places in the Swedish Caledonides, with the most known Nasafjäll, Kvikkjokk and Stekenjokk.

▸ 14.4 Human impact on agricultural land

Modern humans arrived in Europe some 35,000–40,000 years ago. Following the final retreat of the Weichselian ice sheet, humans spread northwards and Scandinavia became colonised about 14,000 years ago, with the first people arriving in eastern Fennoscandia and Finland some 11,000 years ago (Tallavaara et al. 2009). These hunter-gatherers followed herds of prey, and lived mostly as a nomadic people, until the introduction of food production through cultivation of land and domestication of animals. Society depending on agriculture developed in southern Scandinavia during the Mesolithic – Neolithic transition about 4,000 years ago (Sørensen & Karg 2012). This limited form of agriculture and farming was somewhat undeveloped until industrial times, when methods to increase the productivity of cultivated fields were introduced, and farmers realised the need for additives, such as organic matter and nutrients (nitrogen, potash and phosphate). At later stages, the use of chemical fertilisers led to the accumulation of nitrates and phosphates in soil. The spreading of waste sludge on land led to an accumulation of heavy metals, such as lead, cadmium, copper and zinc, in the surface soil layers. With the focus on plant productivity, the use of organic herbicides and pesticides to avoid plant diseases has led to accumulation of toxic residues in soil. Another problem that is prominent in areas of the Baltic Basin, are sulphidic sediments and potential acid sulphate (PAS) soil. By lowering the water table of these sulphidic sediments through ditching, contamination of crop roots is avoided, but the outflow of toxic soil solution to watercourses remains a problem at seasonal events of high groundwater discharge (Joukainen & Yli-Halla 2003).

▸ 14.5 Sampling and methodology

The Geological Surveys of Norway, Sweden and Finland have collected samples of agricultural and grazing land soil, according to the procedure provided to all project members (EGS 2008). Sampling took place during the summer and autumn of 2008. In total, 453 samples (excluding field duplicates) of agricultural soil (Ap, Ap-horizon, 0 – 20 cm) and 350 samples (excluding field duplicates) of grazing land soil (Gr, land under permanent grass cover, topsoil 0 – 10 cm) were collected in Norway, Sweden and Finland. In Sweden, the type of crop was recorded for Ap soil and kind of animal grazing on grassland. In Finland, the field protocol provides extra information about the type of soil sample (sandy, clayey and organic rich).

▸ 14.6 Discussion

14.6.1 Main features of the geochemical landscape of Norway, Sweden and Finland

The complex interpretation of elemental patterns presented on the maps, and with the help of a variety of statistical tools (boxplots, cumulative probability plots), revealed several groups of factors influencing the observed trends in the geochemical patterns in soil of Norway, Sweden and Finland (Table 14.1). The most important factors are: sample type and its composition, sample density, bedrock geology, presence of mineralisation and mining centres, Quaternary history, climate and human activity.

All element distribution maps show clear regional scale features and, for the majority of elements, anomalies that can be explained by the occurrence of mineralisation, distinct bedrock types and alteration. Large-scale differences can be seen in areas underlain by the Archaean domain (northern and north-eastern regions), crystalline rocks of the Fennoscandian Shield, the Caledonides, and younger Mesozoic cover (in southern Sweden). Crystalline bedrock, especially granitic domains, strongly influence the geochemistry of overlying soil, imparting higher concentrations and median values in aqua regia extractable Ag, Ce, Ge, Mo, Na, Nb, P, S, Ti and U in agricultural soil. The presence of underlying pegmatite is marked by positive Li, Nb, Ta, Mo and W anomalies. Mafic and ultramafic rocks (intrusions, dykes, greenstone belts) are outlined by anomalously high Ni, Cr, Fe, Mg and Sc. Some tectonic boundaries can also be observed, e. g., the major tectonic boundary between the Proterozoic and Archaean domains on the Fennoscandian Shield in the Boden region.

The study area is often referred to as Europe's biggest repository for iron, base-, and high-tech metals. Soil geochemistry clearly marks numerous mineralisation sites and ore deposits that have been mined in the past to the present day. Several mining districts and famous ore fields can be recognised, e.g., the Kiruna mining district in northernmost Sweden, the Skellefte mining district with base metals, the Gold Line in northern Sweden, the Bergslagen district in central Sweden, the Outokumpu mining district in central Finland, and the Kongsberg in southern Norway (Fig. 14.1).

Soil type is another important parameter influencing soil chemistry. In Norway, Sweden and Finland several soil types occur, i.e., clay-rich, sandy, minerogenic, and organic-rich (peaty) soil. The dominant soil component (clay, sand, organic) often has a significant influence on the observed distribution of the elements. The Ap samples often show somewhat better (less noisy) results than the Gr samples. An explanation for this feature is that ploughing over many years homogenises soil, which becomes more weathered and finer grained.

The 'Central Scandinavian Clay Belt' (CSCB) is one of the largest and most significant regional features, extending from southern Finland in the east to central Sweden (the Mälaren region) in the west, consisting of soil overlying postglacial marine clay (Fig. 14.2). It constitutes the most prominent geochemical feature in the Scandinavian results. Many elements exhibiting an affinity to clay minerals are visibly enriched in the CSCB, regardless of the type of underlying bedrock (Table 14.1). Clayey soil shows high concentrations of aqua regia extractable potassium, iron and aluminium, and the concentrations of several trace elements (Ag, As, B, Ba, Be, Bi, Cd, Co, Cr, Cs, Cu, Ga, Ge, Hf, In, La, Li, Mg, Na, Nb, Rb, Sc, Th, Ti, Tl, V, Y, Zn and Zr) are relatively high in clay-rich soil. Some of these elements reveal very high concentrations even at the European scale, e.g., Ce, Ga, Ge, Hf, U.

Several elements (e.g., As, Cd, Hg, Pb, S, Se) tend to adsorb to organic matter and, therefore, biogenic (peaty) soil has higher levels of many of these potentially toxic elements. In the Baltic Soil Survey (Reimann et al. 2003), organic soil was collected in eastern and northern Finland, while in the GEMAS data set small-scale anomalies, explained by biogenic soil, were also identified in

Table 14.1: Explanation of some key anomalies in Ap soil samples (aqua regia extraction).

Source of anomaly	Explanation of anomalies	Element
Geology	mafic and ultramafic rocks, greenstone belts	Cd, Cl, Co, Cr, Fe, Mg, Ni, PGE, Sc, Sr, Ti, V,
Geology	pegmatite and evolved granite	Be, Ce, Cs, La, Li, Nb, Sc, Sn,Ta, Th, U, W, Y, Zr
Geology	the Oslo Rift with volcanic rocks, black shale and sulphide mineralisation	B, Ba, Bi, Cd, Ce, Co, Cr, Cu, Ga, In, K, Li, Mn, Mo, Nb, Pb, Sb, Sc, Sr, Te, Tl, U, V, Zn
Geology	black shale	As, B, Cd, Co, Cu, Hg, In, Mn, Mo, Ni, S, Sb, Te, Tl, U, V
Soil type	clayey soil (the CSCB)	Ag, Al, As, B, Ba, Be, Bi, Ce, Co, Cr, Cs, Ga, Ge, Hf, In, K, La, Li, Mg, Na, Ni, Rb, Sc, Sn, Sr, Th, Ti, Tl, U, V, W, Y, Zn, Zr
Soil type	organic matter rich (peaty) soil	Ag, Cd, Ge, Hg, P, Pb, S, Sb, Se, U, V
Mineralisation	polymetallic sulphide ores	Ag, As, Bi, Cd, Co, Cu, Fe, Ga, Ge, In, Mo, Pb, S, Sb, Se, Te, Tl, Zn
Mineralisation	Fe-Ti-V ores	Fe, In, Mn, Mo, P, Ti, V
Mineralisation	other (Sn, W, Au, U, etc.)	Au, Hg, In, La, Mo, PGE, Sb, Sn, Ta, Te, Th, U, W, Y
Climate	high precipitation	Ag, Cl, Ga, Ge, Hg, K, Na, P, Pb, S, Se, Sr
Anthropogenic	contamination	Al, Zn, Hg?
Anthropogenic	addition of fertilisers	B?, Cd, P, Se, Zn, Hg?
Anthropogenic	liming	Ca

south-eastern Finland, where, in general, more minerogenic soil was sampled. Areas of anomalously high values correspond to those detected during the Baltic Soil Survey (Reimann et al. 2003; see also Chapter 4).

Sandy soil usually has low concentrations of metals, and often masks the geological nature of the underlying bedrock and any mineralisation present. In areas with intensive physical erosion and very thin soil cover, e.g., in the Caledonides, certain elemental anomalies are strengthened due to higher contents of lithic clasts and primary minerals (rock-forming and accessory minerals) in minerogenic soil. In sandy soil, collected from river valleys, a higher fraction of heavy minerals influences the elemental distribution pattern.

Most of the collected soil samples have pH < 5 (min. 3.5, max. 7.3), which is low in comparison with the rest of Europe (Reimann et al. 2003) and is characteristic of rather infertile organic rich soil. Only a small percentage of clayey soil samples (the CSCB, eastern coast of Sweden) have slightly higher pH. Most metals become mobile at low pH levels, and some show an increased affinity to bond to organic matter (e.g., Co, Cu, Hg, Ni, Pb, Cd). Clay minerals and Fe-Mn-Al (hydro)oxides in soil also have high adsorption properties and attract many metals. Organic soil (peaty soil) in coastal Norway, Finland and northern regions, and rare calcareous soil (Öland and Gotland), have high cation exchange capacity values (CEC), which indicate high retention of specific cations within the soil horizon, e.g., potentially toxic Zn, Cd and Pb.

The latest Quaternary history has left a significant impact on the spatial distribution of elements in soil. Ice movement and activity of glacial water are responsible for a displacement and dilution of some geochemical anomalies, although the transport distance, up to 5 – 10 km on average in till in northern Scandinavia, is usually not large (Sohlenius et al. 2009).

Climate and distance from the sea are factors that play an important role in the geochemical characteristics of soil in Norway, where high precipitation and a large quantity of marine aerosols in the air contribute to the high metal contents in organic-rich coastal soil (Table 14.1).

Some limitations of the GEMAS project data set are caused by the low sampling density (1 site/2500 km^2), and also the instructions in the field manual not to collect samples near to mining areas, since the objective was to determine the natural geochemical background variation (EGS 2008). This is particularly obvious in a few regions with known ore deposits, e.g., in southern Norway and northern Sweden. The GEMAS data set certainly does not reflect the presence of mining centres in the Skellefte mining district at Västerbotten and the large Cu-mine at Aitik in northern Sweden. The choice of sampling site carries a certain subjectivity burden, which is difficult to avoid due to the sampling scale of the project. This personal human factor is thought, however, to be negligible, and the overall picture still gives reliable information.

A certain number of geochemical anomalies cannot be explained due to insufficient information about the local geological situation, the presence of mineralisation and other factors (e.g., soil type, grain size, precipitation rate, vegetation) that might influence soil chemistry.

14.6.2 Differences between aqua regia extraction and XRF total concentrations

The greatest advantage of the GEMAS project is the availability of analytical data from both aqua regia extraction (leachable elements) and X-ray fluorescence (XRF: total concentrations). The difference between the two methods is visible on elemental maps, and the calculated extractability (Reimann et al. 2011) can be used to extract much useful information, e.g., soil mineralogy, redox conditions, presence of acidic soil, the need for micronutrient fertilisation, climate influence, etc.

Unsurprisingly, the elements with low extractability (see Table 4, p. 24 – 25 in Reimann et al. 2011) show the greatest difference between the aqua regia extraction and XRF elemental patterns (Al, Ba, Bi, Hf, K, Na, Nb, Sc, Sr, Ti and Zr). Leachable fractions of these elements (with the exception of Ba) are commonly enriched in clay-rich samples, especially in the CSCB (e.g., Bi in Fig. 14.3). This is of particular interest because the main host minerals belong to resistant types (heavy minerals) that do not weather easily (e.g., zircon, garnet, columbite, ilmenite, rutile, titanite) and have low solubility. High concentrations of extractable elements indicate their presence as trace elements in the more common rock-forming minerals (e.g., pyroxene, amphibole, mica, feldspar) where they occur at low levels, but can be easily released to the surficial environment during physical and chemical weathering. Another factor, which enhances the higher solubility of these otherwise insoluble elements, is the prevalence of low pH in the studied soil and the presence of organic matter. Organic-rich soil can mask high contents of an element in the underlying parent material, e.g., low K concentrations in peaty soil in Finland overlying felsic igneous rocks, rich in mica and feldspar.

On the other hand, total concentrations (XRF) of elements largely reflect the type of underlying bedrock (e.g., mafic versus felsic, igneous versus sedimentary) and the presence of ore deposits (e.g., Bi in Fig. 14.3). The map of total concentrations of Al, for example, shows a slightly different pattern in grazing land, with higher concentrations in the soil of central and northern Sweden, and in western Norway, related to occurrences of Palaeoproterozoic igneous rocks (Section 11.8 → Part A). Soil from regions underlain by sedimentary rocks (e.g., Dala sandstone and metasediments in the Bothnian Basin) has, by contrast, low Al concentrations. High total concentrations of Ba are characteristic of soil overlying feldspar-rich metamorphic rocks in southern Sweden and Norway. Local high concentrations of Bi (XRF) point to both sulphide (Pb-Zn) and gold mineralisation sites (note, however, the relatively high detection limit of 3 mg/kg for Bi by the XRF technique); (Fig. 14.3). Elevated total XRF concentrations of Nb indicate the Nb-rich parental material, e.g., felsic igneous rocks (with biotite, accessory minerals, such as rutile, titanite and zircon) and metasediments, while high leachable Nb (and Bi) concentrations (aqua regia extraction) occur in clay-rich soil (Fig. 14.4).

Some geographical trends in the distribution of elements can be noted. The regionality of aqua regia extraction and XRF anomalies is related to latitude, altitude, climate, and weathering, together with soil properties (pH, content of organic matter). In the northern regions (north of the polar circle), several elements show low extractable contents, while total concentrations are high, even at the European scale (Al, Ba, Hf, Na, Sc and Sr – see Sections 11.8, 11.12, 11.28, 11.37, 11.49, and 11.53 → Part A). Low concentrations of extractable Na, for example, indicate that Na resides in resistant minerals and that soluble Na, which can

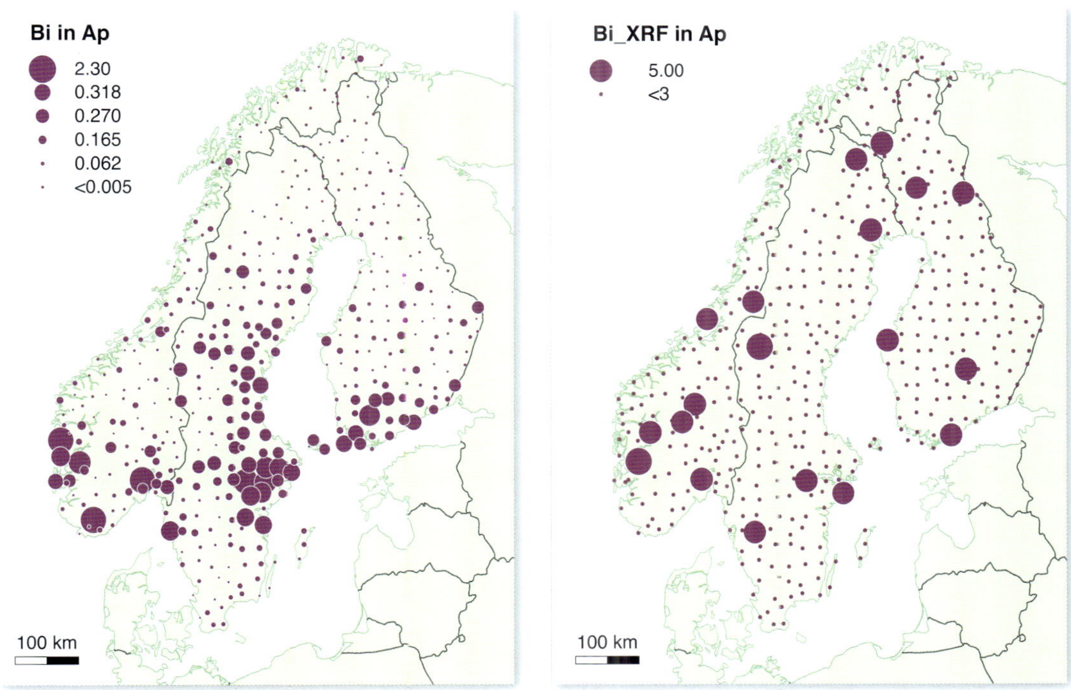

Fig. 14.3: Geochemical maps of aqua regia extractable Bi (left) and Bi total XRF concentrations (right) in agricultural soil of Norway, Sweden and Finland. Even though the detection limit of the XRF method is high, the outliers with high Bi concentrations do not overlap between these two methods.

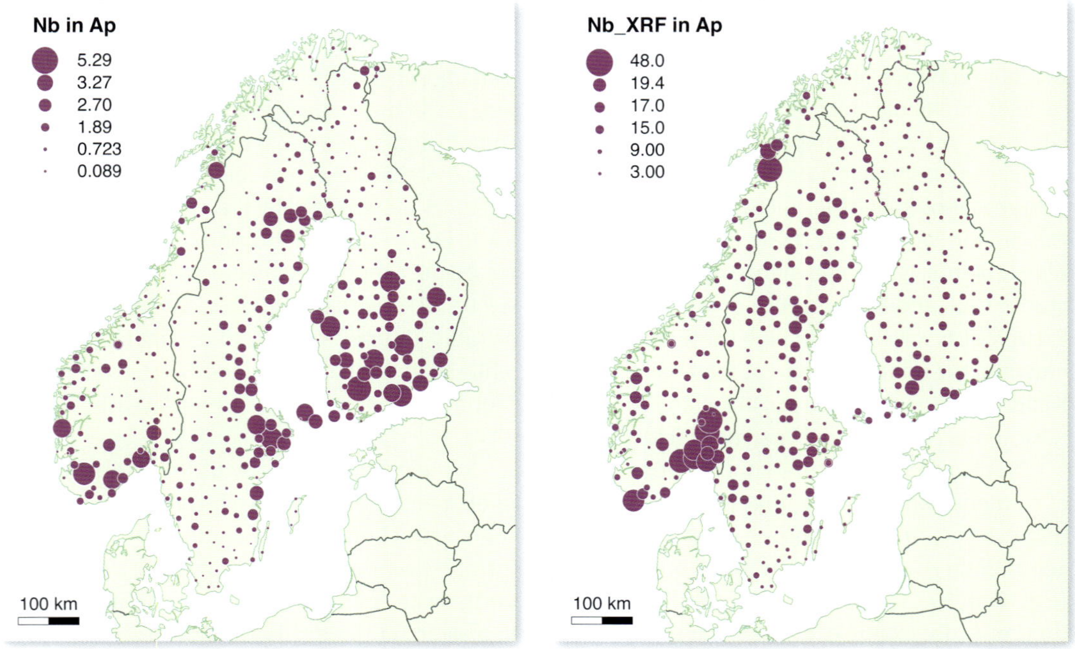

Fig. 14.4: Geochemical maps of aqua regia extractable Nb (left) and Nb total XRF concentrations (right) in agricultural soil of Norway, Sweden and Finland. Niobium enrichment in the CSCB in southern Finland and eastern central Sweden can be observed on the aqua regia extractable Nb map. Total XRF concentrations of Nb are generally related to the underlying bedrock type.

adsorb onto clay particles in soil, is depleted. At the same time, total Na concentrations are high in all three countries compared to the rest of Europe, especially in the north. This particular enrichment can be explained by the widespread presence of hydrothermally altered (e.g., scapolitised) mafic rocks ('greenstone belts'), which host many types of the key mineralisation types (apatite-Fe, base metals) (Bergman et al. 2001). The same pattern is observed for extractable and total Sr and Ba concentrations in Ap soil.

The opposite pattern is observed for Nb, with high extractable concentrations and low total concentrations in northern regions (Fig. 14.4). The enhanced concentrations of extractable Nb may originate from biotite-rich parent material, which can decompose easily in peaty soil. Higher concentrations of Nb in northern peaty soil can also be explained by the fact that this normally insoluble element can, under special conditions (organic soil), become more mobile. Niobium substitutes for Ti in many minerals; therefore, the elemental distribution pattern of both elements is similar (Fig. 14.5). Concentrations of leachable Ti (with leachability as low as 3 %) in all three countries are the highest in Europe. Titanium seems to be enriched in the clay-rich soil of the CSCB, and in soil overlying high-grade metamorphic rocks and both mafic and felsic older igneous rocks. The pattern for total Ti differs slightly, with high values across Norway and in the northern parts of Sweden and Finland with metamorphic and igneous rocks and Ti-V mineralisation.

In this survey, several elements (As, Ce, Co, Cr, Cs, Cu, Fe, La, Mg, Mn, Mo, Ni, P, Pb, Rb, Th, U, V, Y, Zn – see Chapter 11 element descriptions → Part A) show similar patterns for both aqua regia extraction and XRF, indicating good leaching properties for these elements and residence in either common rock-forming minerals and/or sulphides.

14.6.3 Implications for agriculture and soil quality

The GEMAS results from aqua regia extraction can be used as indicators of soil quality and to some extent their bioavailability. The significant difference for some elements between total concentrations and aqua regia extraction (see above) can be an invaluable tool for soil quality assessments. The most important geochemical key parameters for soil quality are pH, CEC and TOC (provided by the GEMAS data set). These three factors govern the chemical behaviour of macro- and micro-nutrients in soil, i.e., P, K, Ca, Mg, S, and B, Cl, Co, Cr, Cu, Fe, I, Mn, Mo, Se, Ni, Zn, respectively.

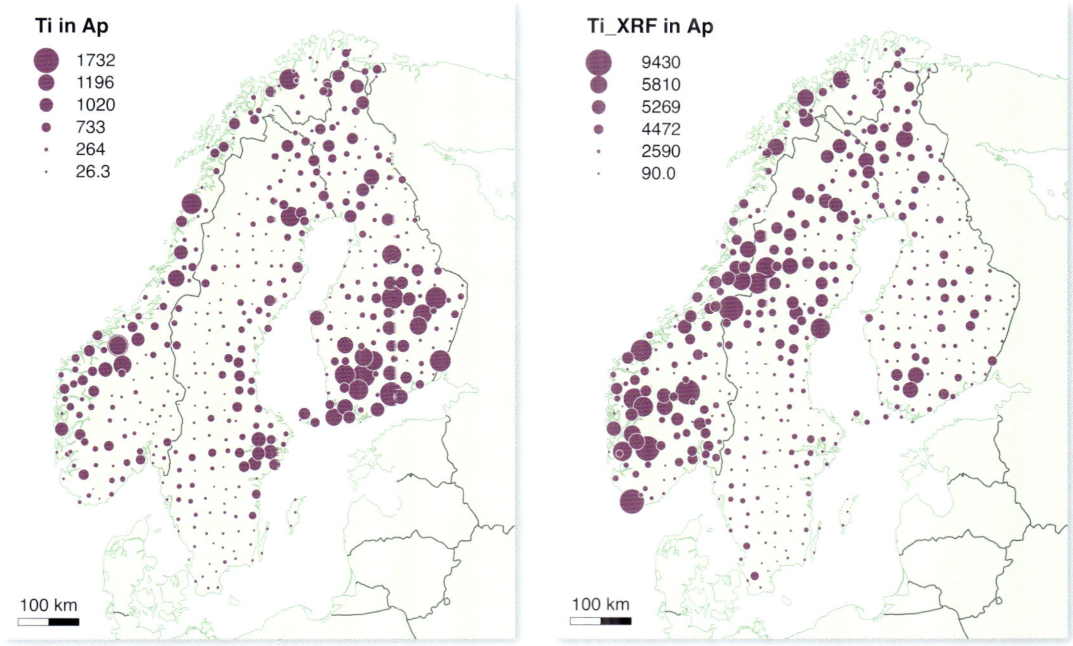

Fig. 14.5: Titanium elemental maps of aqua regia extractable Ti (left) and Ti total XRF concentrations (right) in agricultural soil of Norway, Sweden and Finland. Significant enrichment (in comparison with the rest of the Europe) in Ti resulting from the underlying bedrock geology can be noted.

Norway, Sweden and Finland are generally covered by a relatively thin layer of agricultural soil. Arable land in Norway, Sweden and Finland constitutes only approximately 10 % of the total land area (Reimann et al. 2003). Areas suitable for agriculture are limited due to the lack of suitable soil, cold climate and 'difficult' landscapes, e.g., fjords in Norway, peat bogs and swamps in the north). The most common crops are grass (silage), cereal, oil seed and potatoes. In Sweden, for example, only the southern part of the country produces significant amounts of wheat and vegetables for the country as a whole, while the rest of the area is used for pasture and forestry (Sweden is the biggest wood supplier in Europe).

The weathering rate varies in the studied countries; physical weathering is high in Norway as a whole, locally in Sweden and in south-western Finland, especially in soil overlying marine clay and silt. Erosion usually occurs at a lower rate in sandy and peaty soil (Øygarden et al. 2006; Ulén 2006; Tattari & Rekolainen 2006).

The nutrient status of Ap soil varies between the southern and northern parts of the studied countries. Deficiency in many macro- and micro-nutrients can potentially be responsible for a reduction in crop productivity (which is low anyway), and certain health issues for livestock and humans. Deficiencies can occur in some macronutrients (K, Fe, Mg) and micronutrients (I, Se, Zn), especially in the north (see relevant maps in Chapter 11 → Part A). Generally, a low pH in the studied soil (min. 3.5 in northern Finland and max. 7.3 on Gotland) makes more micronutrients bioavailable, with the exception of Mo, P and Ca. Low pH (< 5.5) in most of the studied soil is, however, an indicator of less fertile soil with limited bioavailability of Ca, Mg and Mo, and potential toxicity (e.g., Al and Mn) (Sims 2000). In coastal areas of Norway, additional soil low fertility originates from the cold climate and high rainfall, which lead to an excess of soil moisture and delayed growth of spring crops (Bradford & Peterson 2000; Reimann et al. 2009a). The clay-rich samples (overlying glacial and marine deposits, e.g., the CSCB) tend to have higher concentrations of essential macro- and micro-elements. Soil developed on black shale (central Sweden, Oslo region) show potentially toxic levels of metals like As, Cd, Mo, Se, U, Zn, Ag, Pb.

The availability of macronutrients varies geographically and between the types of soil. Phosphorus (P), an important nutrient for plant growth (Sharpley 2000) displays low concentrations in agricultural soil in southern Sweden and northern Finland. In northern regions, where P concentrations are high, the uptake by plants can be reduced by soil temperature (cold climate) and low pH. Clayey soil tends to have higher P contents than sandy and organic soil (Sharpley 2000). Potassium (K) is, like phosphorus, an important nutrient for plant growth. Total concentrations of potassium are quite high in agricultural soil, while high concentrations of leachable K is limited to the clay-rich soil of the CSCB and coastal regions in western Norway. The magnesium (Mg) content of soil in southern Norway and Sweden is low, which is a consequence of local geology, the presence of organic matter, low pH (northern regions), and liming activities (southern Sweden and central Finland). Sulphur concentrations are quite low as a result of low pH and low Ca content (exceptions are Gotland and Öland islands). Sulphur retention in soil is generally low (Camberato & Pan 2000), even though the bedrock locally is enriched in sulphide minerals due to mineralisation. In coastal Norway, an additional source of sulphur is from ocean water aerosols. The most common form of sulphur is the sulphide form (e.g., pyrite) and, as such, is less bioavailable. Calcium is low in the studied countries, but this deficiency has not adversely affected agriculture and human health.

There is a complicated pathway for the essential micronutrients, such B, Cl, Cr, Cu, Fe, I, Mn, Mo, Se, Ni, Zn Al, Co, Na, Ni and Si, the last five

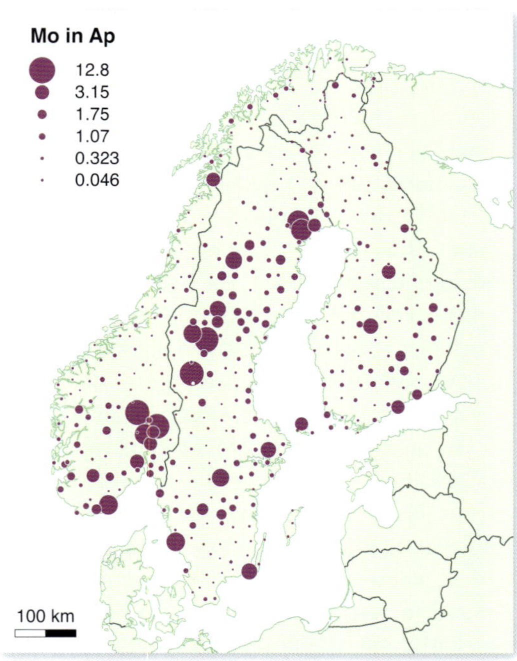

Fig. 14.6: Aqua regia extractable Mo concentrations in agricultural soil of Norway, Sweden and Finland.

known to be beneficial for plant nutrition (Mortvedt 2000; see also Chapter 1). For Fe, Mn, Cu and Zn, bioavailability depends on soil pH, with solubility decreasing as pH increases. Due to the relatively low pH in the studied soil, these elements should be relatively bioavailable (apart from a few exceptions, such as the calcareous soil of Öland and Gotland). The presence of organic-rich soil, however, lowers the bioavailability of Cu and, to a lesser extent, Mn and Zn and their deficiency can occur in northern areas of Norway and Sweden and Finland (high TOC, low pH, widespread peat soil). Reduction reactions, due to low pH, can even lead to toxic levels of Mn for some plant species (Mortvedt 2000). High concentrations of aqua regia extractable Mo occur in Sweden, southern Norway and central-southern Finland (Fig. 14.6). Soil with high pH shows low concentrations of Mo, which is mobile under alkaline conditions, but adsorbs strongly to hydrous Fe oxides and to organic matter. The presence of soluble S and P usually decreases the Mo uptake (e.g., at the Norwegian coast and in northern regions). Although there are high Mo concentrations in the acidic organic soil of Norway, Sweden and Finland, its bioavailability increases with pH increase. Due to liming activities, the Mo bioavailability can be increased, but in practice, this may deplete soil in other nutrients, such as Cu, and cause negative health effects (e.g., molybdenosis and the resultant moose deaths in Sweden in the 1980's; Selinus et al. 2010). The studied soil has rather low boron (B) concentrations, with the exception of the clay-rich soil of the CSCB (sorption to clay minerals and Fe and Al oxides at neutral to higher pH conditions), calcareous soil on Öland, Gotland and southern Sweden, soil overlying younger volcanic rocks (Oslo Rift) and a few locations in Norway (Section 11.11 → Part A). Boron bioavailability is hindered by cold climate and poor aeration of soil. In the studied soil, aqua regia extractable selenium (Se) concentrations are rather low, therefore, Se is an element usually supplied with fertilisers (Section 11.50 → Part A). It is an important micronutrient but can become, at high concentrations, toxic. High concentrations occur at the Norwegian coast, where the uptake can be inhibited by high P concentrations (Wilkinson et al. 2000). Additionally, the bioavailability of Se is hampered in low pH-organic soil (e.g., Norway, southern Finland) due to formation of insoluble complexes with Fe (FORDYCE 2005), and in clayey soil (the CSCB) due to Se adsorption.

Soil quality in Sweden, Norway and Finland is locally influenced by the presence of potential acid sulphate (PAS) soil (JOUKAINEN & YLI-HALLA 2003). Due to post-glacial isostatic rebound, some of these sediments now make up parts of the coastal lowlands of the Baltic Basin, primarily between 0 and 45 m.a.s.l. (BOMAN et al. 2008). If the groundwater table is lowered in PAS soil, oxidation of sulphide minerals, such as pyrite (FeS_2) and metastable iron sulphides (e.g., FeS) begins, and leads to subsequent formation of sulphuric acid (H_2SO_4) in the soil solution. If the pH is lowered to levels below 4, PAS soil become acid sulphate (AS) soil. The acid soil solution induces leaching of metals, mobilising metal ions, such as Fe^{2+} and Al^{2+}, together with trace metals (e.g., Zn, Ni, Co, Mn and Cu). The sulphur content is the main concern of this often cultivated soil. For agricultural purposes, farmers have traditionally limed the top soil layer of their lands in order to neutralise the low pH and induce plant productivity (TOIVONEN & ÖSTERHOLM 2011).

14.6.4 Implications for mineral exploration

Despite the low sample density, the GEMAS data provide (or confirm) information about natural mineral resources in Norway, Sweden and Finland. The location of many base metal deposits can be recognised in the studied samples, especially in the total XRF concentrations results from grazing land soil that does not undergo regular ploughing. Metals, such as Ag, Zn, Pb and As, often indicate the presence of polymetallic mineralisations (Table 14.2). For example, in Sweden alone, over 600 deposits with Ag occur in a variety of deposits scattered all over the country, and the dispersion pattern for Ag in Ap and Gr samples correlates well to these mineralisation types (Chapter 3; see also Section 11.7 → Part A). Some metals can be used as pathfinders, e.g., Ag anomalies correlate well with Mo, U and Cu, and Sb anomalies are usually associated with Pb-Zn-Cu, Bi and Au mineralisation.

At the same time, both sample types due to the low sampling density and the deposit type (completely buried without surface exposure), several economically important active mining centres (with underground mines) are not visible in the GEMAS data set. The most striking 'misses', which are not observed on elemental maps in this study, are the Kiruna mining district (the biggest European iron ore mine) and the copper mine at Aitik in northern Sweden.

Soil type can disturb the element pattern, due to different element mobility in sandy, clayey and peaty soil. Sandy soil usually displays significantly lower levels of economic metals, than clay-rich and organic-rich soil. The Lapland greenstone belt is, for example, not well distinguished in the data set with respect to its presumed elevated Cu contents, as the anomalies are suppressed by the organic-rich nature of the overlying peat-rich soil since, in contrast to Hg, Pb and Se, Cu is somewhat depleted in such soil.

The use of the GEMAS data set in mineral exploration is, therefore, limited, and it should only be viewed as a supporting tool to delineate large metallogenic provinces. High-density sampling is required for detailed exploration-oriented surveys.

14.6.5 Potential anthropogenic input

Anthropogenic impact on soil is to be expected in every geochemical survey. In Norway, Sweden and Finland, however, human activity seems to have had only a minor influence on the soil geochemistry of both agricultural and grazing land. Identified geochemical anomalies, both of aqua regia extraction and total XRF concentrations, related to anthropogenic contamination, are small-scale and scarce in Scandinavian countries. In mining regions, with a natural signal from the mineralisation, it is often difficult to discriminate between the original natural anomaly and any anthropogenic contamination signal. Anthropogenic contamination in Norway, Sweden and Finland becomes visible in the Stockholm area, in southern Sweden and Finland (fertilising, liming activities), around the Zn-smelter at Odda near Bergen in Norway, in Öland and Gotland (fertilising, historical iron and brass manufacturing), and in the Skellefte mining district (mining-related contamination). The results from MMI® (Mobile Metal Ion – see Chapter 13) method may give better insight into possible anthropogenic impact on soil (not discussed here).

Table 14.2: Explanation of some key anomalies related to mineralisation.

Element	Mineralisation
Ag	High Ag concentrations correlate with numerous occurrences of polymetallic mineralisation in Finland, Sweden (e.g., Arjeplog region) and Norway (e.g., Oslo Rift); vein type Ag-Zn-Pb deposits, Ag-bearing sulphides in alum shale, Ni, Cu, Co ore deposits located in mafic metavolcanic rocks (Kemi ore deposits in Finland).
As	High As concentrations occur in the 'Tampere arsenic province' in Finland, in the Oslo Rift (sulphide mineralisation in Permian volcanic rocks and Early Palaeozoic black shale), in 'the copper belt' of known sulphide mineralisation in the Caledonides (central Norway) and at several locations in Sweden (sulphide mineralisation in Bergslagen, Jämtland, Skellefte and the Boden region in Norrbotten).
Au	Au anomalies occur in the vicinity of known gold mineralisation, e.g., Suurikuusikko in northern Finland, and Korpilampi-Pampalo in eastern Finland, the Gold Line in Västerbotten county in northern Sweden, near Silvergruvan mine in central Sweden.
Bi	High Bi concentrations occur in the Oslo Rift, the Kongsberg region (associated with felsic volcanic rocks and the well-known silver ore deposits) and in the Hardanger region. In south-west Norway, in the central Caledonides (also Sweden), south-eastern Sweden Bi anomalies are linked to the occurrence of Pb-Zn ore deposits and Bi, Au, Cu, pyrite mineralisation hosted by quartz-dolomite veins of hydrothermal origin.
Co	High concentrations of Co are related to Zn, Pb, Cu, Ni and PGE mineralisation in mafic and ultramafic rocks (e.g., central Norway, northern Finland, central Sweden). Metamorphosed ultramafic rocks rich in talc-soapstone are also enriched in Co (e.g., Handöl in Sweden). Cobalt mineralisation coinciding with Co anomalies in soil is known at Brattbaken, east from Steinkjer and in northern Norway (Vaddas, Kåfjord). In the Oslo Rift, southern Norway, volcanic rocks and black shale correlate with high Co concentrations in soil.
Cr	Cr anomalies are related to the occurrence of mafic and ultramafic rocks ('greenstone belts', ophiolite rocks in Caledonides, soapstone).
Cu	High Cu concentrations in Norwegian soil are mainly related to known polymetallic ore deposits (VMS deposits) in central and northern Norway. In Jämtland and southern Sweden, Cu anomalies correlate with known polymetallic mineralisation. As a peculiarity, no clear Cu anomalies can be found in regions with large base-metal ore deposits, such as the Skellefte mining district and Bergslagen.
Ge	High Ge concentrations in Västerbotten in Sweden correlate with Ag, Cd, Bi and sulphide mineralisation.
In	Indium anomalies occur over Zn-Cu (-Ag) ore deposits, (e.g., Cu-Zn VMS deposits in the Skellefte region in Sweden, Salo-Issakka in SE Finland, Pyhäsalmi in central Finland), Ni-Cu-Co ores (e.g., northern Norway, Vammala in SW Finland). High In concentrations occur in soil from the Oslo Rift (sulphide mineralisation and black shale) and associated with ore deposits from the Kongsberg region. The large anomaly in southernmost Norway (Gr samples) occurs in the vicinity of V-Fe-Ti ore deposits, e.g., the World class Tellnes anorthosite-hosted Ti deposits. In Jämtland in central Sweden, high In concentrations correlate with known Cu-Zn mineralisation within the Caledonian mountain chain (e.g., Storsjö) and within the Fennoscandian Shield (e.g., Tjärnberget).
Mn	In central and northern Finland (Central Lapland Area) the anomalies coincide with Fe-V and even Mn deposits (the Porkonen-Pahtavaara Fe-Mn zone) hosted by manganiferous iron formations. In Sweden, higher Mn concentrations correlate with known Mn mineralisation that accompanies the Fe and sulphide mineralisation, e.g., in Bergslagen. Smaller anomalies in grazing land soil samples from the Kiruna mining district and the Junosuando region reflect the location of iron ores.
Mo	In central Finland, anomalies coincide with Mo mineralisation at Saviselkä and Mätäsvaara (the only Mo ore mined). The small anomaly in southern Lapland by the Russian border occurs over the Archaean hosted Aittojätvi Mo ore deposits. In Norway, the large Mo anomaly north of Oslo correlates with the Mo ore deposit at Nordli near Hurdal (untouched porphyry Mo ore deposit, one of the largest known, as yet unmined, deposits in the world). The other anomaly in southern Norway correlates with Mo mineralisation at Knaben, a mining district where Mo was exploited from 1885 to 1973. In Sweden, high Mo concentrations occur in Jämtland (alum shale in Myrviken) and Västerbotten (felsic volcanics, limestone), and correlate with sulphide mineralisation (Cu-Zn), Au and U occurring by the Caledonian mountain front. In Västerbotten, Mo soil anomalies follow occurrences of sulphide mineralisation with Mo, Cu, but also W, Au and fluorite. In southern Norrbotten (the Boden region) Mo anomalies coincide with Mo, Cu and Au mineralisation. Enhanced Mo contents in soil follow Fe and sulphide ore deposits in the Kiruna region. In western Bergslagen, an extensive anomaly in agricultural soil is located close to the large Zn-Pb-Ag Zinkgruvan mine.

Table 14.2: continued

Element	Mineralisation
Pb	Locally, Zn-Pb deposits contribute to the high Pb contents in soil, e.g., in the Oslo Rift (skarn type mineralisation), in the vicinity of Stockholm (Pb, Zn, Ag in breccia), Åtvidaberg (Pb, Zn, Cu in felsic volcanic), in south-eastern Scania (alum shale and galena in Cambrian sandstone), and south of the Skellefte mining district.
PGE	Pd and Pt originate mainly from mafic and ultramafic rocks, often hosting Fe-V, Ni, Cr mineralisation. Higher concentrations of PGE in soil correlate with known PGE-Ni-Au mineralisation, e.g., Kukkola (PGE, Cr, Au) in northern Sweden, Notträsk (Au, PGE) at the north-eastern coast, and Kläppsjö (PGE, Au) in central Sweden.
Sb	The Sb anomalies correlate with occurrences of Pb-Zn-Cu vein mineralisation (e.g., Oslo Rift). Locally, high concentrations are found in the vicinity of alum shale (in Jämtland, central Sweden). High Sb concentrations have been found in the Skellefte mining district. North-west of Luleå, both XRF and aqua regia extraction results point to a region with known mineralisation of Sb, Bi, Au and As.
Sn	In south-eastern Finland, Sn anomalies can be related to known Sn deposits on the Russian side of Finland's eastern border, which have been mined for Sn during 1910-1920. In central Sweden, Sn anomalies coincide with known Sn mineralisation in Gävleborg (e.g., Mosisjön), in the Stockholm area (Norrskogen) and in cassiterite-bearing greisens veins.
Ti	Many Fe-Ti-V deposits occur in Norway (e.g., Tellnes in the Rogaland), both in the basement rocks (south-western and northern Norway), and in the central Caledonides. Titanium anomalies occur in soil overlying mafic and ultramafic rocks rich in ilmenite and high-grade metamorphic rocks rich in rutile (e.g., eclogite). In Sweden and Finland, concentrations of Ti are higher in regions with Fe-Ti-V mineralisation, especially in the northern regions.
U	In central and northern Finland, U anomalies follow known U ore deposits and V-Zn mineralisation. In Sweden, elevated concentrations are observed in association with known ore deposits including the Arjeplog–Arvidsjaur epigenetic and the Caledonian black-shale deposits.
V	In general, V follows known Fe(-V-Ti) deposits. High concentrations of V are observed in relation to outcrops of mafic and ultramafic rocks, e.g., Handöl in central Sweden (ophiolites with soapstone), in the Haparanda and Pajala regions in northern Sweden. Known polymetallic mineralisation in Sweden is indicated by V anomalies in Västerbotten, Jämtland (near Östersund) and south-eastern Scania.
W	The W anomalies in south-eastern Finland correlate with U mineralisation and pegmatite occurrences. Tungsten anomalies in central Finland might correlate with the occurrence of Au, Ag, and Li mineralisation in pegmatite and quartz-rich veins. Dispersion patterns show higher concentrations of W in Norrbotten, Bergslagen and along the central coastline in Sweden; these anomalies follow known W mineralisation occurring within granitic rocks (together with U and Au). In the Caledonides of northern Jämtland, in the Gäddede region, high W concentrations (Gr, XRF) in soil correlate with U(+fluorite+haematite) mineralisation hosted by cataclastic breccias within fractured granitic rocks.
Zn	High concentrations of Zn occur in southern Finland and in central part of the country in correlation with clay-rich soil and Zn-Pb-Cu mineralisation. In Sweden, Zn anomalies follow sulphide mineralisation with Zn (+Pb+Cu+Ag), e.g., in northern Dalarna (e.g., Vassbo), Jämtland (e.g., Tjärnberget), Västernorrland (e.g., Rockliden), Bergslagen and Värmland (e.g., Dingelvik). The large Zn anomaly (and Cd) in grazing land soil in south-eastern Scania corresponds to mineralisation with galena in Cambrian sandstone.

14.6.6 Comparison with the rest of Europe

When compared with the entire European data set, GEMAS data from Norway, Sweden and Finland show several interesting features (Table 14.3; see country boxplots on DVD → Part A). Titanium, Nb and U in all three countries show the highest concentrations of these elements in Europe (Fig. 14.7; see also Section 11.38; 11.57; 11.59 → Part A). This feature is related to the presence of large massifs containing igneous rocks of Palaeoproterozoic age. High concentrations of Ti reflect the high abundance of titanomagnetite and ilmenite in igneous rocks and titanite and rutile in metamorphic rocks. A high percentage of Ti originates from rock-forming minerals, e.g., biotite and amphibole, present both in igneous and metamorphic rocks. High Nb, Mo and U levels are related to the presence of highly evolved (fractionated) igneous and meta-igneous rocks. High concentrations of Sr and Na point to overprinting alteration processes (e.g., metasomatism), which play an important role in ore-forming processes. The extreme Sr and Na anomalies, which occur in the northern regions, correlate with large apatite-Fe ore and base metals deposits. Maximum values and outliers often point to known mineralisation; however, they can also be related to the occurrence of organic-rich soil with low pH.

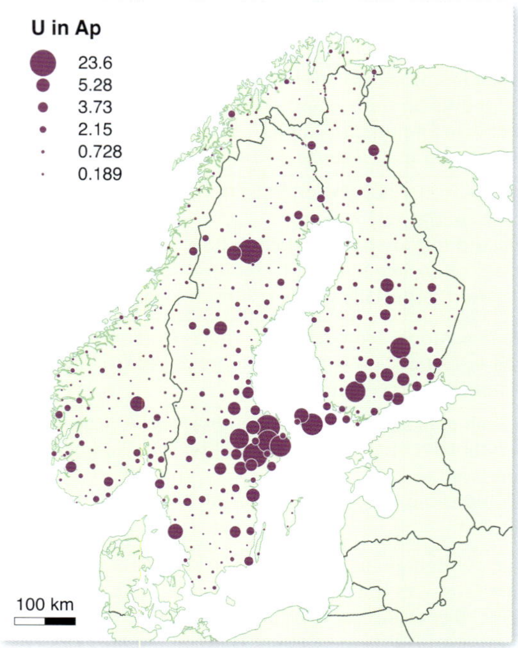

Fig. 14.7: Aqua regia extractable U concentrations in agricultural soil of Norway, Sweden and Finland. One of the highest U values in Europe is present in the CSCB.

Considering the amount of iron ore and base metal sulphide ore deposits in Norway, Sweden and Finland, some metals display a surprisingly low concentration in comparison with the rest of Europe, e.g., Fe, Cu, Pb, As, Cd, Cr, Ni (Chapter 11 → Part A). Several factors are thought to influence this result. The density of GEMAS sampling is low (1 site/ 2500 km^2), therefore, some of the ore deposits/ active mines, where the soil anomaly is localised around the site, could easily have been missed. Certain diluting effects can also influence the results, e.g., sandy soil tends to be leached more efficiently and, therefore, exhibits certain element depletion, while organic soil retains metals even in areas without significant mineralisation in the bedrock.

Table 14.3: Comparison of median and maximum aqua regia extractable element concentrations in Norway, Sweden and Finland with respect to the European data set.

Feature	Norway	Sweden	Finland
High median concentrations in comparison with Europe (in brackets: place on the order list of country boxplots – see country boxplots on attached DVD → Part A)	Ge (5), Na (6), Nb (3), P (2), S (4), Sr (6), Ti (2), U (6)	Ag (3), Mo (4), Na (8), Nb (1), P (9), S (7), Ti (3), U (1)	Ge (3), Na (4), Nb (2), S (12), Ti (1), U (3)
High maximum values of elements in Ap* and Gr soil samples** (including extreme concentrations and outliers)***	Ag (Gr), Ba (Gr, Ap), Bi (Ap), Ce (Gr), Cr (Ap, Gr), Cs (Gr), Fe (Ap), Ge (Ap, Gr), Hg (Gr), La (Ap, Gr), Mn (Ap, Gr), Mo (Gr, Ap), Na (Ap, Gr), Nb (Ap, Gr), P (Ap, Gr), Pb (Gr), Rb (Gr), S (Ap, Gr), Se (Ap, Gr), Sr (Gr), Th (Ap), Tl (Gr), U (Ap, Gr), V (Ap), Zn (Gr)	Ag (Gr), As (Gr), Au (Gr, Ap), B (Gr), Ba (Gr), Be (Gr), Cd (Gr), Ce (Ap, Gr), Fe (Ap), Ge (Gr, Ap), Hg (Gr), La (Ap, Gr), Mn (Gr), Mo (Gr, Ap), Na (Ap, Gr), Nb (Ap, Gr), P (Ap, Gr), Pb (Gr), S (Ap, Gr), Sb (Gr), Se (Gr), Sn (Gr), Th (Ap), U (Ap, Gr), V (Ap, Gr), Y (Gr), Zn (Gr)	Al (Ap), Ba (Ap), Ce (Ap), Co (Ap), Fe (Ap), Ga (Ap), Ge (Ap, Gr), K (Ap), La (Ap, Gr), Li (Ap), Mn (Ap), Mo (Gr, Ap), Na (Ap, Gr), Nb (Ap, Gr), P (Gr), Pb (Gr), Rb (Ap), S (Ap, Gr), Sb (Gr), Sc (Ap), Se (Ap, Gr), Th (Ap, Gr), U (Ap, Gr), V (Ap)

* Agricultural soil ** Grazing land soil *** Based on boxplots for all GEMAS countries

▸ 14.7 Conclusions

The geochemical mapping of agricultural and grazing land soil (the GEMAS project) in Norway, Sweden and Finland, provides an exceptional opportunity to demonstrate regional geochemical trends. When looking at the European data set as a whole (Chapter 11 → Part A), Norway, Sweden and Finland stand out as geochemically distinct, and it was thus considered important to study the data set of these countries as a separate entity.

The interpretation of the elemental maps and statistics (boxplots, cumulative probability plots) identified several groups of factors influencing the observed trends in the geochemical patterns of Norway, Sweden and Finland. The most important factors being bedrock geology, the presence of ore deposits, the soil type and its properties, and climate.

Anthropogenic impact on soil geochemistry seems to be insignificant, both in agricultural and grazing land soil in Norway, Sweden and Finland. Especially in mining regions, it is often difficult to discriminate between the original natural anomaly and any anthropogenic contamination signal.

The GEMAS project is based on a low sampling density (1 site/2500 km^2) and a number of important features in the geochemical landscape do not show up (e.g., main mining districts in northern Sweden). The choice of sampling site carries a certain subjectivity burden, which is difficult to avoid; however, due to the scale of the project, this human factor is thought to be negligible and the overall continental scale picture still gives reliable information. |◂

Anna Ladenberger
Geological Survey of Sweden (SGU);
Uppsala, Sweden
E-mail: anna.ladenberger@sgu.se

Jo Uhlbäck
Geological Survey of Sweden (SGU);
Uppsala, Sweden

Madelen Andersson
Geological Survey of Sweden (SGU);
Uppsala, Sweden

Clemens Reimann
Geological Survey of Norway (NGU);
Trondheim, Norway

Timo Tarvainen
Geological Survey of Finland (GTK);
Espoo, Finland

George Morris
Geological Survey of Sweden (SGU);
Uppsala, Sweden

Martiya Sadeghi
Geological Survey of Sweden (SGU);
Uppsala, Sweden

Mikael Eklund
Geological Survey of Finland (GTK);
Espoo, Finland

Peter Filzmoser
Vienna University of Technology (TU Wien);
Vienna, Austria

Chapter 15

Geochemical Characteristics of Ukrainian Soil Using Landscape-Geochemical Regionalisation Based on the GEMAS Data

Volodymyr Klos, Manfred Birke, Grygorii Akinfiev & Yulia Amashukeli

▸ 15.1 Introduction

Geochemical mapping of European soil exhibits considerable variation in chemical element concentrations across the continent caused, as a rule (in the absence of anthropogenic factors), by natural conditions (climate, topography, soil-forming processes, vegetation and underlying parent materials). The issue of contouring areas covered with soil of 'homogeneous' composition from the geochemical point of view, which are sufficiently extensive to host an adequate number of sampling sites in order to obtain reliable interpolated data, may present certain difficulties at the low sample density geochemical mapping of the GEMAS project (≈ 1 sample site/2500 km²).

To resolve this issue in Ukraine, the methodology of landscape geochemistry was applied. Selecting a geochemical landscape class (the second largest one after a bioclimatic zone) as a regionalisation unit requires an in-depth knowledge of the interaction of the lithosphere with the hydrosphere, atmosphere and biosphere. In landscape-geochemical regionalisation, the class represents physical and chemical conditions of element migration in geochemical landscapes (groups of elemental geographical landscapes), including soil (biogeochemical migration of elements is considered at the bioclimatic zone level).

Landscape geochemistry originated in Russia, largely as a result of the pioneering work by B. B. Polynov, who investigated the migration of major elements in the weathering zone (Polynov 1937). He concieved the idea that during geological time under a given set of climatic constraints, the chemical interactions between the lithosphere, hydrosphere, atmosphere, and biosphere are largely controlled by relationships between the ground (daylight) surface and the water table (or hydrosphere). Research work in this area continued and gradually the 'landscape-geochemical school' was developed and regionalisation principles, based on landscape-geochemical conditions of element migration were extensively studied during the 1950-1970s in the Soviet Union by Polynov (1953), Perelman (1961, 1966, 1972, 1975), Glazovskaya (1963, 1988), Kozlovskiy (1972) and others, including Fortescue (1967, 1974, 1974a, 1975, 1992) in the Western World. The '*geochemical landscape*' is the smallest unit area in landscape-geochemical regionalisation. According to Polynov (1953), the *geochemical landscape* is a paragenetic association of ajoining elementary landscapes, which are related by a specific type of element migration. He suggested soil homogeneity as a major criterion of the elementary landscape contouring: "*... typical elementary landscape shall represent one concrete type of topography made up of the same rock or deposit covered by a certain vegetation community in every moment of its existence. All above conditions develop a well-defined soil variety and confirm the identical interaction between rocks and bionts within the elementary landscape*" (Table 15.1).

Landscape geochemistry therefore classifies the migration of elements in soil according to forms of movement of matter. The great advantage of the landscape geochemistry approach is that it integrates information of several kinds, which otherwise would be displayed independently.

Table 15.1: Classification pattern for elementary landscapes by type of migration of chemical elements (from GLAZOVSKAYA 1963, in FORTESCUE 1992: Table 2, p.13).

Landscape groups after B. B. POLYNOV	**According to type of geochemical integration**	**According to migration cycles of elements in original rocks**		
		Primary (ortho)	**Secondary (para)**	**Superimposed secondary (neo)**
Eluvial	Eluvial *(flat tops, well drained ancient plains)*	Orth - el *(ortho-eluvial on massive igneous rocks)*	Par - el *(para-eluvial on dense sedimentary rocks)*	N - el *(neo-eluvial on loose sediments)*
	Trans - eluvial *(upper parts of slopes)*	Trans - orth - el *(trans-orthoeluvial)*	Trans - par - el *(trans-paraeluvial)*	Trans - n - el *(trans-neoeluvial)*
	Eluvial - accumulative *(parts of slopes and dry gulleys)*	Orth - el - ac *(trans-orthoeluvial-accumulative)*	Par - el - ac *(trans-paraeluvial-accumulative)*	N - el - ac *(neoeluvial-accumulative)*
	Accumulative - eluvial *(local, confined to lower ground with deep groundwater table)*	Orth - ac - el *(ortho-accumulative-eluvial)*	Par - ac - el *(para-accumulative-eluvial)*	N - ac - el *(neo-accumulative-eluvial)*
Superaqual	Trans - superaqual *(trans - hydromorphic)*	Trans - hydro - orth *(trans-ortho-hydromorphic)*	Trans - hydro - p *(para-trans-hydromorphic)*	Trans - hydro - n *(neo-trans-hydromorphic)*
	Superaqual *(confined to lower ground with weak water exchange)*	Hydro - orth *(ortho-superaqual)*	Hydro - p *(para-superaqual)*	Hydro - n *(neo-superaqual)*
Subaqual	Trans - aqual *(streams, flowing lakes)*		Trans - aqual	
	Aqual *(stagnant lakes)*		Aqual	

▸ 15.2 Materials and methods

Migration of elements within a water solution in the humus horizon is taken as a basis for distinguishing geochemical landscape classes, because geochemical processes are the most intense within this particular layer (PERELMAN 1975). The classes are denoted by typomorphic (specific) elements, their ions or complexes (O_2, CO_2, H^+, OH^-, Cl^-, SO_4^{2-}, HCO^{3-}, Ca^{2+}, Mg^{2+}, Na^+, Fe^{2+}, etc.), which govern water solution characteristics as migration media. As a rule, these are elements with high Clarke values. The same element can be typomorphic in one landscape class and non-typomorphic in another. In most cases, an element's role in a landscape is determined not so much by its concentration, but by the intensity of its migration and capacity for accumulation in any landscape.

Distribution of geochemical landscape classes is governed by climate, geological structure and topography.

Based on the map of landscape-geochemical regionalisation of Ukraine at a scale of 1:500,000 (Fig. 15.3) compiled by OLSHEVSKAYA (1995), data on geochemical mapping of the country's soil from the GEMAS project and other sources (Polupan et al. 2005) were selected. Agricultural and grazing land soil samples were selected separately and the sample location classified according to the geochemical landscape class (in view of a bioclimatic zone), and contained information on major soil types. Each geochemical landscape class included a minimum of five samples.

The distribution of elements was estimated within the selected landscape classes, based on all available analytical data, but the compiled summary tables contain results of those analytical methods, which give total element concentrations in soil, because these are more accurate and correspond to their Clarke values in soil and lithosphere (Bowen 1979; Vinogradov 1962), although in some cases the aqua regia results were used.

Fig. 15.1: Agricultural soil sample site (Ukr 68 Ap) in northern Ukraine (Kyiv region, forest steppe, geochemical landscape class: gleyed acidic in combination with calcic and nitric) (photograph by Volodymyr Klos).

Geochemical characteristics of soil samples in each landscape class were determined using the *concentration coefficient* (K_c), *which represents the ratio of an element's background (median) concentration in the selected landscape class to its background (median) concentration in European soil*, estimated separately for agricultural and grazing land soil. Elements with concentration coefficients exceeding 1.5 were considered as accumulating, and the ones with $K_c < 0.7$ were regarded as being leached (lower concentrations) (UkrDGRI 2006).

Accumulation/leaching of elements in agricultural land soil was determined within each landscape class, based on the ratio of an element's background (median) concentration in agricultural land soil, and similarly in the grazing land soil (accumulation if $K_c > 1.5$ and leaching if $K_c < 0.7$).

Fig. 15.2: Profile of agricultural soil sample site (Ukr 76 Ap) in south-eastern Ukraine (Kharkiv region, forest steppe, black soil on loess, geochemical landscape class: calcic) (photograph by Volodymyr Klos).

▸ 15.3 Results and discussion

Of 17 geochemical landscape classes that are most widespread in Ukraine (Fig. 15.3), background concentrations of elements were determined for 10 classes of agricultural land soil (0-20 cm depth) and for 13 classes of grazing land soil (0–10 cm) (Tables 15.2 and 15.3). These geochemical studies covered 99.16 % of the country. Only soil of mountain and seacoast landscape classes of local development and specific processes of element migration in water solutions remained unstudied, namely: calcic brown (Ca^{2+}, Al^{3+}-Fe^{3+} – brown podzolic soil, 0.14 % of Ukraine); acidic gleyic in combination with natric (H^{+}- Fe^{2+}, Na^{+} – solonised meadow bog soil, 0.13 %); natric-chloride-sulphate (Na^{+}, Cl^{-}, SO_4^{2-} – alkaline soil, salt marshes, 0.56 %), and natric (Na^{+} – alkaline soil, 0.01 %).

Geochemical specialisation of soil within the major geochemical landscape classes of Ukraine was determined relative to the overall median value of European soil (Tables 15.4 and 15.5). These tables show that elements (S, B, In, Te, F, Li, Ni, Co, Cr and others) are accumulated in soil of the calcic (Ca^{2+}) landscape classes, the leaching of Mg, Ca, Fe, S, Mn, Li, Ni, V, Co and other elements being observed in soil of the acidic (H^{+}) classes, which is consistent with the fundamental principles of landscape geochemistry. Brown soil of the Carpathian Mountain acidic landscape class is a sole exception where accumulation of Hg, S, Cd, Sb, Ag and other elements are observed as a result of their high primary concentrations in soil-forming and underlying rocks of the Ukrainian Carpathians.

Data on comparison of element concentrations in agricultural and grazing land soil by geochemical landscape classes are presented in Table 15.6. One can see from the Table that the agricultural land soil of the acidic landscape classes has an increased number of elements with elevated concentrations when compared to the same soil of the calcic landscape classes. Most likely, this is caused by the more intense agrochemical treatment of the former. In general, regular accumulation of elements in agricultural land soil was not observed, but Cl, F, Pt, As, Na and other elements were concentrated more often. Decrease in Ca and S concentrations in the agricultural land soil is most probably caused by their biogeochemical migration. Accumulation of S in arable soil of the Carpathian Mountains requires additional studies.

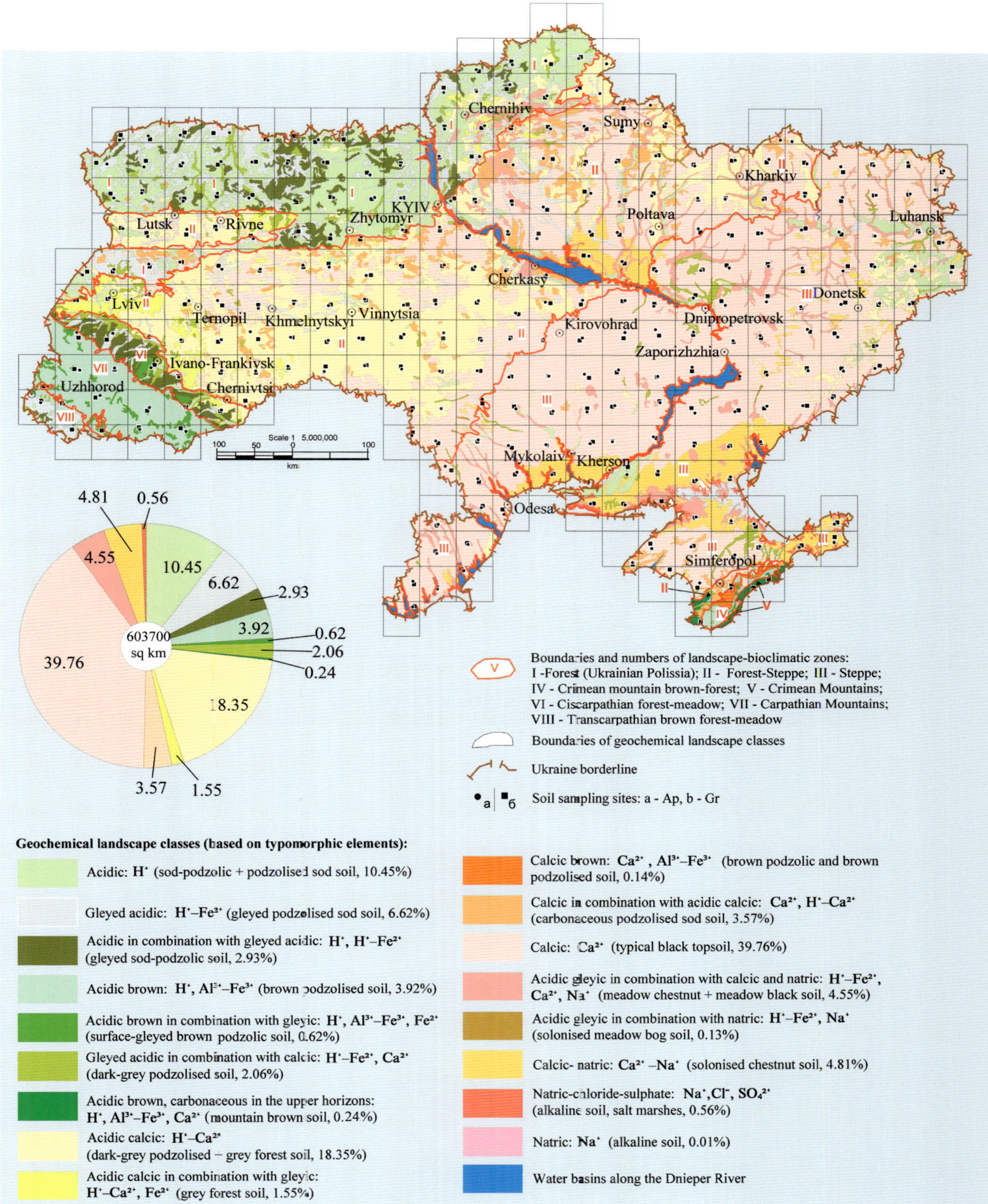

Fig. 15.3: Classes of geochemical landscapes in Ukraine: Overview map (based on OLSHEVSKAYA 1995).

Table 15.2: Element concentrations (mg/kg) in agricultural land soil (Ap-horizon) of Ukraine, according to geochemical landscape classes.

Loc	Geochemical landscape class	Soil type	Analysis:	ICP[1]	XRF[2]	XRF	ICP	ICP
			Element:	Ag	Al	As	Au	B
Z1	acidic: H^+ (47.1 % of the area)	sod-podzolic + podzolised sod soil (N = 26)	*min*	0.001	2435	1.5	0.0001	0.5
			max	0.041	30,114	11	0.0011	4
			median	0.013	13,417	1.5	0.0002	1
	gleyed acidic: H^+- Fe^{2+} (26.5 %); acidic in combination with gleyed acidic: H^+, H^+- Fe^{2+} (12.2 %)	gleyed podzolised sod soil, gleyed sod-podzolic soil (N = 12)	*min*	0.002	7092	1.5	0.0001	0.5
			max	0.022	26,251	1.5	0.0020	2
			median	0.011	11,432	1.5	0.0001	1
	acidic calcic: H^+- Ca^{2+} (4.7%); gleyed acidic in combination with calcic: H^+- Fe^{2+}, Ca^{2+} (3.7 %); calcic in combination with acidic calcic: Ca^{2+}, H^+- Ca^{2+} (1.4 %); acidic calcic in combination with gleyic: H^+-Ca^{2+}, Fe^{2+} (1 %)	dark-grey podzolised + grey forest soil, dark-grey podzolised soil, carbonaceous podzolised sod soil, grey forest soil (N = 7)	*min*	0.015	24,398	1.5	0.0001	2
			max	0.037	35,513	4	0.0005	7
			median	0.019	30,485	3	0.0001	3
Z2	calcic: Ca^{2+} (31.3 %)	typical black topsoil (N = 32)	*min*	0.016	33,925	1.5	0.0001	3
			max	0.058	61,287	11	0.0028	10
			median	0.032	45,674	5	0.0007	5
	acidic calcic: H^+- Ca^{2+} (43.4 %); acidic calcic in combination with gleyic: H^+- Ca^{2+}, Fe^{2+} (3.4 %); calcic in combination with acidic calcic: Ca^{2+}, H^+- Ca^{2+} (2.4 %)	dark-grey podzolised + grey forest soil, grey forest soil, carbonaceous podzolised sod soil (N = 52)	*min*	0.015	21,276	1.5	0.0001	1
			max	0.055	61,446	10	0.0074	8
			median	0.035	46,998	5	0.0006	5
	acidic gleyic in combination with calcic and natric: H^+- Fe^{2+}, Ca^{2+}, Na^+ (4.1 %)	meadow chestnut + meadow black soil, (N = 6)	*min*	0.008	14,290	1.5	0.0001	1
			max	0.028	44,034	5	0.0013	7
			median	0.018	33,714	3.5	0.0002	4.5
Z3	calcic: Ca^{2+} (76 %)	typical black topsoil (N = 87)	*min*	0.020	21,911	1.5	0.0001	2
			max	0.080	70,284	17	0.0122	22
			median	0.046	59,752	9	0.0010	7
	calcic-natric: Ca^{2+}- Na^+ (9.8 %); acidic gleyic in combination with calcic and natric: H^+- Fe^{2+}, Ca^{2+}, Na^+ (6.2 %)	solonised chestnut soil, meadow chestnut + meadow black soil (N = 12)	*min*	0.025	26,886	3	0.0002	5
			max	0.101	68,167	11	0.0013	14
			median	0.058	60,600	9.5	0.0007	9.5
Z4	acidic in combination with gleyed acidic: H^+, H^+- Fe^{2+} (41.6 %)	gleyed sod-podzolic soil (N = 6)	*min*	0.028	41,652	5	0.0006	2
			max	0.088	69,808	11	0.0019	8
			median	0.050	48,189	8.5	0.0016	4
Z5	acidic brown: H^+– Al^{3+}–Fe^{2+} (88.3 %)	brown podzolised soil (N = 8)	*min*	0.038	29,744	5	0.0003	1
			max	0.103	83,304	30	0.0029	7
			median	0.072	57,794	8.5	0.0013	4

Loc Landscape bioclimatic zone **Z1** Forest (Ukrainian Polissia) **Z2** Forest steppe **Z3** Steppe **Z4** Ciscarpathian forest-meadow **Z5** Carpathian Mountains

Loc	Analysis:	XRF	ICP	XRF	XRF	SS	XRF	XRF	XRF	XRF	XRF	ICP	ICP	XRF	ICP	XRF	ICP
	Element:	La	Li	Mg	Mn	Mo	Na	Nb	Ni	P	Pb	Pd	Pt	Rb	Re	S	Sb
Z1	*min*	10	0.2	121	46	0.6	149	2	1.5	172	4	0.005	0.001	7	0.0005	20	0.03
	max	10	4.5	3196	604	16	4235	9	11	1050	17	0.005	0.001	48	0.0010	400	0.21
	median	10	2.35	875	267	1.2	2264	6	4	398	8.5	0.005	0.001	23.5	0.0005	20	0.07
	min	10	0.6	241	85	0.72	1117	2	1.5	255	5	0.005	0.001	11	0.0005	20	0.04
	max	24	4.7	1689	650	2.8	3580	9	11	796	14	0.005	0.001	46	0.0010	160	0.10
	median	10	1.25	603	174	1.4	1572	5.5	4	335	9	0.005	0.001	22	0.0005	20	0.065
	min	10	3.9	1930	310	0.89	2914	8	7	326	9	0.005	0.001	33	0.0005	20	0.07
	max	21	7.2	3679	511	3.3	5295	12	68	1265	15	0.005	0.001	55	0.0005	440	0.16
	median	10	4.5	2231	434	1.2	4460	10	12	484	12	0.005	0.001	48	0.0005	20	0.08
Z2	*min*	10	4.9	2834	395	0.28	2328	10	10	476	11	0.005	0.001	55	0.0005	20	0.08
	max	38	22.4	8322	798	2.1	6183	17	42	1686	25	0.005	0.005	88	0.0005	520	0.52
	median	25.5	10.6	4975	542	1.3	5262	13	22.5	670	16.5	0.005	0.001	69.5	0.0005	80	0.185
	min	10	3.1	1206	217	0.45	3273	9	5	226	9	0.005	0.001	32	0.0005	20	0.06
	max	47	18	8202	937	2.2	7272	18	40	2065	22	0.010	0.004	103	0.0020	360	0.34
	median	26	9.5	4433	631	1.4	5760	15	21	683	19	0.005	0.001	75.5	0.0005	80	0.16
	min	10	2	844	194	0.59	2163	5	4	238	6	0.005	0.001	28	0.0005	20	0.05
	max	37	10.1	4945	519	1.6	5656	13	19	638	16	0.005	0.001	70	0.0005	280	0.15
	median	20.5	6.65	2985	353	1	4571	11.5	11.5	477	12.5	0.005	0.001	54	0.0005	40	0.095
Z3	*min*	10	6.2	2653	286	0.82	1331	6	16	269	8	0.005	0.001	31	0.0005	20	0.13
	max	43	27.7	11,880	1580	2.6	12,026	17	53	1248	44	0.005	0.006	96	0.002	1400	0.90
	median	28	19.6	8202	751	1.7	5321	15	39	680	19	0.005	0.001	87	0.0005	160	0.32
	min	10	7.2	2834	294	1.1	3081	9	14	378	11	0.005	0.001	43	0.0005	20	0.13
	max	38	25	11,760	1138	2.1	6799	17	49	893	23	0.005	0.001	101	0.0005	360	0.39
	median	25	19.3	8141	786	1.7	5914	15	39	668	19	0.005	0.001	86.5	0.0005	80	0.265
Z4	*min*	10	9.4	2714	294	0.87	5169	13	10	610	18	0.005	0.001	63	0.0005	20	0.20
	max	38	21.9	9227	1154	3.2	7100	16	39	1004	32	0.005	0.002	120	0.0005	360	0.35
	median	26.5	11.1	4433	717	1.55	5670	15	25	816	22	0.005	0.001	79.5	0.0005	30	0.27
Z5	*min*	10	8	3377	364	1.5	2391	8	22	557	17	0.005	0.001	53	0.0005	20	0.25
	max	40	33.1	11,036	1874	4.4	10,286	18	59	2004	39	0.005	0.001	131	0.001	520	0.80
	median	18	19	5820	972	2.4	6658	13.5	26	968	31	0.005	0.001	72	0.0005	100	0.465

[1] aqua regia extraction and determination by inductively coupled plasma mass spectrometry (ACME Laboratories, Vancouver, Canada)
[2] X-ray fluorescence (BGR Laboratories, Hannover, Germany)

XRF Ba	ICP Be	ICP Bi	XRF Ca	ICP Cd	XRF Ce	XRF Cl	ICP Co	XRF Cr	XRF Cs	XRF Cu	XRF F	XRF Fe	XRF Ga	SS[3] Ge	XRF Hf	AAS[4] Hg	ICP In	XRF K
48	0.05	0.02	143	0.02	10	10	0.1	7	1.5	2.5	250	839	1	0.5	2.5	0.017	0.01	1367
355	0.4	0.14	150,373	0.25	48	320	4	44	5	6	250	21,752	5	1.5	23	0.033	0.01	13,915
180	0.1	0.045	2212	0.06	24	45	2	26	1.5	2.5	250	5036	2	1.15	9	0.026	0.01	6398
107	0.05	0.01	615	0.03	10	10	0.4	8	1.5	2.5	250	1888	1	0.84	2.5	0.014	0.01	3762
288	0.4	0.07	13,479	0.13	48	110	4.2	41	5	6	250	10,491	6	2	14	0.037	0.01	10,405
148	0.05	0.035	1923	0.055	15	35	0.9	16	1.5	2.5	250	4757	1.5	1.3	8	0.024	0.01	5341
250	0.1	0.05	3809	0.06	33	10	3.1	42	1.5	2.5	250	9862	5	0.83	6	0.024	0.01	9703
407	0.6	0.11	61,300	0.27	64	70	4.9	143	5	10	250	14,613	7	2	19	0.034	0.01	16,005
348	0.4	0.07	4331	0.07	46	10	4.7	49	1.5	6	250	11,191	6	1.7	15	0.026	0.01	14,138
305	0.3	0.08	5389	0.09	41	10	3.9	51	1.5	5	250	13,569	4	1	2.5	0.023	0.01	12,915
447	1.2	0.28	46,220	0.23	84	180	12.7	108	8	26	1500	33,083	15	1.7	20	0.087	0.04	19,075
404	0.65	0.15	9238	0.16	63	10	7.3	70.5	5.5	12	250	20,144	9.5	1.3	13.5	0.030	0.015	17,023
266	0.2	0.06	2516	0.04	29	10	2.5	31	1.5	2.5	250	6085	4	1	6	0.023	0.01	10,034
465	1	0.29	28,502	0.29	93	490	12.2	100	9	24	900	32,733	14	2.3	28	1.177	0.03	22,294
425	0.65	0.145	6822	0.17	63	40	7.85	75	5.5	12	250	20,423	10	1.5	14	0.032	0.01	18,077
195	0.2	0.05	2030	0.05	10	10	1.7	20	1.5	2.5	250	5176	2	1	12	0.019	0.01	7441
411	0.8	0.11	9591	0.14	63	140	6.8	68	5	10	250	19,934	10	1.9	17	0.032	0.01	17,675
349	0.45	0.08	5571	0.1	45.5	50	4.3	52.5	4	8	250	13,639	6.5	1.25	14	0.024	0.01	14,649
187	0.3	0.08	3066	0.1	10	10	4.7	38	1.5	10	250	12,870	4	1	2.5	0.021	0.01	6247
533	1.4	0.3	231,848	0.74	90	190	23.3	126	10	75	1600	43,295	16	1.9	21	0.074	0.05	23,594
442	0.9	0.22	10,778	0.2	72	10	12.2	94	7	21	700	31,125	13	1.5	9	0.033	0.03	17,641
250	0.4	0.09	3295	0.09	33	10	4.5	40	5	8	250	12,520	6	1.3	2.5	0.023	0.01	10,925
480	1.1	0.27	45,226	0.27	84	10	16.9	107	10	35	1400	36,720	16	1.9	15	0.037	0.03	22,182
429	0.8	0.225	6765	0.185	70	10	12.5	96.5	5.5	20	700	31,265	13	1.6	9	0.030	0.03	19,581
345	0.4	0.14	1987	0.1	46	10	4	69	1.5	8	250	17,975	8	1.3	2.5	0.042	0.01	13,742
554	0.9	0.36	14,087	0.65	82	10	13.7	101	8	24	600	34,902	17	1.9	14	0.056	0.03	21,351
383	0.7	0.215	3913	0.145	59.5	10	9.95	79	7	13.5	250	23,152	10	1.75	11	0.048	0.015	15,875
260	0.5	0.15	1794	0.3	39	10	7.4	52	4	10	250	17,346	6	1.3	2.5	0.049	0.01	9684
513	0.8	0.49	10,027	0.49	78	330	18.7	144	9	31	500	44,344	22	2.2	13	0.135	0.04	22,223
364	0.6	0.275	4474	0.38	51.5	25	12.3	107	6	17.5	250	29,062	13.5	1.65	8	0.077	0.015	14,337

XRF Sc	ICP Se	XRF Si	SS Sn	XRF Sr	XRF Ta	ICP Te	XRF Th	XRF Ti	ICP Tl	ICP U	XRF V	XRF W	XRF Y	SS Yb	XRF Zn	XRF Zr	$CaCl_2$[5] pH	XRF LOI[6]
1	0.05	250,828	0.57	10	2.5	0.01	2.5	432	0.01	0.05	5	2.5	6	0.86	7	107	3.63	1.46
5	0.5	451,079	4.8	317	2.5	0.06	9	2740	0.06	1.1	36	2.5	21	35	35	796	7.45	20.1
1	0.2	419,433	2.7	49	2.5	0.01	2.5	1289	0.04	0.3	19.5	2.5	15	2.45	18.5	270	5.51	3.52
1	0.05	404,662	2.3	17	2.5	0.01	2.5	498	0.01	0.05	2.5	2.5	7	1.1	10	122	4.29	1.65
4	0.2	446,591	4.2	58	2.5	0.03	2.5	2380	0.06	0.4	34	2.5	18	4.2	31	497	7.21	7.32
1	0.2	421,513	3.1	41	2.5	0.01	2.5	1019	0.03	0.2	16	2.5	11.5	2.9	16.5	212	4.97	3.72
4	0.2	311,782	2.3	51	2.5	0.01	2.5	2326	0.05	0.3	29	2.5	21	3	22	332	4.97	4.36
6	0.4	403,587	3.6	198	2.5	0.04	9	3561	0.09	0.6	51	2.5	30	3.8	45	665	7.41	13.5
5	0.3	389,751	3.5	71	2.5	0.01	6	3087	0.07	0.5	35	2.5	24	3.1	25	624	5.44	4.52
5	0.05	289,438	2.7	74	2.5	0.01	2.5	3393	0.07	0.3	36	2.5	25	1.2	29	219	4.85	5.96
12	0.6	375,681	4.6	131	2.5	0.07	13	4622	0.17	0.8	104	2.5	37	5.5	80	651	7.74	16.8
7.5	0.4	334,710	3.35	92.5	2.5	0.02	9.5	3960	0.10	0.5	63	2.5	32	3.05	45	525	7.01	10.3
3	0.05	286,961	2.2	51	2.5	0.01	2.5	2560	0.05	0.4	22	2.5	20	1.7	15	342	4.89	2.78
12	0.6	417,423	5.7	119	2.5	0.08	16	4988	0.17	0.7	103	6	39	5.5	81	748	7.47	15.3
8	0.35	344,737	3.65	88.5	2.5	0.01	10.5	4367	0.115	0.6	61.5	2.5	35	3.3	47	495	6.31	7.89
3	0.05	340,342	2.3	36	2.5	0.01	2.5	1625	0.03	0.3	18	2.5	12	2.1	14	406	4.87	2.97
7	0.6	427,473	3.5	96	2.5	0.08	12	3999	0.11	0.6	51	2.5	34	5.7	39	628	6.26	9.86
6	0.3	374,723	2.85	78.5	2.5	0.01	8.5	3388	0.07	0.5	41	2.5	25.5	3.15	29.5	553	5.90	6.94
4	0.1	112,513	2.7	51	2.5	0.01	2.5	1595	0.07	0.3	42	2.5	16	1.7	32	122	5.87	5.41
16	0.8	399,193	5.5	263	6	0.1	16	4934	0.30	1.2	122	6	39	9.3	129	614	7.95	34.6
11	0.4	297,946	3.7	115	2.5	0.03	9	4508	0.15	0.5	94	2.5	34	3.6	67	355	7.45	13.1
5	0.05	246,761	2.8	64	2.5	0.01	2.5	2410	0.08	0.3	41	2.5	18	2.3	30	215	7.08	4.55
13	0.5	398,445	3.9	165	2.5	0.05	14	4826	0.21	0.7	112	2.5	37	4.3	85	551	7.78	16.6
11.5	0.35	315,849	3.75	115	2.5	0.03	7.5	4559	0.165	0.55	94.5	2.5	35	3.4	68.5	348	7.41	9.49
7	0.4	283,362	3.6	82	2.5	0.01	6	3657	0.12	0.6	56	2.5	26	2.3	39	215	4.40	6.48
13	1	361,845	5.3	413	2.5	0.04	13	4898	0.17	1.1	122	2.5	37	5.4	80	552	6.89	12.9
8	0.7	344,479	4.6	93	2.5	0.035	11.5	4553	0.135	0.75	75	2.5	29.5	4.45	49	417	5.31	8.80
6	0.4	273,966	3.9	63	2.5	0.01	5	2548	0.10	0.4	58	2.5	20	2.7	50	176	4.37	6.50
14	0.9	381,804	20	171	2.5	0.05	13	7224	0.22	1.1	155	2.5	32	8.7	125	407	6.79	14.0
8.5	0.6	310,380	5.9	87.5	2.5	0.035	6	3975	0.155	0.55	100	2.5	23	3.4	87.5	193	6.13	11.8

[3] semiquantitative spectral analysis (Pivnichgeologiya NSRGC, Kyiv, Ukraine)

[4] flameless atomic absorption spectrometry (BGR Laboratories, Hannover, Germany)

[5] pH determined in $CaCl_2$ solution

[6] loss on ignition

Table 15.3: Element concentrations (mg/kg) in grazing land soil of Ukraine, according to geochemical landscape classes.

Landscape bioclimatic Zone	Geochemical landscape class	Soil type	Analysis:	ICP[1]	XRF[2]	XRF	ICP
			Element:	Ag	Al	As	Au
Forest (Ukrainian Polissia)	acidic: H^+ (47.1%)	sod-podzolic + podzolised sod soil (N=24)	*min*	0.001	2911	1.5	0.0001
			max	0.021	25,510	3	0.0011
			median	0.010	11,697	1.5	0.0001
	gleyed acidic: H^+-Fe^{2+} (26.5%); acidic in combination with gleyed acidic: H^+, H^+-Fe^{2+} (12.2%)	gleyed podzolised sod soil, gleyed sod-podzolic soil (N=14)	*min*	0.001	5663	1.5	0.0001
			max	0.035	27,151	5	0.0014
			median	0.012	11,961	1.5	0.0005
	acidic calcic: H^+-Ca^{2+} (4.7%); gleyed acidic in combination with calcic: H^+-Fe^{2+}, Ca^{2+} (3.7%); calcic in combination with acidic calcic: Ca^{2+}, H^+-Ca^{2+} (1.4%); acidic calcic in combination with gleyic: H^+-Ca^{2+}, Fe^{2+} (1%)	dark-grey podzolised + grey forest soil, dark-grey podzolised soil, carbonaceous podzolised sod soil, grey forest soil (N=8)	*min*	0.009	9050	1.5	0.0001
			max	0.031	35,989	14	0.0006
			median	0.016	26,304	1.5	0.0001
Forest steppe	calcic: Ca^{2+} (31.3%)	typical black topsoil (N=39)	*min*	0.012	25,616	1.5	0.0001
			max	0.047	61,340	12	0.0019
			median	0.026	42,287	4	0.0008
	acidic calcic: H^+-Ca^{2+} (43.4%); acidic calcic in combination with gleyic: H^+-Ca^{2+}, Fe^{2+} (3.4%); calcic in combination with acidic calcic: Ca^{2+}, H^+-Ca^{2+} (2.4%)	dark-grey podzolised + grey forest soil, grey forest soil, carbonaceous podzolised sod soil (N=53)	*min*	0.008	10,162	1.5	0.0001
			max	0.045	62,610	10	0.0028
			median	0.031	45,886	5	0.0006
	acidic gleyic in combination with calcic and natric: H^+-Fe^{2+}, Ca^{2+}, Na^+ (4.1%)	meadow chestnut + meadow black soil (N=5)	*min*	0.008	19,053	1.5	0.0001
			max	0.016	30,326	1.5	0.0004
			median	0.016	26,092	1.5	0.0003
Steppe	calcic: Ca^{2+} (76%)	typical black topsoil (N=77)	*min*	0.008	6351	1.5	0.0001
			max	0.074	82,034	70	0.0039
			median	0.038	54,566	8	0.0010
	calcic-natric: Ca^{2+}-Na^+ (9.8%)	solonised chestnut soil (N=13)	*min*	0.017	28,050	1.5	0.0001
			max	0.091	67,585	17	0.0019
			median	0.040	59,223	10	0.0008
	acidic gleyic in combination with calcic and natric: H^+-Fe^{2+}, Ca^{2+}, Na^+ (6.2%); gleyed acidic in combination with calcic: H^+-Fe^{2+}, Ca^{2+} (1.7%)	meadow chestnut + meadow black soil, dark-grey podzolised soil (N=15)	*min*	0.011	12,967	1.5	0.0001
			max	0.074	69,861	11	0.0085
			median	0.043	54,142	8	0.0010
Ciscarpathian forest-meadow	acidic in combination with gleyed acidic: H^+, H^+-Fe^{2+} (41.6%); gleyed acidic: H^+-Fe^{2+} (18%); acidic: H^+ (5.6%)	gleyed sod-podzolic soil, gleyed podzolised sod soil, sod-podzolic + podzolised sod soil (N=5)	*min*	0.029	39,906	5	0.0008
			max	0.062	56,048	9	0.0070
			median	0.034	47,633	8	0.0014
Carpathian Mountains	acidic brown: H^+-Al^{3+}-Fe^{3+} (88.3%); acidic brown in combination with gleyic: H^+-Al^{3+}-Fe^{3+}, Fe^{2+} (9.8%)	brown podzolised soil, surface-gleyed brown podzolic soil (N=7)	*min*	0.051	28,897	5	0.0005
			max	0.131	74,889	30	0.0023
			median	0.088	53,348	14	0.0016
Transcarpathian brown forest-meadow	acidic brown: H^+-Al^{3+}-Fe^{3+} (54%); acidic brown in combination with gleyic: H^+-Al^{3+}-Fe^{3+}, Fe^{2+} (21.5%)	brown podzolised soil, surface-gleyed brown podzolic soil (N=5)	*min*	0.022	50,808	5	0.0006
			max	0.064	65,945	13	0.0016
			median	0.030	54,883	9	0.0014
Crimean Mountains	acidic brown, carbonaceous in the upper horizons: H^+-Al^{3+}-Fe^{3+}, Ca^{2+} (91%)	mountain brown soil (N=3)	*min*	0.044	59,435	4	0.0012
			max	0.090	92,513	24	0.0017
			median	0.050	84,839	18	0.0014

[1] aqua regia extraction and determination by inductively coupled plasma mass spectrometry (ACME Laboratories, Vancouver, Canada)
[2] X-ray fluorescence (BGR Laboratories, Hannover, Germany)

ICP	XRF	ICP	ICP	XRF	ICP	XRF	XRF	ICP	XRF	XRF	XRF	XRF	XRF	XRF	SS[3]	XRF	AAS[4]
B	Ba	Be	Bi	Ca	Cd	Ce	Cl	Co	Cr	Cs	Cu	F	Fe	Ga	Ge	Hf	Hg
0.5	44	0.05	0.01	100	0.02	10	10	0.05	6	1.5	2.5	250	909	1	0.55	2.5	0.013
2	330	0.3	0.31	20,483	0.18	35	110	4.2	35	4	7	250	14,618	4	3.5	20	0.056
0.5	177	0.07	0.04	1694	0.06	10	30	1.1	19	1.5	2.5	250	4232	1.5	1.2	8	0.025
0.5	99	0.05	0.03	643	0.02	10	10	0.5	12	1.5	2.5	250	2098	1	0.39	2.5	0.015
3	374	0.3	0.09	8083	0.28	47	180	3.8	40	1.5	10	250	12,030	4	1.9	19	0.074
0.75	154	0.1	0.05	2065	0.07	15.5	40	1.05	18	1.5	2.5	250	3672	2	1.2	7	0.027
0.5	105	0.05	0.03	1051	0.04	24	10	0.6	18	1.5	2.5	250	2658	1	0.6	7	0.020
16	375	0.4	0.14	145,513	0.49	66	40	5.8	55	5	14	0.05	17,416	7	3.7	17	0.069
3	296	0.12	0.07	5910.5	0.10	33.5	10	3.6	38	1.5	6.5	250	11,086	5	1.1	11.5	0.029
2	242	0.2	0.06	3716	0.07	37	10	3.1	40	1.5	6	250	10,212	5	0.99	2.5	0.021
20	442	1.2	0.23	54,024	0.23	94	130	12.3	256	8	22	0.06	33,223	16	4.1	21	0.040
6	382	0.6	0.12	9856	0.13	58	10	6.3	64	5	10	250	18,955	8	1.4	14	0.029
2	181	0.1	0.04	579	0.07	10	10	2.7	24	1.5	2.5	250	4966	2	0.71	2.5	0.020
11	444	1.1	0.23	55,768	0.27	83	100	12.1	97	8	23	600	31,544	14	4.4	24	0.053
5	400	0.6	0.15	6983	0.15	63	10	7.4	70	5	11	250	20,144	9	1.5	14	0.032
6	182	0.2	0.08	5417	0.08	10	10	3.5	34	1.5	8	250	11,471	2	0.76	2.5	0.026
12	332	0.4	0.15	50,586	0.11	52	60	4	53	1.5	9	250	14,198	6	3.1	22	0.035
11	290	0.4	0.09	8962	0.10	48	30	3.7	44	1.5	8	250	12,310	5	1.1	14	0.031
0.5	115	0.05	0.03	307	0.02	10	10	0.7	12	1.5	2.5	250	3008	1	0.39	2.5	0.022
18	550	2.4	0.33	176,959	22.5	102	80	20	121	11	50	1000	46,722	20	2.1	17	0.118
8	423	0.8	0.21	14,644	0.18	64	10	11.3	88	6	18	250	28,257	12	1.3	9	0.031
6	253	0.4	0.08	6018	0.05	31	10	4.3	49	4	10	250	13,219	5	1.1	2.5	0.025
36	469	1.1	0.28	20,340	0.24	78	60	22.5	116	9	44	700	47,631	16	2	15	0.060
8	423	1	0.23	10,728	0.20	66	10	13	103	8	23	250	33,573	13	1.5	8	0.033
2	130	0.2	0.05	1429	0.04	10	10	2.5	25	1.5	6	250	6155	1	1.1	2.5	0.017
23	812	1	0.25	73,328	0.22	94	70	19.1	114	8	26	700	43,784	17	1.6	16	0.095
11	368	0.8	0.20	10,592	0.14	67	10	10.8	88	6	17	250	27,558	12	1.4	7	0.034
2	328	0.4	0.11	800	0.10	47	10	8.1	68	5	10	250	16,297	8	1.4	6	0.036
6	417	0.8	0.23	25,129	0.28	75	100	12.8	88	6	22	500	35,111	13	1.9	14	0.061
3	386	0.6	0.18	3588	0.21	69	10	11.7	76	5	11	250	21,263	10	1.6	10	0.041
3	193	0.3	0.14	2037	0.25	10	10	6.9	51	4	10	250	15,388	6	1.4	2.5	0.057
5	459	0.8	0.47	44,254	2.18	84	50	17.8	142	11	46	800	42,595	19	2.4	11	0.154
3	298	0.4	0.26	4024	0.42	57	10	11.8	109	7	17	250	30,495	11	1.7	7	0.080
0.5	272	0.5	0.21	1787	0.07	49	10	9.9	77	4	8	250	22,732	11	1.6	2.5	0.034
3	415	1.1	0.32	5432	0.29	77	130	14.1	98	7	17	250	33,363	16	4.5	20	0.080
1	387	0.7	0.23	3416	0.21	72	30	10.3	82	6	11	250	27,278	12	1.7	14	0.064
3	345	0.3	0.12	4510	0.07	35	10	7.9	27	4	22	250	30,635	11	1.4	5	0.032
6	574	1	0.40	74,186	0.10	94	170	17	124	11	57	800	47,212	21	1.9	11	0.348
4	352	0.9	0.39	9012	0.08	67	10	16.9	114	10	37	250	41,896	20	1.9	8	0.071

[3] semiquantitative spectral analysis (Pivnichgeologiya NSRGC, Kyiv, Ukraine)
[4] flameless atomic absorption spectrometry (BGR Laboratories, Hannover, Germany)

to be continued ▷

Table 15.3: continued

LBZ	Analysis:	ICP	XRF	XRF	ICP	XRF	XRF	SS	XRF	XRF	XRF	XRF	XRF	ICP	ICP	XRF	ICP	XRF
	Element:	**In**	**K**	**La**	**Li**	**Mg**	**Mn**	**Mo**	**Na**	**Nb**	**Ni**	**P**	**Pb**	**Pd**	**Pt**	**Rb**	**Re**	**S**
Z1	*min*	0.01	1284	10	0.3	121	31	0.34	149	2	1.5	124	1.5	0.005	0.0005	4	0.0005	20
	max	0.01	13,440	30	7.5	1447	558	5.9	4595	11	8	548	18	0.005	0.0024	52	0.0020	320
	median	0.01	5643	10	1.7	603	264	1.2	1596	5	3	338	8	0.005	0.0005	22.5	0.0005	20
	min	0.01	1916	10	0.8	302	54	0.26	372	3	1.5	197	5	0.005	0.0005	1	0.0005	20
	max	0.01	14,174	10	4.7	2231	728	2.3	4084	12	13	708	15	0.002	0.0015	48	0.0010	200
	median	0.01	5348	10	1.6	573	128	0.93	1615	5	3.5	403	9	0.005	0.0005	19.5	0.0005	20
	min	0.01	4184	10	0.9	422	116	0.67	1295	3	1.5	184	6	0.005	0.0005	14	0.0005	20
	max	0.01	16,139	27	6.4	4824	457	7	5271	11	28	1228	18	0.005	0.0011	65	0.0080	1720
	median	0.01	9629	16	3.9	2111	387	1.3	2779	9	9.5	583	11	0.005	0.0005	44	0.0005	70
Z2	*min*	0.01	8352	10	3.9	2050	310	0.56	2431	9	8	343	10	0.005	0.0005	39	0.0005	20
	max	0.03	18,778	38	20	10,554	720	2.3	6066	15	42	1514	20	0.0126	0.0042	96	0.0005	520
	median	0.01	16,617	20	9.3	4644	472	1.2	4954	13	18	620	15	0.0024	0.0010	68	0.0005	120
	min	0.01	5412	10	1	663	279	0.48	1344	7	3	194	8	0.005	0.0005	18	0.0005	20
	max	0.03	20,262	45	19.3	7779	898	2.6	7430	16	36	1293	37	0.0071	0.0032	96	0.0010	840
	median	0.01	17,159	28	10.6	4704	558	1.4	5354	14	20	608	17	0.005	0.0005	74	0.0005	80
	min	0.01	4798	10	3.5	2473	302	0.99	1132	7	11	487	9	0.005	0.0005	27	0.0005	80
	max	0.01	13,312	24	6.1	7900	379	1.4	4647	10	16	854	14	0.0062	0.0020	53	0.0005	440
	median	0.01	12,282	10	4.4	2653	379	1.19	3747	8	12	830	11	0.0048	0.0005	48	0.0005	320
Z3	*min*	0.01	2777	10	1.3	422	70	0.57	609	4	5	128	6	0.005	0.0005	14	0.0005	20
	max	0.04	21,223	45	30.3	12,423	1649	5.3	9454	17	57	1380	30	0.0135	0.0039	105	0.0010	680
	median	0.02	16,638	27	17.7	7538	689	1.6	5061	14	34	659	19	0.005	0.0010	80	0.0005	280
	min	0.01	11,322	10	7.5	3437	302	1.2	3160	10	15	388	14	0.005	0.0010	43	0.0005	80
	max	0.03	25,837	39	27	13,931	1820	2.1	8759	17	55	1185	58	0.005	0.0029	98	0.0010	1040
	median	0.03	19,757	22	20.6	8503	790	1.7	5675	15	42	732	20	0.005	0.0010	86	0.0005	120
	min	0.01	5148	10	3.7	1327	178	1.2	1343	6	7	199	9	0.005	0.0010	22	0.0005	20
	max	0.05	23,608	42	26	9046	999	3	11,405	19	46	1598	24	0.005	0.0030	91	0.002	680
	median	0.02	18,319	24	16.1	7719	635	1.8	5279	13	34	691	19	0.005	0.0010	80	0.0005	160
Z4	*min*	0.01	14,584	21	8	2895	743	1.4	4861	13	15	615	19	0.005	0.0005	66	0.0005	20
	max	0.03	18,142	37	13.5	5428	1262	3.5	6213	16	35	1305	27	0.005	0.0029	96	0.0030	400
	median	0.03	16,175	26	11.2	4342	1038	2	5786	15	20	715	21	0.005	0.0005	83	0.0010	20
Z5	*min*	0.01	6806	10	7.8	2533	573	1.7	2174	9	10	617	19	0.005	0.0005	44	0.0005	20
	max	0.04	22,666	44	25.8	21,107	3662	16	8151	19	68	2442	43	0.005	0.002	123	0.004	600
	median	0.02	10,989	27	14	5729	1138	2.7	5732	12	25	989	24	0.005	0.0005	66	0.0005	20
Z6	*min*	0.01	11,807	23	7.2	3257	627	0.88	4788	13	16	451	20	0.005	0.0005	59	0.0005	20
	max	0.03	15,088	37	27.6	6453	1254	1.3	8304	20	33	738	31	0.004	0.0005	89	0.0010	40
	median	0.01	12,879	35	13.4	4583	759	1	6435	18	23	585	26	0.005	0.0005	69	0.0005	20
Z7	*min*	0.01	10,629	10	9.2	8322	550	0.8	6433	6	12	577	12	0.005	0.0010	34	0.0005	20
	max	0.05	24,078	26	35.7	8624	906	2.4	11,759	16	68	778	29	0.005	0.0020	126	0.0005	1120
	median	0.04	19,919	24	16.1	8383	813	1.4	7549	15	64	723	29	0.005	0.0010	107	0.0005	400

ICP	XRF	ICP	XRF	SS	XRF	XRF	ICP	XRF	XRF	ICP	ICP	XRF	XRF	XRF	SS	XRF	XRF	$CaCl_2$[5]	XRF
Sb	Sc	Se	Si	Sn	Sr	Ta	Te	Th	Ti	Tl	U	V	W	Y	Yb	Zn	Zr	pH	LOI[6]
0.02	1	0.05	366,659	1.5	7	2.5	0.01	2.5	492	0.01	0.05	2.5	2.5	6	1	7	97	3.56	1.03
0.18	4	0.5	457,296	6.6	69	2.5	0.04	9	2944	0.06	0.6	26	2.5	23	7.6	28	738	7.18	17.9
0.07	1	0.15	427,824	2.7	31	2.5	0.01	2.5	1148	0.03	0.15	13.5	2.5	9	2.3	15	230	4.88	3.34
0.03	1	0.05	268,544	1.4	18	2.5	0.01	2.5	594	0.01	0.05	8	2.5	6	1.2	8	115	3.55	1.69
0.33	5	0.9	447,246	4.3	85	2.5	0.03	9	3327	0.06	0.8	39	2.5	22	14	26	666	6.24	37.5
0.08	1	0.2	423,149	2.7	39.5	2.5	0.01	2.5	1298	0.04	0.2	12.5	2.5	9	2.3	14	224	5.15	4.02
0.04	1	0.05	132,052	1.6	23	2.5	0.01	2.5	875	0.02	0.1	11	2.5	8	1.5	8	149	4.55	2.41
0.29	7	2.3	439,299	3.8	727	2.5	0.19	11	3321	0.08	1.5	45	2.5	27	4.2	54	621	7.47	41.2
0.13	3.5	0.25	373,110	2.7	73	2.5	0.01	7	2306	0.05	0.4	29.5	2.5	19	2.1	24.5	317	6.34	7.21
0.09	5	0.1	285,652	2.6	46	2.5	0.01	2.5	2818	0.06	0.3	30	2.5	20	1.8	28	229	4.95	4.46
0.32	12	0.7	398,165	5.3	165	6	0.09	15	4538	0.16	0.9	106	7	35	5.5	67	654	7.99	17.5
0.16	7	0.3	342,259	3.3	91	2.5	0.01	11	3771	0.10	0.5	54	2.5	29	3.1	42	503	7.04	10.2
0.06	3	0.05	263,495	1.8	27	2.5	0.01	2.5	1697	0.02	0.2	19	2.5	11	1.4	14	249	4.31	1.48
0.35	11	0.8	440,842	5.3	160	6	0.05	18	4772	0.18	0.7	91	2.5	39	5.5	77	714	7.42	17.5
0.20	8	0.3	340,810	3.6	82	2.5	0.01	11	4083	0.11	0.6	61	2.5	31	3.4	46	466	6.67	9.85
0.11	4	0.2	311,782	2.4	61	2.5	0.01	7	1643	0.05	0.3	31	2.5	12	1.7	26	169	6.27	7.78
0.11	7	0.7	381,477	3.1	244	2.5	0.04	11	3237	0.07	1.9	38	2.5	23	3.3	34	531	7.8	14.2
0.11	5	0.4	360,255	2.5	79	2.5	0.01	9	2830	0.05	0.5	38	2.5	20	2.6	31	485	6.71	14.2
0.09	1	0.05	161,173	1.2	17	2.5	0.01	2.5	671	0.04	0.05	13	2.5	6	1.1	15	81	5.04	2.53
11.1	17	0.6	444,581	5.6	575	6	0.08	18	5342	0.21	2.3	127	6	59	7.6	95	690	7.93	30.1
0.29	10	0.3	295,982	3.5	113	2.5	0.03	10	4274	0.14	0.5	83	2.5	32	3.4	62	361	7.52	13.4
0.14	5	0.1	257,699	2.7	59	2.5	0.01	2.5	2536	0.07	0.3	43	2.5	21	2.8	31	204	7	4.94
0.69	14	0.6	392,462	4.3	177	2.5	0.10	17	4760	0.22	0.8	120	2.5	37	4.4	84	657	8.01	16.6
0.33	11	0.3	296,917	3.5	116	2.5	0.02	9	4460	0.15	0.5	102	2.5	33	3.6	72	334	7.38	11.3
0.08	3	0.05	232,083	2.8	30	2.5	0.01	2.5	1553	0.04	0.2	21	2.5	13	1.7	16	190	7.2	2.94
0.36	16	1.2	430,558	4.2	227	2.5	0.05	15	5276	0.17	1	117	2.5	43	5.2	94	525	7.72	19.8
0.29	10	0.3	314,353	3.5	115	2.5	0.02	8	4244	0.13	0.5	84	2.5	31	3.8	63	312	7.47	12.0
0.18	6	0.3	316,223	3.2	58	2.5	0.02	12	4065	0.09	0.6	54	2.5	23	2.1	37	267	4.05	5.17
0.33	9	0.7	374,793	5.1	97	2.5	0.06	17	4712	0.16	1.1	95	2.5	35	4.9	81	506	7.19	11.5
0.24	8	0.5	332,302	4.2	88	2.5	0.04	14	4658	0.14	0.7	67	2.5	26	4	51	375	5.06	10.3
0.23	5	0.4	233,720	3.8	58	2.5	0.01	2.5	2410	0.09	0.4	56	2.5	14	2.4	48	140	4.75	6.62
1.26	14	1.5	379,934	7.5	304	2.5	0.09	18	5467	0.37	1.8	157	2.5	26	6	115	299	7.3	18.1
0.51	9	0.8	325,431	5.8	78	2.5	0.03	11	3387	0.15	0.6	118	2.5	23	4.2	80	207	5.53	12.3
0.26	8	0.2	297,104	4	64	2.5	0.02	11	4089	0.1	0.5	76	2.5	21	3.8	47	204	4.2	6.29
0.41	12	0.7	352,823	5.3	129	6	0.06	19	6403	0.26	1	105	2.5	37	5.6	80	532	6.24	14.1
0.40	9	0.4	346,793	4.9	96	2.5	0.05	12	5881	0.12	0.9	80	2.5	33	5.1	65	452	5.67	7.27
0.23	11	0.2	254,427	3.3	112	2.5	0.03	2.5	3057	0.07	0.2	109	2.5	20	3.3	56	101	6.88	9.2
0.39	16	0.4	282,801	4.6	426	2.5	0.10	11	5228	0.10	0.3	146	15	34	4.3	125	224	7.65	14.2
0.39	14	0.2	272,844	4.5	113	2.5	0.08	9	4808	0.08	0.2	138	2.5	33	3.3	104	192	7.61	11.1

[5] pH determined in $CaCl_2$ solution

[6] loss on ignition

Table 15.4: Geochemical characterisation of agricultural land soil (Ap-horizon) in Ukraine, according to geochemical landscape classes relative to the overall background (median) value of European agricultural soil. Notation: Elements highlighted in brown exhibit accumulation in agricultural soil ($K_c > 1.5$) and elements highlighted in green exhibit leaching ($K_c < 0.7$).

Landscape bioclimatic Zone	Geochemical landscape class	Soil type	Elements				
Forest (Ukrainian Polissia)	acidic: H^+ (47.1 %)	sod-podzolic + podzolised sod soil (N = 26)	Cl	Si	Hf	Zr	W
			2.3	1.3	1.1	1.0	1.0
	gleyed acidic: H^+ - Fe^{2+} (26.5 %); acidic in combination with gleyed acidic: H^+, H^+ - Fe^{2+} (12.2 %)	gleyed podzolised sod soil, gleyed sod-podzolic soil (N = 12)	Cl	Si	F	Hf	In
			1.8	1.3	1.0	1.0	1.0
	acidic calcic: H^+ - Ca^{2+} (4.7 %); gleyed acidic in combination with calcic: H^+ - Fe^{2+}, Ca^{2+} (3.7 %); calcic in combination with acidic calcic: Ca^{2+}, H^+ - Ca^{2+} (1.4 %); acidic calcic in combination with gleyic: H^+ - Ca^{2+}, Fe^{2+} (1 %)	dark-grey podzolised + grey forest soil, dark-grey podzolised soil, carbonaceous podzolised sod soil, grey forest soil (N = 7)	Zr	Hf	Si	Ge	B
			2.4	1.9	1.2	1.2	1.2
Forest steppe	calcic: Ca^{2+} (31.3 %)	typical black topsoil (N = 32)	S	Te	Zr	B	Hf
			2.0	2.0	2.0	2.0	1.7
	acidic calcic: H^+ - Ca^{2+} (43.4 %); acidic calcic in combination with gleyic: H^+ - Ca^{2+}, Fe^{2+} (3.4 %); calcic in combination with acidic calcic: Ca^{2+}, H^+ - Ca^{2+} (2.4 %)	dark-grey podzolised + grey forest soil, grey forest soil, carbonaceous podzolised sod soil (N = 52)	Cl	S	B	Zr	Hf
			2.0	2.0	2.0	1.9	1.8
	acidic gleyic in combination with calcic and natric: H^+ - Fe^{2+}, Ca^{2+}, Na^+ (4.1 %)	meadow chestnut + meadow black soil (N = 6)	Cl	Zr	B	Hf	Si
			2.5	2.1	1.8	1.8	1.2
Steppe	calcic: Ca^{2+} (76 %)	typical black topsoil (N = 87)	S	In	Te	F	B
			4.0	3.0	3.0	2.8	2.7
	calcic-natric: Ca^{2+} - Na^+ (9.8 %); acidic gleyic in combination with calcic and natric: H^+ - Fe^{2+}, Ca^{2+}, Na^+ (6.2 %);	solonised chestnut soil; meadow chestnut + meadow black soil (N = 12)	B	In	Te	F	S
			3.7	3.0	3.0	2.8	2.0
Ciscarpathian forest-meadow	acidic in combination with gleyed acidic: H^+, H^+ - Fe^{2+} (41.6 %)	gleyed sod-podzolic soil (N = 6)	Te	Se	Au	Hg	Zr
			3.5	2.0	1.7	1.6	1.6
Carpathian Mountains	acidic brown: H^+ - Al^{3+} - Fe^{3+} (88.3 %)	brown podzolised soil (N = 8)	Te	Hg	S	Cd	Sb
			3.5	2.6	2.5	2.1	1.9

Landscape bioclimatic Zone	Elements															
Forest (Ukrainian Polissia)	Ba	Mn	La	Ce	Cr	Pb	U	K	Na	B	Ti	Tl	Cd	Ag	Rb	Cs
	0.5	0.4	0.4	0.4	0.4	0.4	0.4	0.4	0.4	0.4	0.3	0.3	0.3	0.3	0.3	0.3
	Y	Sr	B	Ba	K	Cd	Cs	Rb	Mn	Th	Ti	Ag	Na	U	Sb	Zn
	0.4	0.4	0.4	0.4	0.3	0.3	0.3	0.3	0.3	0.3	0.3	0.3	0.3	0.3	0.3	0.3
	Na	Nb	Cr	Be	Mn	Sr	U	Th	Rb	P	Sc	Co	Ni	Tl	Pb	Al
	0.8	0.8	0.8	0.7	0.7	0.7	0.7	0.7	0.6	0.6	0.6	0.6	0.6	0.6	0.6	0.5
Forest steppe	Pd	Pt	Re	Ta	W	Sn	Co	Sc	Ge	Rb	Sr	Na	V	Yb	Li	Mn
	1.0	1.0	1.0	1.0	1.0	1.0	0.9	0.9	0.9	0.9	0.9	0.9	0.9	0.9	0.9	0.9
	Rb	F	In	Pd	Pt	Re	Sc	Ta	Te	W	Na	Se	Yb	Tl	Cd	Mo
	1.0	1.0	1.0	1.0	1.0	1.0	1.0	1.0	1.0	1.0	1.0	1.0	1.0	1.0	0.9	0.9
	Se	Be	Cr	Sn	Cs	Hg	Na	Sr	Ce	Sc	Rb	U	Mo	Ca	P	Al
	0.8	0.8	0.8	0.8	0.8	0.8	0.8	0.8	0.8	0.8	0.7	0.7	0.7	0.6	0.6	0.6
Steppe	Ti	Ce	La	Y	Ag	Rb	Nb	Sr	Mo	Ba	Hf	Se	K	Hg	Cd	Ga
	1.2	1.2	1.2	1.2	1.2	1.2	1.2	1.1	1.1	1.1	1.1	1.1	1.1	1.1	1.1	1.1
	K	Ce	Nb	Rb	Ge	Sr	Mo	Hf	Zn	Cs	Ba	Al	Sb	La	Ga	Sn
	1.2	1.2	1.2	1.2	1.1	1.1	1.1	1.1	1.1	1.1	1.1	1.1	1.1	1.1	1.1	1.1
Ciscarpathian forest-meadow	La	Tl	Sb	Si	V	Rb	P	Y	Pb	Mo	Ce	U	F	Pd	Pt	Re
	1.2	1.1	1.1	1.1	1.1	1.1	1.1	1.1	1.0	1.0	1.0	1.0	1.0	1.0	1.0	1.0
Carpathian Mountains	Cl	Cu	As	Cs	Ge	Fe	Na	Ga	Be	Ti	Sc	Al	Nb	Mg	F	Hf
	1.3	1.3	1.2	1.2	1.2	1.2	1.1	1.1	1.1	1.1	1.1	1.0	1.0	1.0	1.0	1.0

Note: Table shows concentration ratios of chemical elements relative to their overall background contents (medians) in European soil.

Elements																	
Te	Ta	Re	Pt	Pd	In	F	Hg	Ge	Mo	Sn	Yb	Se	Y	P	S	Sr	Nb
1.0	1.0	1.0	1.0	1.0	1.0	1.0	0.9	0.8	0.8	0.8	0.7	0.6	0.5	0.5	0.5	0.5	0.5
Pd	Pt	Re	Ta	Te	W	Mo	Ge	Sn	Yb	Zr	Hg	Se	S	La	P	Pb	Nb
1.0	1.0	1.0	1.0	1.0	1.0	0.9	0.9	0.9	0.9	0.8	0.8	0.6	0.5	0.4	0.4	0.4	0.4
F	In	Pd	Pt	Re	Sn	Ta	Te	W	Yb	Ba	K	Hg	Y	Se	Ti	Mo	Ce
1.0	1.0	1.0	1.0	1.0	1.0	1.0	1.0	1.0	0.9	0.9	0.9	0.9	0.9	0.8	0.8	0.8	0.8
In	Be	Y	Ni	Se	La	Cr	Cs	Ti	Ca	Si	Ce	K	Th	Ba	F	Hg	Nb
1.5	1.2	1.1	1.1	1.1	1.1	1.1	1.1	1.1	1.1	1.1	1.1	1.1	1.1	1.0	1.0	1.0	1.0
Y	Be	Ti	Cr	Th	Nb	La	K	Si	Cs	Ba	Ge	Ce	Hg	Ni	Sn	Mn	Co
1.3	1.2	1.2	1.2	1.2	1.2	1.1	1.1	1.1	1.1	1.1	1.1	1.1	1.1	1.1	1.0	1.0	1.0
F	In	Pd	Pt	Re	S	Ta	Te	W	Th	Yb	Ti	K	Y	Ge	Ba	La	Nb
1.0	1.0	1.0	1.0	1.0	1.0	1.0	1.0	1.0	0.9	0.9	0.9	0.9	0.9	0.9	0.9	0.9	0.9
Ni	Be	Li	Co	Cu	Cr	Mg	Cs	Sc	Zr	V	Sb	Bi	As	Tl	Ca	Fe	Mn
2.0	1.7	1.7	1.6	1.5	1.5	1.4	1.4	1.4	1.3	1.3	1.3	1.3	1.3	1.3	1.2	1.2	1.2
Ni	Li	Co	Cr	Ag	Be	Sc	Mg	Cu	Tl	As	V	Bi	Zr	Mn	Fe	Y	Ti
2.0	1.6	1.6	1.5	1.5	1.5	1.4	1.4	1.4	1.4	1.4	1.4	1.3	1.3	1.3	1.3	1.3	1.2
B	In	Cs	Hf	Sn	Yb	Ag	Be	Co	Th	Bi	Ge	Ni	Cr	Ti	As	Mn	Nb
1.6	1.5	1.4	1.4	1.3	1.3	1.3	1.3	1.3	1.3	1.3	1.3	1.3	1.2	1.2	1.2	1.2	1.2
Ag	Sn	Se	Cr	Bi	Li	Mo	Mn	Co	B	In	Pb	V	Zn	Au	Tl	Ni	P
1.9	1.7	1.7	1.7	1.6	1.6	1.6	1.6	1.6	1.6	1.5	1.5	1.4	1.4	1.4	1.3	1.3	1.3

Elements																	
Zn	Sb	V	Th	Bi	Co	Ca	Al	Au	As	Fe	Ni	Li	Be	Cu	Ga	Mg	Sc
0.3	0.3	0.3	0.3	0.3	0.3	0.3	0.2	0.2	0.2	0.2	0.2	0.2	0.2	0.2	0.2	0.2	0.1
Ce	Tl	Cr	V	Ca	As	Al	Bi	Ni	Fe	Cu	Ga	Sc	Co	Au	Mg	Li	Be
0.3	0.3	0.3	0.2	0.2	0.2	0.2	0.2	0.2	0.2	0.2	0.1	0.1	0.1	0.1	0.1	0.1	0.1
Ca	Cl	Ga	S	V	Ag	Fe	La	As	Cu	Bi	Zn	Mg	Cd	Li	Sb	Cs	Au
0.5	0.5	0.5	0.5	0.5	0.5	0.4	0.4	0.4	0.4	0.4	0.4	0.4	0.4	0.4	0.3	0.3	0.1
Bi	Cd	Mg	P	Mo	Cu	Tl	Al	Ag	Fe	Ga	Pb	Sb	Zn	As	Au	U	Cl
0.9	0.9	0.9	0.9	0.9	0.9	0.8	0.8	0.8	0.8	0.8	0.8	0.8	0.7	0.7	0.7	0.7	0.5
Ag	Pb	P	V	Sr	Cu	Bi	Al	Ga	Fe	Li	U	Ca	Mg	Zn	As	Sb	Au
0.9	0.9	0.9	0.9	0.9	0.9	0.9	0.8	0.8	0.8	0.8	0.8	0.8	0.8	0.8	0.7	0.7	0.6
Pb	Tl	V	Mn	Ni	Cu	Li	Co	Cd	Fe	Ga	Mg	As	Zn	Bi	Ag	Sb	Au
0.6	0.6	0.6	0.6	0.6	0.6	0.6	0.6	0.5	0.5	0.5	0.5	0.5	0.5	0.5	0.5	0.4	0.2
Zn	Au	Al	Ge	Yb	Sn	Pd	Pt	Re	Ta	Th	W	Si	Na	Pb	P	U	Cl
1.1	1.1	1.1	1.1	1.1	1.1	1.0	1.0	1.0	1.0	1.0	1.0	1.0	0.9	0.9	0.9	0.7	0.5
Na	Hg	Cd	Si	Pd	Pt	Re	Ta	W	Yb	Se	Pb	P	Th	Ca	Au	U	Cl
1.0	1.0	1.0	1.0	1.0	1.0	1.0	1.0	1.0	1.0	1.0	0.9	0.9	0.8	0.8	0.8	0.7	0.5
Sc	Ta	W	K	Sr	Na	Ba	Cu	Li	Fe	Al	Ga	Cd	Zn	Mg	S	Cl	Ca
1.0	1.0	1.0	1.0	1.0	1.0	1.0	1.0	0.9	0.9	0.9	0.8	0.8	0.8	0.8	0.8	0.5	0.5
Pd	Pt	Re	Ta	W	Yb	Si	Rb	Ba	K	Ce	Sr	Y	La	U	Zr	Th	Ca
1.0	1.0	1.0	1.0	1.0	1.0	1.0	1.0	0.9	0.9	0.9	0.9	0.8	0.8	0.7	0.7	0.7	0.5

$K_c > 1.5$ – accumulating elements

$K_c < 0.7$ – leached elements

Table 15.5: Geochemical characterisation of grazing land soil in Ukraine, according to geochemical landscape classes relative to the overall background (median) value of European grazing land soil. Notation: Elements highlighted in brown exhibit accumulation in agricultural soil ($K_c > 1.5$) and elements highlighted in green exhibit leaching ($K_c < 0.7$).

Landscape bioclimatic Zone	Geochemical landscape class	Soil type	Elements				
Forest (Ukrainian Polissia)	acidic: H^+ (47.1 %)	sod – podzolic + podzolised sod soil (N = 24)	Cl	Si	F	Hf	In
			3.0	1.4	1.0	1.0	1.0
	gleyed acidic: $H^+ - Fe^{2+}$ (26.5 %); acidic in combination with gleyed acidic: H^+, $H^+ - Fe^{2+}$ (12.2 %)	gleyed podzolised sod soil, gleyed sod-podzolic soil (N = 14)	Cl	Si	F	In	Re
			4.0	1.4	1.0	1.0	1.0
	acidic calcic: $H^+ - Ca^{2+}$ (4.7 %); gleyed acidic in combination with calcic: $H^+ - Fe^{2+}$, Ca^{3+} (3.7 %); calcic in combination with acidic calcic: Ca^{2+}, $H^+ - Ca^{2+}$ (1.4 %); acidic calcic in combination with gleyic: $H^+ - Ca^{2+}$, Fe^{2+} (1 %)	dark-grey podzolised + grey forest soil, dark-grey podzolised soil, carbonaceous podzolised sod soil, grey forest soil (N = 8)	S	Hf	Zr	Si	B
			1.8	1.4	1.3	1.2	1.1
Forest steppe	calcic: Ca^{2+} (31.3 %)	typical black topsoil (N = 39)	S	B	Zr	Hf	Ca
			3.0	2.2	2.0	1.8	1.2
	acidic calcic: $H^+ - Ca^{2+}$ (43.4 %); acidic calcic in combination with gleyic: $H^+ - Ca^{2+}$, Fe^{2+} (3.4 %); calcic in combination with acidic calcic: Ca^{2+}, $H^+ - Ca^{2+}$ (2.4 %)	dark – grey podzolised + grey forest soil, grey forest soil, carbonaceous podzolised sod soil (N = 53)	S	Zr	B	Hf	Y
			2.0	1.9	1.8	1.8	1.2
	acidic gleyic in combination with calcic and natric: $H^+ - Fe^{2+}$, Ca^{2+}, Na^+ (4.1 %)	meadow chestnut + meadow black soil (N = 5)	S	B	Cl	Zr	Hf
			8.0	4.0	3.0	1.9	1.8
Steppe	calcic: Ca^{2+} (76 %)	typical black topsoil (N = 77)	S	B	In	Ca	Ni
			7.0	2.9	2.0	1.8	1.7
	calcic – natric: $Ca^{2+} - Na^+$ (9.8 %)	solonised chestnut soil (N = 13)	In	S	B	Ni	Be
			3.0	3.0	2.9	2.1	1.9
	acidic gleyic in combination with calcic and natric: $H^+ - Fe^{2+}$, Ca^{2+}, Na^+ (6.2 %); gleyed acidic in combination with calcic: $H^+ - Fe^{2+}$, Ca^{2+} (1.7 %)	meadow chestnut + meadow black soil, dark-grey podzolised soil (N = 15)	B	S	In	Ni	Be
			4.0	4.0	2.0	1.7	1.5
Ciscarpathian forest-meadow	acidic in combination with gleyed acidic: H^+, $H^+ - Fe^{2+}$ (41.6 %); gleyed acidic: $H^+ - Fe^{2+}$ (18 %); acidic: H^+ (5.6 %)	gleyed sod – podzolic soil, gleyed podzolised sod soil, sod – podzolic + podzolised sod soil (N = 5)	In	Re	Te	Mn	Co
			3.0	2.0	2.0	1.8	1.6
Carpathian Mountains	acidic brown: $H^+ - Al^{3+} - Fe^{3+}$ (88.3 %); acidic brown in combination with gleyic: $H^+ - Al^{3+} - Fe^{3+}$, Fe^{2+} (9.8 %)	brown podzolised soil, surface-gleyed brown podzolic soil (N = 7)	Hg	Ag	Cd	Se	As
			2.6	2.3	2.1	2.0	2.0
Transcarpathian brown forest-meadow	acidic brown: $H^+ - Al^{3+} - Fe^{3+}$ (54 %); acidic brown in combination with gleyic: $H^+ - Al^{3+} - Fe^{3+}$, Fe^{2+} (21.5 %)	brown podzolised soil, surface-gleyed brown podzolic soil (N = 5)	Cl	Te	Hg	Zr	Hf
			3.0	2.5	2.1	1.8	1.8
Crimean Mountains	acidic brown, carbonaceous in the upper horizons: $H^+ - Al^{3+} - Fe^{3+}$, Ca^{2+} (91 %)	mountain brown soil (N = 3)	S	In	Te	Ni	Cu
			10.0	4.0	3.9	3.2	2.8

Landscape bioclimatic Zone	Elements															
Forest (Ukrainian Polissia)	Se	Pb	Y	Sr	Ti	Na	Cd	Rb	Cr	Cs	Tl	Sb	Zn	Ag	Th	Al
	0.4	0.4	0.4	0.3	0.3	0.3	0.3	0.3	0.3	0.3	0.3	0.3	0.2	0.2	0.2	0.2
	Pb	Nb	Y	K	Ti	Cd	Tl	Na	Ag	Bi	Cs	Sb	Cr	U	B	Ce
	0.4	0.4	0.4	0.4	0.4	0.4	0.3	0.3	0.3	0.3	0.3	0.3	0.3	0.3	0.3	0.3
	Yb	K	Ti	Th	Se	Cr	Rb	Ce	U	Na	Cd	Al	Cu	Pb	Pt	Te
	0.7	0.6	0.6	0.6	0.6	0.6	0.6	0.6	0.6	0.5	0.5	0.5	0.5	0.5	0.5	0.5
Forest steppe	W	Sr	Sn	Na	Yb	Hg	Rb	Ni	Sc	Tl	Mo	Mg	Co	La	P	Mn
	1.0	1.0	1.0	1.0	0.9	0.9	0.9	0.9	0.9	0.9	0.9	0.9	0.8	0.8	0.8	0.8
	Mo	Ni	Rb	Re	Sc	Ta	Th	W	Co	Mn	Tl	Li	Sr	V	Al	Ca
	1.0	1.0	1.0	1.0	1.0	1.0	1.0	1.0	1.0	1.0	0.9	0.9	0.9	0.9	0.9	0.9
	Yb	Ge	Ti	Be	Sn	Na	U	Cr	Mn	Rb	Sc	Cu	Nb	Ni	V	Zn
	0.8	0.8	0.8	0.8	0.7	0.7	0.7	0.7	0.7	0.6	0.6	0.6	0.6	0.6	0.6	0.5
Steppe	Bi	Mo	As	Hf	La	Ce	K	Rb	Nb	Al	Sb	Yb	Sn	Zn	Au	Cl
	1.2	1.1	1.1	1.1	1.1	1.1	1.1	1.1	1.1	1.0	1.0	1.0	1.0	1.0	1.0	1.0
	Ti	Mo	Sb	Zn	Ba	Rb	Ce	Nb	Al	Na	Yb	Ga	Ge	Hg	Sn	Ag
	1.2	1.2	1.2	1.2	1.2	1.2	1.2	1.2	1.1	1.1	1.1	1.1	1.1	1.1	1.0	1.0
	Fe	Tl	Ag	Bi	Hg	Mn	Rb	Sb	Al	Si	Zn	Sn	Ba	Na	Au	Cl
	1.1	1.1	1.1	1.1	1.1	1.1	1.1	1.0	1.0	1.0	1.0	1.0	1.0	1.0	1.0	1.0
Ciscarpathian forest-meadow	Na	Si	B	K	La	Ba	Cd	Y	Cl	Cs	F	Ni	Sc	Ta	W	Bi
	1.1	1.1	1.1	1.1	1.1	1.1	1.1	1.0	1.0	1.0	1.0	1.0	1.0	1.0	1.0	1.0
Carpathian Mountains	Ni	Ge	Li	La	Sc	Na	B	Pb	Si	Mg	Al	Ce	Cl	F	Re	Ta
	1.3	1.2	1.2	1.1	1.1	1.1	1.1	1.1	1.1	1.1	1.0	1.0	1.0	1.0	1.0	1.0
Transcarpathian brown forest-meadow	Cs	Pb	V	Ni	Li	Si	Sc	Fe	Th	Ba	Cd	Zn	Al	Sr	Tl	Se
	1.2	1.2	1.2	1.2	1.1	1.1	1.1	1.1	1.1	1.1	1.1	1.1	1.1	1.0	1.0	1.0
Crimean Mountains	Mn	Sb	Li	Ge	K	Ti	Sn	Y	Pb	Ag	Sr	Ce	Nb	Ca	Cl	F
	1.4	1.4	1.4	1.4	1.3	1.3	1.3	1.3	1.3	1.3	1.2	1.2	1.2	1.1	1.0	1.0

Note: Table shows concentration ratios of chemical elements relative to their overall background contents (medians) in European soil.

Element																	
Re	Ta	W	Zr	Ge	Mo	Hg	Sn	Yb	Pt	S	Ba	Te	Mn	P	La	Nb	K
1.0	1.0	1.0	0.9	0.9	0.9	0.8	0.8	0.7	0.5	0.5	0.5	0.5	0.5	0.4	0.4	0.4	0.4
Ta	W	Zr	Hf	Ge	Hg	Sn	Yb	Mo	P	Se	Pt	S	Te	Au	Sr	Ba	La
1.0	1.0	0.9	0.9	0.9	0.9	0.8	0.7	0.7	0.5	0.5	0.5	0.5	0.5	0.5	0.4	0.4	0.4
Cl	F	In	Re	Ta	W	Hg	Mo	Ba	Sn	Sr	Ge	P	Y	Ca	Nb	La	Mn
1.0	1.0	1.0	1.0	1.0	1.0	0.9	0.9	0.8	0.8	0.8	0.8	0.8	0.8	0.7	0.7	0.7	0.7
Y	Be	Si	K	Ba	Ti	Ce	Cr	Cl	Cs	F	Ge	In	Nb	Pt	Re	Ta	Th
1.2	1.1	1.1	1.1	1.1	1.0	1.0	1.0	1.0	1.0	1.0	1.0	1.0	1.0	1.0	1.0	1.0	1.0
La	K	Be	Ti	Si	Ba	Cr	Ce	Nb	Ge	Sn	Hg	Na	Yb	Cl	Cs	F	In
1.2	1.2	1.1	1.1	1.1	1.1	1.1	1.1	1.1	1.1	1.1	1.0	1.0	1.0	1.0	1.0	1.0	1.0
Si	Ca	P	Se	F	Hg	In	Re	Ta	W	Pd	Sr	Mo	Ce	K	Th	Ba	Y
1.2	1.1	1.1	1.0	1.0	1.0	1.0	1.0	1.0	1.0	1.0	0.9	0.9	0.8	0.8	0.8	0.8	0.8
Be	Li	Co	Te	Zr	Cr	Mg	Cu	Y	Sc	Sr	Tl	V	Cs	Mn	Ti	Ba	Fe
1.5	1.5	1.5	1.5	1.4	1.4	1.4	1.4	1.3	1.3	1.2	1.2	1.2	1.2	1.2	1.2	1.2	1.2
Cu	Li	Co	Cr	Cs	Mg	V	As	Fe	Sc	Mn	Ca	Zr	K	Y	Tl	Bi	Sr
1.8	1.8	1.7	1.6	1.6	1.6	1.5	1.4	1.4	1.4	1.4	1.3	1.3	1.3	1.3	1.3	1.3	1.3
Co	Mg	Cr	Li	Ca	Cu	Mo	Zr	Sc	Sr	Y	K	V	Cs	Ce	Ti	Yb	As
1.4	1.4	1.4	1.4	1.3	1.3	1.3	1.3	1.3	1.3	1.2	1.2	1.2	1.2	1.2	1.2	1.2	1.1
Zr	Mo	Au	Hg	Ti	Th	Se	Hf	Sn	Yb	Ce	Tl	Cr	Nb	Ge	As	Be	Rb
1.5	1.4	1.4	1.3	1.3	1.3	1.3	1.3	1.2	1.2	1.2	1.2	1.2	1.2	1.1	1.1	1.1	1.1
In	Mn	Mo	Sb	Cr	V	Sn	Au	Co	Te	Bi	Cs	Zn	Cu	P	Tl	Yb	Fe
2.0	2.0	1.9	1.8	1.7	1.7	1.7	1.6	1.6	1.5	1.4	1.4	1.3	1.3	1.3	1.3	1.3	1.3
Ti	Yb	La	Sn	Sb	Au	Nb	Co	Be	Y	Mn	Cr	As	U	Bi	Ce	Na	Ge
1.6	1.5	1.5	1.4	1.4	1.4	1.4	1.4	1.3	1.3	1.3	1.3	1.3	1.3	1.3	1.3	1.2	1.2
As	Hg	Co	Bi	Cs	V	Cr	Sc	Fe	Zn	Be	Ga	Al	Mg	B	Na	Rb	Au
2.6	2.3	2.2	2.1	2.0	2.0	1.8	1.8	1.7	1.7	1.7	1.7	1.6	1.5	1.5	1.5	1.4	1.4

Elements																	
Bi	As	Ca	U	V	Cu	B	Ce	Fe	Ni	Co	Li	Be	Ga	Sc	Mg	Au	Pd
0.2	0.2	0.2	0.2	0.2	0.2	0.2	0.2	0.2	0.2	0.1	0.1	0.1	0.1	0.1	0.1	0.1	0.1
Rb	Ca	Al	Zn	Th	Mn	As	Cu	Be	V	Ni	Ga	Fe	Co	Li	Sc	Mg	Pd
0.3	0.3	0.2	0.2	0.2	0.2	0.2	0.2	0.2	0.2	0.2	0.2	0.2	0.1	0.1	0.1	0.1	0.1
Co	Ni	Sb	Fe	Sc	Tl	V	Ga	Ag	Zn	Mg	Bi	Li	Cs	Be	As	Au	Pd
0.5	0.5	0.5	0.5	0.4	0.4	0.4	0.4	0.4	0.4	0.4	0.4	0.3	0.3	0.2	0.2	0.1	0.1
Al	Au	Li	V	Fe	Cu	Se	U	Zn	Pb	Ga	Ag	Cd	Bi	Sb	As	Te	Pd
0.8	0.8	0.8	0.8	0.8	0.8	0.8	0.7	0.7	0.7	0.7	0.7	0.7	0.7	0.6	0.6	0.5	0.5
Mg	Cu	U	Fe	Bi	P	Ag	Pb	Cd	Zn	Se	Ga	Sb	As	Au	Pt	Te	Pd
0.9	0.8	0.8	0.8	0.8	0.8	0.8	0.8	0.8	0.8	0.8	0.8	0.7	0.7	0.6	0.5	0.5	0.1
Cd	Fe	Al	Pb	Pt	Bi	Co	Te	Mg	Tl	Ga	La	Ag	Sb	Li	Au	Cs	As
0.5	0.5	0.5	0.5	0.5	0.5	0.5	0.5	0.5	0.4	0.4	0.4	0.4	0.4	0.4	0.3	0.3	0.2
F	Ga	Hg	Pd	Pt	Re	Ta	W	Si	Na	Ag	Ge	Cd	Th	P	Pb	Se	U
1.0	1.0	1.0	1.0	1.0	1.0	1.0	1.0	1.0	1.0	1.0	0.9	0.9	0.9	0.9	0.9	0.8	0.7
Cd	Cl	F	Hf	Pd	Pt	Re	Ta	W	Te	Si	P	La	Pb	Th	Au	Se	U
1.0	1.0	1.0	1.0	1.0	1.0	1.0	1.0	1.0	1.0	1.0	1.0	0.9	0.9	0.8	0.8	0.8	0.7
F	Ga	Ge	La	Nb	Pd	Pt	Re	Ta	W	Te	P	Hf	Pb	Se	Th	Cd	U
1.0	1.0	1.0	1.0	1.0	1.0	1.0	1.0	1.0	1.0	1.0	0.9	0.9	0.9	0.8	0.7	0.7	0.7
U	V	Li	Sr	Pb	P	Al	Fe	Ag	Sb	Cu	Zn	Ga	Mg	Pt	S	Ca	Pd
1.0	1.0	1.0	1.0	1.0	0.9	0.9	0.9	0.9	0.9	0.8	0.8	0.8	0.8	0.5	0.5	0.4	0.1
Th	W	Ti	Nb	Y	Ga	Rb	Hf	Sr	U	Zr	Ba	Be	K	Ca	Pt	S	Pd
1.0	1.0	0.9	0.9	0.9	0.9	0.9	0.9	0.8	0.8	0.8	0.8	0.8	0.7	0.5	0.5	0.5	0.1
F	Ga	In	Re	Ta	W	Rb	K	Cu	Mg	P	Ag	Mo	Pt	S	Ca	B	Pd
1.0	1.0	1.0	1.0	1.0	1.0	0.9	0.9	0.8	0.8	0.8	0.8	0.7	0.5	0.5	0.4	0.4	0.1
Hf	La	Mo	Pd	Pt	Re	Ta	W	Yb	Ba	P	Si	Th	Zr	Tl	Se	Cd	U
1.0	1.0	1.0	1.0	1.0	1.0	1.0	1.0	1.0	1.0	1.0	0.9	0.8	0.8	0.7	0.5	0.4	0.3

K_c > 1.5 – accumulating elements

K_c < 0.7 – leached elements

Table 15.6: Geochemical characterisation of agricultural land soil (Ap-horizon) relative to grazing land soil in Ukraine, according to geochemical landscape classes. Notation: Elements highlighted in brown exhibit accumulation in agricultural soil ($K_c > 1.5$) and elements highlighted in green exhibit leaching ($K_c < 0.7$).

Landscape bioclimatic Zone	Geochemical landscape class	Soil type	Elements				
Forest (Ukrainian Polissia)	acidic: H^+ (47.1 %)	sod - podzolic + podzolised sod soil (N = 26)	Ce	Au	B	Pt	U
			2.4	2.0	2.0	2.0	2.0
	gleyed acidic: $H^+ - Fe^{2+}$ (26.5 %); acidic in combination with gleyed acidic: H^+, $H^+ - Fe^{2+}$ (12.2 %)	gleyed podzolised sod soil, gleyed sod-podzolic soil (N = 12)	Pt	Mo	Mn	B	Fe
			2.0	1.5	1.4	1.3	1.3
	acidic calcic: $H^+ - Ca^{2+}$ (4.7 %); gleyed acidic in combination with calcic: $H^+ - Fe^{2+}$, Ca^{2+} (3.7 %); calcic in combination with acidic calcic: Ca^{2+}, $H^+ - Ca^{2+}$ (1.4 %); acidic calcic in combination with gleyic: $H^+ - Ca^{2+}$, Fe^{2+} (1 %)	dark-grey podzolised + grey forest soil, dark-grey podzolised soil, carbonaceous podzolised sod soil, grey forest soil (N = 7)	Be	As	Pt	Zr	Na
			3.2	2.0	2.0	2.0	1.6
Forest steppe	calcic: Ca^{2+} (31.3 %)	typical black topsoil (N = 32)	Pd	Te	In	Se	La
			2.1	2.0	1.5	1.3	1.3
	acidic calcic: $H^+ - Ca^{2+}$ (43.4 %); acidic calcic in combination with gleyic: $H^+ - Ca^{2+}$, Fe^{2+} (3.4 %); calcic in combination with acidic calcic: Ca^{2+}, $H^+ - Ca^{2+}$ (2.4 %)	dark - grey podzolised + grey forest soil, grey forest soil, carbonaceous podzolised sod soil (N = 52)	Cl	Pt	Se	Cd	Mn
			4.0	2.0	1.2	1.1	1.1
	acidic gleyic in combination with calcic and natric: $H^+ - Fe^{2+}$, Ca^{2+}, Na^+ (4.1 %)	meadow chestnut + meadow black soil (N = 6)	Cs	As	La	Pt	Cl
			2.7	2.3	2.1	2.0	1.7
Steppe	calcic: Ca^{2+} (76 %)	typical black topsoil (N = 87)	F	In	Se	Ag	Cs
			2.8	1.5	1.3	1.2	1.2
	calcic - natric: $Ca^{2+} - Na^+$ (9.8 %); acidic gleyic in combination with calcic and natric: $H^+ - Fe^{2+}$, Ca^{2+}, Na^+ (6.2 %);	solonised chestnut soil; meadow chestnut + meadow black soil, (N = 12)	F	In	Te	Ag	Cd
			2.8	1.5	1.5	1.3	1.3
Ciscarpathian forest-meadow	acidic in combination with gleyed acidic: $H^+ - Fe^{2+}$ (41.6 %)	gleyed sod-podzolic soil (N = 6)	Pt	S	Ag	Cs	Se
			2.0	1.5	1.5	1.4	1.4
Carpathian Mountains	acidic brown: $H^+ - Al^{3+} - Fe^{3+}$ (88.3 %)	brown podzolised soil (N = 8)	S	Cl	Pt	Be	Li
			5.0	2.5	2.0	1.5	1.4

Landscape bioclimatic Zone	Elements															
Forest (Ukrainian Polissia)	Fe	P	Zr	Al	K	Bi	Hf	Ti	Yb	Pb	Hg	Rb	Ba	Mn	Pd	As
	1.2	1.2	1.2	1.1	1.1	1.1	1.1	1.1	1.1	1.1	1.1	1.0	1.0	1.0	1.0	1.0
	La	Pb	Re	S	Sc	Se	Ta	Te	Th	U	W	K	Si	Na	Ce	Ba
	1.0	1.0	1.0	1.0	1.0	1.0	1.0	1.0	1.0	1.0	1.0	1.0	1.0	1.0	1.0	1.0
	Ba	Al	Li	Mn	Nb	Pb	Rb	Mg	Si	Zn	Fe	Pd	Au	B	Bi	Cl
	1.2	1.2	1.2	1.1	1.1	1.1	1.1	1.1	1.0	1.0	1.0	1.0	1.0	1.0	1.0	1.0
Forest steppe	Mo	P	Al	Sc	Zn	Mg	Fe	Na	Ba	Ti	Zr	Hg	K	Rb	Sr	Sn
	1.1	1.1	1.1	1.1	1.1	1.1	1.1	1.1	1.1	1.1	1.0	1.0	1.0	1.0	1.0	1.0
	Tl	Al	Zn	Rb	Sn	Fe	Si	V	Pd	As	Au	B	Ce	F	Ge	Hf
	1.0	1.0	1.0	1.0	1.0	1.0	1.0	1.0	1.0	1.0	1.0	1.0	1.0	1.0	1.0	1.0
	Mg	Ag	Be	Rb	Fe	V	Pd	Si	Cd	Cu	F	Hf	In	Re	Ta	Te
	1.1	1.1	1.1	1.1	1.1	1.1	1.0	1.0	1.0	1.0	1.0	1.0	1.0	1.0	1.0	1.0
Steppe	Co	Nb	Tl	Cr	Hg	Mo	Y	K	Yb	Sn	Ti	Na	Bi	Ba	La	P
	1.1	1.1	1.1	1.1	1.1	1.1	1.1	1.1	1.1	1.1	1.1	1.1	1.0	1.0	1.0	1.0
	Al	Zr	U	Cr	Zn	Ga	Rb	Ti	Sn	K	Mg	Ce	La	Si	Be	Cl
	1.1	1.1	1.1	1.1	1.1	1.1	1.1	1.1	1.1	1.1	1.1	1.0	1.0	1.0	1.0	1.0
Ciscarpathian forest-meadow	Fe	U	As	Pb	Cr	Si	Mg	La	Al	Pd	Cl	F	Ga	Nb	Sc	Ta
	1.1	1.1	1.1	1.0	1.0	1.0	1.0	1.0	1.0	1.0	1.0	1.0	1.0	1.0	1.0	1.0
Carpathian Mountains	Tl	Cu	Sn	Mg	Pd	F	Re	Ta	W	Y	Cr	P	Ge	Hg	Si	Fe
	1.0	1.0	1.0	1.0	1.0	1.0	1.0	1.0	1.0	1.0	1.0	1.0	1.0	1.0	1.0	1.0

Note: Table shows concentration ratios of chemical elements in agricultural land soil (Ap-horizon, medians) relative to grazing land soil (medians) in Ukraine.

Elements																	
Co	Y	Sr	Cl	Mg	V	Na	Li	Cr	Be	Se	Tl	Ga	Ni	Ag	Ca	Zn	Nb
1.8	1.7	1.6	1.5	1.5	1.4	1.4	1.4	1.4	1.3	1.3	1.3	1.3	1.3	1.3	1.3	1.2	1.2
V	Y	Yb	Zn	Sn	Hf	Ni	Rb	Nb	Ge	Mg	Sr	Pd	As	Cs	Cu	F	In
1.3	1.3	1.2	1.2	1.1	1.1	1.1	1.1	1.1	1.1	1.1	1.0	1.0	1.0	1.0	1.0	1.0	1.0
Ge	K	Yb	Sc	Tl	Ce	Ti	Co	Hf	Sn	Cr	Ni	Y	U	Ga	Se	Ag	V
1.5	1.5	1.4	1.4	1.4	1.4	1.3	1.3	1.3	1.3	1.3	1.3	1.3	1.3	1.2	1.2	1.2	1.2
As	Bi	Ni	Cd	Ag	Cu	Ga	V	Co	Sb	Mn	Li	Y	Cr	Cs	Pb	Ce	Be
1.3	1.3	1.3	1.2	1.2	1.2	1.2	1.2	1.2	1.2	1.1	1.1	1.1	1.1	1.1	1.1	1.1	1.1
Ag	Y	P	Pb	Ga	Cs	Cu	Be	Sr	Na	Cr	Nb	Ti	Ba	Zr	Co	K	Ni
1.1	1.1	1.1	1.1	1.1	1.1	1.1	1.1	1.1	1.1	1.1	1.1	1.1	1.1	1.1	1.1	1.1	1.1
Li	Nb	Tl	Ga	Al	Y	Na	Yb	Ba	Sc	Ti	Cr	K	Co	Zr	Sn	Pb	Ge
1.5	1.4	1.4	1.3	1.3	1.3	1.2	1.2	1.2	1.2	1.2	1.2	1.2	1.2	1.1	1.1	1.1	1.1
Cu	Ge	Ni	V	As	Be	Ce	Cd	Li	Sb	Fe	Sc	Al	Mn	Mg	Rb	Ga	Zn
1.2	1.2	1.1	1.1	1.1	1.1	1.1	1.1	1.1	1.1	1.1	1.1	1.1	1.1	1.1	1.1	1.1	1.1
Hf	Tl	Mn	Li	As	Cu	Se	Ba	Co	Nb	Sc	Ni	Ge	Fe	Y	Bi	V	Na
1.3	1.3	1.2	1.2	1.2	1.2	1.2	1.2	1.2	1.2	1.2	1.1	1.1	1.1	1.1	1.1	1.1	1.1
B	Ni	Cu	Bi	Hg	Be	P	Y	Sb	Sr	V	Yb	Zr	Au	Hf	Sn	Ge	Ca
1.3	1.3	1.2	1.2	1.2	1.2	1.1	1.1	1.1	1.1	1.1	1.1	1.1	1.1	1.1	1.1	1.1	1.1
B	K	Pb	Ga	Ba	Ti	Te	Na	Hf	Nb	Sr	Ca	Zn	Rb	Al	Bi	Ni	Co
1.3	1.3	1.3	1.2	1.2	1.2	1.2	1.2	1.1	1.1	1.1	1.1	1.1	1.1	1.1	1.1	1.0	1.0

Elements																	
Cd	Cs	Cu	F	In	La	Mo	Re	S	Sb	Sc	Sn	Ta	Te	Th	W	Si	Ge
1.0	1.0	1.0	1.0	1.0	1.0	1.0	1.0	1.0	1.0	1.0	1.0	1.0	1.0	1.0	1.0	1.0	1.0
Al	Zr	Ca	Hg	Cr	Ag	Cl	Co	P	Sb	Cd	Ti	Li	Ga	Tl	Bi	Be	Au
1.0	0.9	0.9	0.9	0.9	0.9	0.9	0.9	0.8	0.8	0.8	0.8	0.8	0.8	0.8	0.6	0.5	0.2
Cs	F	In	Re	Ta	Te	W	Sr	Cu	Mo	Hg	Th	P	Ca	Cd	La	Sb	S
1.0	1.0	1.0	1.0	1.0	1.0	1.0	1.0	0.9	0.9	0.9	0.9	0.8	0.7	0.7	0.6	0.6	0.3
Cl	F	Nb	Pt	Re	Ta	Tl	U	W	Yb	Si	Hf	Ca	Ge	Th	B	Au	S
1.0	1.0	1.0	1.0	1.0	1.0	1.0	1.0	1.0	1.0	1.0	1.0	0.9	0.9	0.9	0.8	0.8	0.7
Hg	In	Mo	Re	S	Sc	Ta	Te	U	W	Ca	Yb	Bi	Th	Mg	La	Li	Sb
1.0	1.0	1.0	1.0	1.0	1.0	1.0	1.0	1.0	1.0	1.0	1.0	1.0	1.0	0.9	0.9	0.9	0.8
U	W	Sr	Ni	Zn	Ce	Th	Mn	Bi	Sb	Mo	Hg	Se	Au	Ca	P	B	S
1.0	1.0	1.0	1.0	1.0	0.9	0.9	0.9	0.9	0.9	0.8	0.8	0.8	0.7	0.6	0.6	0.4	0.1
Sr	Si	Au	Cl	Hf	Pb	Pd	Pt	Re	Ta	Te	U	W	Zr	Th	B	Ca	S
1.0	1.0	1.0	1.0	1.0	1.0	1.0	1.0	1.0	1.0	1.0	1.0	1.0	1.0	0.9	0.9	0.7	0.6
Pb	Pd	Pt	Re	Sr	Ta	W	P	Mo	Th	Cs	Sb	Hg	Yb	B	Au	Ca	S
1.0	1.0	1.0	1.0	1.0	1.0	1.0	1.0	0.9	0.9	0.9	0.9	0.9	0.9	0.9	0.7	0.6	0.5
W	Ba	Li	K	Na	Ti	Tl	Zn	Rb	Te	Ce	Co	Th	Mo	Cd	Mn	In	Re
1.0	1.0	1.0	1.0	1.0	1.0	1.0	1.0	1.0	0.9	0.9	0.9	0.8	0.8	0.7	0.7	0.5	0.5
Sc	Zr	U	Sb	Cd	Ce	Mo	Cs	Mn	V	Ag	Au	Yb	In	Se	La	As	Th
0.9	0.9	0.9	0.9	0.9	0.9	0.9	0.9	0.9	0.8	0.8	0.8	0.8	0.8	0.8	0.7	0.6	0.5

$K_c > 1.5$ – accumulating elements $K_c < 0.7$ – leached elements

▸ 15.4 Conclusions

Landscape-geochemical regionalisation was applied to determine element background concentrations in Ukrainian agricultural and grazing land soil. Selection of soil types was carried out within geochemical landscape classes, each one being characterised by the same conditions of element migration in aqueous solutions. The approach allowed the determination of geochemical characteristics of agricultural and grazing land soil classified in 10 and 13 geochemical landscape classes respectively covering over 99.16 % of the country's area.

As a rule, leaching of most chemical elements and their accumulation were observed in soil of respectively acidic and calcic geochemical landscape classes, which is quite consistent with the fundamental principles of landscape geochemistry.

Concentrations of chemical elements in soil are substantially governed by geochemical specialisation of soil-forming processes and underlying parent materials, as well as by physical and chemical conditions of element migration. Anthropogenic impact on the geochemical composition of soil at the regional scale is insignificant in Ukraine, although anomalously high concentrations of certain elements (e.g., Hg, Ag, Cd, Pb), mainly in agricultural land soil, were detected near a few large cities, e.g., Kyiv, Donetsk.

Agricultural land use in Ukraine had no substantial influence on the natural geochemical composition of soil, but it was found, that geochemical landscape acidic classes of agricultural land soil have a higher number of elements exhibiting accumulation than the calcic classes (black topsoil). Most probably, this is caused by more intense agrochemical treatment of the impoverished acidic soil. The comparatively low Ca and S concentrations observed in agricultural land soil is connected with biogeochemical migration of these elements. |◂

Volodymyr Klos
State Geological and Subsurface Survey of Ukraine (SGSSU);
Kyiv, Ukraine
E-mail: ukrgeochim@mail.ru

Manfred Birke
Federal Institute for Geosciences and Natural Resources (BGR);
Hannover; Germany

Grygorii Akinfiev
State Geological and Subsurface Survey of Ukraine (SGSSU);
Kyiv, Ukraine

Yulia Amashukeli
State Geological and Subsurface Survey of Ukraine (SGSSU);
Kyiv, Ukraine

Part B

References, Acronyms and Abbreviations

References

ABFKLÄRV (2012): Klärschlammverordnung (AbfKlärV) vom 15. April 1992 (BGBl. I S. 912), die zuletzt durch Artikel 5 Absatz 12 des Gesetzes vom 24. Februar 2012 (BGBl. I S 212) geändert worden ist. – Bundesgesetzblatt I (24. Februar 2012): 212 – 226.

ABRAMS, M. M., BURAU, R. G. & ZASOSKI, R. J. (1990): Organic selenium distribution in selected California soils. – Soil Science Society of American Journal, **54**: 979 – 982.

ADRIANO, D. C., BOLAN, N. S., VANGRONSVELD, J. & WENZEL, W. W. (2005): Heavy metals. – In: HILLEL, D. (ed.): Encyclopedia of Soils in the Environment: 175 – 182; Amsterdam (Elsevier).

AGGARWAL, J. (1999): Boron.– In: MARSHALL, C. P. & FAIRBRIDGE, R. W. (eds.): Encyclopedia of geochemistry.– 44 – 49; Dordrecht (Kluwer Academic Publishers).

ALIU, M., ŠAJN, R. & STAFILOV, T. (2010): Distribution of zinc in surface soils in K. Mitrovica Region, Kosovo. – International Journal of Pure & Applied Chemistry, **5**(4): 351 – 357.

—, ŠAJN, R. & STAFILOV, T. (2011): Enrichment of some heavy metals in topsoils around a Pb-Zn smelter in Mitrovica, Kosovo.– 1st International Workshop on the UNESCO-IGCP Project "Anthropogenic effects on the human environment in the Neogene basins in the SE Europe" (Štip/FYROM, June 3 – 4, 2011), Proceedings: 24 – 29.

—, ŠAJN, R. & STAFILOV, T. (2011a): A survey of lead Pollution in surface soils in Mitrovica Region, Kosovo.– 2nd International Workshop on the UNESCO-IGCP Project "Anthropogenic effects on the human environment in the Neogene basins in the SE Europe" (Zagreb/Croatia, October 6 – 7, 2011), Proceedings: 61 – 66.

ALLOWAY, B. J. (1995): Heavy metals in soils.– 368 pp.; London (Blackie Academic & Professional).

AMSTUTZ, G. C., GORESY, A. E., FRENZEL, G., KLUTH, C., MOH, G., WAUSCHKUHN, A. & ZIMMERMANN, R. A. (eds.) (1982): Ore genesis: The state of the art. – Society for Geology Applied to Mineral Deposits, Spec. Publ., **2**: 804 pp.; Berlin (Springer).

ANAND, R. R. & GILKES, R. J. (1984): Weathering of ilmenite in a lateritic pallid zone. – Clays & Clay Minerals, **32**: 363 – 374.

ANDERSON, B. G. & BORNS, H. W. (1997): The Ice Age World.– 208 pp.; Oslo (Scandinavian Univ. Press).

ANDERSSON, P. S., WASSERBURG, G. J., CHEN, J. H., PAPANASTASSIOU, D. A. & INGRI, J. (1995): $^{238}U - ^{234}U$ and $^{232}Th - ^{230}Th$ in the Baltic sea and river water.– Earth and Planetary Science Letters, **130**: 217 – 234.

ANTOINE, P., CATT, J. A., LAUTRIDOU, J-P. & SOMMÉ, J. (2003): The loess and coversands of northern France and southern England.– Journal of Quaternary Science, **18**: 309 – 318.

ARCHER, V.E., WAGONER, J. K. & LUNDIN, F.E. (1973): Cancer mortality among uranium mill workers.– Journal of Occupational Medicine, **15**: 11 – 14.

ARNDT, T. (2012): Problematik, Klinik und Beispiele der Spurenelementvergiftung – Cadmium.– Toxichem. Krimtech., **79**(3): 127 – 135.

ARO, A. & ALFTHAN, G. (1995): Effects of Supplementation of Fertilizers on Human Selenium Status in Finland. – Analyst, **120**: 841 – 3.

ARTEMIEVA, I. M. (2007): Dynamic topography of the East European Craton: Shedding light upon the lithospheric structure, composition and mantle dynamics.– Global and Planetary Change, **58**: 411 – 434.

ASCH, K. (2003): The 1 : 5 Million International Geological Map of Europe and Adjacent Areas: Development and Implementation of a GIS-enabled Concept. – Geol. Jb., Sonderhefte, **SA 3:** 190 pp.; Hannover.

— (2005): The 1 : 5 Million International Geological Map of Europe and Adjacent Areas.– 1 map; Hannover (BGR). URL: http://www.bgr.de/karten/igme5000/ (last access 26 December 2012).

ASHMAN, M. R. & PURI, G. (2002): Essential Soil Science.– 208 pp.; Oxford (Blackwell Science).

ATSDR (2008): Draft Toxicological Profile of Cadmium.– US Department of Health and Human Services, Agency for Toxic Substances and Disease Registry. URL: http://www.atsdr.cdc.gov/toxprofiles/tp5-p.pdf (last access 13 January 2013).

— (2011): Toxicological Profile for Uranium. URL: http://www.atsdr.cdc.gov/toxprofiles/tp.asp?id=440&tid=77 (last access 20 October 2012)

AUER, C. (1977): Cadmium in Phosphate Fertilizer Production. – US Environmental Protection Agency, Midwest Research Institute Report: 48 pp.; Washington, DC.

BALDSCHUHN, R., FRISCH, U. & KOCKEL, F. (eds.) (1996): Geotektonischer Atlas von NW-Deutschland 1:300 000.– BGR; Hannover.

BARBIER, O., JACQUILLET, G., TAUC, M., COUGNON, M. & POUJEOL, P. (2005): Effect of heavy metals on, and handling by, the kidney. – Nephron Physiology, **99**(4): 105 – 110.

Baritz, R., Fuchs, M., Hartwich, R., Krug, D. & Richter, S. (2005): Soil Regions of the European Union and Adjacent Countries 1 : 5 000 000 (Version 2.0) – Europaweite thematische Karten und Datensätze. – European Soil Bureau Network: 122 pp.

—, Zirlewagen, D. & van Ranst, E. (2006): Methodical standards to detect forest soil carbon stocks and stock changes related to land use change and forestry – landscape scale effects. – Final report Deliverable 3.5-II. Multi-source inventory methods for quantifying carbon stocks and stock changes in European forests (CarboInvent), Contract No. EU EVK2-2001-00287.

—, Seufert, G., van Ranst, E. & Montanarella, L. (2010): Carbon concentrations and stocks in forest soils of Europe.– Forest Ecology and Management, **260**: 262 – 277.

—, Zirlewagen, D., Jones, R. J. A, Arrouays, D., Hiederer, R., Schrumpf, M. & Riek, W. (2011): Carbon in European soils. – In: Jandl, R., Rodeghiero, M. & Olsson, M. (eds.): Soil Carbon in Sensitive European Ecosystems: From Science to Land Management.– 49 – 84; Oxford (Wiley).

Barth, N., Pälchen, W., Rank, G. & Heilmann, H. (1996): Bodenatlas des Freistaates Sachsen. Teil 1: Hintergrundwerte für Schwermetalle und Arsen in landwirtschaftlich genutzten Böden.– Landesamt für Umwelt und Geologie: 12 – 13.

Baselt, R. C. (2011): Disposition of toxic drugs and chemicals in man, 9th ed. – Biomedical Publications: 1250 pp.; Seal Beach.

Bateman, A. M. (1950): Economic mineral deposits. – 916 pp.; New York (Wiley).

— (1955): Economic geology. – Fiftieth anniversary volume. – 1130 pp.; Lancaster, PA (Economic Geology Pub. Co.).

Bateman, M. D. (1995): Thermoluminescence dating of the British coversand deposits. – Quaternary Science Reviews, **14**: 791 – 798.

Bates, R. L. & Jackson, J. A. (eds.) (1980): Glossary of Geology, 2nd edition. – American Geological Institute: 751 pp.; Falls Church, VA.

BBodSchV (1999): Bundes-Bodenschutz- und Altlastenverordnung (BBodSchV), Bundesgesetzblatt I, 12. Juli 1999: 1554 – 1585.

BEIR IV (1988): Health risks of radon and other internally deposited alpha-emitters. – Committee on the Biological Effects of Ionizing Radiations, National Research Council: 245 – 275; Washington (National Academy Press).

Benbrahim-Tallaa, L., Waterland, R. A., Dill, A. L., Webber, M. M. & Waalkes, M.P. (2007): Tumor suppressor gene inactivation during cadmium-induced malignant transformation of human prostate cells correlates with overexpression of de novo DNA methyltransferase. – Environmental Health Perspectives, **115**(10): 1454 – 1459.

Benek, R., Kramer, W., McCann, T., Scheck, M., Negendank, J. F. W., Korich, D., Huebscher, H. D. & Bayer, U. (1996): Permo-Carboniferous magmatism of the Northeast German Basin. – Tectonophysics, **266** (1 – 4): 379 – 404.

Bergman, S., Kubler, L. & Martinsson, O. (2001): Description of regional geological and geophysical maps of northern Norrbotten County (east of the Caledonian orogen). – Geological Survey Sweden, Ba **56**: 110 pp.

Bertrain, P., Allenet, G., Gé, T., Naughton, F., Poirier, P. & Sanchez Goni, M. F. (2009): Coversand and Pleistocene palaeosols in the Landes region, southwestern France. – Journal of Quaternary Science, **24**: 259 – 269.

Beuge, P. & Ulique, A. (1997): Stoffliche Belastung von Auenböden. – Abschlussbericht, FuE-Vorhaben LfUG, TU Bergakademie Freiberg: 110 pp.

Beven, K. J. & Kirkby, M. (1979): A physically-based, variable contributing area model of basin hydrology. – Hydrological Sciences Bulletin, **24**(1): 43 – 69.

BGR – Bundesanstalt für Geowissenschaften und Rohstoffe (2005): Soil Regions Map of the European Union and Adjacent Countries 1 : 5,000,000 (Version 2.0). – Spec. Publ., Ispra, EU catalogue no. S.P.I. 05.134.

Bingen, B., Nordgulen, Ø. & Viola, G. (2008): A four-phase model for the Sveconorwegian orogeny, SW Scandinavia. – Norwegian Journal of Geology, **88**: 43 – 72; Trondheim.

—, Andersson, J., Söderlund, U. & Möller, C. (2008a): The Mesoproterozoic in the Nordic countries. – Episodes, **31**: 29 – 34.

Birke, M., Rauch, U. & Rentzsch, J. (1995): Environmental Results of a Regional Geochemical Survey in Eastern Germany. – Geol. Jb., **D 94**: 3 – 35; Hannover.

—, Bityukova, L., Grunsky, E., Shogenova, A. & Kiipli, T. (1998): Urban Geochemistry of Tallinn. Final Report. – Scientific and Environmental Affairs Division NATO, ENVIR.CRG. 960910: 67 pp.

—, Rauch, U., Raschka, H., Wehner, H., Kringel, R., Gäbler, H.-E., Kriete, C., Siewers, U. & Kantor, W. (2006): Geochemischer Atlas der Bundesrepublik Deutschland – Verteilung anorganischer und organischer Parameter in Oberflächenwässern und Bachsedimenten.– Vorabexemplar BGR: 641 pp. [unpubl.]

Bityukova, L. & Birke, M. (2011): Urban geochemistry of Tallinn (Estonia): Major and trace-elements distribution in topsoils – In: Johnson, C. C., Demetriades, A., Locutura, J. & Ottesen, R. T. (eds.): Mapping the chemical environment of urban areas: 348 – 363; Chichester (Wiley).

—, Shohenova, A. & Birke, M. (2000): Urban geochemistry: A study of element distributions in the topsoils of Tallinn (Estonia). – Environmental Geochemistry and Health, **22**: 173 – 195.

BLUME, H.-P., BRÜMMER, G.W., HORN, R., KANDELER, E., KÖGEL-KNABNER, I., KRETZSCHMAR, R., STAHR, K., WILKE, B.-M., THIELE-BRUHN, S. & WELP, G. (2010): Scheffer/Schachtschabel Lehrbuch der Bodenkunde. – 570 pp.; Heidelberg (Springer Spektrum).

BLUNDELL, D., FREEMAN, R. & MUELLER, ST. (eds.) (1992): A Continent Revealed – the European Geotraverse. – 275 pp.; Cambridge (Cambridge Univ. Press).

—, ARNDT, N., COBBOLD, P.R. & HEINRICH, C. (eds.) (2005): Geodynamics and ore deposit evolution in Europe. – 349 pp.; Amsterdam (Elsevier).

BØLVIKEN, B., DEMETRIADES, A., HINDEL, R., LOCUTURA, J., O'CONNOR, P., OTTESEN, R.T, PLANT, J., RIDGWAY, J., SALMINEN, R., SALPETEUR, I., SCHERMANN, O. & VOLDEN, T. (eds.) (1990): Geochemical Mapping of Western Europe towards the Year 2000 – Project Proposal.– Western European Geological Surveys (WEGS), Geological Survey of Norway, NGU Report, **90.106**:12 p.; Trondheim.

—, BOGEN, B., DEMETRIADES, A., DE VOS, W., EBBING, J., HINDEL, R., OTTESEN, R.T., SALMINEN, R., SCHERMANN, O. & SWENNEN, R. (1993): Final Report of the Working Group on Regional Geochemical Mapping 1986–93 – Geochemical Mapping of Western Europe towards the Year 2000.– Forum of European Geological Surveys (FOREGS), Geological Survey of Norway, NGU Report, **93.092**: 18 pp., Trondheim.

—, BOGEN, J., DEMETRIADES, A., DE VOS, W., EBBING, J., HINDEL, R., LANGEDAL, M., LOCUTURA, J., O'CONNOR, P., OTTESEN, R.T., PULKKINEN, E., SALMINEN, R., SCHERMANN, O., SWENNEN, R., VAN DER SLUYS, J. & VOLDEN, T. (1996): Regional geochemical mapping of Western Europe towards the year 2000. – Journal of Geochemical Exploration, **56**(2): 141 – 166.

BOGDANOVA, S.V., PASHKEVICH, I.K., GORBATSCHEV, R. & ORLYUK, M.I. (1996): Riphean rifting and major Palaeoproterozoic crustal boundaries in the basement of the East European Craton: Geology and geophysics. – Tectonophysics, **268**: 1 – 21.

—, BINGEN, B., GORBATSCHEV, R., KHERASKOVA, T.N., KOZLOV, V.I., PUCHKOV, V.N. & VOLOZH, YU.A. (2008): The East European Craton (Baltica) before and during the assembly of Rodinia. – Precambrian Research, **160**: 23 – 45.

BOISSONNAS, J. & OMENETTO, P. (eds.) (1988): Mineral deposits within the European Community. – Society for Geology Applied to Mineral Deposits, Spec. Publ., **6**: 558 pp.; Berlin (Springer).

BOMAN, A., ÅSTRÖM, M. & FÖRJDÖ, S. (2008): Sulfur dynamics in boreal soils rich in metastabel iron sulfide – The role of artificial drainage. – Chemical Geology, **255**: 68 – 77.

BOWEN, H.J.M. (1979): Environmental chemistry of the elements. – 333 pp.; London (Academic Press).

—, KVALHEIM, A., HASLAM, H.W. & NOTHOLT, A.J.G. (eds.) (1978): Mineral deposits of Europe: Volume 1 – Northwest Europe. – Institution of Mining and Metallurgy: 454 pp.

BOWIE, S.H.U., KVALHEIM, A., HASLAM, H.W. & NOTHOLT, A.J.G. (eds.) (1978): Mineral deposits of Europe: Volume 1 – Northwest Europe. – Institution of Mining and Metallurgy: 454 pp.

—, DUNNING, F.W. & HASLAM, H.W. (eds.) (1989): Mineral deposits of Europe: Southwest and eastern Europe with Iceland. – Institution of Mining and Metallurgy: 454 pp.

BRADFORD, J.M. & PETERSON, G.A. (2000): Conservation tillage. – In: SUMNER, M.E. (ed.): Handbook of Soil Science: G247 – G270; Boca Raton, FL (CRC Press).

BROADLEY, M.R., WHITE, P.J., BRYSON, R.J., MEACHAM, M.C., BOWEN, H.C., JOHNSON, S.E., HAWKESFORD, M.J., MCGRATH, S.P., ZHAO, F.J., BREWARD, N., HARRIMAN, M. & TUCKE, M. (2006): Biofortification of UK food crops with selenium. – Proceedings of the Nutrition Society, **65**: 169 – 181.

BROECKER, W.S. & VAN DONK, J. (1970): Insolation changes, ice volumes, and the O (super 18) record in deep-sea cores. – Reviews of Geophysics and Space Physics, **8**(1): 169 – 198.

BRUGGLE, B., GLASER, B., ZOELLER, L., HAMBACH, U., MARKOVIC, S., GLASER, I. & GERASIMENKO, N. (2008): Geochemical characterisation and origin of Southeastern and Eastern European loesses (Serbia, Riomania, Ukraine). – Quaternary Science Reviews, **27**: 1058 – 1057.

BUCKMAN, H.O. & BRADY, N.C. (1970): The nature and properties of soils. – 653 pp.; New York (Macmillan).

BUIVYDAITĖ, V.V. (ed.) (2001): Lietuvos dirvožemiai. – "Lietuvos mokslo" redakcija, Lietuvos mokslas, Kn **32:** 1244; Vilnius. [in Lithuanian]

BURT, R., WILSON, M.A., MAYS, M.D. & LEE, C.W. (2003): Major and Trace Elements of Selected Pedons in the USA. – J. Environmental Quality, **32**: 2109 – 2121; Madison.

CABALA, J. (2009): Metale ciężkie w środowisku glebowym olkuskiego rejonu eksploatacji rud Zn-Pb.: 130 pp.; Katowice (Wydawn, Uniwersytetu Śląskiego).

CAMBERATO, J.J. & PAN, W.L. (2000): Bioavailability of Calcium, Magnesium and Sulfur. – In: SUMNER, M.E. (ed.): Handbook of Soil Science: D53 – D69; Boca Raton, FL (CRC Press).

CANADIAN SOIL QUALITY GUIDELINES (1999): Canadian Soil Quality Guidelines for the Protection of Environmental and Human Health. Summary Tables, Update 7.0, September 2007. – Canadian Council of Ministry of the Environment, Excerpt from Publication **1299**: 1 – 6.

CARITAT, P. DE & COOPER, M. (2011): National Geochemical Survey of Australia: The Geochemical Atlas of Australia. – Geoscience Australia Record, **2011/20**: 557 pp. URL: www.ga.gov.au/ngsa (last access 28 August 2012).

— & COOPER, M. (2011a): National Geochemical Survey of Australia: Data Quality Assessment. – Geoscience Australia Record, **2011/21**: 478 pp. URL: www.ga.gov.au/ngsa (last access 28 August 2012).

—, Cooper, M., Mann, A. & Prince, P. (2011b): Lithological signatures and exploration implications from overbank sediment sampling and MMI® Analysis: Preliminary findings from the National Geochemical Survey of Australia.– 25th International Applied Geochemistry Symposium (Rovaniemi/Finland, August 22 – 26, 2011). – Programme & Abstracts: 159.

—, Mann, A. & Prince, P. (2012): The world's first continental-scale partial digestion geochemical dataset: Mobile Metal Ion® analysis of Australian soils. – 34th International Geological Congress (Brisbane/Australia, August 5 – 10, 2012). – Proceedings: 2510.

Cassard, D., Bertrand, G., Maldan, F., Gaàl, G., Kaija, J., Aatos, S., Angel, J. M., Arvanitidis, N., Ballas, D., Billa, M., Christidis, C., Dimitrova, D., Eilu, P., Filipe, A., Gazea, E., Inverno, C., Kauniskangas, E., Maki, T., Matos, J., Meliani, M., Michael, C., Mladenova, V., Navas, J., Niedbal, M., Perantonis, G., Pyra, J., Santana, H., Serafimovski, T., Serrano, J.J., Strengell, J., Tasev, G., Tornos, F. & Tudor G. (2012): ProMine pan-European Mineral Deposit database: A new dataset for assessing primary mineral resources in Europe. – Workshop on Mineral Resources Potential Maps: A Tool for Discovering Future Deposits. (Nancy/France, March 12 – 14, 2012). – Proceedings: 9 – 13.

Catt, J. A. (1977): The contribution of loess to soils in lowland Britain. – In: Limbrey, S. & Evans, J. G. (eds.): The Effect of Man on the Landscape: the Lowland Zone.– Council for British Archaeology, Research Report **21**: 12 – 20; Oxford.

— (1985): Soil particle size distribution and mineralogy as indicators of pedogenic and geomorphic history: examples from the loessial soils of England and Wales. – In: Richard, K. S., Arnet, R. R. & Ellis, S. (eds.): Geomorphology and Soils: 202 – 218; London (G. Allen and Unwin).

— (2001): The agricultural importance of loess. – Earth-Science Reviews, **54** (1–3): 213 – 229.

—, Weir, A. H. & Madgett, P. A. (1973): The loess of eastern Yorkshire and Lincolnshire. – Proceedings of the Yorkshire Geological Society, **40**: 23 – 29.

CEC – Commission of the European Communities (2001): Statistical Information needed for Indicators to monitor the Integration of Environmental concerns into the Common Agricultural Policy Communication from the Commission to the Council and the European Parliament COM(2001): 144.

Chaney, R. L., Ryan, J. A. & Li, Y. M. (1996): Phytoavailability and bio-availability in risk assessment of Cd in agricultural environments. – OECD Proceedings Sources of Cadmium in the Environment (Stockholm/Sweden, October 15 – 22, 1995), OECD: 49 – 79; Paris.

—, Ryan, J. A., Li, Y. M. & Brown, S. L. (1999): Soil cadmium as a threat to human health. – In: McLaughlin, M. J. & Single, B. R. (eds.): Cadmium in soil and plants: 219 – 256; Dordrecht (Kluwer Academic).

—, Reeves, P. G., Ryan, J. A., Simmons, R. W., Welch, R. M. & Angle, J. S. (2004): An improved understanding of soil Cd risk to humans and low cost methods to phytoextract Cd from contaminated soils to prevent soil Cd risk. – Bio Metals, **17**: 549 – 553; Dordrecht (Kluwer Academic).

Cheng, Q., Jing, L. & Panahi, A. (2006): Principal component analysis with optimum order sample correlation coefficient for image enhancement. – International Journal of Remote Sensing, **27**(16): 3387 – 3401.

Chorley, R. J., Stoddart, D. R., Haggett, P. & Slaymaker, H. O. (1966): Regional and local components in the areal distribution of surface sand facies in the Breckland, eastern England. – Journal of Sedimentary Petrology, **36**: 209 – 220.

Cienciala, E., Seufert, G., Blujdea, V., Grassi, G. & Exnerová, Z. (eds.) (2010): Harmonised Methods for Assessing Carbon Sequestration in European Forests: Results of the Project "Study under EEC 2152/2003 Forest Focus regulation on developing harmonized methods for assessing carbon sequestration in European forests".– European Commission, Joint Research Centre, Institute for Environment and Sustainability: 324 pp.; Luxembourg. URL: http://afoludata.jrc.ec.europa.eu/metaed/project_collection/JRC_54744_Mascaref_Full_Report_for_print.pdf (last access 24 January 2013).

Clarke, M. L., Rendell, H. M., Hoare, P. G., Godby, S. P. & Stevenson, C. R. (2001): The timing of coversand deposition in northwest Norfolk, UK: A cautionary tale. – Quaternary Science Reviews, **20**: 705 – 713.

Cohen, K. M., Finney, S. & Gibbard, P.L. (2013): International Chronostratigraphic Chart, v 2013/01.– International Commission on Stratigraphy. http://www.stratigraphy.org/ICSchart/ChronostratChart2013-01.pdf (last access 28 July 2013).

Colwell, J. D. (1967): Soil tests for phosphorous fertilizer requirements. – Journal of Australian Institute for Agricultural Science, **33**: 289 – 295.

— (1967a): The calibration of soil tests.– Journal of Australian Institute for Agricultural Science, **33**: 321 – 323.

Cook, M. E. & Morrow, H. (1995): Anthropogenic Sources of Cadmium to the Canadian Environment. – Workshop Proceedings, National Workshop on Cadmium Transport into Plants, Canadian Network of Toxicology Centres (Ottawa/Canada, June 20 – 21, 1995).

Cooper, M. E. & Kuck, P. H. (2007): Cadmium.– USGS 2005 Minerals Yearbook, US Geological Survey: 15.1 – 15.12.

Coppin, F., Chabroullet, C., Martin-Garin, A., Balesdent, J. & Gaudet, J. P. (2006): Methodological approach to assess the effect of soil ageing on selenium behaviour: First results concerning mobility and solid fractionation of selenium. – Biology and Fertility of Soils, **42**: 379 – 86.

Council of Europe (1994): Lead, Cadmium and Mercury in Food: Assessment of Dietary Intakes and Summary of Heavy Metal Limits of Foodstuffs. – Council of Europe Publishing: 59 pp; Strasbourg.

Darnley, A., Björklund, A., Bølviken, B., Gustavsson, N., Koval, P.V., Plant, J.A., Steenfelt, A., Tauchid, M., Xuejing, X., Garrett, R.G. & Hall, G.E.M. (1995): A global geochemical database for environmental and resource management. Recommendations for international geochemical mapping. – Final report of IGCP-project 259: 122 pp.; Paris (UNESCO Publishing). URL: http://www.globalgeochemicalbaselines.eu/files/Blue_Book_GGD_IGCP259.pdf (last access 11 February 2013).

Darnley, A.G. & Garrett, R.G. (1998): Global Geochemical Baselines – an update. – Episodes, **21**(1): 43 – 44.

de Meeûs, C., Eduljee, G.H. & Hutton, M. (2002): Assessment and management of risk arising from exposure to cadmium in fertilisers.– Science of The Total Environment, **291**: 167-187; Amsterdam (Elsevier).

de Vos, W. & Tarvainen, T. (chief-eds.), Salminen, R., Reeder, S., de Vivo, B., Demetriades, A., Pirc, S., Batista, M.J., Marsina, K., Ottesen, R.T., O'Connor, P.J., Bidovec, M., Lima, A., Siewers, U., Smith, B., Taylor, H., Shaw, R., Salpeteur, I., Gregorauskiene, V., Halamic, J., Slaninka, I., Lax, K., Gravesen, P., Birke, M., Breward, N., Ander, E.L., Jordan, G., Duris, M., Klein, P., Locutura, J., Bel-lan, A., Pasieczna, A., Lis, J., Mazreku, A., Gilucis, A., Heitzmann, P., Klaver, G. & Petersell, V. (2006): Geochemical Atlas of Europe. Part 2 - Interpretation of Geochemical Maps, Additional Tables, Figures, Maps, and Related Publications. – Geological Survey of Finland: 692 pp.; Espoo. URL: http://weppi.gtk.fi/publ/foregsatlas/part2.php (last access 10 January 2014).

—, Gregorauskiene, V., Marsina, K., Salminen, R., Salpeteur, I., Tarvainen, T., O'Connor, P.J., Demetriades, A., Pirc, S., Batista, M.J. & Bidovec, M. (2006a): Distribution of elements in subsoil and topsoil. – In: de Vos, W. & Tarvainen, T. (chief-eds.): Geochemical Atlas of Europe. Part 2 - Interpretation of Geochemical Maps, Additional Tables, Figures, Maps, and Related Publications. – Geological Survey of Finland: 21 – 29; Espoo. URL: http://weppi.gtk.fi/publ/foregsatlas/articles/Soil.pdf (last access 12 February 2013).

—, Demetriades, A., Marsina, K., Ottesen, R.T., Reeder, S., Pirc, S., Salminen, R. & Tarvainen, T. (2006b): Comparison of elements in all sample media, general comments and conclusions. – In: de Vos, W. & Tarvainen, T. (chief-eds.): Geochemical Atlas of Europe. Part 2 - Interpretation of Geochemical Maps, Additional Tables, Figures, Maps, and Related Publications. – Geological Survey of Finland: 45 – 47; Espoo. URL: http://weppi.gtk.fi/publ/foregsatlas/articles/Comparison.pdf (last access 12 February 2013).

de Wit, M.J. & Ashwal, L.D. (eds.) (1997): Greenstone Belts.– Oxford Monographs on Geology and Geophysics, **35**: 809 pp. Oxford (Oxford Univ. Press).

Debackere, M. & Delbeke, F.T. (1978): Fluoride pollution caused by a brickwork in the Flemish countryside of Belgium. – International Journal of Environmental Studies, **11**(4): 245 – 252.

Decree No. V-114 (2004): Decree No. V-114 of Minter of Health of the Republic of Lithuania of 8th March 2004: On the Hygenie Standard HN 60:2004 "Maximum permissible concentrations of hazardous chemical substances in soils": 8.

Decree of the Polish Ministry (2002): Decree of the Polish Ministry of the Environment of 9th September 2002: Permissible limit values in agricultural and grazing land soils. – The Governmental Gazette: 165, Item 1359.

Defra (2012): Technical Guidance Sheet (TGS) on normal levels of contaminants in English soils: Cadmium (Cd). – Technical guidance sheet TGS06 (Soils R&D Project SP1008), Defra: 4 pp. URL: http://nora.nerc.ac.uk/19967/ (last access 10 January 2014).

—, Euras, Eurometaux & ICMM (2007): Metals environmental risk assessment guidance (MERAG). – Fact sheet 2. Exposure assessment. URL : http://www.icmm.com/page/1185/metals-environmental-risk-assessment-guidance-merag (last access 19 November 2012).

Degryse, F., Smolders, E. & Parker, D.R. (2009): Partitioning of metals (Cd, Co, Cu, Ni, Pb, Zn) in soils: Concepts, methodologies, prediction and applications – a review. – European Journal of Soil Science, **60**: 590 – 612.

Deneux-Mustin, S., Roussel-Debet, S., Mustin, C.H., Munier-Lamy, C., Berthelin, J., Garnier-Laplace, J. & Leyval, C. (2003): Mobilité et transfert racinaire des éléments en traces: influence des microorganismes du sol. – **1**: 260 pp.; Lavoisier, Paris (Editions Tec & Doc).

Dercourt, J., Gaetani, M., Vrielynck, B., Barrier, E., Biju-Duval, B., Brunet, M.F., Cadet, J.P., Crasquin, S. & Sandulescu, M. (2000): Peri-Tethys atlas; palaeogeographical maps; explanatory notes. – Commission for the Geologic Map of the World; Paris.

Dewey, J.F. & Strachan, R.A. (2005): Caledonides of Britain and Ireland. – In: Selley, R.C., Cocks, L.R.M. & Plimer, I.R. (eds.): Encyclopaedia of Geology, **2**: 56 – 63; Amsterdam (Elsevier).

Dhillon, K.S. & Dhillon, S.K. (1999): Adsorption-desorption reactions of selenium in some soils of India. – Geoderma, **93**: 19 – 31.

Digital Elevation Data (2013): Digital vectorised contours at 20 m intervals based on 1 : 100 k and

20 k Russian military maps. URL: http://www.viewfinderpanoramas.org/dem3.html (last access 24 January 2013).

Dilek, Y. & Newcomb, S. (eds.) (2003): Ophiolite concept and the evolution of geological thought. – Geol. Soc. Am., Spec. Paper, **373**: 482 pp.; Boulder, CO.

Disnar, J. R. & Sureau, J. F. (1989): Organic matter in ore genesis: Progress and perspectives. – Advances in Organic Geochemistry, **16**: 577 – 599.

Domsch, K. H. (1992): Pestizide im Boden: Mikrobieller Abbau und Nebenwirkungen auf Mikroorganismen.– 575 pp.; Weinheim (VCH).

Doornenbal, J. C. & Stevenson, A. G. (eds.) (2010): Petroleum Geological Atlas of the Southern Permian Basin Area. – 354 pp.; Houten (EAGE Publications).

Dubois, M., Ayt Ougougdal, M., Meere, P., Royer, J.-J., Boiron, M. C. & Cathelineau, M. (1996): Temperature of paleo- to modern self-sealing within a continental rift basin; the fluid inclusion data (Soultz-sous-Forets, Rhine Graben, France). – European Journal of Mineralogy, **8**: 1065 – 1080.

Dunning, F. W. & Evans, A. M. (eds.) (1986): Mineral deposits of Europe. Volume 3 – Central Europe. – Institution of Mining and Metallurgy: 355 pp.

—, Mykura, W. & Slater, D. (eds.) (1982): Mineral deposits of Europe. Volume 2 – Southeast Europe. – Institution of Mining and Metallurgy: 304 pp.

—, Garrard, P., Haslam, H. W. & Ixer, R. A. (eds.) (1989): Mineral deposits of Europe: Volume 4/5 – Southwest and Eastern Europe, with Iceland. – Institution of Mining and Metallurgy and The Mineralogical Society: 454 pp.

EC (1986): Council Directive 86/278/EEC of 12 June 1986 on the protection of the environment, and in particular of the soil, when sewage sludge is used in agriculture. – Official Journal of the European Communities: **L181**: 6 – 12; Luxembourg.

— (1991): Council Directive 91/338/EEC of 18 June 1991 amending for the 10th time Directive 76/769/EEC on the approximation of the laws, regulations and administrative provisions of the Member States relating to restrictions on the marketing and use of certain dangerous substances and preparations. – Official Journal of the European Communities, **L186**: 59 – 63; Luxembourg.

— (2005): Directive 2005/32/EC of the European Parliament and of the Council of 6 July 2005 establishing a framework for the setting of ecodesign requirements for energy-using products and amending Council Directive 92/42/EEC and Directives 96/57/EC and 2000/55/EC of the European Parliament and of the Council. – Official Journal of the European Union, **L 191**: 29 – 58; Luxembourg. URL: http://eur-lex.europa.eu/LexUriServ/LexUriServ.do?uri=OJ:L:2005:191:0029:0029:EN:PDF (last access 14 December 2012).

— (2005a): Soil Atlas of Europe. – European Soil Bureau Network, Office for Official Publications of the European Communities, **L-2995**: 128 pp.; Luxembourg.

— (2006): Regulation (EC) No 1907/2006 of the European Parliament and of the Council of 18 December 2006 concerning the Registration, Evaluation, Authorisation and Restriction of Chemicals (REACH), establishing a European Chemicals Agency, amending Directive 1999/45/EC and repealing Council Regulation (EEC) No 793/93 and Commission Regulation (EC) No 1488/94 as well as Council Directive 76/769/EEC and Commission Directives 91/155/EEC, 93/67/EEC, 93/105/EC and 2000/21/EC. – Official Journal of the European Union, **L396**: 1 – 849; Luxembourg. URL: http://eur-lex.europa.eu/LexUriServ/LexUriServ.do?uri=OJ:L:2006:396:0001:0849:EN:PDF (last access 14 December 2012).

ECB – European Chemicals Bureau (2007): European Union Risk Assessment Report: Cadmium oxide and cadmium metal, Part 1: Environment. – European Commission, Joint Research Centre, **72**: 678 pp.

ECHA – European Chemicals Agency (2010): Guidance on information requirements and chemical safety assessment. – Chapter R.16: Environmental Exposure Estimation: 136 pp.; Helsinki. URL: http://echa.europa.eu/documents/10162/13632/information_requirements_r16_en.pdf (last access 29 December 2012).

Edel, J. B., Schulmann, K. & Rotstein, Y. (2007): The Variscan tectonic inheritance of the Upper Rhine Graben: evidence of reactivations in the Lias, Late Eocene–Oligocene up to the recent. – International Journal of Earth Sciences, **96**(2): 305 – 325.

EEA – European Environmental Agency (1998): Environment in the European Union at the turn of the century – Chapter 3.7: Waste generation and management.– Environmental assessment report, **2**: 24 pp.; Copenhagen.

— (2005): Interpolated nitrogen oxides of the year 2005 included in file 'gis_data_2005.zip. URL: http://www.eea.europa.eu/data-and-maps/data/interpolated-air-quality-data-1 (last access 8 February 2013).

EFSA – European Food Safety Authority (2011): Statement on tolerable weekly intake for cadmium.– EFSA Panel on Contaminants in the Food Chain (CONTAM). – EFSA Journal, **9** (2): 1 – 19.

EGS – EuroGeoSurveys Geochemistry Working Group (2008): EuroGeoSurveys geochemical mapping of agricultural and grazing land in Europe (GEMAS) – Field manual. – NGU Report, **2008.038**: 46 pp.; Trondheim. URL: http://www.ngu.no/upload/Publikasjoner/Rapporter/2008/2008_038.pdf (last access 11 February 2013).

EID (2013): Earth impact database. – Planetary and Space Science Centre. URL: http://www.passc.net/ (last access 5 February 2013).

Eilu, P. (2012): Mineral deposits and metallogeny of Fennoscandia. – Geological Survey of Finland, Special paper, **53**: 403 pp.; Espoo. URL: http://arkisto.gtk.fi/sp/sp53/sp53.pdf; see also http://en.gtk.fi/informationservices/databases/fodd/index.html (last access 7 December 2012).

Eisbacher, G. H. & Brandner, R. (1996): Superposed fold-thrust structures and high-angle faults, northwestern Calcareous Alps, Austria. – Eclogae Geologicae Helvetiae, **89**: 553 – 571.

Elberling, B., Breuning-Madsen, H., Hinge, H. & Asmund, G. (2010): Heavy metals in 3300-year-old agricultural soils used to assess present soil contamination. – European Journal of Soil Science, **61**: 74 – 83.

Environment Agency (2009): Soil guideline values for selenium in soil. – Science Report SC050021/Selenium SGV: 8 pp.; Bristol. URL: http://www.environment-agency.gov.uk/static/documents/Research/SCHO0309BPQH-e-e.pdf (last access 10 January 2014).

— (2009a): Human health toxicological assessment of contaminants in soil. – Science Report, Final SC050021/SR2: 79 pp.; Bristol. URL: http://www.environment-agency.gov.uk/static/documents/Research/TOX_guidance_report_-_final.pdf (last access 10 January 2014).

— (2009b): Soil guideline values for cadmium in soil. – Science Report SC050021/Cadmium SGV: 11 pp.; Bristol. URL: http://www.environment-agency.gov.uk/static/documents/Research/SCHO0709BQRO-e-e.pdf (last access 10 January 2014).

Eriksson, J., Mattsson, L. & Söderström, M. (2010): Tillståndet i svensk åkermark och gröda, data från 2001 – 2007 (Current status of Swedish arable soils and cereal crops. Data from the period 2001 – 2007). – Naturvårdsverket, Report **6349**: 131 pp. URL: http://www.naturvardsverket.se/Documents/publikationer/978-91-620-6349-8.pdf (last access 7 December 2012).

ESBN (2005): Soil atlas of Europe. – European Commission in collaboration with the European Soil Bureau Network, Office for Official Publications of the European Communities: 128 pp.; Luxembourg.

ESDB (2004): The European Soil Database – distribution version V2.0. – European Commission and the European Soil Bureau Network. – CD-ROM. URL: http://eusoils.jrc.ec.europa.eu/ESDB_Archive/ESDB_Data_Distribution/ESDB_data.html (last access 24 January 2013).

EU (2010): The factory of life – why soil biodiversity is so important. – Publications Office of the European Union: 22 pp.; Luxembourg.

Evans, A. M. (1980): An introduction to ore geology. – 231 pp.; New York (Elsevier).

Evans, D. J., Graham, C., Armour, A. & Bathurst, P. (2003): The Millennium Atlas: Petroleum geology of the central and northern North Sea. – Geological Society, London: 389 pp.

Fall, D. A. (2003): The geotechnical and geochemical characterisation of the brickearth of southern England. – PhD Thesis, School of Earth and Environmental Sciences; University of Portsmouth.

FAO (2006): Guidelines for soil description. – Food and Agricultural Organization of the United Nations: 97 pp.; Rome.

Fergusson, J. E. (1990): The heavy metals: Chemistry, environmental impact and health effects. – 614 pp.; New York (Pergamon Press).

Finch, T. F. (1971): Soils of County Clare, Soil. – National Soil Survey of Ireland, Survey Bulletin, **23**.

Fink, J., Haase, G. & Ruske, R. (1977): Bemerkungen zur Lösskarte von Europa 1: 2.5 Mio. – Petermanns Geographische Mitteilungen, **2**: 81 – 93.

Flint, R. F. (1971): Glacial and Quaternary geology.– 892 pp.; New York (Wiley).

Flowers, T. J. (1988): Chloride as a nutrient and as an osmoticum. – Advanced Plant Nutrition, **3**: 55 – 78.

Fordyce, F. (2005): Selenium deficiency and toxicity in the environment. – In: Selinus, O., Alloway, B. J., Centeno, J. A., Finkelman, R. B., Fuge, R., Lindh, U. & Smedley, P. (eds.): Essentials of medical geology – Impacts of the natural environment on public health: 373 – 415; Amsterdam (Elsevier).

Fordyce, F. M., Zhang, G., Green, K. & Liu, X. (2000): Soil, grain and water chemistry and human selenium imbalances in Enshi District, Hubei Province, China. – Applied Geochemistry, **15**(1): 117 – 132.

—, Brereton, N., Hughes, J., Luo, W. & Lewis, J. (2010): An initial study to assess the use of geological parent materials to predict the Se concentration in overlying soils and in five staple foodstuffs produced on them in Scotland. – Science of the Total Environment, **408**(22): 5295 – 5305.

Fortescue, J. A. C. (1967): The role of geochemistry in exploration architecture. – Canadian Institute of Mining and Metallurgy, Bulletin, **60**: 1 – 4.

— (1974): The environment and landscape geochemistry. – Western Miner, March 1974.

— (1974a): Exploration geochemistry and landscape. – Canadian Institute of Mining and Metallurgy, Bulletin, **67**(751): 80 – 87.

— (1975): The use of landscape geochemistry to process exploration geochemical data. – In: Bradshaw, P.M.D. (ed.): Conceptual models in exploration geochemistry – The Canadian Cordillera and Canadian Shield. – Journal of Geochemical Exploration, **4**(1): 3 – 13.

— (1992): Landscape geochemistry: Retrospect and prospect – 1990. – Applied Geochemistry, **7**(1): 1 – 53.

Franke, W. (2000): The mid-European segment of the Variscides: Tectonostratigraphic units, terrane boundaries and plate tectonic evolution. – Geological Society, London, Spec. Publ., **179**: 35 – 61.

— & Engel, W. (1986): Synorogenic sedimentation in the Variscan belt of Europe. – Bulletin de la Société Géologique de France, **82**(1): 25 – 33.

Frechen M., Oches, E.A. & Kohfeld, K.E. (2003): Loess in Europe – mass accumulation rates during the Last Glacial Period. – Quaternary Science Reviews, **22**: 1835 – 1857.

Friedrich, G.H. & Herzig, P.M. (eds.) (1988): Base metal sulfide deposits in sedimentary and volcanic environments. – Society for Geology Applied to Mineral Deposits, Spec. Publ., **5**: 290 pp.; Berlin (Springer).

Fuge, R. (2005): Anthropogenic Sources. – In: Selinus, O., Alloway, B.J., Centeno, J.A., Finkelman, R.B., Fuge, R., Lindh, U. & Smedley, P. (eds.): Essentials of medical geology – Impacts of the natural environment on public health: 43 – 60; Amsterdam (Elsevier).

— (2005a): Soils and iodine deficiency. – In: Selinus, O., Alloway, B.J., Centeno, J.A., Finkelman, R.B., Fuge, R., Lindh, U. & Smedley, P. (eds.): Essentials of medical geology – Impacts of the natural environment on public health: 417 – 433; Amsterdam (Elsevier).

Gao, S., Luo, T.-C., Zhang, B.-R., Zhang, H.-F., Han, Y.-W., Hu, Y.-K. & Zhao, Z.-D. (1998): Chemical composition of the continental crust as revealed by studies in east China. – Geochimica et Cosmochimica Acta, **62**: 1959 – 1975.

Garrett, R.G., MacLaurin, A.I., Gawalko, E.J., Thachuk, R. & Hall, G.E.M. (1996): A precipitation model for estimating the cadmium content of durum wheat soil chemistry. – Journal of Geochemical Exploration, **64** (1 – 3): 101 – 110.

—, Porter, A.R.D., Hunt, P.A. & Lalor, G.C. (2008): The presence of anomalous trace element levels in present day Jamaican soils and the geochemistry of Late-Miocene or Pliocene phosphorites. – Applied Geochemistry, **23**: 822 – 834.

—, Reimann, C., Smith, D.B. & Xie, X. (2008a): From geochemical prospecting to international geochemical mapping: A historical overview. – In: Reimann, C. & Smith, D.B. (eds): Thematic set in honour of Arthur G. Darnley (1930-2006). – Geochemistry: Exploration, Environment, Analysis, **8**(3-4): 205 – 217.

—, Porter, A.R.D. & Hunt, P.A. (2010): An occurrence of cadmium phosphorite soil concretions in Jamaica. – Applied Geochemistry, **25**: 1047 – 1055.

Gee, D.G. (2005): Scandinavian Caledonides (with Greenland). – In: Selley, R.C., Cocks, L.R.M. & Plimer, I.R. (eds.): Encyclopaedia of Geology: 64 – 74; Amsterdam (Elsevier).

— & Stephenson, R.A. (2006): The European lithosphere: An introduction. – Geological Society, London, Memoirs, **32**(1): 1 – 9.

—, Fossen, H., Henriksen, N. & Higgins, A.K. (2008): From the Early Paleozoic Platforms of Baltica and Laurentia to the Caledonide Orogen of Scandinavia and Greenland. – Episodes, **31**: 44 – 51.

Gijssel, K. van (2006): A continent-wide framework for local and regional stratigraphies: Application of generic sequence and event stratigraphy to the Middle Pleistocene terrestrial succession of Northwest and Central Europe. – Ph.D. thesis, Palaeolithic Studies Leiden: Pionier Research group 'Changing Views on Ice Age Foragers', Leiden University, Faculty of Archaeology: 144 pp. URL: https://openaccess.leidenuniv.nl/handle/1887/4985 (last access 8 December 2012).

Glazovskaya, M.A. (1963): On geochemical principles of the classification of natural landscapes. – International Geological Reviews, **5**: 1403 – 1431.

— (1988): Geochemistry of natural and industrial landscapes of the USSR. – Vysshaya Shkola: 328 pp.; Moscow.

GLC 2000 database (2003): Global Land Cover 2000 database. – European Commission, Joint Research Centre, 2003. URL: http://bioval.jrc.ec.europa.eu/products/glc2000/glc2000.php (last access 13 January 2013).

Godt, J., Schneidig, F., Grosse-Siestrup, Ch., Esche, V., Brandenburg, P., Reich, A. & Groneberg, D. A. (2006): The toxicity of cadmium and resulting hazards for human health. – Journal of Occupational Medicine and Toxicology, **1**:22: 6 pp. URL: http://www.occup-med.com/content/pdf/1745-6673-1-22.pdf (last access 13 January 2013).

Görür, N., Monod, O., Okay, A. I., Sengör, A.M. C., Tüysüz, O., Yiğitbaş, E., Sakınç, M. & Akkök, R. (1997): Palaeogeographic and tectonic position of the Carboniferous rocks of the western Pontides (Turkey) in the frame of the Variscan belt. – Bulletin de la Société Géologique de France, **168**(2): 197 – 205.

Goidts, E., van Wesemael, B. & van Oost, K. (2009): Driving forces of soil organic matter evolution at the landscape and regional scale using data from a stratified soil monitoring. – Global Change Biology, **15**: 2981-3000.

Goldschmidt, V.M. (1937): The principles of distribution of chemical elements in minerals and rocks. – Journal of the Chemical Society: 655 – 673; London.

— (1958): Geochemistry.– 730 pp.; London (Oxford Univ. Press).

Gottschalch, U., Birke, M., Stärk, H.-J., Kupsch, H. & Lippold, H. (2007): Characterization of urban NOM in a municipal area with disused toxic waste sites. – Applied Geochemistry, **22**, 2435 – 2455.

Grad, M. & Tiira, T. (2009): The Moho depth map of the European Plate. – Geophysical Journal International, **176**(1): 279 – 292.

GRAHMANN, R. (1932): Der Löss in Europa. – Mitt. Ges. Erdk., **51**: 5 – 24; Leipzig.

GREENWOOD, N. N. & EARNSHAW, A. (1997): Chemistry of the Elements (2nd Edition): 1341 pp.; Amsterdam (Elsevier).

GREGORAUSKIENĖ, V. (2012): Regularities of vertical element distribution within the soil profile in Lithuania. – Summary of doctoral dissertation. Vilnius University, Physical sciences, physical geography: 89 pp.

GRENNE, T., IHLEN, P. M. & VOKES, F. M. (1999): Scandinavian Caledonide Metallogeny in a plate tectonic perspective. – Mineralium Deposita, **34**: 422 – 471.

GRIEVE, R. A. F. (1997): Extraterrestrial impact events: The record in the rocks and the stratigraphic column. – Palaeogeography, Palaeoclimatology, Palaeoecology, **132**(1 – 4): 5 – 23.

GRIEVE, R., RUPERT, J., SMITH, J. & THERRIAULT, A. (1995): The record of terrestrial impact cratering. – GSA Today, **5**: 189 – 196.

GRIFFITH, L. C. (2001): Potassium channels: the importance of transport signals. – Current Biology, **11**(6): 226 – 228.

GRUNSKY, E. & SMEE, B. W. (1999): The differentiation of soil types and mineralization from multi-element geochemistry using multivariate methods and digital topography. – Journal of Geochemical Exploration, **67**: 287 – 299.

GÜNTHER, A., VAN DEN EECKHAUT, M., REICHENBACH, P., HERVÁS, J., MALET, J.-P., FOSTER, C. & GUZZETTI, F. (2013): New developments in harmonized landslide susceptibility mapping over Europe in the framework of the European Soil Thematic Strategy. – Proceedings, Second World Landslide Forum (Rome/Italy, October, 3-7, 2011). – In: MARGOTTINI, C., CANUTI, P. & SASSA, K. (eds.): Landslide Science and Practice, **1**: 297 – 301; Berlin (Springer).

GUILBERT, J. M. & PARK, C. F. (2007): The geology of ore deposits. – Waveland Press: 985 pp.

GUSTAFSSON, J. P. & JOHNSSON, L. (1994): The association between selenium and humic substances in forested ecosystems-laboratory evidence – Applied Organometallic Chemistry, **8**: 141 – 147.

HAASE, D., FINK, J., HAASE, G., RUSKE, R., PECSI, M., RICHTER, H., ALTERMANN, M. & JAEGER, K.-D. (2007): Loess in Europe – its spatial distribution based on a European Loess Map, scale 1 : 2.500.00. – Quaternary Science Reviews, **26**: 1301 – 1312.

HALL, G. E. M. (1998): Analytical perspective on trace element species of interest in exploration. – Geochemistry: Exploration, Environment, Analysis, **61**: 1 – 19.

HALLBERG, A., BERGMAN, T., GONZALEZ, J., LARSSON, D., MORRIS, G., PERDAHL, J-A., RIPA, M., NIIRANEN, T. & EILU, P. (2012): Metallogenic areas in Sweden. – In: EILU, P. (ed.): Mineral deposits and metallogeny of Fennoscandia. – Geological Survey of Finland, Special paper, **53**: 139 – 206; Espoo.

HAMILTON, S. M. (1998): Electrochemical mass transport in overburden; a new model to account for the formation of selective-leach geochemical anomalies in glacial terrain. – Geochemistry: Exploration, Environment, Analysis, **63**: 155 – 172.

HANDRECK, K. A. & BLACK, N. D. (1994): Growing Media for Plants and Turf: 448 pp.; Randwick (Univ. of New South Wales Press).

HANS, J., JÖST, C. & ZAUKE, G.-P. (2010): Trace metal concentrations in Antarctic Benthic Polychaetes, Extended Abstract.– Presented at the 15th International Conference on Heavy Metals in the Environment (Gdansk/Poland, September 19. – 23., 2010): 1 – 5.

HANSKI, E. & HUHMA, H. (2005): Central Lapland greenstone belt. – Developments in Precambrian Geology, **14**: 139 – 193.

HARANGI, S., DOWNES, H. & SEGHEDI, I. (2006): Tertiary-Quaternary subduction processes and related magmatism in the Alpine-Mediterranean region. – In: GEE, D. & STEPHENSON, R. (eds.): European Lithosphere Dynamics. Geological Society, Memoirs, **32**: 167 – 190; London.

HARTWICH, R. (1998): Karte der Bodenregionen Europas 1 : 5 000 000 (Version 1.0); part of Georeferenced Soils Database for Europe. – BGR; Hannover.

—, BARITZ, R., FUCHS, M., KRUG, D. & THIELE, S. (2005): Explanation to the Soil Regions Map of the European Union and Adjacent Countries 1 : 5,000,000 (Version 2.0). – European Soil Bureau, research reports: 122 pp.; Luxembourg.

—, THIELE, S., BARITZ, R., FUCHS, M. & KRUG, D. (2006): Erläuterungen zur Bodenregionenkarte der Europäischen Union und ihrer Nachbarstaaten im Maßstab 1 : 5 000 000 (Version 2.0). BGR; Hannover. URL: http://www.bgr.bund.de (last access 24 January 2013).

HASTIE, T., TIBSHIRANI, R. & FRIEDMAN, J. H. (2009): The Elements of Statistical Learning – Data Mining, Inference and Prediction: 746 pp.; New York (Springer).

HAWKES, H. E. & WEBB, J. S. (1962): Geochemistry in Mineral Exploration.– 415 pp.; New York (Harper & Row).

HAYES, K. F., PAPELIS, C. & O'LECKIE, J. (1988): Modeling ionic strength effects on anion adsorption at hydrous oxide/solution interfaces. – Journal of Colloid and Interface Science, **125**: 717 – 726.

HAYGARTH, P. M. (1994): Global importance and global cycling of selenium. – In: FRANKENBERGER, W. T. & BENSON, S. (eds.): Selenium in the Environment, **1**: 1 – 27; New York (Dekker).

HEDENQUIST, J. W. (2005): Economic geology – One hundredth anniversary volume. – 1136 pp.; Littleton, CO (Economic Geology Pub. Co.).

Herber, R.F.M. (2004): Cadmium. – In: Merian, E., Anke, M., Ihnat, M. & Stoeppler. M. (eds.): Elements and their Compounds in the Environment, **2**: 689 – 708; Weinheim (Wiley-VCH).

Hijmans, R.J., Cameron, S.E., Parra, J.L., Jones, P.G. & Jarvis, A. (2005): Very high resolution interpolated climate surfaces for global land areas. – International Journal of Climatology, **25**: 1965 – 1978.

Hill, R.J. (1996): Risks to humans – An Overview. – In: Proceedings of Sources of Cadmium in the Environment.– Organization for Economic Co-operation and Development: 84 – 94; Paris.

Hodgson, E. (2004): A textbook of modern toxicology: 557 pp.; New York (Wiley).

Högdahl, K., Andersson, U.B. & Eklund, O. (2004): The Transscandinavian igneous belt (TIB) in Sweden; a review of its character and evolution. – Geological Survey of Finland, Special Paper, **37**: 125 pp.; Espoo.

Hölttä, P., Balagansky, V., Garde, A.A., Mertanen, S., Peltonen, P., Slabunov, A., Sorjonen-Ward, P. & Whitehouse, M. (2008): Archaean of Greenland and Fennoscandia. – Episodes, **31**: 13 – 19.

Hoffman, J.E. & King, M.G. (1997): Selenium and selenium compounds. – In: Kroschwitz, J.I. & Howe Grant, M. (eds.): Kirk-Othmer Encyclopedia of Chemical Technology, 4th edition, **21**: 686 – 719; New York (Wiley).

Holmgren, G.G.S., Meyer, M.W., Chaney, R.L. & Daniels, R.B. (1993): Cadmium, lead, zinc, copper, and nickel in agricultural soils in the United States of America. – Journal of Environmental Quality, **22**(2): 335 – 348. URL: http://nature.berkeley.edu/classes/espm-120/Website/Holmgren1993.pdf (last access 13 January 2013).

Hrnčiarová, T., Mackovčin, P. & Zvara, I. (eds.) (2009): Landscape Atlas of the Czech Republic. – Ministry of the Environment of the Czech Republic and Silva Tarouca Research Institute for Landscape and Ornamental Gardening: 332 pp.; Prague.

Hu, Z. & Gao, S. (2008): Upper crustal abundances of trace elements: A revision and update. – Chemical Geology, **253** (3-4): 205 – 221.

Huber, S., Prokop, G., Arrouays, D., Banko, G., Bispo, A., Jones, R.J.A., Kibblewhite, M.G., Lexer, W., Möller, A., Rickson, R.J., Shishkov, T., Stephens, M., Toth, G., van den Akker, J.J.H., Varallyay, G., Verheijen, F.G.A. & Jones, A.R. (2008): Environmental assessment of soils for monitoring – Volume I: Indicators and criteria.– Office for Official Publications of the European Communities, EUR 23490 EN: 339 pp.; Luxembourg.

Hüttl, R.F., Prechtel, A. & Bens, O. (2007): Humusversorgung von Böden in Deutschland.– Research Report for the Environmental Protection Agency, Berlin. Project UFOPLAN No. FKZ 360 13 008. – Publications of the German Environmental Protection Agency: 233 pp.; Berlin.

Hutchison, C.S. (1983): Economic deposits and their tectonic setting. – 365 pp.; New York (Wiley).

Hutton, M. & de Meeûs, C. (2001): Analysis and Conclusions from Member States' Assessment of the Risk to Health and the Environment from Cadmium in Fertilisers. – European Commission – Enterprise DG, ETD/00/503201: 134 pp.

IFA – International Fertilizer Industry Association (2012): Cadmium. URL: http://www.fertilizer.org/ifa/HomePage/SUSTAINABILITY (last access 28 August 2012).

Il'yasova, D. & Schwartz, G.G. (2005): Cadmium and renal cancer. – Toxicology and Applied Pharmacology, **207** (2): 179 – 186.

Iskander, I.K. & Kirkham, M.B. (2001): Trace Elements in Soil. – Bioavailability, Flux and Transfer.– 287 pp.; Boca Raton, FL (CRC Press).

ISO (1995): International Standard ISO 11466: 1995. Soil quality – Extraction of trace elements soluble in aqua regia.

— 11277 (1998): International Organisation for Standardisation (05.98): Soil quality – Determination of particle size distribution in mineral soil material – Method by sieving and sedimentation.

— 11277 (2009): International Organisation for Standardisation: Soil quality – Determination of particle size distribution in mineral soil material – Method by sieving and sedimentation.

IUGS/IAGC TGGGB (2013): Website of the IUGS/IAGC Task Group on Global Geochemical Baselines. URL: http://www.globalgeochemicalbaselines.eu/ (last access 8 February 2013).

IUSS – International Union of Soil Sciences (2006): World reference base for soil resources 2006. – FAO, World Soil Resources Report, **103**: 132 pp.; Rome.

Ivanov, V.V. (1996): Ekologicheskaya geokhimia elementov. – Ekologia, kn. 1-6; Moscow [in Russian].

Jacobs, L.W. (ed.) (1989): Selenium in Agriculture and the Environment. – Soil Science Society of America, Spec. Publ., **23**: 233 pp.

Järup, L., Berglund, M., Elinder, C.G., Nordberg, G. & Vahter, M. (1998): Health effects of cadmium exposure: A review of the literature and a risk estimate. – Scandinavien Journal of Work, Environmental & Health, **24**(1): 1 – 51.

—, Best, N., Toledano, M.B., Wakefield, J. & Elliott, P. (2002): Geographical epidemiology of prostate cancer in Great Britain. – International Journal of Cancer, **97**(5): 695 – 699.

Janik, L.J., Merry, R.H. & Skjemstad, J.O. (1998): Can mid infrared diffuse reflectance analysis replace soil extractions?– Australian Journal of Experimental Agriculture, **38**: 681 – 696.

—, FORRESTER, S.T. & RAWSON, A. (2009): The prediction of soil chemical and physical properties from mid-infrared spectroscopy and combined partial least-squares regression and neural networks (PLS-NN) analysis. – Chemometrics and Intelligent Laboratory Systems, **97**: 179 – 188.

JEFFERY, S., GARDI, C., JONES, A., MONTANARELLA, L., MARMO, L., MIKO, L., RITZ, K., PERES, G., RÖMBKE, J. & VAN DER PUTTEN, W.H. (eds.) (2010): European Atlas of Soil Biodiversity. – Publications Office of the European Union: 128 pp.; Luxembourg.

JENNESS, J. (2005): Topographic Position Index (tpi_jen.avx) extension for ArcView 3.x.; (Jenness Enterprises).

JENNY, H. (1941): Factors of Soil Formation – A System of Quantitative Pedology. – 281 pp.; New York (McGraw Hill Book Company).

JÖST, C. & ZAUKE, G.-P. (2008): Trace-metal concentrations in Antarctic sea spiders (Pycnogonida, Pantopoda). – Marine Pollution Bulletin, **56**: 1396 – 1399.

JOHANSSON, Å. & LARSEN, O. (1989): Radiometric age determinations and Precambrian geology of Blekinge, southern Sweden. – Geologiska Föreningens i Stockholm Förhandlingar, **111**: 35 – 50.

JOHNSON, C.C., GE, X., GREEN, K.A. & LIU, X. (2000): Selenium distribution in the local environment of selected villages of the Keshan Disease belt, Zhangjiakou District, Hebei Province, People's Republic of China. – Applied Geochemistry, **15**: 385 – 401.

—, FORDYCE, F.M. & RAYMAN, M.P. (2009): Symposium on 'Geographical and geological influences on nutrition': Factors controlling the distribution of selenium in the environment and their impact on human health. – Proceedings of the Nutrition Society, **69**: 119 – 132.

JONES, A., MONTANARELLA, L. & JONES, R. (eds.) (2005): Soil Atlas of Europe. – European Soil Bureau Network, European Commission, Office for Official Publications of the European Communities: 128 pp.; Luxembourg.

JONES, R.J.A., HIEDERER, R., RUSCO, E., LOVELAND, P.J. & MONTANARELLA, L. (2004): The map of organic carbon in topsoils in Europe, Version 1.2, September 2003: Explanation of Special Publication Ispra 2004 No.72 (S.P.I.04.72). – European Soil Bureau Research Report No.17, Office for Official Publications of the European Communities, EUR 21209 EN: 26 pp.; Luxembourg.

—, HIEDERER, R., RUSCO, E., LOVELAND, P.J. & MONTANARELLA, L. (2005a): Estimating organic carbon in the soils of Europe for policy support. – European Journal of Soil Science, **56**: 655 – 671.

JOUKAINEN, S. & YLI-HALLA, M. (2003): Environmental impacts and acid loads from deep sulfidic layers of two well-drained acid sulfate soils in western Finland. – Agriculture, Ecosystems and Environment, **95**: 297 – 309.

JURKOVIČ, I. B., GARAŠIĆ, V. & JURKOVIĆ, I.M. (2011): Cobalt, nickel, tungsten, cadmium, selenium, silver and gold-bearing mercurian tetrahedrite from the Saski Rad barite–siderite deposit in the Mid-Bosnian Schist Mts. – Geologia Croatica, **64**(3): 223 – 37.

KABATA-PENDIAS, A. (2001): Trace elements in soils and plants. – 413 pp.; Boca Raton, FL (CRC Press).

— & MUKHERJEE, A.B. (2007): Trace elements from soil to human. – 550 pp.; Berlin (Springer).

KADŪNAS, V., BUDAVIČIUS R., GREGORAUSKIENĖ, V., KATINAS, V., KLIAUGIENĖ, E., RADZEVIČIUS, A. & TARAŠKEVIČIUS, R. (1999): Lietuvos geocheminis atlasas – Geochemical Atlas of Lithuania. – Geologijos institutas, Lietuvos geologijos tarnyba: 230 pp.; Vilnius.

KALBERLAH, F. (1999): Cadmium und seine Verbindungen. – In: EIKMANN et. al. (eds.): Gefährdungsabschätzung von Umweltschadstoffen, Toxikologische Basisdaten und ihre Bewertung, **D172**: 1 – 44; Berlin (Erich Schmidt).

KAPUSTA, P., SZAREK-ŁUKASZEWSKA, G. & GODZIK, B. (2006): Spatio-temporal variation of element accumulation by Moehringia trinervia in a polluted forest ecosystem (South Poland). – Environmental Pollution, **143**(2): 285 – 293.

KARDEL, K. & RANK, G. (2010): Geochemische Übersichtskarte der Böden des Freistaates Sachsen 1 : 400 000. Arsen in Oberböden. (Arsengehalt im Königswasseraufschluss) and Cadmium in Oberböden (Cadmiumgehalte im Königswasseraufschluss). – Landesamt für Umwelt, Landwirtschaft und Geologie: 1 Blatt. URL http://www.umwelt.sachsen.de/umwelt/boden/11646.htm (last access 29 December 2012).

KASSE, C. (1997): Cold-climate aeolian sand-sheet formation in north-western Europe (c.14 – 12.4 ka): A response to permafrost degradation and increased aridity. – Permafrost and Periglacial Processes, **8**: 295 – 311.

— (2002): Sandy aeolian deposits and environments and their relation to climate during the Last Glacial Maximum and Lateglacial in northwest and central Europe. – Progress in Physical Geography, **26**: 507 – 532.

KAZANTZIS, G. (1987): Cadmium. – In: FISHBEIN, L., FURST, A. & MYRON, A. (eds.): Genotoxic and Carcinogenic Metals: Environmental and Occupational Occurrence and Exposure. – Advances in Modern Environmental Toxicology, **11**: 339 pp.; Princeton, NJ (Princeton Scientific Publ.).

KESKINEN, R., RÄTY, M. & YLI-HALLA, M. (2011): Selenium fractions in selenate-fertilized field soils of Finland. – Nutrient Cycling Agroecosystems, **91**: 17 – 29.

KHAN, N.A. & KHAN, S. (eds.) (2006): Cadmium toxicity and tolerance in plants.– 190 pp.; Oxford (Alpha Science International Ltd.).

KIBBLEWHITE, M. G. (2012): Definition of priority areas for soil protection at continental scale. – Soil Use and Management **28**: 128 – 133.

KLEY, J. & VOIGT, T. (2008): Late Cretaceous intraplate thrusting in central Europe: Effect of Africa-Iberia-Europe convergence, not Alpine collision. – Geology, **36**: 839 – 842.

KLOMINSKŹ, J. (1969): Krkonošsko-jizerský granitoidni massiv. – Sbor. geol. Věd, Geol., **15**: 7 – 133.

KOEBERL, C. (1998): Identification of meteoritic components in impactites. – Geological Society, London, Spec. Publ., **140**(1): 133 – 153.

— (2007): The geochemistry and cosmochemistry of impacts. – In: DAVIS, A. M. (ed.): Meteorites, Planets, and Comets – Volume 1: Treatise on Geochemistry: p. 1.28.1 – 1.28.52; Oxford (Elsevier). Doi: 10.1016/B978-008043751-4/00228-5.

KOETHE, R. & LEHMEIER, F. (1996): SARA – System zur Automatischen Relief-Analyse. – User Manual, 2nd Edition. – Dept. of Geography, University of Gottingen; Göttingen [unpubl.].

KOLJONEN, T. (ed.) (1992): The Geochemical Atlas of Finland, Part 2: Till. – Geological Survey of Finland: 218 pp.; Espoo.

KOLSTRUP, E., GRÜN, R., MEJDAHL, V., PACKMAN, S. C. & WINTLE, A. G. (1990): Stratigraphy and thermoluminescence dating of Late Glacial cover sand in Denmark. – Journal of Quaternary Science **5**(3): 207 – 224.

KOSTER, E. A. (2009): The "European Aeolian Sand Belt": Geoconservation of drift sand landscapes. – Geoheritage, **1**: 93 – 110.

KOZLOVSKIY, V. V. (1972): Structural functional and mathematical model of migrational landscape geochemical processes (transl.).– Pochvovedeniye, **4**: 122 – 138.

LÅG, J. & STEINNES, E. (1978): Regional distribution of selenium and arsenic in humus layers of Norwegian forest soils. – Geoderma, **20**: 3 – 14.

LAHTINEN, R., GARDE, A. A. & MELEZHIK, V. A. (2008): Paleoproterozoic evolution of Fennoscandia and Greenland. – Episodes, **31**: 20 – 28.

—, KORJA, A. & NIRONEN, M. (2005): Palaeoproterozoic tectonic evolution. – In: LEHTINEN, M., NURMI, P. & RÄMÖ, O. T. (eds.): Precambrian Geology of Finland – Key to the evolution of the Fennoscandian Shield. – Developments in Precambrian Geology, **14**: 481 – 532.

LALOR, G. C., RATTRAY, R., SIMPSON, P. & VUTCHOV, M. (1998): Heavy metals in Jamaica, Part 3: The distribution of cadmium in Jamaica soils. – Revista internacional de contaminación ambiental, **14**: 7 – 12.

LANGNER, E. B. (2000): Selenium and Selenium Compounds. – In: Ullmann's Encyclopedia of Industrial Chemistry; Weinheim (Wiley-VCH).

LE MAITRE, R. W. (ed.) (2002): Igneous rocks – a classification and glossary of terms. – Recommendations of the International Union of Geological Sciences Subcommission on the Systematics of Igneous Rocks: 236 pp.; Cambridge (Cambridge Univ. Press).

LETTENS, S., VAN ORSHOVEN, J., VAN WESEMAEL, B. & MUYS, B. (2004): Soil organic and inorganic carbon content of landscape units in Belgium for 1950 – 1970. – Soil Use and Management, **20**: 40 – 47.

LEVIN, H. L. (2010): The Earth through Time. – 9th edition: 624 pp.; Hoboken, NJ (Wiley).

LI, H.-F., GRAY, C., MICO, C., ZHAO, F. J. & MCGRATH, S. P. (2009): Phytotoxicity and bioavailability of cobalt to plants in a range of soils. – Chemosphere, **75**: 979 – 986.

LI, K., ZHANG, H., PAN, X. & LI, X. (2005): Determination of Fluorine Content by Selective Ionic Electrode Method. Pp- 226-234. – In: ZHENG, K., YE, J. & JIANG, B. (eds.): Selected analytical methods of 57 elements for multipurpose geochemical survey. – Geological Publishing House; Beijing.

LIDE, D. R. (ed.) (1996): CRC Handbook of Chemistry and Physics, 77th Edition, 1996-1997; Boca Raton, FL (CRC Press).

LIMA, A., ALBANESE, S. & CICCHELLA, D. (2005): Geochemical baselines for the radioelements K, U and Th in the Campania region, Italy: a comparison of stream-sediment geochemistry and gamma-ray surveys. – Applied Geochemistry, **20**: 611 – 625.

LINDGREN, W. (1933): Mineral deposits: 930 pp.; New York (McGraw-Hill).

LINDSAY, W. L. & NORVELL, W. A. (1978): Development of DTPA soil test for zinc, iron, manganese and copper. – Soil Science Society of America Journal, **42**: 421 – 428.

LINNEMANN, U., D'LEMOS, R., DROST, K., JEFFRIES, T., GERDES, A., ROMER, R. L. & SAMSON, S. D. (2008): Cadomian tectonics. – In: MCCANN, T. (ed.): The Geology of central Europe – Volume 1: Precambrian and Palaeozoic. – Geological Society, London: 103 – 354.

LITTKE, R., BAYER, U., GAJEWSKI, D. & NELSKAMP, S. (eds.) (2008): Dynamics of complex intracontinental basins; the Central European Basin system. – 520 pp.; Berlin (Springer).

LOCK, K., WAEGENEERS, N., SMOLDERS, E., CRIEL, P., VAN EECKHOUT, H. & JANSSEN, C. R. (2006): Effect of leaching and aging on the bioavailability of lead to the springtail Folsomia candida. – Environmental Toxicology and Chemistry, **25**: 2006 – 2010.

LOFFREDO, N., MOUNIER, S., THIRY, Y. & COPPIN, F. (2011): Sorption of selenate on soils and pure phases: Kinetic parameters and stabilisation. – Journal of Environmental Radioactivity, **102**: 843 – 51.

LOMBI, E., ZHAO, F. J., DUNHAM, S. J. & MCGRATH, S. P. (2001): Phytoremediation of Heavy Metal-Contaminated Soils: Natural Hyperaccumulation versus Chemically Enhanced Phytoextraction. – Journal of Envirnonmental Quality, **30**: 1919 – 1926.

LORENC, M., OLSZYNSKI, W. & WISZNIEWSKA, J. (1998): Distribution of phosphorus in granites and pegmatites of the Sudetes Mts. and rapakivi granites from Mazury Complex (NE Poland). – Acta Universitatis Carolinae, Geologica, **42**: 69 – 72.

LOUKOLA-RUSKEENIEMI, K., RUSKEENIEMI, T., PARVIAINEN, A. & BACKMAN, B. (eds.) (2007): Arseeni Pirkanmaalla – esiintyminen, riskinarviointi ja riskinhallinta. RAMAS-hankkeen tärkeimmät tulokset (Arsenic in the Pirkanmaa Region in Finland: Occurrence in the Environment, Risk Assessment and Risk Management. Final results of the RAMAS project). – Abstract. – Helsinki University of Technology, Geoenvironmental Technology Section, Spec. Publ.: 156 pages; Espoo.

LUOSTO, U. (1991): Crustal Structures of Eastern Fennoscandia. – Tectonophysics, **189**: 19 – 27.

LUUKKONEN, E.J. (1992): Late Archaean and early Proterozoic structural evolutions in the Kuhmo-Suomussalmi terrain, eastern Finland. Annales Universitatis Turkuensis. Sarja-Ser.A.II. – Biologica-Geographica-Geologica, **78**: 115 pp.

MAGILL, J., PFENNIG, G. & GALY, J. (2006): Chart of the Nuclides, 7th edition.– European Commission; Karlsruhe.

MANN, A., CARITAT, P. DE & PRINCE, P. (2012): Bioavailability of nutrients in Australia from Mobile Metal Ion® analysis of catchment outlet sediment samples: Continental scale patterns and processes. – Geochemistry: Exploration, Environment, Analysis, **12**: 1 – 16.

MANN, A.W. (1983): Hydrogeochemistry on the Yilgarn Block, Western Australia – ferrolysis and heavy metals in continental brines. – Geochimica et Cosmochimica Acta, **47**: 181 – 190.

— (2010): Strong versus weak digestions: Ligand-based soil extraction geochemistry. – Geochemistry: Exploration, Environment, Analysis, **10**: 17 – 26.

— & DEUTSCHER, R.L. (1977): Solution geochemistry of copper in water containing carbonate, sulphate and chloride ions. – Chemical Geology, **19**: 253 – 265.

— & DEUTSCHER, R.L. (1980): Solution geochemistry of lead and zinc in water containing carbonate, sulphate and chloride ions. – Chemical Geology, **29**: 293 – 311.

—, MANN, A.T., HUMPHREYS, D.B., DOWLING, S.E., STALTARI, S. & MYERS, L. (1997): Soil geochemical anomalies – their dynamic nature and interpretation. – Mineral and Energy Research Institute of Western Australia, Report **184**: 132 pp.

—, BIRRELL, R.D., MANN, A.T., HUMPHREYS, D.B. & PERDRIX, J.L. (1998): Application of the mobile metal ion technique to routine geochemical exploration. – Journal of Geochemical Exploration, **61**: 87 – 102.

—, BIRRELL, R.D., FEDIKOW, M.A.F. & DE SOUZA, H.A.F. (2005): Vertical ionic migration: mechanisms, soil anomalies and sampling depth for mineral exploration. – Geochemistry: Exploration, Environment, Analysis, **5**: 201 – 210.

MARMO, L. (2002): Heavy metal limits for soil. – In: Sewage Sludge in the EU, European Commission – DG ENV.A2, Brussels, 8-10 April 2002: 17 pp.

MASSFELDT, T. (2011): Metalle (Chap. 17.3). – In: BLUME, H.-P., HORN, R. & THIELE-BRUHN, S. (eds.): Handbuch des Bodenschutzes.– 4th edition; Weinheim (Wiley-VCH).

MATTE, P. (1991): Accretionary history and crustal evolution of the Variscan belt in Western Europe. – Tectonophysics, **196**: 309 – 337.

MAYLAND, H.F., GOUGH, L.P. & STEWART, K.C. (1991): Selenium mobility in soils and its absorption, translocation, and metabolism in plants. Chapter E. – In: SEVERSON, R.C., FISHER, S.E. JR. & GOUGH, L.P. (eds.): Proceedings of 1990 Billings Land Reclamation Symposium; USA-MT-Billings, 1991/03/25-30: 55 – 64.

MCBRATNEY, A.B., MENDONÇA-SANTOS, M.L. & MINASNY, B. (2003): On digital soil mapping. – Geoderma, **117**: 3 – 52.

—, ODEH, I.O.A., BISHOP, T.F.A., DUNBAR, M.S. & SHATAR, T.M. (2000): An overview of pedometric techniques for use in soil survey. – Geoderma, **97**: 293 – 327.

MCBRIDE, M.B. (1989): Reactions controlling heavy metal solubility in soils. – Advances in Agronomy, **10**: 1 – 56.

MCCANN, T. (ed.) (2008): The Geology of Central Europe – Volumes 1 and 2. Volume 1: Precambrian and Palaeozoic; Volume 2: Mesozoic and Cenozoic. – Geological Society, London: 748 and 736 pp.

— (ed.) (2008a): Introduction & Overview. – In: MCCANN, T. (ed.): The Geology of Central Europe. – Geological Society, London, Vol. **1**: 1 – 20.

—, PASCAL, C., TIMMERMANN, M.J., KRZYWIEC, P., LÓPEZ-GÓMEZ, J., KRAWCZYK, C.M., RIEKE, H. & LAMARCHE, J. (2006): Post-Variscan (End Carboniferous-Early Permian) basin evolution in Western and Central Europe. – In: GEE, D. & STEPHENSON, R.A. (eds.): European Lithosphere Dynamics. – Geological Society, London, Memoir, **32**: 355 – 388.

—, SKOMPSKI, S., POTY, E., DUSAR, M., VOZÁROVÁ, A., SCHNEIDER, J., WETZEL, A., KRAINER, K., KORNPIHL, K., SCHÄFER, A., KRINGS, M., OPLUSTIL, S. & TAIT, J. (2008): Carboniferous . – In: MCCANN, T. (ed.): The Geology of Central Europe – Vol. 1: Precambrian and Palaeozoic. – Geological Society, London: 411 – 530.

—, KIERSNOWSKI, H., KRAINER, K., VOZÁROVÁ, A., PERYT, T., OPLUSTIL, S., STOLLHOFEN, H., SCHNEIDER, J., WETZEL, A., BOULVAIN, F., DUSAR, M., TÖRÖK Á., HAAS J., TAIT, J. & KÖRNER, F. (2008a): Permian. – In: MCCANN, T. (ed.): The Geology of Central Europe – Vol. 1: Precambrian and Palaeozoic. – Geological Society, London: 531 – 598.

MCCUNE, B. & KEON, D. (2002): Equations for potential annual direct incident radiation and heat load index. – Journal of Vegetation Science, **13**(4): 603 – 606.

McGrath, D. (1996): Application of single and sequential extraction procedures to polluted and unpolluted soils. – Science of The Total Environment, **178**: 37 – 44.

McGrath, S. P. & Loveland, P. J. (1992): The soil geochemical atlas of England and Wales: 101 pp.; London (Blackie Academic).

Menzies, N. W., Donn, M. J. & Kopittke, P. M. (2007): Evaluation of extractants for estimation of the phytoavailable trace metals in soils. – Environmental Pollution, **145**: 121 – 130.

Meunier, J. D. (1992): Diagenesis and mechanisms of uranium accumulation by detrital organic matter. – In: Pagel, M. & Leroy, J. L. (eds.): Source, Transport and deposition of metals: 565 – 568; Rotterdam (Balkema).

Miko, S., Peh, Z., Bukovec, Z., Prohij, Z. & Kastmuller, E. (2000): Geochemical baseline mapping and lead pollution assessment of soils on the karst in Western Croatia. – Natura Croatica, **9**: 41 – 59.

Miljøstyrelsen (2010): Liste over kvalitetskriterier i relation til forurenet jord og kvalitetskriterier for drikkevand [in Danish]. http://www.mst.dk/NR/rdonlyres/0AB0AF23-4BD6-4901-BCD9-43F6F6FD6FC1/124958/Kvalitetskriterierjord_og_drikkevandfinaljuniogjul.pdf (last access 10 January 2014).

Miller, D. M. (1984): Reducing transformation bias in curve fitting. – The American Statistician, **38**(2): 124 – 126.

Misra, K. C. (2000): Understanding mineral deposits. – 845 pp.; Dordrecht (Springer).

Mitchell, A. H. G. & Garson, M. S. (1981): Mineral deposits and global tectonic settings. – 405 pp.; New York (Academic Press).

MoE – Ministry of the Environment of the Czech Republic (1994): Regulation No. 13/1994 of the Ministry of the Environment of the Czech Republic, specifying some details of the the agricultural land resources protection: 22 pp.

Montanarella, L., Ernstsen, V., Donezar, M., Inghe, O., Bruneau, P., Kubík, L., Berényi Üveges, J., van Ranst, E., Kayadjanian, M., Wolf, D. & Schamann, M. (2004): Existing soil monitoring systems. – In: van Camp, L., Bujarrabal, B., Gentile, A. R., Jones, R. J. A., Montanarella, L., Olazabal, C. & Selvaradjou, S.-K. (eds.): Reports of the Technical Working Groups established under the Thematic Strategy for Soil Protection. – Volume 5: Monitoring. – Office for Official Publications of the European Communities, EUR 21319 EN/5: 79 pp.; Luxembourg.

Monthel, J., Vadala, P., Leistel, J. M., Cottard, F. with the collaboration of Ilic, M., Strumberger, A., Tosovic, R. & Stepanovic, A. (2002): Mineral deposits and mining districts of Serbia. – Compilation map and GIS database. – BRGM/RC-51448-FR: 67.

Moore, I. D., Grayson, R. B. & Ladson, A. R. (1991): Digital terrain modelling: A review of hydrological, geomorphological and biological applications. – Hydrological Processes, **5**: 3 – 30.

— & Wilson, J. P. (1992): Length-slope factors for the revised universal soil loss equation: Simplified method of estimation. – Journal of Soil and Water Conservation, **47**(5): 423 – 428.

Moore, J. A., Noiva, R. & Wells, T. C. (1984): Selenium concentrations in plasma of patients with arteriographically defined coronary arteriosclerosis. – Clinical Chemistry, **30**: 1171 – 1173.

Morrow, H. (2001): Cadmium and Cadmium Alloys. – In: Kirk-Othmer: Encyclopedia of Chemical Technology. – 5th ed., **4**.: 471 – 507; New York (Wiley).

Morton, L. S., Evans, C. V., Harbottle, G. & Estes, G. O. (2001): Pedogenic Fractionation and Bioavailability of Uranium and Thorium in Naturally Radioactive Spodosols. – Soil Science Society of America Journal, **65**: 1197 – 1203.

Mortvedt, J. J. (2000): Bioavailability of Micronutrients. – In: Sumner, M.E. (ed.): Handbook of Soil Science: D69 – D88; Boca Raton, FL (CRC Press).

Mücher, C. A. (ed.) (2000): PELCOM project: Development of a consistent methodology to derive land cover information on a European scale from remote sensing for environmental modelling. – Final Report: 31 pp. URL: http://www.geo-informatie.nl/projects/pelcom/public/ (last access 24 January 2013).

Muecke, R., Klotz, T., Giedl, J., Buentzel, J., Kundt, G., Kisters, K., Prott, F. J. & Micke, O. (2009): Whole blood selenium levels (WBSL) in patients with prostate cancer (PC), benign prostatic hyperplasia (BPH) and healthy male inhabitants (HMI) and prostatic tissue selenium levels (PTSL) in patients with PC and BPH. – Acta Oncologica, **48**: 452 – 456.

Muhs, D. R., Budahn, J., Avila, A., Skipp, G., Freeman, J. & Patterson, D. (2010): The role of African dust in the formation of Quaternary soils on Mallorca, Spain and implications for the genesis of Red Mediterranean soils. – Quarternary Science Review, **29**: 2518 – 2543.

Navarro-Alarcon, M. & Cabrera-Vique, C. (2008): Selenium in food and the human body: A review. – Science of The Total Environment, **400**: 115 – 41.

Neuendorf, K. K. E., Mehl, J. P. & Jackson, J. A. (eds.) (2011): Glossary of geology. – American Geosciences Institute: 800 pp.; Alexandria, VA.

Nikishin, A. M., Ziegler, P. A., Stephenson, R. A., Cloetingh, S. A. P. L., Fume, A. V., Fokin, P. A., Ershov, A. V., Bolotov, S. N., Korotaev, M. V., Alekseev, A. S., Gorbachev, V. I., Shipilov, E. V., Lankreijer, A., Bembinova, E. Yu. & Shalimov, I. V. (1996): Late Precambrian to Triassic history of the East European Craton: Dynamics of sedimentary basin evolution. – Tectonophysics, **268**: 23 – 63.

Nirei, H., Furuno, K. & Kusuda, T. (2010): Medical Geology in Japan. – In: Selinus, O., Finkelman, R. B. & Centeno, J. A. (eds.): Medical geology: A regional synthesis: 329 – 354; Berlin (Springer).

NRIAGU, J. O. & PACYNA, J. M. (1988): Quantitative assessment of worldwide contamination of air, water and soils by trace metals. – Nature, **333**: 134 – 139; London.

NYSTUEN, J. P., ANDRESEN, A., KUMPULAINEN, R.A. & SIEDLECKA, A. (2008): Neoproterozoic basin evolution in Fennoscandia, East Greenland and Svalbard. – Episodes, **31**: 35 – 43.

OCZLON, M. S. (2006): Terrane Map of Europe. – Gaea Heidelbergensis, **15**. URL: http://www.searchanddiscovery.com/abstracts/pdf/2010/kiev/abstracts/ndx_Oczlon02.pdf (last access 13 January 2013).

ODEH, I. O. A., MCBRATNEY, A. B. & CHITTLEBOROUGH, D. J. (1995): Further results on prediction of soil properties from terrain attributes: heterotropic cokriging and regression-kriging. – Geoderma, **67**: 215 – 226.

OECD (2000): Test No. 106: Adsorption – Desorption Using a Batch Equilibrium Method; OECD Guidelines for the Testing of Chemicals, Section 1: Physical-Chemical properties. – OECD Publishing: 44 pp.

ÖNORM L 1075 (2004): Grundlagen für die Bewertung der Gehalte ausgewählter Elemente in Böden. – Österreichisches Normungsinstitut; Wien. URL: http://www.on-norm.at (last access 13 January 2013).

ØYGARDEN, L., LUNDEKVAM, H., ARNOLDUSSEN, A. H. & BØRRESEN, T. (2006): Norway – In: BOARDMAN, J. & POESEN, J. (eds.): Soil erosion in Europe: 1 – 15; Chichester (Wiley).

OFFICIAL GAZETTE (2010): Ordinance on the protection of agricultural land against pollution. – NN32/2012. Government of the Republic of Croatia; Zagreb. URL: http://narodne-novine.nn.hr/default.aspx (last access 16 December 2012).

OKAY, A. I. & TÜYSÜZ, O. (1999): Tethyan sutures of northern Turkey. – In: DURAND, B., JOLIVET, L., HORVÁTH, F. & SÉRANNE, M. (eds.): The Mediterranean Basins: Tertiary Extension within the Alpine Orogen. – Geological Society, London, Spec. Publ., **156**: 475 – 515.

—, ŞENGÖR, A. M. C. & GOROR, N. (1994): Kinematic history of the opening of the Black Sea and its effect on the surrounding regions. – Geology, **22**: 267 – 270.

—, SATIR, M. & SIEBEL, W. (2006): Pre-Alpide Palaeozoic and Mesozoic orogenic events in the Eastern Mediterranean region. – Geological Society, London, Memoirs, **32**: 389-405.

OLSHEVSKAYA, Y. (1995): Landscape-ecogeochemical mapping of recharge and discharge areas of the main water-bearing horizons in Ukraine of 1 : 500 000 scale. – Geoprognoz SGC: 200 pp.; Kyiv.

OORTS, K., GHESQUIERE, U., SWINNEN, K. & SMOLDERS, E. (2006): Soil properties affecting the toxicity of $CuCl_2$ and $NiCl_2$ for soil microbial processes in freshly spiked soils. – Environmental Toxicology and Chemistry, **25**: 836 – 844.

—, BRONCKAERS, H. & SMOLDERS, E. (2006a): Discrepancy of the microbial response to elevated Cu between freshly spiked and long-term contaminated soils. – Environmental Toxicology and Chemistry, **25**: 845 – 853.

OPEN UNIVERSITY (1974) Mineral deposits. – 140 pp.; Walton Hall, Milton Keynes (Open Univ. Press).

OSINSKI, G. R., SPRAY, J. G. & GRIEVE, R. A. F. (2008): Impact melting in sedimentary target rocks: An synthesis. – In: EVANS, K., HORTON, W., KING, D., MORROW, J. & WARME, J. (eds.): The Sedimentary Record of Meteorite Impacts. – Geol. Soc. Am., Spec. Papers, **437**: 1 – 18; Boulder, CO.

OSMOND, J. K. & IVANOVICH, M. (1992): Uranium series mobilization and surface hydrology. – In: IVANOVICH, M. & HARMON, R. S. (eds.): Uranium Series Disequilibrium; Applications to Earth, Marine, and Environmental Sciences: 259 – 289; Oxford (Clarendon Press).

PAGE, A. L., BINGHAM, F. T. & CHANG, A. C. (1981): Cadmium. – In: LEPP, N.W. (ed.): Effect of heavy metal pollution on plants. Volume 1: Effects of trace metals on plant function – Applied Science: 77 – 109; London.

—, CHANG, A. C. & EL-AMANY, M. (1987): Cadmium levels in soils and crops in the United States. – In: HUTCHINSON, T.C. & MEEMA, K. H. (eds.): Lead, mercury, cadmium and arsenic in the environment. – SCOPE, **31**: 119 – 146; New York (Wiley).

PÁLFY, J. (2004): Did the Puchezh-Katunki impact trigger an extinction?– In: DYPVIK, H., BURCHELL, M.J. & CLAEYS, P. (eds.): Cratering in Marine Environments and on Ice. – 135 – 148; Berlin (Springer).

PAN, J., PLANT, J. A., VOULVOULIS, N., OATES, CHR. J. & IHLENFELD, CHR. (2010): Cadmium levels in Europe: implications for human health. – Journal of Environmental Geochemistry and Health, **32**: 1 – 12.

PANAGOS, P. (2006): The European soil database.– GEO: connexion, **5**(7): 32 – 33.

—, VAN LIEDEKERKE, M., JONES, A. & MONTANARELLA, L. (2012): European Soil Data Centre: Response to European policy support and public data requirements. – Land Use Policy, **29**(2): 329 – 338.

—, HIEDERER, R., VAN LIEDEKERKE, M. & BAMPA, F. (2013): Estimating soil organic carbon in Europe based on data collected through an European network. – Ecological Indicators, **24**: 439 – 450.

PARK, C. F. JR & MACDIARMID, R. A. (1975): Ore deposits: 530 pp.; New York (Freeman).

PARKS, D. A. & RENDELL, H M. (1992): TL geochronology of brickearths from south-east England. – Quaternary Science Reviews, **11**: 7 – 12.

PERELMAN, A. I. (1961): Landscape Geochemistry. – Geografgis: 388 pp.; Moscow [in Russian].

— (1966): Landscape Geochemistry. – Vysshaya Shkola: 388 pp.; Moscow [in Russian].

— (1972): Landscape Geochemistry. – Vysshaya Shkola, (Translation No. 676, Geological Survey of Canada, Ottawa): 388 pp.; Moscow.

— (1975): Landscape geochemistry. – Vysshaya Shkola: 392 pp.; Moscow [in Russian].

— (1977): Geochemistry of Elements in the Supergene Zone: 266 pp.; Jerusalem (Keter Publ. House).

Perrin, R. M. S., Davies, H. & Fysh, M. D. (1974): Distribution of Late Pleistocene aeolian deposits in eastern and southern England.– Nature, **248**: 320 – 324.

Petersell, V. (1997): Dictyonema argillite. – In: Geology and mineral resources of Estonia: 327 – 331; Tallinn (Estonian Academy Publishers).

—, Ressar, H., Carlsson, M., Mõttus, V., Enel, M., Mardla, A. & Täht, K. (1997): The geochemical atlas of the humus horizon of Estonian soil – Explanatory text: 75 pp.; Tallinn.

Pharaoh, T. C. (1999): Palaeozoic terranes and their lithospheric boundaries within the Trans-European Suture Zone (TESZ): A review. – Tectonophysics, **314**: 17 – 41.

— & Gibbons, W. (1994): Precambrian rocks in England and Wales south of the Menai Strait Fault System. – In: Gibbons, W. & Harris, A.L. (eds.): A revised correlation of Precambrian rocks in the British Isles. – Geological Society, London: 85 – 97.

—, England, R. W., Verniers, J. & Zelazniewicz, A. (1997): Introduction; geological and geophysical studies in the Trans-European suture zone. – Geological Magazine, **134**(5): 585 – 590; London (Cambridge Univ. Press).

—, Winchester, J. A., Verniers, J., Lassen, A. & Seghedi, A. (2006): The Western Accretionary Margin of the East European Craton: An overview. – Geological Society, Memoirs, **32**: 291 – 311; London.

Pieńkowski, G., Schudack, M. E., Bosák, P., Enay, R., Feldman-Olszewska, A., Golonka, J., Gutowski, J., Herngreen, G. F. W., Jordan, P., Krobicki, M., Lathuilière, B., Leinfelder, R. R., Michalik, J., Mönnig, E., Noe-Nygaard, N., Pálfy, J., Pint, A., Rasser, M.W., Reisdorf, A. G., Schmid, D. U., Schweigert, G., Surlyk, F., Wetzel, A. & Wong, T. E. (2008): Jurassic. – In: McCann, T. (ed.): The Geology of central Europe – Volume 2: Mesozoic and Cenozoic. – Geological Society, London: 823 – 922.

Piteå Kommun (2005): Arsenik i dricksvatten i Piteå kommun. – Lägesrapport, Augusti 2005: 4 pp. [in Swedish] URL: http://www.pitea.se/Sidmappar/36895/Arsenik.pub.7437.pdf (last access 29 December 2012).

Plant, J. A., Klaver, G., Locutura, J., Salminen, R., Vrana, K. & Fordyce, F. (1996): Forum of European Geological Surveys (FOREGS), Geochemistry Task Group 1994 – 1996 Report. A contribution to IUGS Continental Geochemical Baselines. – British Geological Survey, BGS Technical Report, WP/95/14: 52 pp.; Keyworth.

—, Klaver, G., Locutura, J., Salminen, R., Vrana, K. & Fordyce, F. (1997): The Forum of European Geological Surveys, Geochemistry Task Group inventory 1994 – 1996. – Journal of Geochemical Exploration, **59**: 123-146.

—, Kinniburgh, D. G., Smedley, P. L., Fordyce, F. M. & Klinck, B. A. (2003): Arsenic and Selenium. – Treatise on Geochemisty, **9**: 17 – 66

—, Reeder, S., Salminen, R., Smith, D. B., Tarvainen, T., de Vivo, B. & Petterson, M. G. (2003a): The distribution of uranium over Europe: Geological and environmental significance. – Applied Earth Science; Trans. Inst. Min. Metall. B, **112**: B221 – B238.

—, Whittaker, A., Demetriades, A., de Vivo, B. & Lexa, J. (2005): The geological and tectonic framework of Europe. – In: Salminen, R. (ed.): FOREGS – Geochemical Atlas of Europe. Part 1: Background Information, Methodology and Maps. – Geological Survey of Finland: 23 – 42; Espoo. http://weppi.gtk.fi/publ/foregsatlas/ (last access 10 January 2014).

Plein, E., Bachmann, G. H., Gast, R.E., Gebhardt, U., Graf, R. & Helmuth, H. J. (1995): New Rotliegend stratigraphy (Permian) of the North German Basin. – Panstwowy Instytut Geologiczny: 115 pp.; Poland.

Polednak, A. P., Stehney, A. F. & Lucas, H. F. (1983): Mortality among male workers at a thorium-processing plant.– Health Physics, **44** (Suppl. 1): 239 – 251.

Polupan, M., Solovei, V. & Velichko V. (2005): Ukraine soils classification.– 300 pp.; Kyiv (Agrarna Nauka).

Polynov, B. B. (1937): The cycle of weathering (translated by A. Muir). – 250 pp.; London (T. Murby).

— (1953): Concept of Landscapes. – Geography Issues, Collected Volume, **33**: 120 – 285.

Pourbaix, M. J. N. (1963): Atlas d'Equilibres Electrochimiques. – 250 pp.; Paris (Gauthier-Villars).

ProMine (2013): Nano-particle products from new mineral resources in Europe. – Seventh Framework Programme, Nanosciences, Nanotechnologies, Materials and New Production Technologies. URL: http://promine.gtk.fi/ (last access 26 January 2013).

Puchtel, I. S., Hofmann, A. W., Jochum, K. P., Mezger, K., Shchipansky, A. A. & Samsonov, A. V. (1997): The Kostomuksha greenstone belt, NW Baltic shield: Remnant of a late Archaean oceanic plateau?– Terra Nova, **9**(2): 87 – 90.

Pye, K. (1995): The nature, origin and accumulation of loess. – Quaternary Science Reviews, **14** (7-8): 653 – 667.

Rajmon, D. (2009): Impact database 2010.1. URL: http://impacts.rajmon.cz (last access 20 January 2013).

Rank, G., Kardel, K., Pälchen, W. & Weidensdörfer, H. (1999): Bodenatlas des Freistaates Sachsen – Teil 3: Bodenmessprogramm. – Landesamt für Umwelt und Geologie: 24 – 29.

Rasmussen, E. S., Heilmann-Clausen, C., Waagstein, R. & Eidvin, T. (2008): The Tertiary of Norden.– Episodes, **31**: 66 – 72.

RAUDSEP, R. (1997): Phosphorite. – In: Geology and mineral resources of Estonia: 331 – 336: Tallinn (Estonian Academy Publ.).

RAYMAN, M. P. (2000): The importance of selenium to human health. – Lancet, **356**: 233 – 241.

— (2008): Food-chain selenium and human health: emphasis on intake. – British Journal of Nutrition, **100**: 254 – 268.

REEDER, S., TAYLOR, H., SHAW, R. A. & DEMETRIADES, A. (2006): Introduction to the chemistry and geochemistry of the elements. – In: DE VOS, W. & TARVAINEN, T. (eds.): Geochemical Atlas of Europe, Part 2 – Interpretation of Geochemical Maps, Additional Tables, Figures, Maps, and Related Publications. – Geological Survey of Finland: 692 pp.; Espoo. URL: http://weppi.gtk.fi/publ/foregsatlas/ (last access 13 January 2013).

— (2007): Global Geochemical Baselines. – Episodes, **30**(1): 69 – 72.

REEVES III, J. B., MCCARTY, G. W. & REEVES, V. B. (2001): Mid-infrared diffuse reflectance spectroscopy for the quantitative analysis of agricultural soils. – Journal of Agricultural and Food Chemistry, **49**: 766 – 772.

REILLY, C. (2006): Selenium in Food and Health. 2nd Edition.– 222 pp.; New York (Springer Science).

REIMANN, C. & CARITAT, P. DE (1998): Chemical elements in the environment: Factsheets for the geochemist and environmental scientist.– 398 pp.; Berlin (Springer).

— & GARRETT, R. G. (2005): Geochemical background - concept and reality. – Science of the Total Environment, **350**: 12 – 27.

— & BIRKE, M. (eds.) (2010): Geochemistry of European Bottled Water. – 268 pp.; Stuttgart (Borntraeger).

—, SIEWERS, U., TARVAINEN, T., BITYUKOVA, L., ERIKSSON, J., GILUCIS, A., GREGORAUSKIENĖ, V., LUKASHEV, V. K., MATINIAN, N. N. & PASIECZNA, A. (2003): Agricultural Soils in Northern Europe: A Geochemical Atlas. – Geol. Jb., Sonderhefte, **SD 5**: 279 pp.; Hannover.

—, FILZMOSER, P., GARRETT, R. G. & DUTTER, R. (2008): Statistical Data Analysis Explained – Applied Environmental Statistics with R. – 343 pp.; Chichester (Wiley).

—, DEMETRIADES, A., EGGEN, O. A., FILZMOSER, P. & EUROGEOSURVEYS GEOCHEMISTRY EXPERT GROUP (2009): The EuroGeoSurveys Geochemical Mapping of Agricultural and Grazing Land Soils Project (GEMAS) – Evaluation of Quality Control Results of Aqua Regia Extraction Analysis – Geological Survey of Norway, NGU Report, **2009.049**: 94 pp.; Trondheim. URL: http://www.ngu.no/upload/Publikasjoner/Rapporter/2009/2009_049.pdf (last access 11 February 2013).

—, ENGLMAIER, P., GOUGH, L., LAMOTHE, P., NORDGULEN, Ø. & SMITH, D. (2009a): Geochemical gradients in O-horizon soils of southern Norway: Natural or anthropogenic?– Applied Geochemistry, **24**: 62 - 76.

—, DEMETRIADES, A., EGGEN, O. A., FILZMOSER, P. & EUROGEOSURVEYS GEOCHEMISTRY EXPERT GROUP (2011): The EuroGeoSurveys Geochemical mapping of agricultural and grazing land soils project (GEMAS) – Evaluation of quality control results of total C and S, total organic carbon (TOC), cation exchange capacity (CEC), XRF, pH, and particle size distribution (PSD) analyses. – Geological Survey of Norway, NGU Report, **2011.043**: 90 pp.; Trondheim. URL: http://www.ngu.no/upload/Publikasjoner/Rapporter/2011/2011_043.pdf (last access 11 February 2013).

—, DEMETRIADES, A., BIRKE, M., EGGEN, O. A., FILZMOSER, P., KRIETE, C. & EUROGEOSURVEYS GEOCHEMISTRY EXPERT GROUP (2012): The EuroGeoSurveys Geochemical Mapping of Agricultural and grazing land Soils project (GEMAS) – Evaluation of quality control results of particle size estimation by MIR prediction, Pb-isotope and MMI® extraction analyses and results of the GEMAS ring test for the standards Ap and Gr. – Geological Survey of Norway, NGU Report, **2012.051**: 136 pp.; Trondheim. URL: http://www.ngu.no/upload/Publikasjoner/Rapporter/2012/2012_051.pdf (last access 11 February 2013).

—, CARITAT, P. DE, THE GEMAS PROJECT TEAM & THE NGSA PROJECT TEAM (2012a): New soil composition data for Europe and Australia: demonstrating comparability, identifying continental-scale processes and learning lessons for global geochemical mapping. – Science of the Total Environment, **416**: 239 – 252.

—, FILZMOSER, P., FABIAN, K., HRON, K., BIRKE, M., DEMETRIADES, A., DINELLI, E., LADENBERGER, A. & THE GEMAS PROJECT TEAM (2012b): The concept of compositional data analysis in practice – Total major element concentrations in agricultural and grazing land soils of Europe. – Science of the Total Environment, **426**: 196 – 210.

—, FLEM, B., FABIAN, K., BIRKE, M., LADENBERGER, A., NEGREL, P., DEMETRIADES, A., HOOGEWERFF, J. & GEMAS PROJECT TEAM (2012c): Lead and lead isotopes in agricultural soils of Europe – The continental perspective. – Applied Geochemistry, **27**(3): 532 – 542.

REIMOLD, W. U., KOEBERL, C., GIBSON, R. L. & DRESSLER, B. O. (2005): Economic Mineral Deposits in Impact Structures: A Review. – Impact Studies, **6**: 479 – 552; Berlin (Springer).

RILEY, S. J., DE GLORIA, S. D. & ELLIOT, R. (1999): A Terrain Ruggedness that Quantifies Topographic Heterogeneity.– Intermountain Journal of Science, **5**(1-4): 23 – 27.

RING, U., RATSCHBACHER, L. & FRISCH, W. (1988): Plate-boundary kinematics in the Alps: Motion in the Arosa suture zone. – Geology, **16**: 696 – 698.

ROBBERECHT, H., DEELSTRA, H., VANDEN BERGHE, D. & VAN GRIEKEN, R. (1983): Metal pollution and selenium distributions in soils and grass near a non-ferrous plant. – Science of the Total Environment, **29**(3): 229 – 41.

ROGERS, C. D. F., DIJKSTRA, T. A. & SMALLEY, I. J. (1994): Hydroconsolidation and subsidence of loess: Studies from China, Russia, North America and Europe: In memory of Jan Sajgalik. – Engineering Geology, **37**(2), 83 – 113.

ROGERS, P., ARORA, S., FLEMING, G., CRINION, R. & MCLAUGHLIN, J. (1990): Selenium toxicity in farm animals: treatment and prevention. – Irish Veterinary Journal, **43**: 151 – 153.

ROSHOLT, J. N. (1982): Isotopic composition of uranium and thorium in crystalline rocks. – In: Archean Geochemistry Field Conference; Seninoe Mountains and Hartville Uplift, Wyoming; Part II, Program.

RUDNICK, R. L. & GAO, S. (2003): Composition of the Continental Crust. – In: RUDNICK, R. L. (ed.): The Crust. Vol. 3. – In: HOLLAND, H. D. & TUREKIAN, K. K. (eds.): Treatise on Geochemistry: 1 – 64; Oxford (Elsevier - Pergamon).

SABY, N. P. A., BELLAMY, P. H., MORVAN, X., ARROUAYS, D., JONES, R. J. A., VERHEIJEN, F. G. A., KIBBLEWHITE, M. G., VERDOOT, A., ÜVEGES, J. B., FREUDENSCHUSS, A. & SIMOTA, C. (2008): Will European soil monitoring networks be able to detect changes in topsoil organic carbon content?– Global Change Biology, **14**: 2432 – 2442.

SAET, Y. E., REVICH, B. A., YANIN, E. P., SMIRNOVA, R. S., BASHARKEVICH, I. L., ONISHCHENKO, T. L., PAVLOVA, L. N., TREFILOVA, N. Y., ACHKASOV, A. I. & SARKISIAN, S. SH. (1990): Geokhimia okruzhayushchei sredy. – Environmental Geochemistry, 335; Moscow, Nedra [in Russian].

SÆTHER, O. M., ROBERTS, D. & REIMANN, C. (2005): Structural control of Sr-concentrations in stream sediments: An example from major fault zones in Central Norway. – Geochemistry, Exploration, Environment, Analysis, **5**: 183 – 188.

—, XIE, R., AAGAARD, P., ENDRE, E., LØKEN, T. & RUDOLPH-LUND, K. (2010): Main and trace element content of shales from Anderskogen (Hamar) and Øvre Slottsgt. (Oslo), Norway.– Geological Survey of Norway, Report No. **2009.63**: 128 pp.; Trondheim.

SALMINEN, R., TARVAINEN, T., DEMETRIADES, A., DURIS, M., FORDYCE, F. M., GREGORAUSKIENE, V., KAHELIN, H., KIVISILLA, J., KLAVER, G., KLEIN, H., LARSON, J.O., LIS, J., LOCUTURA, J., MARSINA, K., MJARTANOVA, H., MOUVET, C., O'CONNOR, P., ODOR, L., OTTONELLO, G., PAUKOLA, T., PLANT, J. A., REIMANN, C., SCHERMANN, O., SIEWERS, U., STEENFELT, A., VAN DER SLUYS, J. & WILLIAMS, L. (1998): FOREGS – Geochemical Mapping Field Manual. – Geological Survey of Finland, Guide **47**: 36 pp.; Espoo. URL: http://arkisto.gsf.fi/op/op47/op47.pdf (last access 11 February 2013).

— (chief-ed.), BATISTA, M. J., BIDOVEC, M., DEMETRIADES, A., DE VIVO, B., DE VOS, W., DURIS, M., GILUCIS, A., GREGORAUSKIENE, V., HALAMIC, J., HEITZMANN, P., LIMA, A., JORDAN, G., KLAVER, G., KLEIN, P., LIS, J., LOCUTURA, J., MARSINA, K., MAZREKU, A., O'CONNOR, P. J., OLSSON, S.A., OTTESEN, R.-T., PETERSELL, V., PLANT, J. A., REEDER, S., SALPETEUR, I., SANDSTRÖM, H., SIEWERS, U., STEENFELT, A. & TARVAINEN, T. (2005): FOREGS – Geochemical Atlas of Europe. Part 1: Background Information, Methodology, and Maps. – Geological Survey of Finland: 690 pp.; Espoo. URL: http://weppi.gtk.fi/publ/foregsatlas/ (last access 12 February 2013).

SALONEN, J. T., ALFTHAN, G., HUTTUNEN, J. K. & PUSKA, P. (1982): Association between cardiovascular death and myocardial infraction and serum selenium in matched-pair longitudinal study. – Lancet, **2**: 175 – 9.

—, ALFTHAN, G., HUTTUNEN, J. K. & PUSKA, P. (1984): Association between serum selenium and the risk of cancer. – American Journal of Epidemiology, **120**: 342 – 349.

SANDSTAD, J. S., BJERKGÅRD, T., BOYD, R., OHLEN, P., KORNELIUSSEN, A., NILSSON, L. P., OFTEN, M., EILU, P. & HALLBERG, A. (2012): Metallogenic areas in Norway. – In: EILU, P. (ed.): Mineral deposits and metallogeny of Fennoscandia. – Geological Survey of Finland, Special paper, **53**: 35 – 138; Espoo.

SANDSTRÖM, H., REEDER, S., BARTHA, A., BIRKE, M., BERGE, F., DAVIDSEN, B., GRIMSTVEDT, A., HAGEL-BRUNNSTRÖM, M-L., KANTOR, W., KALLIO, E., KLAVER, G., LUCIVJANSKY, P., MACKOVYCH, D., MJARTANOVA, H., OS, B. VAN, PASLAWSKI, P., POPIOLEK, E., SIEWERS, U., VARGA-BARNA, ZS., VILSTEREN, E. VAN & ØDEGÅRD, M. (2005): Sample Preparation and Analysis. – In: SALMINEN, R. (chief-ed.): FOREGS – Geochemical Atlas of Europe. Part 1: Background Information, Methodology, and Maps. Geological Survey of Finland: 81 – 94; Espoo. URL: http://weppi.gtk.fi/publ/foregsatlas/articles/Analysis.pdf (last access 12 February 2013).

SAPPINGTON, J. M., LONGSHORE, K. M. & THOMPSON, D. B. (2007): Quantifying Landscape Ruggedness for Animal Habitat Analysis: A Case Study Using Bighorn Sheep in the Mojave Desert. – Journal of Wildlife Management, **71**(5): 1419 – 1426.

SAS INSTITUTE INC. (2004): SAS OnlineDoc® 9.1.3. – Cary, NC: SAS Institute Inc.

SAUVÉ, S., HENDERSHOT, W. & ALLEN, H. E. (2000): Solid-solution partitioning of metals in contaminated soils: Dependence on pH, total metal burden, and organic matter. – Environmental Science and Technology, **34**: 1125 – 1131.

SAWKINS, F. J. (1984): Metal deposits in relation to plate tectonics.– 325 pp.; Berlin (Springer).

SCHEIB, A. J. & LEE, J. (2010): Mapping Late Pleistocene and Holocene aeolian sediments in East Anglia, UK: The application of regional-scale geochemical data. – Quaternary Newsletter, **120**: 5 – 14.

—, BIRKE, M., DINELLI, E. & THE GEMAS PROJECT TEAM (2014): Geochemical evidence of aeolian deposits in European soils. – Boreas, **43**: 175–192. DOI 10.1111/bor.12029 (last access 10 January 2014).

SCHILS, R., KUIKMAN, P., LISKI, J., VAN OIJEN, M., SMITH, P., WEBB, J., ALM, J., SOMOGYI, Z., VAN DEN AKKER, J., BILLETT, M., EMMETT, B., EVANS, C., LINDNER, M.,

Palosuo, T., Bellamy, P., Jandl, R. & Hiederer, R. (2008): Review of existing information on the interrelations between soil and climate change (CLIMSOIL). – European Commission, Final report: 208 pp.; Brussels. URL: http://ec.europa.eu/environment/soil/pdf/climsoil_report_dec_2008.pdf (last access 24 January 2013).

Schmid, S. M., Pfiffner, O. A., Froitzheim, N., Schönborn, G. & Kissling, E. (1996): Geophysical-geological transect and tectonic evolution of the Swiss-Italian Alps. – Tectonics, **15**: 1036 – 1064.

Schmidt, F. & Persson, A. (2003): Comparison of DEM data capture and topographic wetness indices. – Precision Agriculture, **4**(2): 179 – 192.

Schneider, H.-J. (ed.) (1983): Mineral deposits of the Alps and of the Alpine Epoch in Europe. – Society for Geology Applied to Mineral Deposits, Spec. Publ., **3**: 402 pp.; Berlin (Springer).

Schroeder, D. & Blum, W. E. H. (1992): Bodenkunde in Stichworten (Hirts Stichwortbücher). – 175 pp.; Stuttgart (Borntraeger).

Scotese, C. R. (2010): The PALEOMAP paleoatlas (ArcGIS). Abstracts with Programs – Geol. Soc. Am., **42**(5): 457 – 457.

Selinus, O., Finkelman, R. B. & Centeno, J. A. (eds.) (2010): Medical Geology: A regional synthesis. – 392 pp.; Dordrecht (Springer).

Shamberger, R. J. (1970): Relationship of selenium to cancer: Inhibitory effect of selenium on carcinogenesis. – Journal of the National Cancer Institute, **44**(4): 931 – 936.

Sharpley, A. (2000): Phosphorus availability. – In: Sumner, M. E. (ed.): Handbook of Soil Science. – D18 – D38; Boca Raton, FL (CRC Press).

Shaw, D. M., Reilly, G. A., Muysson, J. R., Pattenden, G. E. & Campbel, F.E. (1967): An estimate of the chemical composition of the Canadian Precambrian shield. – Canadian Journal of Earth Sciences, **4**: 829 – 853.

—, Dostal, J. & Keays, R. R. (1976): Additional estimates of continental surface Precambrian shield composition in Canada. – Geochimica et Cosmochimica Acta, **40**: 73 – 83.

Siewers, U. & Herpin, U. (1998): Schwermetalleinträge in Deutschland. Moos-Monitoring 1995/96. – Geol. Jb., Sonderhefte, **SD 2**: 199 pp.; Hannover.

Sims, T. (2000): Soil fertility evaluation. – In: Sumner, M.E. (ed.): Handbook of Soil Science. – D113 – D153; Boca Raton, FL (CRC Press).

Singer, D. A. & Menzie, W.D. (2010): Quantitative mineral resource assessments – An integrated approach: 219 pp.; Oxford (Oxford Univ. Press).

Singhvi, A. K., Bluszcz, A., Bateman, M. D. & Someshwar Rao, M. (2001): Luminescence dating of loess-paleosol sequences–methodological aspects and paleoclimatic implication. – Earth-Science Reviews, **54**: 193 – 211.

Skinner, B. J. (1981): Economic geology – Seventy-fifth anniversary volume. – 964 pp.; Lancaster, PA (Economic Geology Pub. Co.).

Smalley, I. J., Jefferson, I. F., Dijkstra, T. A. & Derbyshire, E. (2001): Some major events in the development of the scientific study of loess. – Earth-Science Reviews, **54**: 5 – 18.

Smirnov, V. I. (1976): Geology of mineral deposits. – 520 pp.; Moscow (MIR Publishers).

Smith, D. B., Wang, X., Reeder, S. & Demetriades, A. (2012): The IUGS/IAGC Task Group on Global Geochemical Baselines. – Earth Science Frontiers, **19**(3): 1 – 6.

Smith, D. R. & Flegal, A. R. (1995): Lead in the biosphere: recent trends. – Ambio, **24**: 21 – 23.

Smolders, E., Oorts, K., van Sprang, P., Schoeters, I., Janssen, C. R., McGrath, S. P. & McLaughlin, M. J. (2009): Toxicity of Trace Metals in Soil as Affected by Soil Type and Aging after Contamination: Using Calibrated Bioavailability Models to Set Ecological Soil Standards. – Environmental Toxicology and Chemistry, **28**: 1633 – 1642.

Sørensen, L. & Karg, S. (2012): The expansion of agrarian societies towards the north a new evidence for agriculture during the Mesolithic/Neolithic transition in Southern Scandinavia. – Journal of Archaeological Sciences [in press].

Sohlenius, G. & Öborn, I. (2004): Geochemistry and partitioning of trace metals in acid sulphate soils in Sweden and Finland before and after sulphide oxidation. – Geoderma, **122**: 167 – 175.

—, Lax, K. & Ladenberger, A. (2009): Kan SGUs data användas för att uppskatta moränens transportlängd?– SGU-report, **1009**:26: 25 pp.

Soper, N. J., Webb, B. C. & Woodcock, N. H. (1987): Late Caledonian (Acadian) transpression in north-west England: Timing, geometry and geotectonic significance. – Yorkshire Geological Society, Proceedings, **46**: 175 – 192.

Spadoni, M., Voltaggio, M., Carcea, M., Coni, E., Raggi, A. & Cubadda, F. (2007): Bioaccessible selenium in Italian agricultural soils: Comparison of the biogeochemical approach with a regression model based on geochemical and pedoclimatic variables. – Science of the Total Environment, **376**: 160 – 77.

Spielhagen, R. F., Baumann, K.-H., Erlenkeuser, H., Nowaczyk, N. R., Norgaard-Pedersen, N., Vogt, C. & Weiel, D. (2004): Arctic Ocean deep-sea record of northern Eurasian ice sheet history. – Quaternary Science Reviews, **23**(11 – 13): 1455 – 1483.

SRTM (2013): V4 Digital elevation data from the CGIAR-CSI website. URL: http://srtm.csi.cgiar.org/ (last access 24 January 2013).

Stafilov, T., Šajn, R., Pančevski, Z., Beov, B., Frontasyeva, M. V. & Strelakova, L. P. (2010): Heavy metal contamination of topsoils around a lead and zinc smelter in the Former Yugoslavian Republic of

Macedonia. – Journal of Hazardous Materials, **175**: 896 – 914.

—, Aliu, M. & Šajn, R. (2010a): Arsenic in Surface Soils Affected by Mining and Metallurgical Processing in K. Mitrovica Region, Kosovo. – International Journal of Environmental Research and Public Health, **7**: 4050 – 4061.

Stampfli, G. M. & Borel, G. D. (2004): The TRANSMED transects in space and time; constraints on the paleotectonic evolution of the Mediterranean domain. – In: Cavazza, W., Roure, F. M., Spakman, W., Stampfli, G. M. & Ziegler, P. A. (eds.): The TRANSMED atlas – The Mediterranean region from crust to mantle: 53 – 80; Berlin (Springer).

— & Kozur, H. W. (2006): Europe from the Variscan to the Alpine cycles. – Geological Society, London, Memoirs, **32**: 57 – 82.

Stehney, A. F., Polednak, A. P., Rundo, J. et al. (1980): Health status and body radioactivity of former thorium workers. – Argonne, IL, Argonne Nat1 Laboratory, **80 – 37**: 44.

Steiger, R. H. & Jäger, E. (1977): Subcommission on geochronology: Convention on the use of decay constants in geo- and cosmochronology. – Earth and Planetary Science Letters, **36**(3): 359 – 362.

Stückrad, S., Sabel, K.-J. & Wilcke, W. (2010): Contributions of different parent materials in soils developed from peri-glacial cover beds. – European Journal of Soil Science, **61**: 844 – 853.

Su, C. & Suarez, D. L. (2000): Selenate and selenite sorption on iron oxides: An infrared and electrophoretic study. – Soil Science Society of American Journal, **64**: 101 – 111.

Sucharova, J., Suchara, I., Hola, M., Reimann, C., Boyd, R., Filzmoser, P. & Englmaier, P. (2011): Linking chemical elements in forest floor humus in the Czech Republic to contamination sources. – Environmental Pollution, **159**: 1205 – 1214.

Svendsen, J. I., Alexanderson, H., Astakhov, V. I., Demidov, I., Dowdeswell, J. A., Funder, S., Gataullin, V., Henriksen, M., Hjort, C. & Houmark-Nielsen, N. (2004): Late Quaternary ice sheet history of northern Eurasia. – Quaternary Science Reviews, **23**(11 – 13): 1229 – 1271.

Syers, J. K., Mackay, A. D., Brown, M. W. & Currie, C. D. (1986): Chemical and physical characteristics of phosphate rock materials of varying reactivity. – Journal of the Science of Food and Agriculture, **37**: 1057 – 1064.

Szalay, A. (1964): Cation exchange properties of humic acids and their importance in the geochemical enrichment of UO_2^{2+} and other cations. – Geochimica et Cosmochimica Acta, **28**: 1605 – 1614.

Sźkorová, I., Pešek, J., Mizera, J., Havelcová, M., Matysová, P. & Vašíček, M. (2007): Variation in petrology and geochemistry of Tertiary coals in the deposits of the Czech Republic – influence of depositional environment.– Abstracts Book, Unconventional Petroleum Systems and Advances in Organic petrology and Geochemistry, CSCOP-TSOP-ICCP (University of Victoria, August 19 – 25, 2007): 44; Victoria, British Columbia, Canada.

Tallavaara, M., Pesonen, P. & Oinen, M. (2009): Prehistoric population history in eastern Fennoscandia. – Journal of Archaeological Science, **37**: 251 – 260.

Tan Jianan (chief ed.) (1989): The atlas of endemic diseases and their environments in the People's Republic of China. – 193 pp.; Beijing (Science Press) [in Chinese].

—, Wenyu Zhu, Wuyi Wang, Ribang Li, Shaofan Hou, Dacheng Wang & Linsheng Yang (2002): Selenium in soil and endemic diseases in China. – Science of the Total Environment, **284**: 227 – 35.

Tarvainen, T., Albanese, S., Birke, M., Poňavič, M., Reimann, C. & the GEMAS Project Team (2013): Arsenic in agricultural and grazing land soils of Europe. – Applied Geochemistry, **28**(1): 2 – 10.

Tattari, S. & Rekolainen, S. (2006): Finland. – In: Boardman, J. & Poesen, J. (eds.): Soil erosion in Europe: 27 – 32; New York (Wiley).

Taylor, G. N., Mays, C. W., Lloyd, R. D., Jones, C. W., Rojas, J., Wrenn, M. E., Ayoroa, G., Kaul, A. & Riedel, W. (1986): Liver cancer induction by ^{241}Am and Thorotrast in deer mice and grasshopper mice. – In: Gössner, W., Berber, G. B., Hagen, U. & Luz, A. (eds.): The radiobiology of radium and thorotrast: 172 – 177; Munich (Urban & Schwarzenberg).

Taylor, M. D. & Percipal, H. J. (2001): Cadmium in soils solutions from a transect of soils away from a fertilizer bin. – Environmental Pollution, **113**: 35 – 40.

Taylor, S. R. (1964): Abundance of chemical elements in the continental crust: A new table. – Geochimica et Cosmochimica Acta, **28**: 1273 – 1284.

— (1966): Application of trace element data to problems in petrology. – Physics and Chemistry of the Earth, **6**: 135 – 213.

— & McLennan, S. M. (1995): The geochemical evolution of the continental crust. – Reviews of Geophysics, **33**(2): 241 – 265.

—, McLennan, S. M. & McCulloch, M.T. (1983): Geochemistry of loess, continental crustal composition and crustal model ages. – Geochimica et Cosmochimica Acta, **47**: 1897 – 1905.

Terytze, K., Nestler, A., Ricking, M. & Schulze, T. (2000): Einfluss von Bodenparametern auf die mikrobielle Aktivität und den Abbau organischer Schadstoffe in Böden (Effect of soil parameters on microbial activity and decomposition of soil organic matter). – Literature review under the project "Konkretisierung fachlicher Grundlagen für Vorsorgeaspekte im betrieblichen Bodenschutz hinsichtlich tolerierbarer

Einträge", contracted by Deutsche Bundesstiftung Umwelt, Free University Berlin, Faculty of Geosciences: 108 pp.

Thompson, M. & Howarth, R.J. (1978): A new approach to the estimation of analytical precision. – Journal of Geochemical Exploration, **9**: 23 – 30.

Thornton, I. (1991): Metal contamination of soils in urban areas. – In: Bullock, P. & Gregory, P.J. (eds.): Soils in the urban environment: 47 – 75; Oxford (Blackwell).

— (1992): Sources and pathways of cadmium in the environment. – In: Nordberg, G.F., Herber, R.F.M. & Alessio, L. (eds.): Cadmium in the human environment: Toxicity and carcinogenicity. – International Agency for Research on Cancer (IARC), Scientific Publications, **118**: 149 – 162; Lyon, France.

Timmerman, M.J., Heeremans, M., Kirstein, L.A., Larsen, B.T., Spencer-Dunworth, E.-A. & Sundvoll, B. (2009): Linking changes in tectonic style with magmatism in northern Europe during the Late Carboniferous to latest Permian. – Tectonophysics, **473** (3-4): 375 – 390.

Toivonen, J. & Österholm, P. (2011): Characterization of acid sulfate soils and assessing their impact on a humic boreal lake. – Journal of Geochemical Exploration, **110**: 107 – 117.

Tylecote, R.F. (1992): A History of Metallurgy. Second Edition. – The Institute of Materials, Minerals and Mining: 217 pp.; London.

Újvári, G., Varga, A. & Balogh-Brunstad, Z. (2008): Origin, weathering and geochemical composition of loess in southwestern Hungary. – Quaternary Research, **69**: 421 – 437.

UkrDGRI (2006): Methodological recommendations for construction of geochemical maps (Precambrian basement geological formations geochemical specialisation and prognostic-geochemical maps) of 1 : 200,000 and 1 : 50,000 scales in reference to conditions of the Ukrainian Shield. – UkrDGRI (Ukrainian State Geological Exploration Institute) Methodological Recommendations: 96 pp.; Kyiv.

Ulén, B. (2006): Sweden. – In: Boardman, J. & Poesen, J. (eds.): Soil erosion in Europe: 17 – 26; New York (Wiley).

UNEP – United Nations Environment Programme (2006): Report of the first meeting of the Lead and Cadmium Working Group. – United Nations, Geneva, 18 – 22 September 2006.

— (2010): Chemicals Branch, DTIE. Final review of scientific information on cadmium. – Version of December 2010: 201 pp.

— (2010a): Chemicals Branch, DTIE. Appendix: Overview of existing and future national actions, including legislation, relevant to cadmium.– Companium document to the final review of scientific information on cadmium. – Version of December 2010: 118 pp.

USGS (2010): Mineral Commodity Summaries–. Mineral Resource Program. URL: http://minerals.usgs.gov/ (last access 20 October 2012).

— (2011): 2010 Minerals Yearbook, Cadmium (Advanced Release).– U.S. Geological Survey. URL: http://minerals.usgs.gov/minerals/pubs/commodity/cadmium/myb1-2010-cadmi.pdf (last access13 January 2013).

Van Asche, F.J. (1998): A Stepwise Model to Quantify the Relative Contribution of Different Environmental Sources to Human Cadmium Exposure. – Paper to be presented at NiCad'98, Prague, Czech Republic, September 21 – 22, 1998.

van Gestel, C.A.M., Borgman, E., Verweij, R.A. & Ortiz, M.D. (2011): The influence of soil properties on the toxicity of molybdenum to three species of soil invertebrates. – Ecotoxicology and Environmental Safety, **75**: 1 – 9.

—, McGrath, S.P., Smolders, E., Ortiz, M.D., Borgman, E., Verweij, R.A., Buekers, J. & Oorts, K. (2012): Effect of long-term equilibration on the toxicity of molybdenum to soil organisms.– Environmental Pollution, **162**: 1 – 7

VBBO (1998): Verordnung 814.12 über Belastungen des Bodens (VBBO) vom 1. Juli 1998 (Stand am 1. Juni 2012). Der Schweizerische Bundesrat: 12 pp.

Viets, J.G., Leach, D.L., Lichte, F.E., Hopkins, R.T., Gent, C.A. & Powell, J.W. (1996): Paragenetic and minor- and trace-element studies of Mississippi Valley-type ore deposits of the Silesian-Cracow district, Poland. – Prace – Panstwowego Instytutu Geologicznego, **154**: 36-71.

Vinogradov, A.P. (1954): Geochemie seltener und nur in Spuren vorhandener chemischer Elemente in Böden.– 249 pp.; Berlin (Akademie).

— (1957): Geokhimia redkih i rassejanych elementov v pochvah. – Geochemistry of rare and scattered elements in soils: 157; Moskva [in Russian].

— (1962): Average contents of chemical elements in the principal types of igneous rocks of the Earth's crust. – Geochemistry International, **7**: 641 – 664.

Waggott, A. (1969): An investigation of the potential problem of increasing boron concentrations in rivers and water courses. – Water Research, **3**: 749 – 765.

Walker, W. (1976): Metallogeny and global tectonics: 413 pp.; Dowden (Hutchinson & Ross).

Wallnoefer, P.R. & Engelhardt, G. (1995): Schadstoffe aus dem Boden. – In: Hock, B. & Elstner, E.F. (eds.): Schadwirkungen auf Pflanzen, 3. Auflage: 118 – 140; Heidelberg (Spektrum).

Wang, J.F. & Liu, Z. (1999): Effect of vanadium on the growth of soy bean seedlings. – Plant and Soil, **216**(1): 47 – 51.

Weaver, B.L. & Tarney, J. (1984): Empirical approach to estimating the composition of the continental crust. – Nature, **310**: 575 – 577.

Wedepohl, K.H. (1968): Origin and distribution of elements. – In: Ahrens, L.H. (ed.): International Symposium on the Origin and Distribution of the Elements: 909; New York (Pergamon Press).

— (1995): The composition of the continental crust. – Geochimica et Cosmochimica Acta, **59**(7): 1217–1232.

Weigand, H., Mansfeld, T., Baeumler, R., Schneckenburger, D., Wessel-Bothe, S. & Marb, C. (2010): Arsenic release and speciation in a degraded fen as affected by soil redox potential at varied moisture regime. – Geoderma, **159**: 371 – 378.

Weigel, F. (1983): Uranium and uranium compounds. – In: Grayson, M. & Kirk-Othmer, V. (eds.): Encyclopedia of chemical technology, Vol. 23, 3rd ed.: 502 – 547; New York (Wiley).

Weihed, P., Eilu, P., Larsen, R.B., Stendal, H. & Tontti, M. (2008): Metallic mineral deposits in the Nordic countries. – Episodes, **31**: 125 – 132.

Welch, R.M. & Norvell, W.A. (1999): Mechanisms of cadmium uptake, translocation and deposition in plants. – In: McLanghlin, M.J. & Singh, B.R. (eds.): Cadmium in soils and plants, **85**: 125 – 150; Dordrecht (Kluwer Publications).

Wendling, L.A., Ma, Y., Kirby, J.M. & McLaughlin, M.J. (2008): A predictive model of the effects of aging on cobalt fate and behaviour in soil. – Environmental Science & Technology, **42**: 140 – 146.

WHO (1987): Air Quality Guidelines for Europe. World Health Organization, Regional Office for Europe. – European Series: 23; Copenhagen.

— (1992): Cadmium. – World Health Organization, Geneva, Switzerland. – Environmental Health Criteria, **134**: 280 pp.

— (2003): Selenium in drinking-water. – Background document for development of WHO Guidelines for Drinking-Water Quality. – World Health Organization; Geneva.

Wilkinson, S.R., Grunes, D.L. & Sumner, M.E. (2000): Nutrient interactions in soil and plant nutrition. – In: Sumner, M. E. (ed.): Handbook of Soil Science: D89 – D112; Boca Raton, FL (CRC Press).

Wikipedia (2012): Abundance of elements in Earth's crust. – Retrieved from URL: http://en.wikipedia.org/wiki/Abundance_of_elements_in_Earth%27s_crust (last access 18 December 2012).

Wohlfarth, B., Björck, S., Funder, S., Houmark-Nielsen, M., Ingólfsson, Ó., Lunkka, J.-P., Mangerud, J., Saarnisto, M. & Vorren, T. (2008): Quaternary of Norden. – Episodes, **31**: 73 – 81.

Wood, S.A. (1996): The role of humic substances in the transport and fixation of metals of economic interest (Au, Pt, Pd, U, V). – Ore Geology Reviews, **11**: 1 – 31.

World Bank Group (1998): Cadmium. – In: Pollution Prevention Abatement Handbook. – World Bank Group: 212 – 214.

Wright, J.B. (1977): Mineral deposits, continental drift and plate tectonics. – 417 pp.; Dowden (Hutchinson & Ross).

Yan, R., Gauthier, D., Flamant, G. & Peraudeau, G. (2001): Fate of selenium in coal combustion: volatilization and speciation in the flue gas. – Environmental Science & Technology, **35**: 1406 – 1410.

Yao, W., Xie, X. & Wang, X. (2011): Comparison of results analyzed by Chinese and European laboratories for FOREGS geochemical baseline mapping samples. – Geoscience Frontiers **2**(2): 247 – 259.

Zech, W. & Hintermaier-Erhard, G. (2002): Böden der Welt – ein Bildatlas. – 120 pp.; Berlin (Spektrum).

Zeeberg, J.J. (1998): The European sand belt in Eastern Europe – and comparison of Late Glacial dune orientation with GCM simulation results. – Boreas, **27**: 127 – 139.

Zhao, F.J., Rooney, C.P., Zhang, H. & McGrath, S.P. (2006): Comparison of soil solution speciation and diffusive gradients in thin-films measurement as an indicator of copper bioavailability to plants. – Environmental Toxicology and Chemistry, **25**: 733 – 742.

Zheng, K.L., Ye, J.Y. & Jiang, B.L. (2005): Selected Analytical Methods of 57 Elements for Multi-purpose Geochemical Survey. – Geological Publishing House: 401 pp.; Beijing [in Chinese with English abstract].

Ziegler, P.A. (1990): Geological atlas of Western and Central Europe. – Shell Internationale Petroleum Maatschappij B.V.: 239 pp.; The Hague.

— & Dezes, P. (2006): Crustal evolution of Western and Central Europe. – Geological Society, London, Memoirs, **32**: 43 – 56.

Zirlewagen, D. & von Wilpert, K. (2010): Upscaling of environmental information – support of land-use management decisions by spatio-temporal regionalization approaches. – Environmental Management, **46**: 878 – 893.

Acronyms and Abbreviations

AAS	Atomic Absorption Spectrometry
AAT	Annual Accumulated Temperature
AES	Atomic Emission Spectroscopy, *also Optical Emission Spectroscopy (OES)*
ANOVA	Analysis of variance
Ap, A_p	Agricultural soil
AR	Aqua regia extraction
ArcGIS	Geographical Information System
ATSDR	Agency for Toxic Substances and Disease Registry
BIF	Banded Iron Formation
BoTe	Boreal to temperate climate
BSS	Baltic Soil Survey
bw	Body weight in kilograms
CCA	Chromated copper arsenate
CEC	Cation Exchange Capacity
CIA	Chemical Index of Alteration
clr	Centred log-ratio transformation
CP	Cumulative Probability
CSCB	Central Scandinavian Clay Belt
DDT	Dichlorodiphenyltrichloroethane
DIN	Deutsches Institut für Normung *(German Institute for Standardisation)*
DIN EN ISO	German, European, and International standards
DL	Detection Limit
DMEER	Digital Map of European Ecological Regions
EC	European Commission
EC10	Effective Concentration yielding 10% inhibition response
eCEC	effective Cation Exchange Capacity
ECHA	European Chemicals Agency
EDA	Exploratory Data Analysis
EEC	East European Craton *or* European Economic Community
EFSA	European Food Safety Authority *(Expert Committee on Food Activities)*
Eh	Redox potential
EID	Earth Impact Database
ES	Emission Spectroscopy
EU	European Union
FAO	Food and Agricultural Organisation of the United Nations
FOREGS	Forum of European Geological Surveys *(Geochemical Baseline Mapping Programme)*
FYROM	The former Yugoslav Republic of Macedonia
GC-MS	Gas chromatography–Mass spectrometry
Gr	Grazing land soil
GTN	Global Terrestrial Network
HC	Hebridian Craton
HCV	Health critical value
HF	Hydrofluoric acid
HFSE	High-field-strength elements
IARC	International Agency for Research on Cancer
ICP-MS	Inductively Coupled Plasma emission–Mass Spectrometer
ICP-OES	Inductively Coupled Plasma–Optical Emission Spectrometry
IDL	Instrumental Detection Limit
IGCP	International Geological Correlation Programme
IOCG	Iron oxide-copper-gold deposit
IR	Infrared
ISE	Ion Selective Electrode
K_c	Concentration coefficient
K_d	Solid-solution partitioning coefficient
K-T	Cretaceous-Tertiary boundary
L/A	Leaching-ageing factor
LDL	Laboratory Detection Limit

LLD Lower Limit of Detection
LOI Loss On Ignition
LREE Light Rare Earth Elements, *see Rare Earth Elements (REE)*
LUCAS Land Use/Cover Area frame statistical Survey

MAD Median absolute deviation
MAR Mass accumulation rate
MIR Mid-Infrared spectroscopy
MIS Marine (oxygen) Isotope Stages
MMC Midlands Microcraton
MMI Mobile Metal Ion
MOHO Mohorovičić discontinuity
MOP Muriate of Potash
MS Magnetic susceptibility

NOEC No Observed Effect Concentrations
NWCB North-West Carboniferous Basin

OM Organic matter

PAS Potential Acid Sulphate
PC Principal component
PCA Principal component analysis
PEC Predicted Environmental Concentration *(Exposure Concentration)*
PEC_{local} Predicted Environmental Concentration at the local scale
$PEC_{RWC,regional}$ Reasonable worst-case Predicted Environmental Concentration at the regional scale
PELCOM Pan-European Land Cover Monitoring project
PGE Platinum Group Elements
PLS Partial least-squares
PNEC Predicted No Effect Concentration
$PNEC_{generic}$ Generic (reasonable worst-case) Predicted No Effect Concentration
$PNEC_{RWC}$ Reasonable worst-case Predicted No Effect Concentration
ProMine Pan-European GIS-based database containing the known and predicted metalliferous and non-metalliferous resources
PSD Particle size distribution

QC Quality Control

RCR Risk Characterisation Ratio
RCR_{local} Local Risk Characterisation Ratio
$RCR_{RWC,regional}$ Regional reasonable worst-case Risk Characterisation Ratio
REACH Registration, Evaluation, Authorisation and Restriction of Chemicals
REE Rare Earth Elements
RMSE Root mean square error
RNCP Roman-Neapolitan Comagmatic Province
RPD Residual predictive deviation
RSD Relative standard deviation
RWC Reasonable worst-case

SGVs Soil Guideline Values
SLF Slope Length Factor
SOP Sulphate of Potash
SPB Southern Permian Basin
SPI Stream Power Index
StDev Standard deviation
STU Soil Typological Unit

TBP Tributyl phosphate
TC Total Carbon
TESZ Trans European Suture Zone
TIB Transscandinavian Igneous Belt
TOC Total Organic Carbon
TPI Topographic Position Index
TRI Terrain Ruggedness Index
TTZ Tornquist-Teisseyre Zone
TWI Tolerable weekly intake *or* Topographic Wetness Index

UCC Upper Continental Crust
UFZ Umweltforschungszentrum
ULD Upper Limit of Detection
USM Upper Silesian Massif

VS Vardar Suture

WBSF West Black Sea Fault
WCF Crimean Fault
WHO World Health Organisation

XRF X-ray fluorescence
XRFS X-ray fluorescence spectrometry

Part B

Appendices
A, B, C

Appendix A

Soil Guideline Values (SGVs) for Agricultural and Grazing Land Soil in Different European Countries

Appendix A

		EU Directive 86/278/EEC [1]		AUSTRIA		BELGIUM [5] (Brussels)			
						Guideline values [6]		Intervention values [7]	
Parameter	unit	min.	max.	Agricultural soil [4]	Grazing land soil [4]	Agricultural soil	Grazing land soil	Agricultural soil	Grazing land soil
Ag		–	–	–	–	–	–	–	–
As		–	–	20	30	35	35	58	58
B		–	–	–	–	–	–	–	–
Ba		–	–	–	–	–	–	–	–
Be		–	–	–	–	–	–	–	–
Bi		–	–	–	–	–	–	–	–
Cd		1	3	0.5	1	1.2	1.2	2	2
Co		–	–	50	50	–	–	–	–
Cr (total) [a]		(2)	(2)	100	100	91	91	130	130
Cr (VI)		–	–	–	–	–	–	–	–
Cu		50 (3)	140 (3)	60	100	72	72	120	120
F		–	–	–	–	–	–	–	–
Hg	[mg/kg]	1	1.5	0.5	0.5	1.7	1.7	2.9	2.9
Mn		–	–	–	–	–	–	–	–
Mo		–	–	2.5	2.5	–	–	–	–
NO_3		–	–	–	–	–	–	–	–
Ni		30 (3)	75 (3)	60	60	56	56	93	93
Pb		50	300	100	100	120	120	200	200
S		–	–	–	–	–	–	–	–
Sb		–	–	–	–	–	–	–	–
Se		–	–	2	2	–	–	–	–
Tl		–	–	1	1	–	–	–	–
U		–	–	–	–	–	–	–	–
V		–	–	100	100	–	–	–	–
Zn		150 (2)	300 (2)	150	300	200	200	333	333
pH	–	6	7	–	–	–	–	–	–

a Cr (III) + Cr (VI)

1 Council Directive 86/278/EEC of 12 June 1986 on the protection of the environment, and in particular of soil, when sewage sludge is used in agriculture. Member States may permit the limit values they fix to be exceeded in the case of the use of sludge on land which at the time of notification of this Directive is dedicated to the disposal of sludge, but on which commercial food crops are being grown exclusively for animal consumption. Member States must inform the Commission of the number and type of sites concerned. They must also seek to ensure that there is no resulting hazard to human health or the environment. All Member States have incorporated in their national legislation this directive.

2 It is not possible at this stage to fix limit values for chromium. The Council will fix these limit values later on the basis of proposals to be submitted by the Commission, within one year following notification of this Directive. Member States may permit the limit values they fix to be exceeded in respect of these parameters on soil with a pH consistently higher than 7. The maximum authorised concentrations of these heavy metals must in no case exceed those values by more than 50 %. Member States must also seek to ensure that there is no resulting hazard to human health or the environment and in particular to groundwater.

3 Methods of analysis for heavy metals must be carried out following strong acid digestion. The reference method of analysis must be that of atomic absorption spectrometry and the limit of detection for each metal should be no greater than 10 % of the appropriate limit value.

4 Grain size <2 mm; aqua regia digestion

5 Concentrations are based on dry soil sieved over 4 mm, aqua regia (or similar concentrated acid) destruction. All values are for a standard soil: pH KCl 5, 2 % organic matter and 10 % clay. In the Flemish legislation ('Vlarebo') equations are reported to normalise the values for standard soil towards the pH, clay and organic matter content of the site of interest. *Guideline values:* concentrations at which risks for human health and the environment are considered negligible. At higher concentrations site-specific investigations must be initiated. These are also the threshold concentrations that must be reached in case soil remediation is necessary. *Reference values:* predicted background concentrations in absence of anthropogenic emissions, target concentration for soil remediation projects. *Intervention values:* above these concentrations an intervention (remediation or safety measures) is required.

6 Saneringsnormen

7 Interventienormen

BELGIUM (Flanders)						BELGIUM (Wallonia)					
Guideline values [8]		Reference values [9]		Intervention values [10]		Guideline values [11]		Reference values [12]		Intervention values [13]	
Agricultural soil	Grazing land soil	Agricultural soil	Grazing land soil	Agricultural soil	Grazing land soil	Agricultural soil	Grazing land soil	Agricultural soil	Grazing land soil	Agricultural soil	Grazing land soil
–	–	–	–	–	–	–	–	–	–	–	–
35	35	16	16	58	58	30	30	12	12	265	265
–	–	–	–	–	–	–	–	–	–	–	–
–	–	–	–	–	–	–	–	–	–	–	–
–	–	–	–	–	–	–	–	–	–	–	–
–	–	–	–	–	–	–	–	–	–	–	–
1.2	1.2	0.7	0.7	2	2	1	1	0.2	0.2	10	10
–	–	–	–	–	–	–	–	–	–	–	–
91	91	62	62	130	130	85	85	34	34	175	175
–	–	–	–	–	–	4	4	2.5	2.5	40	40
72	72	20	20	120	120	50	50	14	14	145	145
–	–	–	–	–	–	–	–	–	–	–	–
1.7	1.7	0.1	0.1	2.9	2.9	1	1	0.05	0.05	6	6
–	–	–	–	–	–	–	–	–	–	–	–
–	–	–	–	–	–	–	–	–	–	–	–
–	–	–	–	–	–	–	–	–	–	–	–
56	56	16	16	93	93	65	65	24	24	200	200
120	120	31	31	200	200	200	200	25	25	400	400
–	–	–	–	–	–	–	–	–	–	–	–
–	–	–	–	–	–	–	–	–	–	–	–
–	–	–	–	–	–	–	–	–	–	–	–
–	–	–	–	–	–	–	–	–	–	–	–
–	–	–	–	–	–	–	–	–	–	–	–
–	–	–	–	–	–	–	–	–	–	–	–
200	200	77	77	333	333	155	155	67	67	300	300
–	–	–	–	–	–	–	–	–	–	–	–

to be continued

8 Richtwaarden
9 Streefwarden
10 Bodemsaneringsnormen
11 Valeur seuil
12 Valeur de reference
13 Valeur d'intervention

Appendix A (continued)

Parameter	unit	BOSNIA and HERZEGOVINA		CROATIA		CZECH REPUBLIC		DENMARK	
		Agricultural soil	Grazing land soil	Agricultural soil [14,15]	Grazing land soil	Agricultural soil	Grazing land soil	Agricultural soil	Grazing land soil
Ag	[mg/kg]	–	–	–	–	–	–	50	50
As		20	20	–	–	30	30	20	20
B		30 – 50	30 – 50	–	–	40	40	–	–
Ba		–	–	–	–	–	–	100	100
Be		–	–	–	–	7	7	–	–
Bi		0.5–1	0.5–1	–	–	–	–	–	–
Cd		–	–	0.5 – 1.0	0.5 – 1.0	1	1	0.5	0.5
Co		30 – 60	30 – 60	–	–	50	50	–	–
Cr (total) [a]		50 – 100	50 – 100	40 – 80	40 – 80	200	200	500	500
Cr (VI)		50 – 80	50 – 80	–	–	–	–	20	20
Cu		–	–	60 – 90	60 – 90	100	100	500	500
F		150 – 350	150 – 350	–	–	–	–	20	20
Hg		0.5–1.5	0.5–1.5	0.5–1.0	0.5–1.0	0.8	0.8	1	1
Mn		–	–	–	–	–	–	–	–
Mo		30 – 60	30 – 60	–	–	5	5	5	5
NO_3		–	–	–	–	–	–	–	–
Ni		30 – 50	30 – 50	30 – 50	30 – 50	–	–	30	30
Pb		–	–	50 – 100	50 – 100	140	140	40	40
S		300 – 500	300 – 500	–	–	–	–	–	–
Sb		–	–	–	–	–	–	–	–
Se		–	–	–	–	–	–	20	20
Tl		0.5–1	0.5–1	–	–	–	–	1	1
U		–	–	–	–	–	–	–	–
V		–	–	–	–	220	220	–	–
Zn		100 – 200	100 – 200	60 – 150	60 – 150	200	200	500	500
pH		–	–	–	–	–	–	–	–

14 The same limits for both agricultural and grazing land soil

15 Precaution values for soil type loam soil

GERMANY		HUNGARY		LITHUANIA		MONTENEGRO		POLAND	
Agricultural soil	**Grazing land soil (24)**	**Agricultural soil (4)**	**Grazing land soil (26)**	**Agricultural soil (27)**	**Grazing land soil (27)**	**Agricultural soil**	**Grazing land soil**	**Agricultural soil (4)**	**Grazing land soil (4)**
–	–	2	2	2	2	–	–	–	–
200 (16)	50	15	15	10	10	20	20	20	20
–	–	1000	1000	50	50	5	5	–	–
–	–	250	250	600	600	–	–	200	200
–	–	–	–	10	10	–	–	–	–
–	–	–	–	–	–	–	–	–	–
0.04/0.1 (17)	20	1	1	3	3	2	2	4	4
–	–	30	30	30	30	50	50	20	20
30–100 (18)	–	75	75	100	100	50	50	150	150
–	–	1	1	–	–	–	–	–	–
20–60 (19)	1300 (25)	75	75	100	100	100	100	150	150
–	–	–	–	200	200	300	300	–	–
5.0 (20)	2.0	0.5	0.5	1.5	1.5	1.5	1.5	2	2
–	–	–	–	1500	1500	–	–	–	–
–	–	7	7	5	5	10	10	10	10
–	–	–	500	–	–	–	–	–	–
15–70 (21)	1.9	40	40	75	75	50	50	100	100
0.1 (22)	1.2	100	100	100	100	50	50	100	100
–	–	–	–	160	160	–	–	–	–
–	–	–	5	10	10	–	–	–	–
–	–	1	1	5	5	–	–	–	–
0.1 (22)	15	–	–	–	–	–	–	–	–
–	–	–	–	20	20	–	–	–	–
–	–	–	–	150	150	–	–	–	–
60–200 (23)	–	200	200	300	300	300	300	300	300
–	–	–	–	–	–	–	–	–	–

to be continued

[16] In case of soil with occasionally decreasing conditions, a trigger value of 50 mg/kg shall be applied.

[17] Action value for ammonium nitrate extraction; on areas that are used for growing bread wheat or strongly cadmium-accumulating vegetables an action value of 0.04 mg/kg shall be applied; otherwise, the action value is 0.1 mg/kg.

[18] Precaution values for soil type sand (30 mg/kg) and clay (100 mg/kg)

[19] Precaution values for soil type sand (20 mg/kg) and clay (60 mg/kg)

[20] Trigger value for AR

[21] Precaution values for soil type sand (15 mg/kg) and clay (70 mg/kg)

[22] Trigger value for ammonium nitrate extraction

[23] Precaution values for soil type sand (60 mg/kg) and clay (200 mg/kg)

[24] Action values for the pollutant transition soil-food plant on grassland

[25] In case sheep are kept on grassland, an action value of 200 mg/kg shall be applied.

[26] Geological medium except arable (agricultural) land

[27] Grain size <1 mm; real total content

Appendix A (continued)

		PORTUGAL	SERBIA	SLOVENIA		SWITZERLAND		UNITED KINGDOM	
Parameter	unit	Agricultural soil [28]	Agricultural soil [29]	Agricultural soil	Grazing land soil	Agricultural soil [30]	Grazing land soil [30]	Agricultural soil [39]	Grazing land soil [39]
Ag		–	–	–	–	500 [31]	500 [31]	–	–
As		–	25	20	20	50 [31]	50 [31]	43	43
B		–	50	–	–	–	–	–	–
Ba		–	–	–	–	–	–	–	–
Be		–	–	–	–	–	–	–	–
Bi		–	–	–	–	–	–	–	–
Cd		20	3	1	1	0.8 [32]	0.8 [32]	1.8	1.8
Co		–	–	20	20	–	–	–	–
Cr (total) [a]		1000	–	100	100	50	50	130	130
Cr (VI)		–	100	25	25	100 [31]	100 [31]	–	–
Cu		1000	100	60	60	40 [33]	40 [33]	–	–
F		–	300	–	–	700 [34]	700 [34]	–	–
Hg	[mg/kg]	16	–	0.8	0.8	0.5 [35]	0.5 [35]	80 [40]; 8 [41]	80 [40]; 8 [41]
Mn		–	–	–	–	–	–	–	–
Mo		–	–	10	10	5	5	–	–
NO_3		–	–	–	–	–	–	–	–
Ni		300	–	50	50	50 [36]	50 [36]	230	230
Pb		750	–	85	85	50 [37]	50 [37]	450	450
S		–	–	–	–	–	–	–	–
Sb		–	–	–	–	50 [31]	50 [31]	–	–
Se		–	–	–	–	–	–	120	120
Tl		–	–	–	–	–	–	–	–
U		–	–	–	–	–	–	–	–
V		–	–	–	–	–	–	–	–
Zn		2500	–	200	200	150 [38]	150 [38]	–	–
pH	–	–	–	–	–	–	–	–	–

28 Limits when adding manure in soil

29 Maximum allowed concentration in soil

30 Guideline values for the total content in dried samples, sieved <2mm, soil up to 15 % humus. Sampling depth 0–20 cm; playgrounds 0–5 cm. Extraction with 2 molar nitric acid. *Terminology:* French Valeur indicative = Guideline value; Seul d'investigation = Trigger value; Valeur d'assainissement = Action value

31 As, Ag, Cr(VI), Sb: Action values for gardens and playgrounds.

32 Cd: Trigger value 2 mg/kg; Action value 30 mg/kg for agricultural soil, 20 mg/kg for gardens and playgrounds.

33 Cu: Trigger value 150 mg/kg in soil for forage crops; Action value 1000 mg/kg for agricultural soil in general, gardens and playgrounds.

34 Similar to (30), but NaOH fusion.

35 Hg: Action value 5 mg/kg for gardens and playgrounds.

36 Ni: Action value 1000 mg/kg for gardens and playgrounds.

37 Pb: Trigger value 200 mg/kg; Action value 2000 mg/kg for agricultural soil, 1000 mg/kg for gardens and playgrounds.

38 Zn: Action value 2000 mg/kg for agricultural soil, gardens and playgrounds.

39 Soil Guideline Values for the UK are released by the UK Environment Agency covering soil of England and Wales and are for allotments rather than agricultural or grazing land. Revised contaminated land Statutory Guidance published in April 2012 will see new tools released to replace SGVs for use in the indentification of contaminated land and health risk assessments. Revised SGVs were published in 2009 for some elements, though for Pb and Cr the original SGVs have been withdrawn and not replaced.

40 Inorganic Hg^{2+}

41 Methyl Hg^{4+}

References to EU and National Legislation

Austria
ÖNORM L 1075 (2004): Grundlagen für die Bewertung der Gehalte ausgewählter Elemente in Böden. Medieninhaber und Hersteller: Österreichisches Normungsinstitut, 1020 Wien
http://www.on-norm.at

Belgium (Brussels)
17 December 2009. – Arrêté du Gouvernement de la Région de Bruxelles-Capitale déterminant les normes d'intervention et les normes d'assainissement
http://www.bruxellesenvironnement.be/Templates/Professionnels/Informer.aspx?id=32685&langtype=2060

Belgium (Flanders)
Decree of 27 October 2006 on soil remediation and soil protection. This decree only describes the regulatory framework. The threshold values are reported in "Vlarebo 2008":
Vlaams Reglement rond de Bodemsanering (Order of the Flemish Government establishing the Flemish regulation on soil remediation and soil protection. 14 December 2007).
http://www.ovam.be/jahia/Jahia/pid/1858
http://www.ovam.be/jahia/Jahia/pid/1859

Belgium (Wallonia)
5 décembre 2008 – Décret relatif à la gestion des sols
http://environnement.wallonie.be/legis/solsoussol/sol003.htm
http://environnement.wallonie.be/legis/legdecw8.htm#SOL

Croatia
Narodne Novine (Official Gazette) (2010): Pravilnik o zaštiti poljoprivrednog zemljišta od onečišćenja (Ordinance on the protection of agricultural land against pollution).-
Ministry of agriculture, fishery and rural development, 032/2010, Zagreb, 6 pp.
http://narodne-novine.nn.hr/default.aspx

Czech Republic
Regulation No. 13/1994 of the Ministry of the Environment of the Czech Republic, determining certain details of the protection of the agricultural land resources.

Denmark
Miløstyrelsen (2010): Liste over kvalitetskriterier i relation til forurenet jord og kvalitetskriterier for drikkevand. Danish EPA, 13 pp.

European Union
EU directive 86/278/EEC of 12 June 1986 on the protection of the environment, and in particular of the soil, when sewage sludge is used in agriculture. Official Journal of the European Communities, 4/7/1986, L181, 6 – 12.
http://eur-lex.europa.eu/LexUriServ/LexUriServ.do?uri=CELEX:31986L0278:EN:NOT
http://eur-lex.europa.eu/LexUriServ/LexUriServ.do?uri=OJ:L:1986:181:0006:0012:EN:PDF

FYROM
Regulation on the manner and procedure for the use of sludge, the maximum concentration values of the of heavy metals in the soil that the sludge is used, values of concentrations of heavy metals in sludge, in accordance with its purpose and the maximum annual quantities of heavy metals that may be entered in soil. Official gazette of RM, No. 73/11, 31.05.2011.

Germany
Federal Soil Protection and Contaminated Sites Ordinance (BBodSchV), dated 12 July 1999.
Bundes-Bodenschutz- und Altlastenverordnung (BBodSchV), Bundesgesetzblatt I, 12. Juli 1999, S. 1554.

Hungary
Contamination limiting values and the measuring method can be found in 6/2009. (IV. 14.) KvVM-EüM-FVM joint regulation (valid in 2012.10.15.):
Environmental protection. Testing of soil.
Sampling MSZ 21470-1: 1998.
Sampling soil for management purposes in agriculture
MSZ 08-0202: 1977.
Soil quality – Sampling – Part 2: Guidance on sampling techniques
ISO 10381-2: 2002.
http://www.complex.hu/jr/gen/hjegy_doc.cgi?docid=A0900006.KVV

Lithuania
Decree No. V-114 of Minister of Health of the Republic of Lithuania of 8th March 2004, On the Hygiene Standard HN 60:2004 - Maximum permissible concentrations of hazardous chemical substances in soil. Lietuvos higienos norma HN 60:2004. Kenksmingų cheminių medžiagų didžiausios leidžiamos koncentracijos dirvožemyje (Valstybės žinios, 2004, No. 41-1357).

Montenegro
Montenegro: Official document, 1996. Pravilnik o dozvoljenim količinama opasnih i štetnih materija u zemljištu i metodama za njihovo ispitivanje. Službeni list Republike Crne Gore 18/97.

Poland
Permisible limit values in agricultural and grazing land soil. Decree of the Polish Ministry of the Environment of 9th September 2002

Portugal
Diário da República, Decreto-Lei n.º 276/2009 de 2 de Outubro
http://dre.pt/pdf1s/2009/10/19200/0715407165.pdf

Serbia
Official Gazette SRY 23/94. Regulations on allowed concentrations of hazardous and harmful substances in agricultural soil and water for irrigation.

Slovenia
Official Gazette RS, 1996. Uredba o mejnih, opozorilnih in kritičnih imisijskih vrednostih nevarnih snovi v tleh (in Slovenian). Uradni list Republike Slovenije 68/96, 5773 – 5774.
Off. Gaz. RS 68/96 is in accordance with The New Dutchlist
Limit values denote the concentration of individual dangerous substances in the soil at which soil quality and fertility are not yet affected by pollution.
'Warning values' represent the concentration of individual dangerous substances in the soil that probably cause harmful effects or influence human health and environment.
Critical values represent the concentration of individual dangerous substances in the soil that have harmful effects and influences on human health and environment. Polluted soil is not adequate for plant growing and animal keeping and for water retention and filtration.

Switzerland
SR 814.12
Verordnung vom 1. Juli 1998 über Belastungen des Bodens (VBBo)
Ordonnance du 1er juillet 1998 sur les atteintes portées aux sols (OSol)
Ordinanza del 1° luglio 1998 contro il deterioramento del suolo (O suolo)
SR 814.680
Verordnung vom 26. August 1998 über die Sanierung von belasteten Standorten (Altlasten-Verordnung, AltlV)
Ordonnance du 26 août 1998 sur l'assainissement des sites pollués (Ordonnance sur les sites contaminés, OSites)
Ordinanza del 26 agosto 1998 sul risanamento dei siti inquinati (Ordinanza sui siti contaminati, OSiti)

Appendix B

GEMAS Project: Statistical Parameters for All of Europe (Ap and Gr samples)

Appendix B

Parameter	Material	Method	N	Unit	DL	% < DL	min.	Q2
Altitude	Ap		2108	m a.s.l.			**-5**	4
Altitude	Gr		2023	m a.s.l.			**-4**	4
CEC	Ap	silver-thiourea	2108	meq/100 g	0.1	0	**1.8**	4.7
CEC	Gr	silver-thiourea	2024	meq/100 g	0.1	0	**2.54**	6.85
CIA	Ap	calculated	2108	%			**4.25**	36.3
CIA	Gr	calculated	2020	%			**2.35**	28.9
Clay	Ap	MIR	2108	%	2	10	**1**	1
Clay	Gr	MIR	2022	%	2	13	**1**	1
LOI	Ap	gravimetry	2108	wt%	0.01	0	**1.46**	2.59
LOI	Gr	gravimetry	2024	wt%	0.01	0	**1.03**	2.73
MS	Ap	susceptibility-meter	2097	10^{-6} m^3/kg		0	**< 0.0005**	0.028
pH_$CaCl_2$	Ap	potentiometry	2108	pH	0.1	0	**3.32**	3.95
pH_$CaCl_2$	Gr	potentiometry	2023	pH	0.1	0	**3.26**	3.81
TOC	Ap	IR spectroscopy	2095	wt%	0.1	0	**0.4**	0.6
TOC	Gr	IR spectroscopy	2023	wt%	0.1	0	**0.41**	0.694

Element	Material	Method	N	Unit	DL	% < DL	min.	Q2
Ag - Silver								
Ag	Ap	MMI®	2108	mg/kg	0.001	4.9	**<0.001**	<0.001
Ag	Ap	ICP_MS, ACME 1F_15	2108	mg/kg	0.002	0.4	**<0.002**	0.007
Ag	Gr	ICP_MS, ACME 1F_15	2024	mg/kg	0.002	0.4	**<0.002**	0.006
Al - Aluminium								
Al	Ap	MMI®	2108	mg/kg	1 / 300	0.33	**<1**	7
Al	Ap	ICP_MS, ACME 1F_15	2108	mg/kg	50	0	**329**	1536
Al	Gr	ICP_MS, ACME 1F_15	2024	mg/kg	50	0	**495**	1432
Al	Ap	XRF	2108	mg/kg	265	0	**1958**	10,479
Al	Gr	XRF	2024	mg/kg	265	0	**1535**	7777
Al_2O_3	Ap	XRF	2108	wt%	0.05	0	**0.37**	2.0
Al_2O_3	Gr	XRF	2024	wt%	0.05	0	**0.29**	1.5
As - Arsenic								
As	Ap	MMI®	2108	mg/kg	0.01	29	**<0.01**	<0.01
As	Ap	ICP_MS, ACME 1F_15	2108	mg/kg	0.05	0.1	**<0.05**	0.42
As	Gr	ICP_MS, ACME 1F_15	2024	mg/kg	0.05	0.05	**<0.05**	0.40
As	Ap	XRF	2108	mg/kg	3	24	**<3**	<3
As	Gr	XRF	2024	mg/kg	3	23	**<3**	<3
Au - Gold								
Au	Ap	MMI®	2108	mg/kg	0.0001	17	**<0.0001**	<0.0001
Au	Ap	ICP_MS, ACME 1F_15	2108	mg/kg	0.0002	15	**<0.0002**	<0.0002
Au	Gr	ICP_MS, ACME 1F_15	2024	mg/kg	0.0002	13	**<0.0002**	<0.0002
B - Boron								
B	Ap	ICP_MS, ACME 1F_15	2108	mg/kg	0.5	7.4	**<0.5**	<0.5
B	Gr	ICP_MS, ACME 1F_15	2024	mg/kg	0.5	7	**<0.5**	<0.5

Q5	Q10	Q25	Q50	Q75	Q90	Q95	Q98	max.	Powers
10	24	72	**159**	348	657	829	1000	**1644**	3.2
11	27	80	**177**	395	715	889	1057	**2400**	3.4
6.24	8	11.1	**16.4**	23.2	30.3	34.7	39.2	**48.3**	1.4
8.18	9.71	12.8	**18.7**	26.9	33.9	38.1	42.8	**49.9**	1.3
49.9	54.3	60.7	**66.8**	73.9	79.6	82.3	85.4	**92.6**	1.3
47.7	53.1	60.2	**67.5**	74.9	80.4	83.2	85.9	**97.1**	1.6
1	1	7	**15**	21	27	30	34	**63**	1.8
1	1	5	**13**	19	25	29	32	**69**	1.8
3.11	3.91	5.87	**8.57**	12.8	20.5	27.4	36	**95.3**	1.8
3.98	5	7.74	**11.6**	16.6	25.9	33.2	49	**93.9**	2
0.044	0.065	0.11	**0.21**	0.45	1.03	1.87	3.59	**40**	4.9
4.18	4.44	4.96	**5.77**	7.03	7.36	7.45	7.53	**7.98**	0.4
4.04	4.26	4.69	**5.45**	6.89	7.33	7.45	7.54	**8.06**	0.4
0.7	0.9	1.2	**1.8**	2.6	3.9	5.4	10	**46**	2.1
0.921	1.1	1.7	**2.7**	4.3	6.8	10	23.6	**49**	2.1

Q5	Q10	Q25	Q50	Q75	Q90	Q95	Q98	max.	Powers
0.001	0.002	0.006	**0.014**	0.025	0.04	0.053	0.083	**1.34**	3.4
0.01	0.014	0.023	**0.038**	0.060	0.096	0.13	0.19	**2.95**	3.5
0.010	0.014	0.023	**0.040**	0.065	0.10	0.14	0.22	**3.91**	3.6
10	14	25	**61**	155	253	>300	>300	**>300**	3
2445	3540	6683	**10,993**	15,809	20,756	24,091	29,610	**65,090**	2.3
2213	3220	6146	**10,746**	15,966	20,740	25,314	32,491	**62,542**	2.1
15,510	21,699	36,347	**55,068**	67,876	77,001	83,603	90,857	**143,850**	1.9
12,809	18,896	34,087	**51,898**	66,493	77,421	83,621	91,280	**141,429**	2
2.9	4.1	6.9	**10**	13	15	16	17	**27**	1.9
2.4	3.6	6.4	**9.8**	13	15	16	17	**27**	2
<0.01	<0.01	<0.01	**0.02**	0.03	0.05	0.07	0.13	**0.53**	2
0.72	1.2	2.5	**5.5**	9.5	15	21	34	**666**	4.4
0.76	1.3	2.7	**5.6**	9.8	16	25	40	**292**	4.1
<3	<3	3	**7**	11	18	25	40	**695**	2.7
<3	<3	3	**7**	12	20	29	44.5	**291**	2.3
<0.0001	<0.0001	0.0001	**0.0003**	0.0006	0.0012	0.0017	0.003	**0.024**	2.7
<0.0002	<0.0002	0.000395	**0.00093**	0.0018	0.0030	0.0045	0.0078	**0.083**	2.9
<0.0002	<0.0002	0.000447	**0.001**	0.0019	0.0031	0.0047	0.0079	**0.23**	3.4
<0.5	0.61	1.23	**2.42**	4.68	7.64	10	13.8	**49**	2.3
<0.5	0.67	1.27	**2.58**	5.14	8.38	11.4	15.8	**41**	2.2

to be continued

Appendix B (continued)

Parameter	Material	Method	N	Unit	DL	% < DL	min.	Q2
Ba - Barium								
Ba	Ap	MMI®	2108	mg/kg	0.01	0.09	**<0.01**	0.12
Ba	Ap	ICP_MS, ACME 1F_15	2108	mg/kg	0.5	0	**2.6**	10
Ba	Gr	ICP_MS, ACME 1F_15	2024	mg/kg	0.5	0	**3.8**	10
Ba	Ap	XRF	2108	mg/kg	5	0	**16**	107
Ba	Gr	XRF	2024	mg/kg	5	0	**22**	92
Be - Beryllium								
Be	Ap	ICP_MS, ACME 1F_15	2108	mg/kg	0.05	1.8	**<0.05**	0.056
Be	Gr	ICP_MS, ACME 1F_15	2024	mg/kg	0.05	2.4	**<0.05**	<0.05
Bi - Bismuth								
Bi	Ap	MMI®	2108	mg/kg	0.001	88	**<0.001**	<0.001
Bi	Ap	ICP_MS, ACME 1F_15	2108	mg/kg	0.005	0.05	**<0.005**	0.031
Bi	Gr	ICP_MS, ACME 1F_15	2024	mg/kg	0.005	0	**0.011**	0.035
Bi	Ap	XRF	2108	mg/kg	3	94	**<3**	<3
Bi	Gr	XRF	2024	mg/kg	3	97	**<3**	<3
C - Carbon								
TC	Ap	IR spectroscopy	2108	wt%	0.07	0	**0.08**	0.641
TC	Gr	IR spectroscopy	2024	wt%	0.07	0	**0.27**	0.73
Ca - Calcium								
Ca	Ap	MMI®	2108	mg/kg	10	0.4	**<10**	50
Ca	Ap	ICP_MS, ACME 1F_15	2108	mg/kg	100	0.2	**<100**	323
Ca	Gr	ICP_MS, ACME 1F_15	2024	mg/kg	100	0.4	**<100**	306
Ca	Ap	XRF	2108	mg/kg	36	0	**86**	773
Ca	Gr	XRF	2024	mg/kg	36	0	**50**	746
CaO	Ap	XRF	2108	wt%	0.005	0	**0.012**	0.11
CaO	Gr	XRF	2024	wt%	0.005	0	**0.007**	0.10
Cd - Cadmium								
Cd	Ap	MMI®	2108	mg/kg	0.001	0.19	**<0.001**	0.007
Cd	Ap	ICP_MS, ACME 1F_15	2108	mg/kg	0.01	0.2	**<0.01**	0.031
Cd	Gr	ICP_MS, ACME 1F_15	2024	mg/kg	0.01	0.3	**<0.01**	0.031
Ce - Cerium								
Ce	Ap	MMI®	2108	mg/kg	0.005	6.2	**<0.005**	<0.005
Ce	Ap	ICP_MS, ACME 1F_15	2108	mg/kg	0.1	0	**1.6**	5.31
Ce	Gr	ICP_MS, ACME 1F_15	2024	mg/kg	0.1	0	**1.7**	4.7
Ce	Ap	XRF	2108	mg/kg	20	6.5	**<20**	<20
Ce	Gr	XRF	2024	mg/kg	20	8.7	**<20**	<20
Cl - Chlorine								
Cl	Ap	XRF	2108	wt%	0.002	45	**<0.002**	<0.002
Cl	Gr	XRF	2024	wt%	0.002	59	**<0.002**	<0.002

Q5	Q10	Q25	Q50	Q75	Q90	Q95	Q98	max.	Powers
0.19	0.31	0.57	**1.1**	2.0	3.0	4.2	6.3	**52**	4
15	22	36	**62**	95	142	180	252	**818**	2.5
15	21	37	**63**	98	143	184	258	**2043**	2.7
146	188	271	**384**	505	641	739	918	**3350**	2.3
133	170	251	**354**	467	611	703	883	**5410**	2.4
0.094	0.14	0.28	**0.52**	0.81	1.2	1.4	2.0	**11**	2.6
0.091	0.13	0.27	**0.51**	0.83	1.2	1.6	2.4	**9.5**	2.6
<0.001	<0.001	<0.001	**<0.001**	<0.001	0.0010	0.0020	0.0040	**0.20**	2.6
0.043	0.058	0.094	**0.16**	0.26	0.33	0.49	0.76	**3.2**	3.1
0.050	0.065	0.10	**0.18**	0.28	0.42	0.55	0.87	**18**	3.2
<3	<3	<3	**<3**	<3	<3	4	4	**9**	0.8
<3	<3	<3	**<3**	<3	<3	<3	4	**17**	1.1
0.824	1.01	1.41	**2.21**	3.48	5.83	7.85	11.2	**48.6**	2.8
1.03	1.31	2	**3.29**	5.1	8.02	11.4	23.2	**50.1**	2.3
100	150	320	**595**	902	1160	1340	1539	**2490**	2.7
571	906	1635	**3034**	8088	71,146	131,930	192,637	**347,847**	3.8
529	803	1547	**3141**	9853	64,691	132,133	184,761	**328,124**	3.8
1310	1949	3745	**8494**	18,802	74,293	139,027	202,489	**378,148**	3.6
1202	1896	3495	**7662**	17,953	69,761	134,239	196,074	**362,803**	3.9
0.18	0.27	0.52	**1.2**	2.6	10	20	28	**53**	3.6
0.17	0.27	0.49	**1.1**	2.5	9.8	19	27	**51**	3.9
0.011	0.018	0.032	**0.054**	0.085	0.13	0.17	0.27	**2.8**	3.7
0.046	0.065	0.11	**0.18**	0.28	0.45	0.61	0.98	**7.5**	3.2
0.047	0.069	0.12	**0.20**	0.33	0.57	0.89	1.4	**6.0**	3.1
<0.005	0.009	0.033	**0.14**	0.37	0.85	1.3	2.1	**23**	4
8.1	11.9	18.5	**28.4**	39.5	51.4	62.3	81.3	**265**	2.2
7.8	10.5	17.5	**27.1**	37.9	48.4	61.3	82.4	**272**	2.2
<20	25	40.8	**58**	75	92	103	125	**317**	1.5
<20	23	37	**56**	73	89	103	125	**315**	1.5
<0.002	<0.002	<0.002	**0.003**	0.007	0.013	0.018	0.026	**0.060**	1.8
<0.002	<0.002	<0.002	**<0.002**	0.004	0.007	0.009	0.012	**0.042**	1.6

to be continued

Appendix B (continued)

Parameter	Material	Method	N	Unit	DL	% < DL	min.	Q2
Co – Cobalt								
Co	Ap	MMI®	2108	mg/kg	0.005	0.33	**<0.005**	0.018
Co	Ap	ICP_MS, ACME 1F_15	2108	mg/kg	0.1	0.05	**<0.1**	0.54
Co	Gr	ICP_MS, ACME 1F_15	2024	mg/kg	0.1	0.1	**<0.1**	0.50
Co	Ap	XRF	2108	mg/kg	3	14	**<3**	<3
Co	Gr	XRF	2024	mg/kg	3	16	**<3**	<3
Cr – Chromium								
Cr	Ap	MMI®	2108	mg/kg	0.001	0.38	**<0.001**	0.007
Cr	Ap	ICP_MS, ACME 1F_15	2108	mg/kg	0.2	0	**0.4**	2.6
Cr	Gr	ICP_MS, ACME 1F_15	2024	mg/kg	0.2	0	**0.64**	2.4
Cr	Ap	XRF	2108	mg/kg	4	0.05	**<4**	10
Cr	Gr	XRF	2024	mg/kg	4	0.3	**<4**	10
Cs – Caesium								
Cs	Ap	MMI®	2108	mg/kg	0.0005	40	**<0.0005**	<0.0005
Cs	Ap	ICP_MS, ACME 1F_15	2108	mg/kg	0.02	0	**0.029**	0.17
Cs	Gr	ICP_MS, ACME 1F_15	2024	mg/kg	0.02	0	**0.036**	0.13
Cs	Ap	XRF	2108	mg/kg	3	25	**<3**	<3
Cs	Gr	XRF	2024	mg/kg	3	28	**<3**	<3
Cu – Copper								
Cu	Ap	MMI®	2108	mg/kg	0.01	0.05	**<0.01**	0.19
Cu	Ap	ICP_MS, ACME 1F_15	2108	mg/kg	0.01	0	**0.3**	1.8
Cu	Gr	ICP_MS, ACME 1F_15	2024	mg/kg	0.01	0	**0.62**	1.8
Cu	Ap	XRF	2108	mg/kg	5	16	**<5**	<5
Cu	Gr	XRF	2024	mg/kg	5	20	**<5**	<5
Dy – Dysprosium								
Dy	Ap	MMI®	2108	mg/kg	0.001	0.95	**<0.001**	0.002
Er – Erbium								
Er	Ap	MMI®	2108	mg/kg	0.001	0.38	**<0.0005**	0.0013
Eu – Europium								
Eu	Ap	MMI®	2108	mg/kg	0.001	5.1	**<0.0005**	<0.0005
F – Fluorine								
F	Ap	XRF	2108	wt%	0.05	77	**<0.05**	<0.05
F	Gr	XRF	2024	wt%	0.05	80	**<0.05**	<0.05
Fe – Iron								
Fe	Ap	MMI®	2108	mg/kg	1	0.14	**<1**	5
Fe	Ap	ICP_MS, ACME 1F_15	2108	mg/kg	100	0	**392**	1959
Fe	Gr	ICP_MS, ACME 1F_15	2024	mg/kg	100	0	**380**	1976
Fe	Ap	XRF	2108	mg/kg	70	0	**839**	3507
Fe	Gr	XRF	2024	mg/kg	70	0	**909**	3179
Fe_2O_3	Ap	XRF	2108	wt%	0.01	0	**0.12**	0.501
Fe_2O_3	Gr	XRF	2024	wt%	0.01	0	**0.13**	0.455

Q5	Q10	Q25	Q50	Q75	Q90	Q95	Q98	max.	Powers
0.027	0.043	0.077	**0.18**	0.43	0.94	1.4	2.0	**8.9**	3.6
0.96	1.7	3.7	**7.5**	12	17	22	29	**126**	3.4
0.96	1.5	3.3	**7.2**	12	17	21	28	**255**	3.7
<3	<3	5	**9**	14	20	25	33	**144**	2
<3	<3	5	**9**	14	19	24	31	**271**	2.3
0.011	0.017	0.033	**0.065**	0.12	0.20	0.26	0.39	**1.3**	3.4
4	5.8	11	**20**	33	48	70	107	**696**	3.2
3.7	5.4	10	**20**	32	48	69	103	**577**	3
14	19	34	**62**	89	125	213	349	**6133**	3.5
13	18	32	**61**	90	128	199	360	**7552**	3.6
<0.0005	<0.0005	<0.0005	**0.0007**	0.002	0.0054	0.0086	0.018	**0.695**	3.4
0.24	0.36	0.62	**1.1**	1.7	3.0	4.3	7.0	**43**	3.2
0.21	0.31	0.59	**1.1**	1.7	3.0	4.5	7.7	**29**	2.9
<3	<3	<3	**5**	7	9	11	16	**42**	1.4
<3	<3	<3	**5**	8	9.7	12.8	16	**56**	1.6
0.28	0.40	0.68	**1.3**	2.7	5.2	8.3	16	**62**	4.1
3.2	4.6	8.3	**15**	24	38	50	78	**395**	3.1
2.7	4.1	7.5	**15**	25	37	48	68	**373**	2.8
<5	<5	6	**13**	22	35	48	78	**391**	2.2
<5	<5	6	**13**	22	33	45	64	**261**	2
0.003	0.006	0.015	**0.035**	0.075	0.15	0.21	0.31	**2.2**	3.6
0.0025	0.0043	0.0093	**0.020**	0.045	0.088	0.14	0.20	**1.2**	3.7
<0.0005	0.001	0.0028	**0.0075**	0.016	0.031	0.044	0.064	**0.21**	2.9
<0.05	<0.05	<0.05	**<0.05**	<0.05	0.07	0.09	0.1	**0.21**	0.9
<0.05	<0.05	<0.05	**<0.05**	<0.05	0.07	0.08	0.1	**0.23**	1
7	9	13	**31**	88	173	239	384	**2200**	3.6
3095	5053	9890	**17,200**	25,067	32,159	36,897	45,769	**133,926**	2.5
3174	5127	9610	**17,004**	25,188	32,647	38,310	44,949	**94,759**	2.4
5386	8323	15,527	**24,550**	34,902	42,896	48,376	59,136	**154,715**	2.3
5106	7973	14,478	**23,989**	34,690	43,293	49,367	58,438	**114,422**	2.1
0.77	1.19	2.22	**3.51**	4.99	6.13	6.92	8.45	**22.1**	2.3
0.73	1.14	2.07	**3.43**	4.96	6.19	7.06	8.36	**16.4**	2.1

to be continued

Appendix B (continued)

Parameter	Material	Method	N	Unit	DL	% < DL	min.	Q2
Ga - Gallium								
Ga	Ap	MMI®	2108	mg/kg	0.001	34	**<0.001**	<0.001
Ga	Ap	ICP_MS, ACME 1F_15	2108	mg/kg	0.05	0	**0.14**	0.57
Ga	Gr	ICP_MS, ACME 1F_15	2024	mg/kg	0.05	0	**0.19**	0.51
Ga	Ap	XRF	2108	mg/kg	2	3.3	**<2**	<2
Ga	Gr	XRF	2024	mg/kg	2	4.7	**<2**	<2
Gd - Gadolinium								
Gd	Ap	MMI®	2108	mg/kg	0.001	1.6	**<0.001**	0.001
Ge - Germanium								
Ge	Ap	ICP_MS, ACME 1F_15	2108	mg/kg	0.02	26	**<0.02**	<0.02
Ge	Gr	ICP_MS, ACME 1F_15	2024	mg/kg	0.02	27	**<0.02**	<0.02
Hf - Hafnium								
Hf	Ap	ICP_MS, ACME 1F_15	2108	mg/kg	0.01	16	**<0.01**	<0.01
Hf	Gr	ICP_MS, ACME 1F_15	2024	mg/kg	0.01	16	**<0.01**	<0.01
Hf	Ap	XRF	2108	mg/kg	5	20	**<5**	<5
Hf	Gr	XRF	2024	mg/kg	5	16	**<5**	<5
Hg - Mercury								
Hg	Ap	MMI®	2108	mg/kg	0.001	75	**<0.001**	<0.001
Hg	Ap	ICP_MS, ACME 1F_15	2108	mg/kg	0.003	0.8	**<0.003**	0.0056
Hg	Gr	ICP_MS, ACME 1F_15	2024	mg/kg	0.003	0.35	**<0.003**	0.0072
In - Indium								
In	Ap	MMI®	2108	mg/kg	0.0005	96	**<0.0005**	<0.0005
In	Ap	ICP_MS, ACME 1F_15	2108	mg/kg	0.003	3.8	**<0.003**	<0.003
In	Gr	ICP_MS, ACME 1F_15	2024	mg/kg	0.003	4.1	**<0.003**	<0.003
K - Potassium								
K	Ap	MMI®	2108	mg/kg	0.1	0.33	**<0.1**	14
K	Ap	ICP_MS, ACME 1F_15	2108	mg/kg	20	0	**40**	166
K	Gr	ICP_MS, ACME 1F_15	2024	mg/kg	20	0	**43**	141
K	Ap	XRF	2108	mg/kg	42	0	**241**	3841
K	Gr	XRF	2024	mg/kg	42	0	**240**	2860
K_2O	Ap	XRF	2108	wt%	0.005	0	**0.029**	0.463
K_2O	Gr	XRF	2024	wt%	0.005	0	**0.029**	0.345
La - Lanthanum								
La	Ap	MMI®	2108	mg/kg	0.001	10	**<0.001**	<0.001
La	Ap	ICP_MS, ACME 1F_15	2108	mg/kg	0.1	0	**1**	2.8
La	Gr	ICP_MS, ACME 1F_15	2024	mg/kg	0.1	0	**0.93**	2.6
La	Ap	XRF	2108	mg/kg	20	43	**<20**	10
La	Gr	XRF	2024	mg/kg	20	40	**<20**	<20
Li - Lithium								
Li	Ap	MMI®	2108	mg/kg	0.005	56	**<0.005**	<0.005
Li	Ap	ICP_MS, ACME 1F_15	2108	mg/kg	0.1	0	**0.16**	1.0
Li	Gr	ICP_MS, ACME 1F_15	2024	mg/kg	0.1	0.05	**<0.1**	0.70

Q5	Q10	Q25	Q50	Q75	Q90	Q95	Q98	max.	Powers
<0.001	<0.001	<0.001	**0.002**	0.005	0.013	0.021	0.033	**0.15**	2.5
0.85	1.2	2.2	**3.4**	4.9	6.4	7.6	9.1	**20**	2.1
0.77	1.1	2.0	**3.4**	4.8	6.3	7.6	9.3	**23**	2.1
2	4	7	**12**	16	19	21	23	**36**	1.6
2	4	7	**12**	16	19	21	23	**38**	1.6
0.003	0.006	0.015	**0.036**	0.081	0.16	0.22	0.32	**2.2**	3.6
<0.02	<0.02	<0.02	**0.037**	0.058	0.073	0.094	0.13	**0.21**	1.3
<0.02	<0.02	<0.02	**0.034**	0.056	0.077	0.095	0.12	**0.31**	1.5
<0.01	<0.01	0.0178	**0.0488**	0.104	0.178	0.232	0.333	**2.97**	2.8
<0.01	<0.01	0.0167	**0.0458**	0.106	0.182	0.239	0.334	**6.73**	3.1
<5	<5	6	**8**	11	14	16	19	**28**	1
<5	<5	6	**8**	12	14	16	19	**24**	1
<0.001	<0.001	<0.001	**<0.001**	0.001	0.002	0.002	0.003	**0.053**	2
0.0085	0.012	0.018	**0.030**	0.048	0.076	0.10	0.19	**1.6**	3
0.0099	0.013	0.020	**0.035**	0.059	0.094	0.13	0.21	**3.1**	3.3
<0.0005	<0.0005	<0.0005	**<0.0005**	<0.0005	<0.0005	<0.0005	0.0007	**0.0022**	0.9
0.00366	0.00573	0.0102	**0.0176**	0.026	0.0357	0.0422	0.0547	**0.348**	2.4
0.00346	0.00553	0.0103	**0.0177**	0.0276	0.0379	0.0465	0.0595	**0.206**	2.1
19	25	40	**64**	105	161	207	291	**749**	4.2
230	342	691	**1250**	2120	3137	3869	5084	**26,739**	2.8
198	296	609	**1126**	1965	3112	3937	5361	**27,957**	2.8
5681	7577	11,373	**15,818**	20,027	24,421	27,963	33,986	**79,221**	2.5
4494	6530	10,348	**14,780**	19,272	24,007	28,217	34,282	**50,020**	2.3
0.684	0.913	1.37	**1.91**	2.41	2.94	3.37	4.09	**9.54**	2.5
0.541	0.787	1.25	**1.78**	2.32	2.89	3.4	4.13	**6.03**	2.3
<0.001	<0.001	0.006	**0.049**	0.15	0.33	0.53	0.85	**12**	4.4
4.3	6.1	9.5	**14**	20	26	32	42	**109**	2
3.9	5.4	8.9	**14**	19	25	33	42	**230**	2.4
10	10	10	**23**	32	43	49	60	**155**	1.2
<20	<20	<20	**24**	34	43	50	64.5	**269**	1.4
<0.005	<0.005	<0.005	**<0.005**	0.018	0.052	0.075	0.11	**0.85**	2.5
1.8	2.8	5.8	**11**	19	28	33	45	**136**	2.9
1.4	2.3	5.1	**11**	19	28	37	50	**153**	3.5

to be continued

Appendix B (continued)

Element	Material	Method	N	Unit	DL	%<DL	min.	Q2
Mg – Magnesium								
Mg	Ap	MMI®	2108	mg/kg	1 / 500	0.19	**<1**	4
Mg	Ap	ICP_MS, ACME 1F_15	2108	mg/kg	50	0.05	**<50**	186
Mg	Gr	ICP_MS, ACME 1F_15	2024	mg/kg	50	0.05	**<50**	159
Mg	Ap	XRF	2108	mg/kg	302	0.7	**<302**	482
Mg	Gr	XRF	2024	mg/kg	302	0.8	**<302**	422
MgO	Ap	XRF	2108	wt%	0.01	0	**0.02**	0.08
MgO	Gr	XRF	2024	wt%	0.01	0	**0.02**	0.07
Mn – Manganese								
Mn	Ap	MMI®	2108	mg/kg	0.01	0.05	**<0.01**	1.8
Mn	Ap	ICP_MS, ACME 1F_15	2108	mg/kg	1	0	**2**	50
Mn	Gr	ICP_MS, ACME 1F_15	2024	mg/kg	1	0	**7**	40.5
Mn	Ap	XRF	2108	mg/kg	8	0	**39**	116
Mn	Gr	XRF	2024	mg/kg	8	0	**31**	85
MnO	Ap	XRF	2108	wt%	0.001	0	**0.005**	0.015
MnO	Gr	XRF	2024	wt%	0.001	0	**0.004**	0.011
Mo – Molybdenum								
Mo	Ap	MMI®	2108	mg/kg	0.005	13	**<0.005**	<0.005
Mo	Ap	ICP_MS, ACME 1F_15	2108	mg/kg	0.01	0	**0.026**	0.077
Mo	Gr	ICP_MS, ACME 1F_15	2024	mg/kg	0.01	0	**0.033**	0.078
Mo	Ap	XRF	2108	mg/kg	2	95	**<2**	<2
Mo	Gr	XRF	2024	mg/kg	2	91	**<2**	<2
Na – Sodium								
Na	Ap	ICP_MS, ACME 1F_15	2108	mg/kg	2	0.2	**<2**	7.71
Na	Gr	ICP_MS, ACME 1F_15	2024	mg/kg	2	0.15	**<2**	7.8
Na	Ap	XRF	2108	mg/kg	74	0	**74**	519
Na	Gr	XRF	2022	mg/kg	74	0.1	**148**	445
Na_2O	Ap	XRF	2108	wt%	0.01	0	**0.01**	0.07
Na_2O	Gr	XRF	2024	wt%	0.01	0.1	**<0.01**	0.06
Nb – Niobium								
Nb	Ap	MMI®	2108	mg/kg	<0.0005	39	**0.00025**	<0.0005
Nb	Ap	ICP_MS, ACME 1F_15	2108	mg/kg	0.02	0.09	**<0.02**	0.084
Nb	Gr	ICP_MS, ACME 1F_15	2024	mg/kg	0.02	0.05	**<0.02**	0.081
Nb	Ap	XRF	2108	mg/kg	2	0.05	**<2**	4
Nb	Gr	XRF	2024	mg/kg	2	0.15	**<2**	3
Nd – Neodymium								
Nd	Ap	MMI®	2108	mg/kg	0.001	2.9	**<0.001**	<0.001
Ni – Nickel								
Ni	Ap	MMI®	2108	mg/kg	0.005	0.05	**<0.005**	0.067
Ni	Ap	ICP_MS, ACME 1F_15	2108	mg/kg	0.1	0.05	**<0.1**	1.2
Ni	Gr	ICP_MS, ACME 1F_15	2024	mg/kg	0.1	0	**0.28**	1.1
Ni	Ap	XRF	2108	mg/kg	3	4.9	**<3**	<3
Ni	Gr	XRF	2024	mg/kg	3	4.2	**<3**	<3

Q5	Q10	Q25	Q50	Q75	Q90	Q95	Q98	max.	Powers
8	13	26	**52**	103	172	230	340	**>500**	3.2
365	613	1370	**2860**	5447	9040	13,037	22,253	**116,908**	3.7
314	534	1232	**2820**	5096	8750	12,386	21,180	**124,783**	3.7
784	1387	3015	**5488**	9407	14,853	19,897	29,283	**126,638**	2.9
724	1206	2714	**5186**	8804	13,990	18,684	28,388	**145,202**	3
0.13	0.23	0.5	**0.91**	1.6	2.5	3.3	4.9	**21**	3
0.12	0.2	0.45	**0.86**	1.5	2.3	3.1	4.7	**24.1**	3.1
3.5	5.5	11	**20**	31	45	55	72	**180**	4.6
74.4	124	241	**445**	701	1080	1415	1831	**14,969**	3.9
73	115	232	**435**	715	1089	1416	1839	**5453**	2.9
170	232	379	**604**	867	1249	1580	1967	**17,425**	2.7
132	201	349	**573**	883	1222	1562	1987	**5483**	2.2
0.022	0.03	0.049	**0.078**	0.112	0.161	0.204	0.254	**2.25**	2.7
0.017	0.026	0.045	**0.074**	0.114	0.158	0.202	0.257	**0.708**	2.2
<0.005	<0.005	0.008	**0.015**	0.027	0.045	0.064	0.088	**0.53**	2.3
0.11	0.15	0.25	**0.42**	0.72	1.2	1.7	3.1	**14**	2.7
0.11	0.15	0.26	**0.42**	0.75	1.3	2.0	3.5	**13**	2.6
<2	<2	<2	**<2**	<2	<2	<2	3	**17**	1.2
<2	<2	<2	**<2**	<2	<2	3	6	**33**	1.5
11.6	16.9	29.9	**48**	81.7	144	201	361	**7237**	3.9
12	18	31	**50**	84	138	209	404	**12,285**	4.1
890	1536	3264	**6009**	12,537	18,621	20,920	23,814	**33,977**	2.7
816	1261	2819	**5267**	10,090	16,099	19,289	22,566	**29,231**	2.3
0.12	0.21	0.44	**0.81**	1.7	2.5	2.8	3.2	**4.6**	2.7
0.11	0.17	0.38	**0.71**	1.4	2.2	2.6	3.0	**3.9**	2.9
<0.0005	<0.0005	<0.0005	**0.0009**	0.0031	0.0077	0.012	0.019	**0.086**	2.5
0.12	0.15	0.26	**0.48**	0.93	1.82	2.57	3.88	**13.1**	3.1
0.12	0.16	0.30	**0.52**	1.0	1.9	3.0	4.5	**21.1**	3.3
5	6	9	**13**	15	19	21	28	**126**	2.1
4	6	8	**12**	15	18	22	29	**174**	2.2
0.002	0.005	0.021	**0.10**	0.26	0.58	0.89	1.2	**11**	4.4
0.10	0.14	0.29	**0.65**	1.5	2.8	4.0	5.8	**30**	4.1
1.9	3.1	6.8	**15**	27	43	66	139	**2475**	4.7
1.8	2.8	6.3	**14**	28	43	66	131	**2466**	3.9
3	5	9	**20**	35	53	83.6	167	**2535**	3.2
3	5	9	**20**	34	57	84.8	163	**2480**	3.2

to be continued

Appendix B (continued)

Element	Material	Method	N	Unit	DL	%<DL	min.	Q2
P - Phosphorus								
P	Ap	MMI®	2108	mg/kg	0.1	0.38	**<0.1**	0.3
P	Ap	ICP_MS, ACME 1F_15	2108	mg/kg	10	0	**24**	211
P	Gr	ICP_MS, ACME 1F_15	2024	mg/kg	10	0	**46**	144
P	Ap	XRF	2108	mg/kg	4	0	**61**	279
P	Gr	XRF	2024	mg/kg	4	0	**87**	223
P_2O_5	Ap	XRF	2108	wt%	0.001	0	**0.014**	0.064
P_2O_5	Gr	XRF	2024	wt%	0.001	0	**0.02**	0.051
Pb - Lead								
Pb	Ap	MMI®	2108	mg/kg	0.01	0.24	**<0.01**	0.04
Pb	Ap	ICP_MS, ACME 1F_15	2108	mg/kg	0.01	0	**1.6**	3.5
Pb	Gr	ICP_MS, ACME 1F_15	2024	mg/kg	0.01	0	**1.8**	4
Pb	Ap	XRF	2108	mg/kg	3	1	**<3**	6
Pb	Gr	XRF	2024	mg/kg	3	1.7	**<3**	4
Pd - Palladium								
Pd	Ap	MMI®	2108	mg/kg	0.001	99.8	**<0.001**	<0.001
Pd	Ap	ICP_MS, ACME 1F_15	2108	mg/kg	0.001	59	**<0.001**	<0.001
Pd	Gr	ICP_MS, ACME 1F_15	2024	mg/kg	0.001	65	**<0.001**	<0.001
Pr - Praesodymium								
Pr	Ap	MMI®	2108	mg/kg	0.001	13	**<0.001**	<0.001
Pt - Platinum								
Pt	Ap	MMI®	2108	mg/kg	0.001	100	**<0.001**	<0.001
Pt	Ap	ICP_MS, ACME 1F_15	2108	mg/kg	0.001	70	**<0.001**	<0.001
Pt	Gr	ICP_MS, ACME 1F_15	2024	mg/kg	0.001	73	**<0.001**	<0.001
Rb - Rubidium								
Rb	Ap	MMI®	2108	mg/kg	0.005	0.43	**<0.005**	0.01
Rb	Ap	ICP_MS, ACME 1F_15	2108	mg/kg	0.1	0	**0.21**	1.9
Rb	Gr	ICP_MS, ACME 1F_15	2024	mg/kg	0.1	0	**0.24**	1.3
Rb	Ap	XRF	2108	mg/kg	2	0.4	**<2**	15
Rb	Gr	XRF	2024	mg/kg	2	0.4	**<2**	13
Re - Rhenium								
Re	Ap	ICP_MS, ACME 1F_15	2108	mg/kg	0.0005	77	**<0.0005**	<0.0005
Re	Gr	ICP_MS, ACME 1F_15	2024	mg/kg	0.0005	73	**<0.0005**	<0.0005
S - Sulphur								
S	Ap	MMI®	2108	mg/kg	0.5	0.24	**<0.5**	7.2
S	Ap	ICP_MS, ACME 1F_15	2108	mg/kg	5	1.3	**<5**	36
S	Gr	ICP_MS, ACME 1F_15	2024	mg/kg	5	1.2	**<5**	35.5
S	Ap	IR spectroscopy	2108	wt%	0.01	24	**<0.01**	<0.01
S	Gr	IR spectroscopy	2024	wt%	0.01	19	**<0.01**	<0.01
S	Ap	XRF	2108	wt%	0.004	46	**<0.004**	<0.004
S	Gr	XRF	2024	wt%	0.004	43	**<0.004**	<0.004
SO_3	Ap	XRF	2108	wt%	0.01	46	**<0.01**	<0.01
SO_3	Gr	XRF	2024	wt%	0.01	43	**<0.01**	<0.01

Q5	Q10	Q25	Q50	Q75	Q90	Q95	Q98	max.	Powers
0.4	0.6	1.7	**4**	10	25	42	62	**188**	3.6
268	328	466	**653**	908	1250	1581	2162	**3881**	2.2
222	290	424	**642**	927	1310	1639	2248	**6898**	2.2
349	432	576	**786**	1104	1494	1871	2469	**4417**	1.9
305	385	532	**777**	1123	1531	1879	2472	**8754**	2
0.08	0.099	0.13	**0.18**	0.25	0.34	0.43	0.57	**1.0**	1.9
0.07	0.088	0.12	**0.18**	0.26	0.35	0.43	0.57	**2.0**	2
0.06	0.09	0.16	**0.32**	0.73	1.6	2.6	4.3	**58**	4.1
4.9	6.3	9.9	**16**	24	35	44	73	**1309**	2.9
5.6	7.2	11	**18**	28	41	53	80	**1090**	2.8
8	11	15	**21**	30	40	52	87.9	**1287**	2.9
7	10	16	**22**	32	45	57	89.6	**1057**	2.8
<0.001	<0.001	<0.001	**<0.001**	<0.001	<0.001	<0.001	<0.001	**0.002**	0.6
<0.001	<0.001	<0.001	**<0.001**	0.00207	0.00437	0.0061	0.00967	**0.148**	2.5
<0.001	<0.001	<0.001	**<0.001**	0.00168	0.00397	0.00558	0.00879	**0.0328**	1.8
<0.001	<0.001	0.003	**0.02**	0.054	0.12	0.184	0.274	**2.98**	3.8
<0.001	<0.001	<0.001	**<0.001**	<0.001	<0.001	<0.001	<0.001	**<0.001**	
<0.001	<0.001	<0.001	**<0.001**	0.0012	0.0019	0.0025	0.0035	**0.027**	1.7
<0.001	<0.001	<0.001	**<0.001**	0.0011	0.0019	0.0025	0.0035	**0.038**	1.9
0.016	0.022	0.039	**0.069**	0.13	0.25	0.38	0.65	**2.9**	3.1
3.1	4.9	9.0	**14**	22	33	44	62	**329**	3.2
2.2	4.0	8.0	**14**	21	31	42	61	**196**	2.9
24	33	51	**73**	98	123	143	192	**404**	2.6
21	30	50	**74**	98	126	148	205	**562**	2.7
<0.0005	<0.0005	<0.0005	**<0.0005**	<0.0005	0.000887	0.0012	0.0018	**0.0052**	1.3
<0.0005	<0.0005	<0.0005	**<0.0005**	0.00056	0.0011	0.0015	0.0024	**0.0095**	1.6
9.8	12	17	**24**	35	53	69	129	**2620**	4
64	89	137	**207**	322	487	729	1526	**68,226**	4.4
69	106	176	**295**	484	795	1169	2799	**98,190**	4.6
<0.01	<0.01	0.01	**0.016**	0.026	0.043	0.070	0.13	**9.1**	3.3
<0.01	<0.01	0.012	**0.023**	0.039	0.065	0.11	0.24	**14**	3.5
<0.004	<0.004	<0.004	**0.004**	0.02	0.044	0.056	0.080	**8.31**	3.6
<0.004	<0.004	<0.004	**0.004**	0.028	0.052	0.072	0.10	**15**	3.9
<0.01	<0.01	<0.01	**0.01**	0.05	0.11	0.14	0.2	**21**	3.6
<0.01	<0.01	<0.01	**0.01**	0.07	0.13	0.18	0.26	**36**	3.9

to be continued

Appendix B (continued)

Element	Material	Method	N	Unit	DL	%<DL	min.	Q2
Sb – Antimony								
Sb	Ap	MMI®	2108	mg/kg	0.001	15	**<0.001**	<0.001
Sb	Ap	ICP_MS, ACME 1F_15	2108	mg/kg	0.02	0.7	**<0.02**	0.029
Sb	Gr	ICP_MS, ACME 1F_15	2024	mg/kg	0.02	0.4	**<0.02**	0.041
Sb	Ap	XRF	2108	mg/kg	5	96	**<5**	<5
Sb	Gr	XRF	2024	mg/kg	5	95	**<5**	<5
Sc – Scandium								
Sc	Ap	MMI®	2108	mg/kg	0.005	6.6	**<0.005**	<0.005
Sc	Ap	ICP_MS, ACME 1F_15	2108	mg/kg	0.05	0.05	**<0.05**	0.26
Sc	Gr	ICP_MS, ACME 1F_15	2024	mg/kg	0.05	0.05	**<0.05**	0.23
Sc	Ap	XRF	2108	mg/kg	2	8.5	**<2**	<2
Sc	Gr	XRF	2024	mg/kg	2	9.3	**<2**	<2
Se – Selenium								
Se	Ap	MMI®	2108	mg/kg	0.01	39	**<0.01**	<0.01
Se	Ap	ICP_MS, ACME 1F_15	2108	mg/kg	0.05	2.2	**<0.05**	0.025
Se	Gr	ICP_MS, ACME 1F_15	2024	mg/kg	0.05	1.7	**<0.05**	0.055
Si – Silicon								
Si	Ap	XRF	2108	mg/kg	467	0	**11,499**	104,731
Si	Gr	XRF	2024	mg/kg	467	0	**7058**	79,605
SiO_2	Ap	XRF	2108	wt%	0.1	0	**2.46**	22.4
SiO_2	Gr	XRF	2024	wt%	0.1	0	**1.51**	17
Sm – Samarium								
Sm	Ap	MMI®	2108	mg/kg	0.001	4.4	**<0.001**	<0.001
Sn – Tin								
Sn	Ap	MMI®	2108	mg/kg	0.001	75	**<0.001**	<0.001
Sn	Ap	ICP_MS, ACME 1F_15	2108	mg/kg	0.4	0.38	**<0.1**	0.17
Sn	Gr	ICP_MS, ACME 1F_15	2024	mg/kg	0.3	0.35	**<0.1**	0.17
Sn	Ap	XRF	2108	mg/kg	4	91	**<4**	<4
Sn	Gr	XRF	2024	mg/kg	4	90	**<4**	<4
Sr – Strontium								
Sr	Ap	MMI®	2108	mg/kg	0.01	0.19	**<0.01**	0.15
Sr	Ap	ICP_MS, ACME 1F_15	2108	mg/kg	0.5	0	**0.8**	2.5
Sr	Gr	ICP_MS, ACME 1F_15	2024	mg/kg	0.5	0	**0.6**	2.5
Sr	Ap	XRF	2108	mg/kg	2	0	**7**	25
Sr	Gr	XRF	2024	mg/kg	2	0	**7**	23
T – Tantalum								
Ta	Ap	MMI®	2108	mg/kg	0.001	98	**<0.001**	<0.001
Ta	Ap	ICP_MS, ACME 1F_15	2108	mg/kg	0.005	94	**<0.005**	<0.005
Ta	Gr	ICP_MS, ACME 1F_15	2024	mg/kg	0.005	92	**<0.005**	<0.005
Ta	Ap	XRF	2108	mg/kg	5	99	**<5**	<5
Ta	Gr	XRF	2024	mg/kg	5	99	**<5**	<5

Q5	Q10	Q25	Q50	Q75	Q90	Q95	Q98	max.	Powers
<0.001	<0.001	0.001	**0.002**	0.004	0.007	0.009	0.013	**0.071**	2.2
0.050	0.072	0.12	**0.23**	0.42	0.70	1.0	1.8	**17**	3.2
0.059	0.085	0.14	**0.28**	0.49	0.82	1.2	2.0	**25**	3.4
<5	<5	<5	**<5**	<5	<5	<5	6	**36**	1.2
<5	<5	<5	**<5**	<5	<5	<5	6	**33**	1.1
<0.005	0.006	0.008	**0.016**	0.046	0.10	0.15	0.21	**0.90**	2.6
0.36	0.58	1.2	**2.2**	3.4	4.8	5.9	7.4	**21**	2.9
0.34	0.51	1.0	**2.0**	3.4	4.8	5.9	7.5	**22**	2.9
<2	3	5	**8**	12	15	17	20	**42**	1.6
<2	3	5	**8**	12	15	17	20	**42**	1.6
<0.01	<0.01	<0.01	**0.01**	0.02	0.03	0.03	0.04	**0.17**	1.5
0.10	0.16	0.24	**0.35**	0.49	0.68	0.88	1.2	**3.8**	2.2
0.11	0.17	0.28	**0.40**	0.57	0.83	1.1	1.4	**6.8**	2.4
166,392	217,270	275,532	**313,932**	359,087	401,315	418,543	431,066	**451,079**	1.6
144,370	198,486	262,048	**302,968**	348,856	392,406	413,551	427,495	**457,257**	1.8
35.6	46.5	58.9	**67.2**	76.8	85.9	89.5	92.2	**96.5**	1.6
30.9	42.5	56.1	**64.8**	74.6	84	88.5	91.5	**97.8**	1.8
0.001	0.003	0.009	**0.029**	0.068	0.14	0.21	0.30	**2.0**	3.6
<0.001	<0.001	<0.001	**<0.001**	0.0006	0.003	0.005	0.007	**0.03**	1.8
0.23	0.29	0.46	**0.72**	1.1	1.7	2.4	4.3	**30**	2.8
0.25	0.31	0.49	**0.81**	1.3	2.0	2.6	4.2	**158**	3.5
<4	<4	<4	**<4**	<4	<4	6	11	**129**	1.8
<4	<4	<4	**<4**	<4	5	6	11	**809**	2.6
0.26	0.41	0.81	**1.7**	3.1	5.2	7.4	12	**90**	4.3
4.1	5.9	10	**18**	38	104	204	404	**3777**	3.7
4	5.8	9.7	**18**	38	105	220	449	**4247**	3.8
34	45	68	**101**	173	259	342	540	**4564**	2.8
33	42	63	**90**	151	252	368	613	**8072**	3.1
<0.001	<0.001	<0.001	**<0.001**	<0.001	<0.001	<0.001	<0.001	**0.029**	1.8
<0.005	<0.005	<0.005	**<0.005**	<0.005	<0.005	0.0055	0.0079	**0.034**	1.1
<0.005	<0.005	<0.005	**<0.005**	<0.005	<0.005	0.006	0.0088	**0.048**	1.3
<5	<5	<5	**<5**	<5	<5	<5	<5	**10**	0.6
<5	<5	<5	**<5**	<5	<5	<5	<5	**10**	0.6

to be continued

Appendix B (continued)

Element	Material	Method	N	Unit	DL	%<DL	min.	Q2
Tb - Terbium								
Tb	Ap	MMI®	2108	mg/kg	0.001	11	**<0.001**	<0.001
Te - Tellurium								
Te	Ap	MMI®	2108	mg/kg	0.01	99.9	**<0.01**	<0.01
Te	Ap	ICP_MS, ACME 1F_15	2108	mg/kg	0.02	56	**<0.02**	<0.02
Te	Gr	ICP_MS, ACME 1F_15	2024	mg/kg	0.02	50	**<0.02**	<0.02
Th - Thorium								
Th	Ap	MMI®	2108	mg/kg	0.0005	1.3	**<0.0005**	0.001
Th	Ap	ICP_MS, ACME 1F_15	2108	mg/kg	0.02	0	**0.03**	0.18
Th	Gr	ICP_MS, ACME 1F_15	2024	mg/kg	0.02	0	**0.032**	0.13
Th	Ap	XRF	2108	mg/kg	5	24	**<5**	<5
Th	Gr	XRF	2024	mg/kg	5	14	**<5**	<5
Ti - Titanium								
Ti	Ap	MMI®	2024	mg/kg	0.003	0.62	**<0.003**	0.005
Ti	Ap	ICP_MS, ACME 1F_15	2108	mg/kg	5	0.1	**<5**	12
Ti	Gr	ICP_MS, ACME 1F_15	2024	mg/kg	5	0.2	**<5**	12
Ti	Ap	XRF	2108	mg/kg	6	0	**90**	763
Ti	Gr	XRF	2024	mg/kg	6	0	**114**	596
TiO_2	Ap	XRF	2108	wt%	0.001	0	**0.015**	0.13
TiO_2	Gr	XRF	2024	wt%	0.001	0	**0.019**	0.10
Tl - Thallium								
Tl	Ap	MMI®	2108	mg/kg	0.0005	45	**<0.0005**	<0.0005
Tl	Ap	ICP_MS, ACME 1F_15	2108	mg/kg	0.0005	0.2	**<0.0005**	0.022
Tl	Gr	ICP_MS, ACME 1F_15	2024	mg/kg	0.0005	0.1	**<0.0005**	0.024
U - Uranium								
U	Ap	MMI®	2108	mg/kg	0.001	0.14	**<0.001**	0.014
U	Ap	ICP_MS, ACME 1F_15	2108	mg/kg	0.04	0.05	**<0.04**	0.20
U	Gr	ICP_MS, ACME 1F_15	2024	mg/kg	0.04	0	**0.047**	0.18
U	Ap	XRF	2108	mg/kg	3	63	**<3**	<3
U	Gr	XRF	2024	mg/kg	3	77	**<3**	<3
V - Vanadium								
V	Ap	MMI®	2108	mg/kg	0.001	0.09	**<0.001**	0.004
V	Ap	ICP_MS, ACME 1F_15	2108	mg/kg	0.2	0	**1**	4.4
V	Gr	ICP_MS, ACME 1F_15	2024	mg/kg	0.2	0	**0.7**	3.5
V	Ap	XRF	2108	mg/kg	5	0.14	**<5**	12
V	Gr	XRF	2024	mg/kg	5	0.15	**<5**	11
W - Tungsten								
W	Ap	MMI®	2108	mg/kg	0.001	47	**<0.001**	<0.001
W	Ap	ICP_MS, ACME 1F_15	2108	mg/kg	0.01	1.7	**<0.01**	0.012
W	Gr	ICP_MS, ACME 1F_15	2024	mg/kg	0.01	1.3	**<0.01**	0.013
W	Ap	XRF	2108	mg/kg	5	94	**<5**	<5
W	Gr	XRF	2024	mg/kg	5	96	**<5**	<5

Q 5	Q 10	Q 25	Q 50	Q 75	Q 90	Q 95	Q 98	max.	Powers
<0.001	<0.001	0.002	**0.005**	0.012	0.023	0.0326	0.048	**0.33**	2.8
<0.01	<0.01	<0.01	**<0.01**	<0.01	<0.01	<0.01	<0.01	**0.03**	0.8
<0.02	<0.02	<0.02	**<0.02**	0.035	0.060	0.082	0.12	**0.27**	1.4
<0.02	<0.02	<0.02	**0.020**	0.040	0.073	0.11	0.17	**0.95**	2
0.0019	0.0038	0.0092	**0.022**	0.057	0.12	0.18	0.28	**2.8**	4.1
0.35	0.68	1.6	**2.9**	4.6	6.3	7.9	10	**63**	3.3
0.24	0.45	1.2	**2.5**	4.3	6.0	7.4	10	**56**	3.2
<5	<5	5	**9**	12	16	18	22	**84**	1.5
<5	<5	7	**11**	14	18	21	25	**80**	1.5
0.006	0.009	0.015	**0.038**	0.22	0.73	1.3	1.9	**13**	3.9
17	24	43	**86**	300	711	974	1365	**7860**	3.5
15	21	36	**74**	211	595	859	1217	**10,421**	3.6
1109	1565	2440	**3621**	4592	5386	5957	7048	**24,556**	2.4
895	1360	2337	**3566**	4537	5364	6030	7096	**23,756**	2.3
0.19	0.26	0.41	**0.60**	0.77	0.90	0.99	1.2	**4.1**	2.4
0.15	0.23	0.39	**0.60**	0.76	0.90	1.0	1.2	**4.0**	2.3
<0.0005	<0.0005	<0.0005	**0.0006**	0.0011	0.00	0.0026	0.0039	**0.017**	1.8
0.033	0.046	0.075	**0.12**	0.18	0.29	0.38	0.62	**2.5**	3
0.033	0.044	0.072	**0.12**	0.18	0.30	0.44	0.69	**2.5**	3
0.019	0.026	0.045	**0.084**	0.16	0.29	0.47	0.77	**5.3**	4
0.30	0.37	0.52	**0.77**	1.2	2.2	3.4	5.2	**24**	3.1
0.25	0.34	0.51	**0.74**	1.2	2.2	3.3	5.3	**73**	3.2
<3	<3	<3	**<3**	3	5	6	8	**25**	1.2
<3	<3	<3	**<3**	<3	4	5	8	**94**	1.8
0.005	0.007	0.013	**0.027**	0.057	0.12	0.21	0.42	**13**	4.4
6.3	9.5	16	**25**	37	51	63	85	**188**	2.3
5.7	8.6	16	**26**	38	54	68	97	**552**	2.9
17	24	42	**68**	96	122	139	171	**323**	2.1
16	22	40	**68**	97	124	145	184	**706**	2.5
<0.001	<0.001	<0.001	**0.001**	0.002	0.005	0.008	0.016	**0.067**	2.1
0.020	0.028	0.042	**0.073**	0.13	0.23	0.35	0.66	**5.2**	3
0.020	0.027	0.044	**0.077**	0.15	0.28	0.45	0.93	**19**	3.6
<5	<5	<5	**<5**	<5	<5	6	6	**25**	1
<5	<5	<5	**<5**	<5	<5	<5	6	**37**	1.2

to be continued

Appendix B (continued)

Element	Material	Method	N	Unit	DL	%<DL	min.	Q2
Y – Yttrium								
Y	Ap	MMI®	2108	mg/kg	0.005	0.81	**<0.005**	0.01
Y	Ap	ICP_MS, ACME 1F_15	2108	mg/kg	0.01	0	**0.23**	0.94
Y	Gr	ICP_MS, ACME 1F_15	2024	mg/kg	0.01	0	**0.23**	0.91
Y	Ap	XRF	2108	mg/kg	3	0	**6**	9
Y	Gr	XRF	2024	mg/kg	3	0	**3**	7
Yb – Ytterbium								
Yb	Ap	MMI®	2108	mg/kg	0.001	1.3	**<0.001**	0.001
Zn – Zinc								
Zn	Ap	MMI®	2108	mg/kg	0.02	0.14	**<0.02**	0.09
Zn	Ap	ICP_MS, ACME 1F_15	2108	mg/kg	0.1	0	**2.8**	8.5
Zn	Gr	ICP_MS, ACME 1F_15	2024	mg/kg	0.1	0	**1.5**	7
Zn	Ap	XRF	2108	mg/kg	3	0.2	**<3**	14
Zn	Gr	XRF	2024	mg/kg	3	0.1	**<3**	11
Zr – Zirconium								
Zr	Ap	MMI®	2108	mg/kg	0.005	4.6	**<0.005**	<0.005
Zr	Ap	ICP_MS, ACME 1F_15	2108	mg/kg	0.1	2.4	**<0.1**	0.05
Zr	Gr	ICP_MS, ACME 1F_15	2024	mg/kg	0.1	2.3	**<0.1**	<0.1
Zr	Ap	XRF	2108	mg/kg	3	0	**4**	70
Zr	Gr	XRF	2024	mg/kg	3	0	**4**	57

Isotope Ratio	Material	Method	N				min.	Q2
$^{208}Pb/^{207}Pb$	Ap	SF_ICP_MS	2106				**2.4**	2.45
$^{207}Pb/^{208}Pb$	Ap	SF_ICP_MS	2106				**0.287**	0.373
$^{208}Pb/^{206}Pb$	Ap	SF_ICP_MS	2106				**1.48**	1.77
$^{206}Pb/^{208}Pb$	Ap	SF_ICP_MS	2106				**0.37**	0.474
$^{206}Pb/^{207}Pb$	Ap	SF_ICP_MS	2106				**1.12**	1.17
$^{207}Pb/^{206}Pb$	Ap	SF_ICP_MS	2106				**0.579**	0.681

Q5	Q10	Q25	Q50	Q75	Q90	Q95	Q98	max.	Powers
0.019	0.035	0.082	**0.18**	0.39	0.78	1.2	1.8	**16**	3.8
1.5	2.3	3.9	**6.7**	9.9	13	16	22	**65**	2.4
1.5	2.2	3.8	**6.5**	9.7	14	19	27	**77**	2.5
10	13	20	**27**	33	38	41	46	**111**	1.3
9	12	18	**25**	31	37	41	50	**118**	1.6
0.002	0.004	0.008	**0.016**	0.038	0.074	0.12	0.19	**0.79**	3.2
0.12	0.18	0.34	**0.76**	1.7	3.6	5.3	9.1	**36**	3.6
12	17	27	**45**	65	87	104	129	**1396**	2.7
12	17	26	**46**	70	96	110	131	**642**	2.6
19	26	39	**61**	85	110	129	158	**1413**	3
16	25	39	**61**	88	116	134	160	**742**	2.7
<0.005	0.008	0.02	**0.051**	0.10	0.20	0.29	0.44	**1.9**	2.9
0.15	0.27	0.75	**1.8**	3.8	6.4	9.0	16	**165**	3.5
0.15	0.26	0.69	**1.6**	3.7	6.4	9.2	16	**376**	3.9
104	135	192	**258**	346	463	547	610	**963**	2.4
93	123	178	**245**	323	427	520	603	**871**	2.3

Q5	Q10	Q25	Q50	Q75	Q90	Q95	Q98	max.	Powers
2.46	2.46	2.47	**2.48**	2.51	2.57	2.6	2.68	**3.48**	0.2
0.384	0.39	0.399	**0.403**	0.405	0.406	0.407	0.408	**0.417**	0.2
1.84	1.92	2.03	**2.07**	2.08	2.09	2.1	2.11	**2.7**	0.3
0.477	0.478	0.481	**0.484**	0.492	0.52	0.544	0.564	**0.677**	0.3
1.17	1.18	1.19	**1.2**	1.24	1.32	1.41	1.47	**1.73**	0.2
0.707	0.757	0.805	**0.832**	0.84	0.847	0.851	0.857	**0.896**	0.2

Appendix C

GEMAS Project: Statistical Parameters of Northern and Southern Europe (Ap samples)

Appendix C

Parameter	Material	Method	N	Unit	DL	% < DL	min.	Q2
Altitude	Ap-north		814	m a.s.l.			**0**	4
Altitude	Ap-south		1294	m a.s.l.			**-5**	4.9
CEC	Ap-north	silver-thiourea	814	meq	0.1		**1.8**	3.5
CEC	Ap-south	silver-thiourea	1294	meq	0.1		**2.2**	6.1
CIA	Ap-north	calculated	814	perc			**15**	48
CIA	Ap-south	calculated	1294	perc			**4.3**	27
Clay	Ap-north	MIR	814	%	2	21.4	**<2**	<2
Clay	Ap-south	MIR	1294	%	2	3.4	**<2**	<2
LOI	Ap-north	gravimetry	814	wt%	0.01	0	**1.5**	2.2
LOI	Ap-south	gravimetry	1294	wt%	0.01	0	**1.5**	2.9
MS	Ap-north	susceptibility-meter	814	10^{-6} m³/kg			**0.0052**	0.023
MS	Ap-south	susceptibility-meter	1294	10^{-6} m³/kg			**0.017**	0.04
pH_$CaCl_2$	Ap-north	potentiometry	814	pH	0.1	0	**3.3**	3.9
pH_$CaCl_2$	Ap-south	potentiometry	1294	pH	0.1	0	**0.5**	4
TOC	Ap-north	IR spectroscopy	814	wt%	0.1	0	**0.4**	0.7
TOC	Ap-south	IR spectroscopy	1294	wt%	0.1	0	**0.41**	0.5

Element	Material	Method	N	Unit	DL	% < DL	min.	Q2
Ag – Silver								
Ag	Ap-north	MMI®	814	mg/kg	0.001	6.9	**<0.001**	<0.001
Ag	Ap-south	MMI®	1294	mg/kg	0.001	3.6	**<0.001**	<0.001
Ag	Ap-north	ICP_MS, ACME 1F_15	814	mg/kg	0.002	0.74	**<0.002**	0.0059
Ag	Ap-south	ICP_MS, ACME 1F_15	1294	mg/kg	0.002	0.15	**<0.002**	0.00799
Al – Aluminium								
Al	Ap-north	MMI®	814	mg/kg	1	0.2	**<1**	12
Al	Ap-south	MMI®	1294	mg/kg	1	0.4	**<1**	6
Al	Ap-north	ICP_MS, ACME 1F_15	814	mg/kg	50	0	**329**	1219
Al	Ap-south	ICP_MS, ACME 1F_15	1294	mg/kg	50	0	**352**	2142
Al	Ap-north	XRF	814	mg/kg	265	0	**1958**	7886
Al	Ap-south	XRF	1294	mg/kg	265	0	**2752**	12,801
Al_2O_3	Ap-north	XRF	814	wt%	0.05	0	**0.37**	1.5
Al_2O_3	Ap-south	XRF	1294	wt%	0.05	0	**0.52**	2.4
As – Arsenic								
As	Ap-north	MMI®	814	mg/kg	0.01	27.6	**<0.01**	<0.01
As	Ap-south	MMI®	1294	mg/kg	0.01	30.1	**<0.01**	<0.01
As	Ap-north	ICP_MS, ACME 1F_15	814	mg/kg	0.05	0.25	**<0.05**	0.24
As	Ap-south	ICP_MS, ACME 1F_15	1294	mg/kg	0.05	0	**0.191**	1.5
As	Ap-north	XRF	814	mg/kg	3	52	**<3**	<3
As	Ap-south	XRF	1294	mg/kg	3	6.0	**<3**	<3

Q5	Q10	Q25	Q50	Q75	Q90	Q95	Q98	max.	Powers
8	18	54	**124**	203	308	401	580	**956**	3
13	25	87	**218**	465	774	931	1052	**1644**	3.2
5.1	6.4	9.1	**12**	19	26	32	40	**48**	1.4
7.6	9.9	13.7	**19**	26	32	36	39	**48**	1.3
51	53	57	**61**	66	69	72	74	**90**	0.8
49	58	65	**72**	77	82	84	87	**93**	1.3
<2	<2	2	**8**	16	21.7	25	27	**36**	1.6
3	7	12.2	**18**	24	29	31.3	36	**63**	1.8
2.7	3.2	4.8	**7.1**	9.6	14	21	56	**94**	1.8
3.8	4.7	6.7	**9.6**	15	23	28	34	**95**	1.8
0.032	0.047	0.082	**0.16**	0.29	0.65	1.3	2.6	**5.4**	3
0.06	0.084	0.13	**0.25**	0.6	1.2	2.2	4.2	**40**	3.4
4.1	4.3	4.7	**5.2**	5.9	6.8	7.1	7.2	**7.5**	0.3
4.3	4.6	5.3	**6.3**	7.3	7.4	7.5	7.6	**8**	0.4
0.8	1	1.3	**2**	2.9	4.7	9	29	**46**	2
0.7	0.8	1.1	**1.6**	2.4	3.5	4.6	6.3	**46**	2.1

Q5	Q10	Q25	Q50	Q75	Q90	Q95	Q98	max.	Powers
<0.001	0,002	0.004	**0.01**	0.016	0.025	0.032	0.041	**0.114**	2.4
0.001	0.004	0.009	**0.018**	0.03	0.047	0.065	0.102	**1.34**	3.4
0.0088	0.012	0.020	**0.031**	0.051	0.088	0.12	0.14	**0.26**	2.4
0.012	0.017	0.027	**0.042**	0.063	0.098	0.14	0.23	**2.95**	3.5
17	24	54	**134**	215	298	450	450	**450**	3
9	12	20	**37**	97.8	183	248	450	**450**	3
1843	2604	4394	**7524**	11,509	15,780	18,951	22,373	**51,351**	2.2
3601	5637	9142	**12,996**	17,772	22,519	26,194	34,437	**65,090**	2.3
12,893	16,433	29,373	**49,961**	63,444	70,390	74,326	78,368	**92,513**	1.7
19,701	28,119	40,488	**57,662**	70,642	80,922	87,366	95,142	**143,850**	1.7
2.4	3.1	5.6	**9.4**	12	13	14	15	**18**	1.7
3.7	5.3	7.7	**10.9**	13	15	17	18	**27**	1.7
<0.01	<0.01	<0.01	**0.02**	0,03	0.05	0.07	0.1	**0.33**	1,8
<0.01	<0.01	<0.01	**0.02**	0.03	0.06	0.08	0.15	**0.53**	2
0.42	0.65	1.2	**2.3**	3.7	5.6	7.2	10	**37**	3,2
2.3	3.3	5.4	**8**	12	19	27	44	**666**	3,5
<3	<3	<3	**<3**	5	7	9	11	**43**	1,5
<3	4	7	**10**	15	23	33	57.1	**695**	2,7

to be continued

Appendix C (continued)

Element	Material	Method	N	Unit	DL	%<DL	min.	Q2
Au – Gold								
Au	Ap-north	MMI®	814	mg/kg	0.0001	36.5	**<0.0001**	<0.0001
Au	Ap-south	MMI®	1294	mg/kg	0.0001	5.2	**<0.0001**	<0.0001
Au	Ap-north	ICP_MS, ACME 1F_15	814	mg/kg	0.0002	27	**<0.0002**	<0.0002
Au	Ap-south	ICP_MS, ACME 1F_15	1294	mg/kg	0.0002	8.0	**<0.0002**	<0.0002
B – Boron								
B	Ap-north	ICP_MS, ACME 1F_15	814	mg/kg	0.5	11	**<0.5**	<0.5
B	Ap-south	ICP_MS, ACME 1F_15	1294	mg/kg	0.5	4.9	**<0.5**	<0.5
Ba – Barium								
Ba	Ap-north	MMI®	814	mg/kg	0.01	0.2	**<0.01**	0.11
Ba	Ap-south	MMI®	1294	mg/kg	0.01	0	**0.02**	0.12
Ba	Ap-north	ICP_MS, ACME 1F_15	814	mg/kg	0.5	0	**3.9**	9.4
Ba	Ap-south	ICP_MS, ACME 1F_15	1294	mg/kg	0.5	0	**2.6**	12
Ba	Ap-north	XRF	814	mg/kg	5	0	**27**	110
Ba	Ap-south	XRF	1294	mg/kg	5	0	**16**	106
Be – Beryllium								
Be	Ap-north	ICP_MS, ACME 1F_15	814	mg/kg	0.05	3.9	**<0.05**	<0.05
Be	Ap-south	ICP_MS, ACME 1F_15	1294	mg/kg	0.05	0.46	**<0.05**	0.11
Bi – Bismuth								
Bi	Ap-north	MMI®	814	mg/kg	0.001	83	**<0.001**	<0.001
Bi	Ap-south	MMI®	1294	mg/kg	0.001	91	**<0.001**	<0.001
Bi	Ap-north	ICP_MS, ACME 1F_15	814	mg/kg	0.005	0.1	**<0.005**	0.023
Bi	Ap-south	ICP_MS, ACME 1F_15	1294	mg/kg	0.005	0	**0.009**	0.055
Bi	Ap-north	XRF	814	mg/kg	3	97	**<3**	<3
Bi	Ap-south	XRF	1294	mg/kg	3	91	**<3**	<3
C – Carbon								
TC	Ap-north	IR spectroscopy	814	wt%	0.07	0	**0.2**	0.74
TC	Ap-south	IR spectroscopy	1294	wt%	0.07	0	**0.08**	0.57
Ca – Calcium								
Ca	Ap-north	MMI®	814	mg/kg	10	0.6	**<10**	32.6
Ca	Ap-south	MMI®	1294	mg/kg	10	0.2	**<10**	80
Ca	Ap-north	ICP_MS, ACME 1F_15	814	mg/kg	100	0.2	**<100**	223
Ca	Ap-south	ICP_MS, ACME 1F_15	1294	mg/kg	100	0.2	**<100**	423
Ca	Ap-north	XRF	814	mg/kg	36	0	**143**	869
Ca	Ap-south	XRF	1294	mg/kg	36	0	**86**	650
CaO	Ap-north	XRF	814	wt%	0.005	0	**0.02**	0.12
CaO	Ap-south	XRF	1294	wt%	0.005	0	**0.012**	0.091
Cd – Cadmium								
Cd	Ap-north	MMI®	814	mg/kg	0.001	0.1	**<0.001**	0.007
Cd	Ap-south	MMI®	1294	mg/kg	0.001	0.2	**<0.001**	0.0069
Cd	Ap-north	ICP_MS, ACME 1F_15	814	mg/kg	0.01	0.2	**<0.01**	0.026
Cd	Ap-south	ICP_MS, ACME 1F_15	1294	mg/kg	0.01	0.2	**<0.01**	0.037

Q5	Q10	Q25	Q50	Q75	Q90	Q95	Q98	max.	Powers
<0.0001	<0.0001	<0.0001	**0.0001**	0.0002	0.0003	0.0005	0.0006	**0.0015**	2.7
<0.0001	0.0001	0.0003	**0.0005**	0.0009	0.0015	0.0023	0.00381	**0.0237**	2.7
<0.0002	<0.0002	<0.0002	**0.00049**	0.00095	0.0015	0.002	0.003	**0.083**	2.9
<0.0002	<0.0002	<0.0002	**0.0013**	0.0023	0.0040	0.0061	0.010	**0.054**	3.4
<0.5	<0.5	0.89	**1.7**	3.3	5.6	7.0	8.1	**16**	1.8
0.51	0.89	1.6	**3.0**	5.6	8.9	11	16	**49**	2.3
0.17	0.28	0.5	**0.9**	1.56	2.3	2.92	3.88	**7.86**	3.2
0.21	0.34	0.65	**1.25**	2.21	3.46	4.71	6.98	**51.6**	3.4
12	17	26	**43**	71	100	130	177	**426**	2
20	28	48	**74**	108	162	209	303	**818**	2.5
159	206	308	**421**	564	677	778	895	**2675**	2
144	180	255	**363**	459	598	715	943	**3350**	2.3
0.061	0.092	0.16	**0.28**	0.46	0.70	0.88	1.120	**2.0**	1.9
0.18	0.29	0.45	**0.68**	0.98	1.310	1.690	2.390	**11**	2.6
<0.001	<0.001	<0.001	**<0.001**	<0.001	0.002	0.003	0.004	**0.173**	2.5
<0.001	<0.001	<0.001	**<0.001**	<0.001	<0.001	0.002	0.004	**0.199**	2.6
0.033	0.041	0.060	**0.093**	0.14	0.22	0.28	0.36	**2.3**	3
0.077	0.10	0.15	**0.22**	0.31	0.45	0.61	0.94	**3.2**	2.5
<3	<3	<3	**<3**	<3	<3	<3	4	**5**	0.5
<3	<3	<3	**<3**	<3	<3	4	4	**9**	0.8
0.87	1.0	1.5	**2.2**	3.3	5.1	8.2	27	**48**	2.4
0.79	0.99	1.4	**2.2**	3.7	6.1	7.7	9.5	**49**	2.8
70	100	200	**370**	660	970	1123	1357	**2490**	2.7
150	250	470	**730**	990	1247	1400	1584	**2280**	2.7
428	725	1371	**2364**	4001	6752	10,255	21,723	**228,238**	3.7
669	996	1973	**3924**	26,711	107,122	164,425	218,830	**347,847**	3.8
1346	2082	4417	**9520**	14,160	21,241	25,249	34,396	**231,849**	3.2
1222	1903	3490	**7536**	33,366	109,177	170,695	227,209	**378,148**	3.6
0.19	0.29	0.62	**1.3**	2.0	3.0	3.5	4.8	**32**	3.2
0.17	0.27	0.49	**1.1**	4.7	15	24	32	**53**	3.6
0.012	0.019	0.028	**0.043**	0.068	0.0967	0.124	0.172	**0.404**	2.9
0.011	0.018	0.037	**0.061**	0.096	0.145	0.196	0.341	**2.81**	3.7
0.038	0.050	0.079	**0.12**	0.18	0.26	0.35	0.48	**1.1**	2.4
0.061	0.090	0.148	**0.22**	0.34	0.53	0.74	1.2	**7.5**	3.2

to be continued

Appendix C (continued)

Element	Material	Method	N	Unit	DL	%<DL	min.	Q2
Ce – Cerium								
Ce	Ap-north	MMI®	814	mg/kg	0.005	1.5	**<0.005**	0.00626
Ce	Ap-south	MMI®	1294	mg/kg	0.005	9.2	**<0.005**	<0.005
Ce	Ap-north	ICP_MS, ACME 1F_15	814	mg/kg	0.1	0	**1.6**	5
Ce	Ap-south	ICP_MS, ACME 1F_15	1294	mg/kg	0.1	0	**2**	6.4
Ce	Ap-north	XRF	814	mg/kg	20	10	**<20**	<20
Ce	Ap-south	XRF	1294	mg/kg	20	3.9	**<20**	<20
Cl – Chlorine								
Cl	Ap-north	XRF	814	wt%	0.002	36	**<0.002**	<0.002
Cl	Ap-south	XRF	1294	wt%	0.002	51	**<0.002**	<0.002
Co – Cobalt								
Co	Ap-north	MMI®	814	mg/kg	0.005	0.5	**<0.005**	0.0143
Co	Ap-south	MMI®	1294	mg/kg	0.005	0.2	**<0.005**	0.025
Co	Ap-north	ICP_MS, ACME 1F_15	814	mg/kg	0.1	0	**0.12**	0.42
Co	Ap-south	ICP_MS, ACME 1F_15	1294	mg/kg	0.1	0.1	**<0.1**	0.98
Co	Ap-north	XRF	814	mg/kg	3	24	**<3**	<3
Co	Ap-south	XRF	1294	mg/kg	3	7.4	**<3**	<3
Cr – Chromium								
Cr	Ap-north	MMI®	814	mg/kg	0.001	0.2	**<0.001**	0.011
Cr	Ap-south	MMI®	1294	mg/kg	0.001	0.5	**<0.001**	0.005
Cr	Ap-north	ICP_MS, ACME 1F_15	814	mg/kg	0.2	0	**0.6**	1.9
Cr	Ap-south	ICP_MS, ACME 1F_15	1294	mg/kg	0.2	0	**0.4**	4.1
Cr	Ap-north	XRF	814	mg/kg	4	0	**5**	8.3
Cr	Ap-south	XRF	1294	mg/kg	4	0.1	**<4**	12
Cs – Caesium								
Cs	Ap-north	MMI®	814	mg/kg	0.0005	38	**<0.0005**	<0.0005
Cs	Ap-south	MMI®	1294	mg/kg	0.0005	41	**<0.0005**	<0.0005
Cs	Ap-north	ICP_MS, ACME 1F_15	814	mg/kg	0.02	0	**0.029**	0.14
Cs	Ap-south	ICP_MS, ACME 1F_15	1294	mg/kg	0.02	0	**0.07**	0.27
Cs	Ap-north	XRF	814	mg/kg	3	43	**<3**	<3
Cs	Ap-south	XRF	1294	mg/kg	3	14	**<3**	<3
Cu – Copper								
Cu	Ap-north	MMI®	814	mg/kg	0.01	0	**0.03**	0.143
Cu	Ap-south	MMI®	1294	mg/kg	0.01	0.1	**<0.01**	0.23
Cu	Ap-north	ICP_MS, ACME 1F_15	814	mg/kg	0.01	0	**0.65**	1.5
Cu	Ap-south	ICP_MS, ACME 1F_15	1294	mg/kg	0.01	0	**0.3**	3.2
Cu	Ap-north	XRF	814	mg/kg	5	31	**<5**	<5
Cu	Ap-south	XRF	1294	mg/kg	5	6.9	**<5**	<5
Dy – Dysprosium								
Dy	Ap-north	MMI®	814	mg/kg	0.001	0.1	**<0.001**	0.006
Dy	Ap-south	MMI®	1294	mg/kg	0.001	1.5	**<0.001**	0.001

Q 5	Q 10	Q 25	Q 50	Q 75	Q 90	Q 95	Q 98	max.	Powers
0.0197	0.0413	0.132	**0.266**	0.55	1.09	1.59	2.33	**5.49**	3.3
<0.005	0.006	0.018	**0.0675**	0.23	0.613	1.05	1.76	**23.1**	4
6.7	9.6	17	**26**	38.2	52	64	85	**158**	2
9.8	13	20	**30**	40	51	62	78	**265**	2.1
<20	<20	30	**47**	63	81	93	112	**195**	1.3
24	33	48	**63**	80	94	107	129	**317**	1.5
<0.002	<0.002	<0.002	**0.004**	0.008	0.015	0.022	0.031	**0.06**	1.8
<0.002	<0.002	<0.002	**<0.002**	0.006	0.012	0.017	0.021	**0.043**	1.6
0.021	0.03	0.053	**0.101**	0.2	0.351	0.508	0.889	**2.15**	2.9
0.0427	0.059	0.12	**0.288**	0.613	1.18	1.69	2.4	**8.94**	3.6
0.67	0.97	2.1	**4.1**	7.1	11	13	17	**66**	2.7
1.8	3.3	6.1	**10**	14	20	26	32	**126**	3.4
<3	<3	4	**6**	10	14	18	21	**65**	1.6
<3	4	7	**11**	16	23	28	36	**144**	2
0.016	0.021	0.037	**0.0695**	0.123	0.217	0.281	0.502	**0.92**	3.3
0.009	0.014	0.029	**0.0605**	0.114	0.192	0.254	0.338	**1.3**	3.4
2.8	4.0	7.2	**13**	23	36.9	47.5	62.8	**221**	2.6
6.2	9.1	16	**25**	36	57	89	142	**696**	3.2
12	15	24	**39**	66	88	108	158	**318**	1.8
18	26	49	**75**	100	163	270	512	**6133**	3.5
<0.0005	<0.0005	<0.0005	**0.0008**	0.0021	0.0049	0.007	0.00945	**0.0552**	2.3
<0.0005	<0.0005	<0.0005	**0.0007**	0.0019	0.0061	0.0104	0.0245	**0.695**	3.4
0.17	0.24	0.39	**0.66**	1.2	1.9	2.6	3.3	**6.5**	2.4
0.45	0.60	0.88	**1.3**	2.1	3.8	5.8	9.1	**43**	2.8
<3	<3	<3	**4**	5	7	8	8	**11**	0.9
<3	<3	5	**7**	8	10	13	17	**42**	1.4
0.21	0.313	0.492	**0.91**	1.64	2.65	3.53	5.03	**9.66**	2.5
0.36	0.49	0.92	**1.81**	3.64	7	12.8	21.4	**62.1**	4.1
2.3	3.3	5.4	**9.4**	16	23	30	39	**78**	2.1
4.7	7.2	12	**19**	29	44	61	100	**395**	3.1
<5	<5	<5	**7**	13	19	25	34	**66**	1.4
<5	6	10	**18**	27	42	58	105	**391**	2.2
0.01	0.014	0.027	**0.0505**	0.0918	0.179	0.26	0.369	**1.29**	3.4
0.002	0.004	0.011	**0.025**	0.061	0.129	0.184	0.251	**2.17**	3.6

to be continued

Appendix C (continued)

Element	Material	Method	N	Unit	DL	%<DL	min.	Q2
Er - Erbium								
Er	Ap-north	MMI®	814	mg/kg	0.0005	0.1	**<0.0005**	0.00403
Er	Ap-south	MMI®	1294	mg/kg	0.0005	0.5	**<0.0005**	0.001
Eu - Europium								
Eu	Ap-north	MMI®	814	mg/kg	0.0005	0.9	**<0.0005**	0.00113
Eu	Ap-south	MMI®	1294	mg/kg	0.0005	7.8	**<0.0005**	<0.0005
F - Fluorine								
F	Ap-north	XRF	814	wt%	0.05	85	**<0.05**	<0.05
F	Ap-south	XRF	1294	wt%	0.05	71	**<0.05**	<0.05
Fe - Iron								
Fe	Ap-north	MMI®	814	mg/kg	1	0.1	**<1**	9
Fe	Ap-south	MMI®	1294	mg/kg	1	0.2	**<1**	5
Fe	Ap-north	ICP_MS, ACME 1F_15	814	mg/kg	100	0	**392**	1656
Fe	Ap-south	ICP_MS, ACME 1F_15	1294	mg/kg	100	0	**405**	2959
Fe	Ap-north	XRF	814	mg/kg	70	0	**839**	2816
Fe	Ap-south	XRF	1294	mg/kg	70	0	**909**	5166
Fe_2O_3	Ap-north	XRF	814	wt%	0.01	0	**0.12**	0.40
Fe_2O_3	Ap-south	XRF	1294	wt%	0.01	0	**0.13**	0.74
Ga - Gallium								
Ga	Ap-north	MMI®	814	mg/kg	0.001	17.7	**<0.001**	<0.001
Ga	Ap-south	MMI®	1294	mg/kg	0.001	43.6	**<0.001**	<0.001
Ga	Ap-north	ICP_MS, ACME 1F_15	814	mg/kg	0.05	0	**0.14**	0.50
Ga	Ap-south	ICP_MS, ACME 1F_15	1294	mg/kg	0.05	0	**0.19**	0.73
Ga	Ap-north	XRF	814	mg/kg	2	5	**<2**	<2
Ga	Ap-south	XRF	1294	mg/kg	2	2.2	**<2**	<2
Gd - Gadolinium								
Gd	Ap-north	MMI®	814	mg/kg	0.001	0.1	**<0.001**	0.005
Gd	Ap-south	MMI®	1294	mg/kg	0.001	2.6	**<0.001**	<0.001
Ge - Germanium								
Ge	Ap-north	ICP_MS, ACME 1F_15	814	mg/kg	0.02	2.2	**<0.02**	<0.02
Ge	Ap-south	ICP_MS, ACME 1F_15	1294	mg/kg	0.02	30	**<0.02**	<0.02
Hf - Hafnium								
Hf	Ap-north	ICP_MS, ACME 1F_15	814	mg/kg	0.01	23	**<0.01**	<0.01
Hf	Ap-south	ICP_MS, ACME 1F_15	1294	mg/kg	0.01	11	**<0.01**	<0.01
Hf	Ap-north	XRF	814	mg/kg	5	20	**<5**	<5
Hf	Ap-south	XRF	1294	mg/kg	5	19	**<5**	<5
Hg - Mercury								
Hg	Ap-north	MMI®	814	mg/kg	0.001	80.2	**<0.001**	<0.001
Hg	Ap-south	MMI®	1294	mg/kg	0.001	71.3	**<0.001**	<0.001
Hg	Ap-north	ICP_MS, ACME 1F_15	814	mg/kg	0.003	0.74	**<0.003**	0.01
Hg	Ap-south	ICP_MS, ACME 1F_15	1294	mg/kg	0.003	0.77	**<0.003**	0.01

Q5	Q10	Q25	Q50	Q75	Q90	Q95	Q98	max.	Powers
0.00563	0.0079	0.0149	**0.0296**	0.0562	0.108	0.17	0.271	**0.966**	3.6
0.0019	0.00303	0.0072	**0.0157**	0.0366	0.073	0.113	0.161	**1.23**	3.7
0.00207	0.00313	0.00592	**0.0111**	0.0195	0.0332	0.0473	0.0667	**0.145**	2.8
<0.0005	0.0006	0.0017	**0.0048**	0.0138	0.0299	0.0419	0.0625	**0.208**	2.9
<0.05	<0.05	<0.05	**<0.05**	<0.05	0.06	0.08	0.097	**0.19**	0.9
<0.05	<0.05	<0.05	**<0.05**	0.05	0.08	0.09	0.11	**0.21**	0.9
11	14	30	**80**	152	245	361	655	**2200**	3.6
6	8	11	**19**	44	94.7	146	215	**855**	3.2
2212	3156	6422	**10,942**	17,982	24,960	30,862	36,569	**112,947**	2.5
5188	8673	14,157	**20,714**	27,570	34,706	39,056	49,193	**133,926**	2.5
4102	5547	10,631	**18,220**	29,988	39,924	44,998	51,895	**120,932**	2.2
8508	12,660	19,322	**27,977**	36,580	44,344	50,268	63,101	**154,715**	2.2
0.59	0.79	1.5	**2.6**	4.3	5.7	6.4	7.4	**17**	2.2
1.22	1.8	2.8	**4.0**	5.2	6.3	7.2	9.0	**22**	2.2
<0.001	<0.001	0.001	**0.005**	0.011	0.021	0.0303	0.043	**0.154**	2.5
<0.001	<0.001	<0.001	**0.001**	0.003	0.006	0.011	0.0181	**0.141**	2.5
0.72	0.93	1.6	**2.7**	4.1	5.6	6.7	7.9	**17**	2.1
1.2	1.7	2.7	**3.9**	5.2	6.7	8.1	9.7	**20**	2
<2	3	6	**10**	15	18	19	20	**29**	1.5
3.7	5	8	**13**	16	20	22	24	**36**	1.6
0.011	0.015	0.0282	**0.052**	0.096	0.181	0.251	0.353	**0.817**	3.2
0.002	0.004	0.01	**0.026**	0.066	0.146	0.201	0.293	**2.21**	3.6
<0.02	<0.02	<0.02	**0.035**	0.059	0.082	0.10	0.14	**0.21**	1.3
<0.02	<0.02	0.022	**0.039**	0.057	0.076	0.090	0.11	**0.21**	1.3
<0.01	<0.01	0.011	**0.027**	0.065	0.15	0.20	0.27	**0.44**	1.9
<0.01	<0.01	0.029	**0.066**	0.12	0.13	0.25	0.37	**3.0**	2.8
<5	<5	6	**8**	12	14	16	19	**28**	1
<5	<5	6	**8**	11	14	16	19	**23**	1
<0.001	<0.001	<0.001	**<0.001**	<0.001	0.001	0.002	0.002	**0.022**	1.6
<0.001	<0.001	<0.001	**<0.001**	0.001	0.002	0.002	0.004	**0.053**	2
0.01	0.011	0.016	**0.024**	0.035	0.051	0.066	0.103	**0.914**	2.8
0.01	0.013	0.021	**0.036**	0.058	0.088	0.126	0.252	**1.56**	3

to be continued

Appendix C (continued)

Element	Material	Method	N	Unit	DL	%<DL	min.	Q2
In - Indium								
In	Ap-north	MMI®	814	mg/kg	0.0005	93	**<0.0005**	<0.0005
In	Ap-south	MMI®	1294	mg/kg	0.0005	97.2	**<0.0005**	<0.0005
In	Ap-north	ICP_MS, ACME 1F_15	814	mg/kg	0.003		**<0.003**	<0.003
In	Ap-south	ICP_MS, ACME 1F_15	1294	mg/kg	0.003		**<0.003**	<0.003
K - Potassium								
K	Ap-north	MMI®	814	mg/kg	0.1	0.2	**<0.1**	11.8
K	Ap-south	MMI®	1294	mg/kg	0.1	0.4	**<0.1**	17
K	Ap-north	ICP_MS, ACME 1F_15	814	mg/kg	20	0	**49**	137
K	Ap-south	ICP_MS, ACME 1F_15	1294	mg/kg	20	0	**40**	260
K	Ap-north	XRF	814	mg/kg	42	0	**390**	3746
K	Ap-south	XRF	1294	mg/kg	42	0	**241**	3993
K_2O	Ap-north	XRF	814	wt%	0.005	0	**0.047**	0.45
K_2O	Ap-south	XRF	1294	wt%	0.005	0	**0.029**	0.48
L - Lanthanum								
La	Ap-north	MMI®	814	mg/kg	0.001	1.7	**<0.001**	0.002
La	Ap-south	MMI®	1294	mg/kg	0.001	15.4	**<0.001**	<0.001
La	Ap-north	ICP_MS, ACME 1F_15	814	mg/kg	0.1	0	**1**	2.7
La	Ap-south	ICP_MS, ACME 1F_15	1294	mg/kg	0.1	0	**1.1**	3.3
La	Ap-north	XRF	814	mg/kg	20	57	**<20**	<20
La	Ap-south	XRF	1294	mg/kg	20	33	**<20**	<20
Li - Lithium								
Li	Ap-north	MMI®	814	mg/kg	0.005	74.1	**<0.005**	<0.005
Li	Ap-south	MMI®	1294	mg/kg	0.005	45.1	**<0.005**	<0.005
Li	Ap-north	ICP_MS, ACME 1F_15	814	mg/kg	0.1	0	**0.16**	0.78
Li	Ap-south	ICP_MS, ACME 1F_15	1294	mg/kg	0.1	0	**0.17**	1.8
Mg - Magnesium								
Mg	Ap-north	MMI®	814	mg/kg	1	0.4	**<1**	3
Mg	Ap-south	MMI®	1294	mg/kg	1	0.1	**<1**	8
Mg	Ap-north	ICP_MS, ACME 1F_15	814	mg/kg	50	0	**59**	138
Mg	Ap-south	ICP_MS, ACME 1F_15	1294	mg/kg	50	0.1	**<50**	302
Mg	Ap-north	XRF	814	mg/kg	302	0.1	**<302**	362
Mg	Ap-south	XRF	1294	mg/kg	302	0.1	**<302**	655
MgO	Ap-north	XRF	814	wt%	0.01	0	**0.02**	0.06
MgO	Ap-south	XRF	1294	wt%	0.01	0	**0.02**	0.11
Mn - Manganese								
Mn	Ap-north	MMI®	814	mg/kg	0.01	0.1	**<0.01**	1.37
Mn	Ap-south	MMI®	1294	mg/kg	0.01	0	**0.11**	3.18
Mn	Ap-north	ICP_MS, ACME 1F_15	814	mg/kg	1	0	**12**	36
Mn	Ap-south	ICP_MS, ACME 1F_15	1294	mg/kg	1	0	**2**	77
Mn	Ap-north	XRF	814	mg/kg	8	0	**46**	89
Mn	Ap-south	XRF	1294	mg/kg	8	0	**39**	138
MnO	Ap-north	XRF	814	wt%	0.001	0	**0.006**	0.012
MnO	Ap-south	XRF	1294	wt%	0.001	0	**0.005**	0.018

Q5	Q10	Q25	Q50	Q75	Q90	Q95	Q98	max.	Powers
<0.0005	<0.0005	<0.0005	**<0.0005**	<0.0005	<0.0005	0.0006	0.0008	**0.0019**	0.9
<0.0005	<0.0005	<0.0005	**<0.0005**	<0.0005	<0.0005	<0.0005	0.0006	**0.0022**	0.9
<0.003	0.00	0.01	**0.012**	0.018	0.026	0.033	0.041	**0.073**	1.7
0.01	0.01	0.014	**0.021**	0.029	0.039	0.047	0.063	**0.348**	2.4
15.8	20	30.4	**51.3**	82.1	121	168	234	**749**	4.2
23.2	29.7	45.2	**73**	119	180	220	312	**600**	4.1
185	224	378	**822**	1524	2434	3352	4339	**6224**	2.1
455	616	972	**1561**	2386	3396	4339	5560	**26,739**	2.8
5821	7568	11,688	**16,836**	21,250	24,310	26,338	28,466	**34,310**	1.9
5630	7590	11,224	**15,204**	19,382	24,491	31,312	36,639	**79,221**	2.5
0.70	0.91	1.4	**2.0**	2.6	2.9	3.2	3.4	**4.1**	1.9
0.68	0.91	1.4	**1.8**	2.3	3.0	3.8	4.4	**9.5**	2.5
0.006	0.017	0.051	**0.108**	0.234	0.469	0.667	1	**3.34**	3.8
<0.001	<0.001	0.002	**0.017**	0.082	0.215	0.371	0.71	**12.1**	4.4
3.6	5.1	8.9	**14**	20	27	34	44	**93**	2
4.9	6.7	9.9	**15**	20	25	30	40	**109**	2
<20	<20	<20	**<20**	27	36	43	58	**96**	1
<20	<20	<20	**26**	36	44	50	62	**155**	1.2
<0.005	<0.005	<0.005	**<0.005**	0.005	0.025	0.05	0.0717	**0.173**	1.8
<0.005	<0.005	<0.005	**0.007**	0.027	0.062	0.0853	0.13	**0.846**	2.5
1.2	1.8	3.4	**6.4**	12	20	24	28	**58**	2.6
3.2	5.0	9.1	**15**	22	31	38	52	**136**	2.9
5	7	17	**37.5**	75	132	173	222	**750**	3.2
14	19	34	**64**	121	195	259	366	**750**	3.2
241	401	898	**1836**	3639	6190	7482	9866	**20,797**	2.5
501	812	1914	**3514**	6370	11,315	17,331	30,664	**116,908**	3.7
543	923	2111	**4312**	7598	12,706	14,895	18,401	**31,539**	2.4
1266	1869	3679	**6272**	9995	16,505	23,901	38,997	**126,638**	3
0.09	0.15	0.35	**0.72**	1.3	2.1	2.5	3.1	**5.2**	2.4
0.21	0.31	0.61	**1.0**	1.7	2.7	4.0	6.5	**21**	3
2.42	3.66	7.26	**14.6**	24.2	37.9	47.9	66.1	**121**	4.4
5.42	8.68	14.6	**23.5**	35.2	48.1	59	73.9	**180**	3.2
56	76	154	**280**	437	64[illegible]	781	1100	**14,969**	3.1
131	196	354	**584**	846	1296	1611	2022	**14,605**	3.9
139	201	318	**465**	674	875	1087	1486	**13,003**	2.5
194	271	457	**705**	999	1433	1727	2146	**17,425**	2.7
0.018	0.026	0.041	**0.060**	0.087	0.11	0.14	0.19	**1.68**	2.4
0.025	0.035	0.059	**0.091**	0.13	0.19	0.22	0.28	**2.25**	2.7

to be continued

Appendix C (continued)

Element	Material	Method	N	Unit	DL	% < DL	min.	Q2
Mo – Molybdenum								
Mo	Ap-north	MMI®	814	mg/kg	0.005	9.5	**<0.005**	<0.005
Mo	Ap-south	MMI®	1294	mg/kg	0.005	14.5	**<0.005**	<0.005
Mo	Ap-north	ICP_MS, ACME 1F_15	814	mg/kg	0.01	0	**0.031**	0.077
Mo	Ap-south	ICP_MS, ACME 1F_15	1294	mg/kg	0.01	0	**0.026**	0.083
Mo	Ap-north	XRF	814	mg/kg	2	95	**<2**	<2
Mo	Ap-south	XRF	1294	mg/kg	2	96	**<2**	<2
Na – Sodium								
Na	Ap-north	ICP_MS, ACME 1F_15	814	mg/kg	2	0.5	**<2**	5.1
Na	Ap-south	ICP_MS, ACME 1F_15	1294	mg/kg	2	0	**3.6**	12
Na	Ap-north	XRF	814	mg/kg	74	0	**148**	1058
Na	Ap-south	XRF	2022	mg/kg	74	0	**74**	445
Na_2O	Ap-north	XRF	814	wt%	0.01	0	**0.02**	0.14
Na_2O	Ap-south	XRF	1294	wt%	0.01	0	**0.01**	0.06
Nb – Niobium								
Nb	Ap-north	MMI®	814	mg/kg	0.001	13.4	**<0.0005**	<0.0005
Nb	Ap-south	MMI®	1294	mg/kg	0.001	54.3	**<0.0005**	<0.0005
Nb	Ap-north	ICP_MS, ACME 1F_15	814	mg/kg	0.02	0.1	**<0.02**	0.082
Nb	Ap-south	ICP_MS, ACME 1F_15	1294	mg/kg	0.02	0.1	**<0.02**	0.088
Nb	Ap-north	XRF	814	mg/kg	2	0.1	**<2**	3
Nb	Ap-south	XRF	1294	mg/kg	2		**2**	4
Nd – Neodymium								
Nd	Ap-north	MMI®	814	mg/kg	0.001	0.2	**<0.001**	0.007
Nd	Ap-south	MMI®	1294	mg/kg	0.001	4.5	**<0.001**	<0.001
Ni – Nickel								
Ni	Ap-north	MMI®	814	mg/kg	0.005	0	**0.005**	0.063
Ni	Ap-south	MMI®	1294	mg/kg	0.005	0.1	**<0.005**	0.079
Ni	Ap-north	ICP_MS, ACME 1F_15	814	mg/kg	0.1	0	**0.32**	0.96
Ni	Ap-south	ICP_MS, ACME 1F_15	1294	mg/kg	0.1	0.1	**<0.1**	1.9
Ni	Ap-north	XRF	814	mg/kg	3	9.3	**<3**	<3
Ni	Ap-south	XRF	1294	mg/kg	3	2.2	**<3**	<3
P – Phosphorus								
P	Ap-north	MMI®	814	mg/kg	0.1	0	**0.3**	0.9
P	Ap-south	MMI®	1294	mg/kg	0.1	0.6	**<0.1**	0.2
P	Ap-north	ICP_MS, ACME 1F_15	814	mg/kg	10	0	**106**	228
P	Ap-south	ICP_MS, ACME 1F_15	1294	mg/kg	10	0	**24**	198
P	Ap-north	XRF	814	mg/kg	4	0	**148**	289
P	Ap-south	XRF	1294	mg/kg	4	0	**61**	279
P_2O_5	Ap-north	XRF	814	wt%	0.001	0	**0.034**	0.066
P_2O_5	Ap-south	XRF	1294	wt%	0.001	0	**0.014**	0.064

Q5	Q10	Q25	Q50	Q75	Q90	Q95	Q98	max.	Powers
<0.005	0.005	0.008	**0.014**	0.024	0.0377	0.0563	0.0877	**0.529**	2.3
<0.005	<0.005	0.008	**0.015**	0.029	0.047	0.0673	0.0901	**0.239**	2
0.097	0.13	0.19	**0.32**	0.66	1.3	2.0	3.6	**13**	2.6
0.14	0.20	0.31	**0.46**	0.74	1.2	1.6	2.7	**14**	2.7
<2	<2	<2	**<2**	<2	<2	3	4	**14**	1,1
<2	<2	<2	**<2**	<2	<2	<2	3	**17**	1,2
8.5	12	22	**41**	76	133	166	237	**543**	2.7
17	23	34	**51**	84	153	238	517	**7237**	3.3
2003	2767	4396	**9051**	17,582	21,143	23,443	25,798	**33,977**	2.4
742	1187	2745	**5453**	8680	13,724	16,866	20,327	**30,565**	2.6
0.27	0.37	0.59	**1.2**	2.4	2.9	3.2	3.5	**4.6**	2.4
0.1	0.16	0.37	**0.74**	1.2	1.9	2.3	2.7	**4.1**	2.6
<0.0005	<0.0005	0.001	**0.0028**	0.0064	0.0123	0.016	0.0248	**0.0861**	2.5
<0.0005	<0.0005	<0.0005	**<0.0005**	0.0013	0.00367	0.0063	0.0107	**0.0681**	2.4
0.11	0.15	0.28	**0.59**	1.4	2.2	2.8	3.6	**5.3**	2.7
0.12	0.15	0.25	**0.43**	0.71	1.4	2.2	4.0	**13.1**	3.1
4	6	7	**10**	14	17	18	20	**48**	1.7
6	8	10	**13**	17	20	24	34	**126**	1.8
0.019	0.039	0.09	**0.19**	0.36	0.719	1.04	1.48	**4.67**	4
0.001	0.002	0.01	**0.047**	0.185	0.427	0.734	1.09	**11.2**	4.4
0.086	0.115	0.181	**0.348**	0.782	1.81	2.61	3.79	**6.27**	3.1
0.126	0.193	0.434	**0.954**	1.83	3.37	4.59	6.9	**30.1**	4.1
1.4	1.9	3.9	**8.2**	15	24	30	37	**99**	2.5
3.3	5.4	12	**21**	34	56	101	193	**2475**	4.7
<3	3	6	**11**	20	31	41	55	**122**	1.9
5	8	15	**27**	42	68	121	214	**2535**	3.2
1.3	2	3.62	**8.2**	18.2	39.8	58.8	77.1	**188**	2.8
0.3	0.4	1	**2.7**	6.2	13.3	22.8	42.4	**122**	3.4
296	341	457	**630**	884	1222	1576	2121	**3620**	1.5
252	319	473	**663**	926	1270	1579	2208	**3881**	2.2
352	425	572	**772**	1117	1484	1858	2403	**4417**	1.5
348	436	580	**799**	1100	1496	1873	2541	**4369**	1.9
0.081	0.097	0.13	**0.18**	0.26	0.34	0.43	0.55	**1.0**	1.5
0.080	0.10	0.13	**0.18**	0.25	0.34	0.43	0.58	**1.0**	1.9

to be continued

Appendix C (continued)

Element	Material	Method	N	Unit	DL	%<DL	min.	Q2
Pb – Lead								
Pb	Ap-north	MMI®	814	mg/kg	0.01	0.4	**<0.01**	0.033
Pb	Ap-south	MMI®	1294	mg/kg	0.01	0.2	**<0.01**	0.04
Pb	Ap-north	ICP_MS, ACME 1F_15	814	mg/kg	0.01	0	**1.6**	2.9
Pb	Ap-south	ICP_MS, ACME 1F_15	1294	mg/kg	0.01	0	**2.1**	6.1
Pb	Ap-north	XRF	814	mg/kg	3	1.6	**<3**	4
Pb	Ap-south	XRF	1294	mg/kg	3	0.7	**<3**	6
Pd – Palladium								
Pd	Ap-north	MMI®	814	mg/kg	0.001	99.9	**<0.001**	<0.001
Pd	Ap-south	MMI®	1294	mg/kg	0.001	99.7	**<0.001**	<0.001
Pd	Ap-north	ICP_MS, ACME 1F_15	814	mg/kg	0.001	66	**<0.001**	<0.001
Pd	Ap-south	ICP_MS, ACME 1F_15	1294	mg/kg	0.001	55	**<0.001**	<0.001
Pr – Praesodymium								
Pr	Ap-north	MMI®	814	mg/kg	0.001	2	**<0.001**	0.001
Pr	Ap-south	MMI®	1294	mg/kg	0.001	20.1	**<0.001**	<0.001
Pt – Platinium								
Pt	Ap-north	MMI®	814	mg/kg	0.001	100	**<0.001**	<0.001
Pt	Ap-south	MMI®	1294	mg/kg	0.001	100	**<0.001**	<0.001
Pt	Ap-north	ICP_MS, ACME 1F_15	814	mg/kg	0.001	79	**<0.001**	<0.001
Pt	Ap-south	ICP_MS, ACME 1F_15	1294	mg/kg	0.001	64	**<0.001**	<0.001
Rb – Rubidium								
Rb	Ap-north	MMI®	814	mg/kg	0.005	0.2	**<0.005**	0.01
Rb	Ap-south	MMI®	1294	mg/kg	0.005	0.5	**<0.005**	0.01
Rb	Ap-north	ICP_MS, ACME 1F_15	814	mg/kg	0.1	0	**0.21**	1.6
Rb	Ap-south	ICP_MS, ACME 1F_15	1294	mg/kg	0.1	0	**0.61**	3.0
Rb	Ap-north	XRF	814	mg/kg	2	0.9	**<2**	14
Rb	Ap-south	XRF	1294	mg/kg	2	0.2	**<2**	16
Re – Rhenium								
Re	Ap-north	ICP_MS, ACME 1F_15	814	mg/kg	0.001	81	**<0.0005**	<0.0005
Re	Ap-south	ICP_MS, ACME 1F_15	1294	mg/kg	0.001	74	**<0.0005**	<0.0005
S – Sulphur								
S	Ap-north	MMI®	814	mg/kg	0.5	0.2	**<0.5**	9.6
S	Ap-south	MMI®	1294	mg/kg	0.5	0.2	**<0.5**	6.79
S	Ap-north	ICP_MS, ACME 1F_15	814	mg/kg	5	0	**25**	72
S	Ap-south	ICP_MS, ACME 1F_15	1294	mg/kg	5	2	**<5**	<5
S	Ap-north	IR spectroscopy	814	wt%	0.01	19	**<0.01**	<0.01
S	Ap-south	IR spectroscopy	1294	wt%	0.01	27	**<0.01**	<0.01
S	Ap-north	XRF	814	wt%	0.004	65	**<0.004**	<0.004
S	Ap-south	XRF	1294	wt%	0.004	34	**<0.004**	<0.004
SO_3	Ap-north	XRF	814	wt%	0.01	65	**<0.01**	<0.01
SO_3	Ap-south	XRF	1294	wt%	0.01	34	**<0.01**	<0.01

Q5	Q10	Q25	Q50	Q75	Q90	Q95	Q98	max.	Powers
0.05	0.08	0.14	**0.3**	0.71	1.42	1.98	2.91	**11.9**	3.4
0.07	0.1	0.18	**0.33**	0.74	1.73	3.03	4.97	**58.3**	4.1
3.6	4.7	6.7	**9.7**	14	18	22	28	**52**	1.5
8.4	11	15	**21**	29	41	54	97	**1309**	2.8
7	9	13	**16**	20	25	31	34	**73**	1.7
10	14	19	**26**	34	46	64	110	**1287**	2.9
<0.001	<0.001	<0.001	**<0.001**	<0.001	<0.001	<0.001	<0.001	**0.002**	0.6
<0.001	<0.001	<0.001	**<0.001**	<0.001	<0.001	<0.001	<0.001	**0.002**	0.6
<0.001	<0.001	<0.001	**<0.001**	0.0017	0.0035	0.0047	0.0076	**0.0186**	1.6
<0.001	<0.001	<0.001	**<0.001**	0.0025	0.0049	0.0074	0.0113	**0.15**	2.5
0.003	0.008	0.019	**0.039**	0.078	0.161	0.224	0.331	**1.11**	3.3
<0.001	<0.001	0.001	**0.008**	0.034	0.083	0.149	0.23	**2.98**	3.8
<0.001	<0.001	<0.001	**<0.001**	<0.001	<0.001	<0.001	<0.001	**<0.001**	
<0.001	<0.001	<0.001	**<0.001**	<0.001	<0.001	<0.001	<0.001	**<0.001**	
<0.001	<0.001	<0.001	**<0.001**	<0.001	0.0015	0.0021	0.0028	**0.0057**	1.1
<0.001	<0.001	<0.001	**<0.001**	0.0013	0.0021	0.0028	0.0040	**0.027**	1.7
0.015	0.022	0.039	**0.068**	0.121	0.197	0.252	0.349	**1.15**	2.7
0.016	0.023	0.039	**0.07**	0.143	0.307	0.462	0.806	**2.85**	3.1
2.2	3.4	5.9	**11**	17	30	39	50	**82**	2.6
5.4	7.6	11	**16**	24	35	47	70	**329**	2.7
21	27	44	**62**	85	110	128	138	**156**	2.2
27	38	59	**81**	104	130	163	234	**404**	2.6
<0.0005	<0.0005	<0.0005	**<0.0005**	<0.0005	0.00080	0.0011	0.0017	**0.0052**	1.3
<0.0005	<0.0005	<0.0005	**<0.0005**	0.00	0.00094	0.0013	0.0018	**0.0047**	1.3
11.9	13.6	18.1	**25.9**	36.2	52.7	66.8	102	**296**	3.1
8.9	11.4	16.7	**23.2**	34.7	52.8	71.3	162	**2620**	4
93	106	144	**216**	323	513	949	1974	**5512**	2.3
50	75	130	**202**	320	482	649	1201	**68,226**	4.4
<0.01	<0.01	0.011	**0.017**	0.026	0.048	0.081	0.17	**0.46**	2
<0.01	<0.01	<0.01	**0.016**	0.027	0.043	0.060	0.11	**9.1**	3.3
<0.004	<0.004	<0.004	**<0.004**	0.008	0.02	0.032	0.052	**0.12**	1.8
<0.004	<0.004	<0.004	**0.008**	0.032	0.048	0.064	0.1	**8.3**	3.6
<0.01	<0.01	<0.01	**<0.01**	0.02	0.05	0.08	0.13	**0.31**	1.8
<0.01	<0.01	<0.01	**0.02**	0.08	0.12	0.16	0.25	**21**	3.6

to be continued

Appendix C (continued)

Element	Material	Method	N	Unit	DL	%<DL	min.	Q2
Sb - Antimony								
Sb	Ap-north	MMI®	814	mg/kg	0.001	21.1	**<0.001**	<0.001
Sb	Ap-south	MMI®	1294	mg/kg	0.001	10.8	**<0.001**	<0.001
Sb	Ap-north	ICP_MS, ACME 1F_15	814	mg/kg	0.02	1.8	**<0.02**	0.021
Sb	Ap-south	ICP_MS, ACME 1F_15	1294	mg/kg	0.02	0	**0.022**	0.075
Sb	Ap-north	XRF	814	mg/kg	5	99	**<5**	<5
Sb	Ap-south	XRF	1294	mg/kg	5	94	**<5**	<5
Sc - Scandium								
Sc	Ap-north	MMI®	814	mg/kg	0.005	4.5	**<0.005**	<0.005
Sc	Ap-south	MMI®	1294	mg/kg	0.005	7.8	**<0.005**	<0.005
Sc	Ap-north	ICP_MS, ACME 1F_15	814	mg/kg	0.05	0.1	**<0.05**	0.21
Sc	Ap-south	ICP_MS, ACME 1F_15	1294	mg/kg	0.05	0	**0.056**	0.37
Sc	Ap-north	XRF	814	mg/kg	2	13	**<2**	<2
Sc	Ap-south	XRF	1294	mg/kg	2	5.5	**<2**	<2
Se - Selenium								
Se	Ap-north	MMI®	814	mg/kg	0.01	43	**<0.01**	<0.01
Se	Ap-south	MMI®	1294	mg/kg	0.01	36	**<0.01**	<0.01
Se	Ap-north	ICP_MS, ACME 1F_15	814	mg/kg	0.05	3.3	**<0.05**	<0.05
Se	Ap-south	ICP_MS, ACME 1F_15	1294	mg/kg	0.05	1.6	**<0.05**	0.071
Si - Silicon								
Si	Ap-north	XRF	814	mg/kg	467	0	**15,005**	130,389
Si	Ap-south	XRF	1294	mg/kg	467	0	**11,499**	100,807
SiO_2	Ap-north	XRF	814	wt%	0.1	0	**3**	28
SiO_2	Ap-south	XRF	1294	wt%	0.1	0	**2**	22
Sm - Samarium								
Sm	Ap-north	MMI®	814	mg/kg	0.001	0.5	**<0.001**	<0.001
Sm	Ap-south	MMI®	1294	mg/kg	0.001	6.8	**<0.001**	<0.001
Sn - Tin								
Sn	Ap-north	MMI®	814	mg/kg	0.001	60.1	**<0.001**	<0.001
Sn	Ap-south	MMI®	1294	mg/kg	0.001	84.4	**<0.001**	<0.001
Sn	Ap-north	ICP_MS, ACME 1F_15	814	mg/kg	0.1	0.4	**<0.1**	0.12
Sn	Ap-south	ICP_MS, ACME 1F_15	1294	mg/kg	0.1	0.4	**<0.1**	0.23
Sn	Ap-north	XRF	814	mg/kg	4	99	**<4**	<4
Sn	Ap-south	XRF	1294	mg/kg	4	85	**<4**	<4
Sr - Strontium								
Sr	Ap-north	MMI®	814	mg/kg	0.01	0.4	**<0.01**	0.1
Sr	Ap-south	MMI®	1294	mg/kg	0.01	0.1	**<0.01**	0.19
Sr	Ap-north	ICP_MS, ACME 1F_15	814	mg/kg	0.5	0	**0.8**	0
Sr	Ap-south	ICP_MS, ACME 1F_15	1294	mg/kg	0.5	0	**0.8**	2.9
Sr	Ap-north	XRF	814	mg/kg	2	0	**10**	25
Sr	Ap-south	XRF	1294	mg/kg	2	0	**7**	25

Q5	Q10	Q25	Q50	Q75	Q90	Q95	Q98	max.	Powers
<0.001	<0.001	0.001	**0.002**	0.003	0.005	0.007	0.009	**0.019**	1.6
<0.001	<0.001	0.002	**0.003**	0.005	0.008	0.01	0.014	**0.071**	2.2
0.030	0.045	0.073	**0.11**	0.16	0.22	0.28	0.34	**1.0**	2
0.12	0.15	0.24	**0.35**	0.55	0.91	1.4	2.5	**17**	2.9
<5	<5	<5	**<5**	<5	<5	<5	<5	**8**	0.5
<5	<5	<5	**<5**	<5	<5	5	7	**36**	1.2
0.005	0.0063	0.011	**0.034**	0.07	0.119	0.173	0.225	**0.593**	2.4
<0.005	0.005	0.008	**0.012**	0.027	0.0767	0.126	0.203	**0.899**	2.6
0.27	0.37	0.74	**1.4**	2.4	3.8	4.8	6.3	**11**	2.7
0.57	0.99	1.7	**2.7**	3.9	5.3	6.2	7.9	**21**	2.6
<2	<2	4	**7**	10	14	16	18	**34**	1.5
<2	3	7	**9**	12	15	18	20	**42**	1.6
<0.01	<0.01	<0.01	**0.01**	0.02	0.03	0.03	0.04	**0.09**	1.3
<0.01	<0.01	<0.01	**0.01**	0.02	0.03	0.03	0.04	**0.17**	1.5
0.070	0.12	0.20	**0.31**	0.43	0.63	0.77	1.1	**3.5**	2.2
0.14	0.19	0.27	**0.39**	0.52	0.73	0.93	1.2	**3.8**	2.2
237,081	270,268	297,080	**333,822**	386,012	418,302	427,351	435,464	**451,079**	1.5
147,437	197,712	261,462	**302,807**	340,155	382,599	400,974	419,206	**445,563**	1.6
51	58	64	**71**	83	90	91	93	**97**	1.5
32	42	56	**65**	73	82	86	90	**95**	1.6
0.00665	0.012	0.023	**0.047**	0.0875	0.173	0.236	0.328	**0.861**	3.2
<0.001	0.002	0.005	**0.016**	0.053	0.12	0.172	0.27	**2.02**	3.6
<0.001	<0.001	<0.001	**<0.001**	0.002	0.004	0.006	0.009	**0.023**	1.7
<0.001	<0.001	<0.001	**<0.001**	<0.001	0.002	0.003	0.00614	**0.03**	1.8
0.19	0.23	0.32	**0.48**	0.76	1	2	2	**8**	2.2
0.33	0.44	0.63	**0.87**	1	2	3	5	**30**	2.8
<4	<4	<4	**<4**	<4	<4	<4	<4	**20**	1
<4	<4	<4	**<4**	<4	6	8	14.1	**129**	1.8
0.21	0.35	0.67	**1.42**	2.66	4.23	5.88	9.47	**17.5**	3.5
0.307	0.47	0.952	**1.8**	3.41	5.82	7.87	13.6	**90.2**	4.3
3.2	5.5	10	**16**	26	42	59	97	**303**	2.6
4.4	6.2	10	**20**	55	170	313	564	**3777**	3.7
34	43	70	**113**	192	249	290	375	**726**	1.9
35	46	68	**96**	153	270	378	711	**4564**	2.8

to be continued

Appendix C (continued)

Element	Material	Method	N	Unit	DL	%<DL	min.	Q2
Ta – Tantalum								
Ta	Ap-north	MMI®	814	mg/kg	0.001	96	**<0.001**	<0.001
Ta	Ap-south	MMI®	1294	mg/kg	0.001	99.7	**<0.001**	<0.001
Ta	Ap-north	ICP_MS, ACME 1F_15	814	mg/kg	0.005	92	**<0.005**	<0.005
Ta	Ap-south	ICP_MS, ACME 1F_15	1294	mg/kg	0.005	95	**<0.005**	<0.005
Ta	Ap-north	XRF	814	mg/kg	5	99.6	**<5**	<5
Ta	Ap-south	XRF	1294	mg/kg	5	99.1	**<5**	<5
Tb – Terbium								
Tb	Ap-north	MMI®	814	mg/kg	0.001	2.9	**<0.001**	<0.001
Tb	Ap-south	MMI®	1294	mg/kg	0.001	16.5	**<0.001**	<0.001
Te – Tellurium								
Te	Ap-north	MMI®	814	mg/kg	0.01	100	**<0.01**	<0.01
Te	Ap-south	MMI®	1294	mg/kg	0.01	99.9	**<0.01**	<0.01
Te	Ap-north	ICP_MS, ACME 1F_15	814	mg/kg	0.02		**<0.02**	<0.02
Te	Ap-south	ICP_MS, ACME 1F_15	1294	mg/kg	0.02		**<0.02**	<0.02
Th – Thorium								
Th	Ap-north	MMI®	814	mg/kg	0.0005	0.1	**<0.0005**	0.00283
Th	Ap-south	MMI®	1294	mg/kg	0.0005	2.1	**<0.0005**	<0.0005
Th	Ap-north	ICP_MS, ACME 1F_15	814	mg/kg	0.02	0	**0.03**	0.14
Th	Ap-south	ICP_MS, ACME 1F_15	1294	mg/kg	0.02	0	**0.071**	0.33
Th	Ap-north	XRF	814	mg/kg	5	41	**<5**	<5
Th	Ap-south	XRF	1294	mg/kg	5	14	**<5**	<5
Ti – Titanium								
Ti	Ap-north	MMI®	814	mg/kg	0.003	0	**0.004**	0.008
Ti	Ap-south	MMI®	1294	mg/kg	0.003	1	**<0.003**	0.004
Ti	Ap-north	ICP_MS, ACME 1F_15	814	mg/kg	5	0	**6.4**	20
Ti	Ap-south	ICP_MS, ACME 1F_15	1294	mg/kg	5	0.2	**<5**	11
Ti	Ap-north	XRF	814	mg/kg	6	0	**90**	609
Ti	Ap-south	XRF	1294	mg/kg	6	0	**204**	950
TiO_2	Ap-north	XRF	814	wt%	0.001	0	**0.02**	0.10
TiO_2	Ap-south	XRF	1294	wt%	0.001	0	**0.03**	0.16
Tl – Thallium								
Tl	Ap-north	MMI®	814	mg/kg	0.001	39.8	**0.0025**	<0.0005
Tl	Ap-south	MMI®	1294	mg/kg	0.001	47.8	**0.0025**	<0.0005
Tl	Ap-north	ICP_MS, ACME 1F_15	814	mg/kg	0.001	0	**0.0025**	0.019
Tl	Ap-south	ICP_MS, ACME 1F_15	1294	mg/kg	0.001	0	**0.0025**	0.028

Q5	Q10	Q25	Q50	Q75	Q90	Q95	Q98	max.	Powers
<0.001	<0.001	<0.001	**<0.001**	<0.001	<0.001	<0.001	0.001	**0.029**	1.8
<0.001	<0.001	<0.001	**<0.001**	<0.001	<0.001	<0.001	<0.001	**0.003**	0.8
<0.005	<0.005	<0.005	**<0.005**	<0.005	<0.005	<0.005	0.008	**0.03**	1.1
<0.005	<0.005	<0.005	**<0.005**	<0.005	<0.005	<0.005	0.008	**0.02**	0.8
<5	<5	<5	**<5**	<5	<5	<5	<5	**7**	0.4
<5	<5	<5	**<5**	<5	<5	<5	<5	**10**	0.6
0.002	0.002	0.002	**0.008**	0.015	0.027	0.037	0.0565	**0.144**	2.5
<0.001	<0.001	<0.001	**0.004**	0.01	0.021	0.029	0.042	**0.334**	2.8
<0.01	<0.01	<0.01	**<0.01**	<0.01	<0.01	<0.01	<0.01	**<0.01**	
<0.01	<0.01	<0.01	**<0.01**	<0.01	<0.01	<0.01	<0.01	**0.03**	0.8
<0.02	<0.02	<0.02	**<0.02**	<0.02	0.03	0.04	0.06	**0.11**	1
<0.02	<0.02	<0.02	**0.03**	0.05	0.07	0.10	0.14	**0.27**	1.4
0.00683	0.0101	0.0189	**0.0422**	0.0791	0.144	0.195	0.283	**1.66**	3.8
0.00147	0.00253	0.0066	**0.0143**	0.0389	0.0935	0.158	0.279	**2.84**	4.1
0.22	0.37	0.87	**2**	3	5	7	9	**14**	2.7
0.71	1	2	**4**	5	7	8	11	**63**	2.9
<5	<5	<5	**6**	9	11	14	18	**26**	1
<5	<5	7	**11**	14	17	20	25	**84**	1.5
0.0107	0.015	0.0412	**0.222**	0.675	1.42	1.9	3.05	**6.03**	3.2
0.006	0.007	0.012	**0.023**	0.057	0.201	0.372	0.713	**13.2**	3.9
26	36	64	**172**	525	859	1043	1363	**1732**	2.4
15	20	35	**68**	161	474	867	1345	**7860**	3.5
909	1191	2052	**2977**	4106	4844	5348	5966	**9430**	2
1391	1894	2836	**3957**	4808	5569	6455	7410	**24,556**	2.1
0.15	0.20	0.34	**0.50**	0.69	0.81	0.89	1.0	**1.6**	2
0.23	0.32	0.47	**0.66**	0.80	0.93	1.1	1.2	**4.1**	2.1
<0.0005	<0.0005	<0.0005	**0.0006**	0.0011	0.0016	0.00193	0.0028	**0.0064**	1.4
<0.0005	<0.0005	<0.0005	**0.0005**	0.0011	0.0021	0.0028	0.00433	**0.0172**	1.8
0.026	0.034	0.054	**0.086**	0.14	0.24	0.30	0.38	**0.58**	2.4
0.047	0.061	0.092	**0.14**	0.20	0.31	0.46	0.74	**2.5**	3

to be continued

Appendix C (continued)

Element	Material	Method	N	Unit	DL	%<DL	min.	Q2
U - Uranium								
U	Ap-north	MMI®	814	mg/kg	0.001	0.1	**<0.001**	0.015
U	Ap-south	MMI®	1294	mg/kg	0.001	0.2	**<0.001**	0.013
U	Ap-north	ICP_MS, ACME 1F_15	814	mg/kg	0.04	0.1	**<0.04**	0.15
U	Ap-south	ICP_MS, ACME 1F_15	1294	mg/kg	0.04	0	**0.071**	0.25
U	Ap-north	XRF	814	mg/kg	3	75	**<3**	<3
U	Ap-south	XRF	1294	mg/kg	3	56	**<3**	<3
V - Vanadium								
V	Ap-north	MMI®	814	mg/kg	0.001	0	**0.002**	0.005
V	Ap-south	MMI®	1294	mg/kg	0.001	0.2	**<0.001**	0.003
V	Ap-north	ICP_MS, ACME 1F_15	814	mg/kg	0.2	0	**1**	3.1
V	Ap-south	ICP_MS, ACME 1F_15	1294	mg/kg	0.2	0	**1.7**	6.5
V	Ap-north	XRF	814	mg/kg	5	0.3	**<5**	9
V	Ap-south	XRF	1294	mg/kg	5	0.1	**<5**	16
W - Tungsten								
W	Ap-north	MMI®	814	mg/kg	0.001	42.4	**<0.001**	<0.001
W	Ap-south	MMI®	1294	mg/kg	0.001	49.2	**<0.001**	<0.001
W	Ap-north	ICP_MS, ACME 1F_15	814	mg/kg	0.01	1.5	**<0.01**	0.012
W	Ap-south	ICP_MS, ACME 1F_15	1294	mg/kg	0.01	1.8	**<0.01**	0.011
W	Ap-north	XRF	814	mg/kg	5	99	**<5**	<5
W	Ap-south	XRF	1294	mg/kg	5	90	**<5**	<5
Y - Yttrium								
Y	Ap-north	MMI®	814	mg/kg	0.005	0.1	**<0.005**	0.0335
Y	Ap-south	MMI®	1294	mg/kg	0.005	1.2	**<0.005**	0.008
Y	Ap-north	ICP_MS, ACME 1F_15	814	mg/kg	0.01	0	**0.23**	0.78
Y	Ap-south	ICP_MS, ACME 1F_15	1294	mg/kg	0.01	0	**0.3**	1.4
Y	Ap-north	XRF	814	mg/kg	3	0	**6**	7
Y	Ap-south	XRF	1294	mg/kg	3	0	**6**	9.9
Yb - Ytterbium								
Yb	Ap-north	MMI®	814	mg/kg	0.001	0.1	**<0.001**	0.004
Yb	Ap-south	MMI®	1294	mg/kg	0.001	2	**<0.001**	0.00093
Zn - Zinc								
Zn	Ap-north	MMI®	814	mg/kg	0.02	0.1	**<0.02**	0.11
Zn	Ap-south	MMI®	1294	mg/kg	0.02	0.2	**<0.02**	0.08
Zn	Ap-north	ICP_MS. ACME 1F_15	814	mg/kg	0.1	0	**2.9**	7.2
Zn	Ap-south	ICP_MS, ACME 1F_15	1294	mg/kg	0.1	0	**2.8**	11
Zn	Ap-north	XRF	814	mg/kg	3	0.4	**<3**	10
Zn	Ap-south	XRF	1294	mg/kg	3	0.1	**<3**	19

Q5	Q10	Q25	Q50	Q75	Q90	Q95	Q98	max.	Powers
0.021	0.027	0.043	**0.084**	0.182	0.379	0.637	1.23	**5.32**	4
0.019	0.025	0.046	**0.084**	0.147	0.242	0.369	0.579	**2.03**	3.6
0.23	0.33	0.46	**0.68**	1.3	2.7	4.0	5.9	**24**	3.1
0.33	0.42	0.56	**0.81**	1.2	1.8	2.6	4.2	**21**	2.5
<3	<3	<3	**<3**	<3	4	6	8.7	**22**	1.2
<3	<3	<3	**<3**	4	5	6	8	**25**	1.2
0.00765	0.011	0.0192	**0.037**	0.08	0.173	0.272	0.461	**12.9**	3.8
0.005	0.006	0.011	**0.022**	0.045	0.094	0.155	0.333	**1.58**	3.5
4.6	6.2	11	**19**	31	43	51	63	**104**	2
9.8	14	20	**29**	40	57	72	95	**188**	2
13	18	29	**50**	78	106	122	138	**253**	2
24	35	55	**79**	104	128	148	184	**323**	2.1
<0.001	<0.001	<0.001	**0.001**	0.002	0.004	0.007	0.011	**0.036**	1.9
<0.001	<0.001	<0.001	**0.001**	0.003	0.006	0.00935	0.017	**0.067**	2.1
0.020	0.026	0.038	**0.058**	0.11	0.18	0.21	0.28	**0.71**	2.2
0.019	0.029	0.047	**0.083**	0.15	0.28	0.45	0.97	**5.2**	3
<5	<5	<5	**<5**	<5	<5	<5	<5	**6**	0.4
<5	<5	<5	**<5**	<5	5	6	7	**25**	1
0.055	0.0723	0.136	**0.256**	0.466	0.944	1.51	2.23	**7.6**	3.5
0.013	0.024	0.059	**0.134**	0.311	0.67	1.07	1.55	**16.2**	3.8
1.1	1.5	3.0	**5**	8.2	12	14	18	**38**	2.2
2.3	3.2	4.9	**7.6**	11	14	17	23	**65**	2.3
9	11	16	**23**	30	35	37	39	**78**	1.1
13	17	23	**29**	35	39	43	48	**111**	1.3
0.005	0.006	0.012	**0.024**	0.047	0.102	0.163	0.243	**0.741**	3.2
0.002	0.003	0.006	**0.013**	0.03	0.062	0.0927	0.134	**0.788**	3.2
0.14	0.21	0.45	**1.16**	2.41	4.54	6.93	10.4	**23.8**	3.4
0.11	0.16	0.31	**0.62**	1.33	2.8	4.26	7.64	**36.2**	3.6
9.7	13	20	**30**	48	73	87	110	**139**	1.7
16	23	38	**53**	72	94	112	139	**1396**	2.7
14	19	30	**43**	65	92	105	126	**178**	2.1
27	35	52	**71**	93	119	139	174	**1413**	3

to be continued

Appendix C (continued)

Element	Material	Method	N	Unit	DL	%<DL	min.	Q2
Zr - Zirconium								
Zr	Ap-north	MMI®	814	mg/kg	0.005	0.1	**<0.005**	0.0143
Zr	Ap-south	MMI®	1294	mg/kg	0.005	7.4	**<0.005**	<0.005
Zr	Ap-north	ICP_MS, ACME 1F_15	814	mg/kg	0.1	3.3	**<0.1**	<0.1
Zr	Ap-south	ICP_MS, ACME 1F_15	1294	mg/kg	0.1	1.9	**<0.1**	0.11
Zr	Ap-north	XRF	814	mg/kg	3	0	**4**	77
Zr	Ap-south	XRF	1294	mg/kg	3	0	**10**	67

Isotope Ratio	Material	Method	N	Unit	DL	%<DL	min.	Q2
$^{206}Pb/^{207}Pb$	Ap-north	SF_ICP_MS	814				**1.14**	1.18
	Ap-south		1294				**1.12**	1.17
$^{206}Pb/^{208}Pb$	Ap-north	SF_ICP_MS	814				**0.37**	0.473
	Ap-south		1294				**0.444**	0.475
$^{207}Pb/^{206}Pb$	Ap-north	SF_ICP_MS	814				**0.579**	0.657
	Ap-south		1294				**0.721**	0.798
$^{207}Pb/^{208}Pb$	Ap-north	SF_ICP_MS	814				**0.287**	0.361
	Ap-south		1294				**0.379**	0.396
$^{208}Pb/^{206}Pb$	Ap-north	SF_ICP_MS	814				**1.48**	1.72
	Ap-south		1294				**1.82**	2
$^{208}Pb/^{207}Pb$	Ap-north	SF_ICP_MS	814				**2.42**	2.45
	Ap-south		1294				**2.4**	2.45

Q5	Q10	Q25	Q50	Q75	Q90	Q95	Q98	max.	Powers
0.0197	0.025	0.041	**0.072**	0.136	0.233	0.31	0.452	**1.85**	2.9
<0.005	0.006	0.012	**0.033**	0.08	0.165	0.262	0.422	**1.64**	2.8
0.1	0.18	0.47	**1.2**	2.8	5.8	7.9	11	**18**	2.6
0.23	0.41	1.0	**2.2**	4.1	6.9	9.8	20	**165**	3.5
122	158	212	**269**	340	489	573	625	**963**	2.4
96	126	183	**251**	347	454	524	589	**839**	1.9

Q5	Q10	Q25	Q50	Q75	Q90	Q95	Q98	max.	Powers
1.19	1.2	1.22	**1.26**	1.33	1.43	1.47	1.52	**1.73**	0.2
1.17	1.18	1.19	**1.19**	1.2	1.21	1.22	1.25	**1.39**	0.1
0.479	0.482	0.487	**0.496**	0.52	0.552	0.564	0.583	**0.677**	0.3
0.476	0.477	0.48	**0.482**	0.484	0.487	0.492	0.499	**0.55**	0.1
0.681	0.7	0.755	**0.796**	0.82	0.835	0.843	0.849	**0.875**	0.2
0.818	0.824	0.832	**0.837**	0.843	0.849	0.854	0.857	**0.896**	0.1
0.372	0.383	0.39	**0.397**	0.402	0.405	0.407	0.408	**0.414**	0.2
0.399	0.401	0.402	**0.404**	0.405	0.406	0.407	0.408	**0.417**	
1.77	1.81	1.92	**2.02**	2.05	2.08	2.09	2.11	**2.7**	0.3
2.03	2.05	2.07	**2.08**	2.08	2.09	2.1	2.11	**2.25**	0.1
2.46	2.47	2.49	**2.52**	2.57	2.61	2.69	2.77	**3.48**	0.2
2.45	2.46	2.47	**2.48**	2.48	2.5	2.51	2.52	**2.64**	

Photo Credits

Cover:	North Yorkshire, England (Photo: © mauritius images / Photononstop)
Page 19:	Cereal field on marl-rich soils near to Seville, Spain (Photo: Juan Locutura)
Page 79:	View south-eastward along the upper Aare Valley, Switzerland (Photo: Peter Schwab)
Page 181:	Sampling a pasture after a heavy snow fall, Auvergne, France (Photo: Anne Sophie Audion)
Page 233:	Mowed oat field in the northern part of Ukraine (Photo: Oleksiy Maksymenko)
Page 271:	Pastureland in the Mecklenburg-Brandenburg Basin, Germany (Photo: Rainer Hoffmann)
Page 297:	Grazing land soil sample site in central Slovenia (Photo: Mateja Gosar)

Imprint

Geol. Jb.
B 103

Chemistry of Europe's Agricultural Soils
Part B – General Background Information and Further Analysis of the GEMAS Data Set

Published by: BUNDESANSTALT FÜR GEOWISSENSCHAFTEN UND ROHSTOFFE (BGR)
[FEDERAL INSTITUTE FOR GEOSCIENCES AND NATURAL RESOURCES]

Bundesanstalt für Geowissenschaften und Rohstoffe (BGR)
Referat für Schriftenpublikationen und Öffentlichkeitsarbeit
Stilleweg 2, 30655 Hannover (Germany)
Phone: +49 (0)511 643–0
www.bgr.bund.de

Edited by: Dr. Thomas Schubert,
K. Otremba with the assistance of H.-J. Sturm and Th. Oertel

Layout and Prepress: BGR and schreiberVIS, Bickenbach

Printed by: Werbedruck GmbH Horst Schreckhase, Spangenberg
Printed on FSC-certified paper

Available from: E. Schweizerbart'sche Verlagsbuchhandlung
(Nägele u. Obermiller)
Johannesstraße 3 A, 70176 Stuttgart
Telefon: (07 11) 35 14 56–0
Telefax: (07 11) 35 14 56–99
E-mail: mail@schweizerbart.de
www.schweizerbart.de

Chemistry of Europe's Agricultural Soils
Part B: **ISBN 978-3-510-96847-3**
Information on this title: www.schweizerbart.com/9783510968473

Part A: ISBN 978-3-510-96846-6 (Geol. Jb. B 102)
Information on this title: www.schweizerbart.com/9783510968466

Part A+B: ISBN 978-3-510-96848-0
ISSN 0341-6402

BGR www.bgr.bund.de